D1229925

Organometallic compounds of arsenic, antimony, and bismuth

This is a volume in the series
THE CHEMISTRY OF ORGANOMETALLIC COMPOUNDS

THE CHEMISTRY OF ORGANOMETALLIC COMPOUNDS

A Series of Monographs

Dietmar Seyferth, *editor*

Department of Chemistry,
Massachusetts Institute of Technology
Cambridge, Massachusetts

PUBLISHED:

Chemistry of the iron group metallocenes
Part One: ferrocene, ruthenocene, osmocene
By Myron Rosenblum

The organic compounds of lead
By Hymin Shapiro and F. W. Frey

The organic chemistry of tin
By Wilhelm P. Neumann

IN PRESS

The organic compounds of germanium
By M. Lesbre, P. Mazerolles and J. Satge

Organometallic compounds of arsenic, antimony, and bismuth

G. O. DOAK
and
LEON D. FREEDMAN

Department of Chemistry
North Carolina State University
Raleigh, North Carolina

Wiley-Interscience
a Division of
JOHN WILEY & SONS, INC.

NEW YORK · LONDON · SYDNEY · TORONTO

Library of Congress Catalogue Card Number: 75-120703

ISBN 0 471 21650-X

Printed in the United States of America

10 9 8 7 6 5 4 3 2 1

To Bee and Myrle

Preface

The first organoarsenic compounds were prepared by L. C. Cadet de Gassicourt in 1760. Although it was not until many years later that the actual chemical structures of Cadet's compounds were elucidated, his discovery is often cited as the first synthesis of an organometallic compound. During the second half of the nineteenth century, when synthetic organic chemistry was new and exciting, a host of new organoarsenic compounds were prepared and their properties studied. The names of Michaelis, Hofmann, Bunsen, Kolbe, and Baeyer are among the prominent chemists associated with the early history of organoarsenic chemistry. However, it was not until 1910, when Ehrlich announced that a synthetic arsenical was an effective cure for syphilis, that organoarsenic chemistry became one of the favorite research areas of the synthetic organic chemist. It has been estimated that by 1932 a total of 12,500 organic derivatives of arsenic had been synthesized and tested for biological activity. Among the famous twentieth-century chemists who have worked on organoarsenic compounds are Dehn, Karrer, Sir Gilbert Morgan, Roger Adams, Wieland, Nesmeyanov, and F. G. Mann. The area of searching for new arsenicals for the treatment of syphilis came to an abrupt end in 1943, when it was shown that the disease was readily cured by penicillin. For a time the volume of research on organoarsenicals decreased, but in recent years interest has revived and a large number of papers are published each year. Whereas formerly emphasis was on the use of arsenicals as medicinal agents, recent research has been more concerned with the structure and stereochemistry of arsenic compounds or their use as ligands in coordination chemistry and as analytical reagents.

Although organoantimony chemistry cannot boast of as early a start as organoarsenic chemistry, the nineteenth century saw a great deal of research on the synthesis of organic derivatives of antimony. Löwig, Landolt, and Friedländer were among the first chemists working in this field. Again the name of Michaelis appears on many publications. In the early part of this century Sir Gilbert Morgan did much pioneer work in the synthesis of new organoantimonials. Although antimony compounds have found a place in the treatment of disease, particularly of such tropical diseases as schistosomiasis and leishmaniasis, there has not been the incentive for the synthesis of new antimonials that was provided by the use of arsenicals in the treatment of syphilis. This fact, coupled with the decreased stability of organoantimony compounds compared with their arsenic analogs, has resulted in the synthesis of fewer organoantimony compounds. Nevertheless, considerable research is being carried on with these compounds, and interest appears to be increasing as judged by the number of papers on organoantimony chemistry that have been abstracted in succeeding years in *Annual Surveys of Organometallic Chemistry* (now *Organometallic Chemistry Reviews, Section B*). A part of this interest lies in the use of tertiary stibines as ligands in coordination chemistry. Another potentially important and as yet largely unexplored area of research involves Mössbauer spectroscopy of organoantimony compounds. Finally, the chemistry of antimonials has been found to differ sufficiently from that of arsenicals to justify studying them for their own inherent interest.

In contrast to the large number of organic derivatives of arsenic and antimony, relatively few organobismuth compounds have been prepared. The history of organobismuth chemistry began in 1850 with the synthesis of triethylbismuth by Löwig and Schweizer. Michaelis and his students made important contributions to this area during the second half of the nineteenth century. In 1913 Frederick Challenger entered the field with the aim of preparing an asymmetric bismuthonium compound. Although he never succeeded in this objective, his work has added greatly to our knowledge of bismuth chemistry. The important work of Gilman and of Wittig should also be emphasized. As a class, organobismuth compounds are relatively unstable and are difficult to prepare. These properties have, of course, hampered their investigation, and much remains to be learned about their chemistry and structure. For example, we still do not know whether the bismuthonium compounds contain tetra- or pentacovalent bismuth. We do not understand why triphenylbismuth dihalides readily rearrange to form tetraphenylbismuthonium compounds. And the deep purple color of pentaphenylbismuth is a complete mystery. We hope that the section on bismuth compounds in the present volume will help to stimulate research in this area.

We take pleasure in acknowledging the advice of Dr. G. G. Long concerning the structure and physical properties of organoantimony compounds. Thanks are also due to our wives, Beatrice Doak and Myrle N. Freedman, for their patience and care in correcting the manuscript. The first draft of the book was typed by Mrs. Beatrice Doak, and the final draft was typed by Mrs. Rosemary P. Scholl.

G. O. DOAK
LEON D. FREEDMAN

March 1970
Raleigh, North Carolina

Contents

I Introduction

1. SCOPE OF THIS BOOK

A number of books and reviews have been written on organic compounds of arsenic, antimony, and bismuth. Although many of these compilations are now out-of-date, they still possess considerable value, since some of the synthetic methods have not changed in the ensuing years. Sir Gilbert Morgan (1) was the author of a book on organoarsenic and -antimony compounds in 1918. In 1923 Raiziss and Gavron (2) wrote a detailed monograph on organoarsenic compounds, which is still a classic in its field. *Organic Derivatives of Antimony* by W. G. Christiansen (3), published in 1925, was equally valuable for antimony compounds. In 1930 a large and extensive monograph on organoarsenic compounds by A. E. Goddard (4) appeared, and was followed in 1936 by a second monograph, by the same author (5), dealing with organophosphorus, -antimony, and -bismuth compounds. The monographs by Raiziss and Gavron, by Christiansen, and by Goddard describe in considerable detail the preparation and chemical properties of all organoarsenic, -antimony, and -bismuth compounds known at the time these books were published. In 1937 the literature on these compounds was thoroughly summarized by Krause and von Grosse (6) in *Die Chemie der metall-organischen Verbindungen.* A recent noncritical compilation by Dub (7) covers the synthesis, reactions, physical constants, and uses of all such compounds prepared between 1937 and 1964. A monograph by F. G. Mann (8) covering the heterocyclic compounds of arsenic, antimony, and bismuth appeared in 1950; a second edition is now in press. In addition to the above books, a number of chapters in monographs and in textbooks, as well as several review articles in journals, have described certain phases of the organic chemistry of these elements (9–22).

The present volume makes no attempt to list all known organic compounds of arsenic, antimony, and bismuth or all the literature references to such compounds. Nor does it describe in detail the preparation of specific compounds of these elements. Instead we have attempted to give examples of all the methods that have been used to prepare the various types of organoarsenic, -antimony, and -bismuth compounds. We have emphasized those procedures that appear to be of most general application, and we have described the recent work in this area in greater detail. The chemical reactions of each class of these compounds have also been covered. The possible reaction mechanisms involved in the synthesis and reactions have been discussed whenever sufficient data were available. The structure, spectra, and other physical properties of organoarsenic, -antimony, and -bismuth compounds have been reviewed in considerable detail, and in some cases new interpretations are offered. We have attempted to cover all the pertinent literature through the end of 1967, and a few important papers published in 1968 and 1969 have also been included. Beginning with 1967, the literature on organoarsenic, -antimony, and -bismuth compounds is being periodically reviewed in annual surveys published in *Organometallic Chemistry Reviews, Section B.*

2. NOMENCLATURE

The International Union of Pure and Applied Chemistry (IUPAC) has not as yet issued detailed rules for the naming of organometallic compounds, although the authors have been informed that tentative rules are under consideration. Most chemists working in the field of organoarsenic, -antimony, and -bismuth compounds follow *Chemical Abstracts* rules. For the most part we have done this in the present volume. The 1945 *C.A.* rules for arsenic, antimony, and bismuth compounds were based on IUPAC rule 34 (23). The *C.A.* rules were changed considerably in 1962 (24)—not always for the better. Thus the names arsonic and stibonic acids have been retained for compounds of the type $RAsO_3H_2$ and $RSbO_3H_2$, but compounds of the type R_2AsO_2H and R_2SbO_2H are named as hydroxydiaryl (or dialkyl) arsine oxides and hydroxydiaryl (or dialkyl) stibine oxides. This change makes it difficult to compare two classes of compounds (i.e., arsonic with arsinic, stibonic with stibinic acids), and accordingly we have chosen to disregard these changes in the 1945 *C.A.* rules. In a similar manner we have retained the names arsonous and arsinous acids with the full realization that such compounds are very weak acids indeed. In conformity with *C.A.*, we have used the terms arsenobenzene and antimonobenzene although such compounds do not contain As=As or Sb=Sb linkages. We have chosen to name such compounds as $(C_6H_5)_3Bi$ and $(CH_3)_3Bi$ as triphenylbismuth and trimethylbismuth, respectively, rather than as bismuthines. This is in accord with common practice, but not with current *C.A.* usage.

In *C.A.*, substituents on arsenic, antimony, and bismuth are given in alphabetical order. Hence, CH_3AsCl_2 is dichloromethylarsine, but $C_2H_5AsI_2$ is ethyldiiodoarsine. This practice leads to difficulty in comparing compounds of similar structure, and we have chosen to name such compounds as a class, i.e., dichloroarsines and diiodoarsines, regardless of the hydrocarbon substituent attached to the arsenic.

The nomenclature used in the present volume is summarized in Table 1-1.

3. COMPARISON OF ARSENIC, ANTIMONY, AND BISMUTH COMPOUNDS

Arsenic, antimony, and bismuth, together with nitrogen and phosphorus, comprise the representative elements of group V. All possess the following electronic configuration:

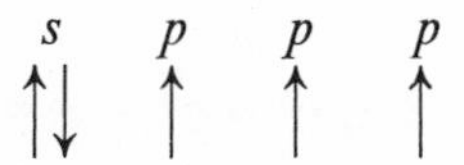

However, since nitrogen readily forms double bonds and does not possess low-energy *d* orbitals, it differs considerably in its chemistry from the other group V elements. There is a gradual increase in metallic properties and usually a decrease in the strength of the covalent bond as one descends from nitrogen to bismuth. A great many of the physical properties of these elements also vary progressively from nitrogen to bismuth. Davies (25) showed that when the atomic radii were plotted against the period numbers of the group V elements (nitrogen excepted), a straight line was obtained. Similar results were obtained by plotting atomic refractivities, atomic parachors, and the dipole moments of the triphenyl compounds against the period numbers. Certain other physical properties, notably bond energies and electronegativities, however, may vary irregularly in going from phosphorus to bismuth.

A. Types of compounds

All of the group V elements form trivalent compounds of the type R_3M, R_2MX, RMX_2, and MX_3 where R is an organic group and X is an inorganic group which is usually, but not always, a halogen. In such compounds of bismuth, the 6*s* electrons almost always remain in the *s* orbital; this pair of electrons is then referred to as an inert pair. By contrast the 2*s* and 3*s* electrons of nitrogen and phosphorus occupy sp^3 hybrid orbitals; the nonbonding electron pair is then referred to as the lone pair. According to

Table 1-1 Nomenclature of Organoarsenic, -Antimony, and -Bismuth Compounds

Radical	Suffix	Prefix	Example
—AsH_2	arsine		CH_3AsH_2 methylarsine
		arsino	p-$H_2AsC_6H_4COOH$ p-arsinobenzoic acid
=AsH	arsine		$(CH_3)_2AsH$ dimethylarsine
		arsylene	p-$HAs(C_6H_4COOH)_2$ p-arsylenedibenzoic acid
≡As	arsine	—	$(CH_3)_3As$ trimethylarsine
—AsO	—	arsenoso	$(C_6H_5AsO)_n$ arsenosobenzene
=As—O—As=	arsenic oxide	—	$[(C_6H_5)_2As]_2O$ bis(diphenylarsenic) oxide
=As—S—As=	arsenic sulfide	—	$[(C_6H_5)_2As]_2S$ bis(diphenylarsenic) sulfide
—AsS	—	thioarsenoso	C_6H_5AsS thioarsenosobenzene
—$As(OH)_2$	arsonous acid	—	p-$NO_2C_6H_4As(OH)_2$ p-nitrobenzenearsonous acid
—$As(SH)_2$	thioarsonous acid	—	(known only as esters, $RAs(SR')_2$)
=AsOH	arsinous acid	—	$(C_6H_5)_2AsOH$ diphenylarsinous acid
=AsSH	thioarsinous acid	—	(known only as esters R_2AsSR')
—AsX_2	dihaloarsine		$C_6H_5AsCl_2$ phenyldichloroarsine
		dihaloarsino	p-$HOOCC_6H_4AsCl_2$ p-dichloroarsinobenzoic acid
=AsX	haloarsine	—	$(CH_3)_2AsCl$ dimethylchloroarsine
—As=As—	—	arseno	$(C_6H_5As)_6$ arsenobenzene
=As—As=	diarsine	—	$[(C_6H_5)_2As]_2$ tetraphenyldiarsine
—AsX_4	arsenic tetrahalide	—	$C_6H_5AsCl_4$ phenylarsenic tetrachloride
=AsX_3	arsenic trihalide	—	$(C_6H_5)_2AsCl_3$ diphenylarsenic trichloride
≡AsX_2	arsenic dihalide	—	$(C_6H_5)_3AsCl_2$ triphenylarsenic dichloride
≡AsXY		—	$(C_6H_5)_3As(OH)NO_3$ triphenylarsenic hydroxynitrate
≡AsO	arsine oxide	—	$(C_6H_5)_3AsO$ triphenylarsine oxide

$\equiv AsNH$	arsine imide	—	$(C_6H_5)_3AsNH$ triphenylarsine imide
$\equiv As^{\oplus}$	arsonium	—	$(CH_3)_4AsBr$ tetramethylarsonium bromide
$—AsO(OH)_2$	arsonic		$C_6H_5AsO(OH)_2$ benzenearsonic acid
		arsono	p-$HOOCC_6H_4AsO(OH)_2$ *p*-arsonobenzoic acid
$\equiv As(OH)X$	arsenic hydroxyhalide	—	$(C_6H_5)_3As(OH)Cl$ triphenylarsenic hydroxychloride
$\equiv As(NH_2)X$	arsenic aminohalide	—	$(C_6H_5)_3As(NH_2)Cl$ triphenylarsenic aminochloride
$=AsO(OH)$	arsinic		$(C_6H_5)_2AsO(OH)$ diphenylarsinic acid
		arsinico	$(p$-$HOOCC_6H_4)_2AsO(OH)$ *p*-arsinicodibenzoic acid
$\equiv As=$	arsenic	—	$(C_6H_5)_5As$ pentaphenylarsenic
$—SbH_2$	stibine		CH_3SbH_2 methylstibine
		stibino	p-$HOOCC_6H_4SbH_2$ *p*-stibinobenzoic acid
$=SbH$	stibine		$(CH_3)_2SbH$ dimethylstibine
		stibylene	$(p$-$HOOCC_6H_4)_2SbH$ *p*-stibylenedibenzoic acid
$\equiv Sb$	stibine	—	$(CH_3)_3Sb$ trimethylstibine
$—SbO$	—	stiboso	$(C_6H_5SbO)_n$ stibosobenzene
$=Sb—O—Sb=$	antimony oxide	—	$[(C_6H_5)_2Sb]_2O$ bis(diphenylantimony) oxide
$—SbS$	—	thiostiboso	C_6H_5SbS thiostibosobenzene
$—SbX_2$	dihalostibine	—	$C_6H_5SbCl_2$ phenyldichlorostibine
$=SbX$	halostibine	—	$(CH_3)_2SbCl$ dimethylchlorostibine
$—Sb=Sb—$	—	antimono	$(C_6H_5Sb)_n$ antimonobenzene
$=Sb—Sb=$	distibine	—	$[(C_6H_5)_2Sb]_2$ tetraphenyldistibine
$—SbX_4$	antimony tetrahalide	—	$C_6H_5SbCl_4$ phenylantimony tetrachloride
$=SbX_3$	antimony trihalide	—	$(C_6H_5)_2SbCl_3$ diphenylantimony trichloride

Table 1-1 (Continued)

Radical	Suffix	Prefix	Example
$(\text{—SbX})_2O$	antimony halide	—	$[(C_6H_5)_3SbCl]_2O$ oxybis(triphenylantimony) chloride
$\equiv SbX_2$	antimony dihalide	—	$(C_6H_5)_3SbCl_2$ triphenylantimony dichloride
$\equiv Sb^{\oplus}$	stibonium	—	$(C_6H_5)_4SbCl$ tetraphenylstibonium chloride
$\text{—}SbO(OH)_2$	stibonic		$C_6H_5SbO(OH)_2$ phenylstibonic acid
		stibono	p-$HOOCC_6H_4SbO(OH)_2$ p-stibonobenzoic acid
$=SbO(OH)$	stibinic		$(C_6H_5)_2SbO(OH)$ diphenylstibinic acid
		stibinico	$(p\text{-}HOOCC_6H_4)_2SbO(OH)$ p-stibinicodibenzoic acid
$\equiv Sb=$	antimony	—	$(C_6H_5)_5Sb$ pentaphenylantimony
$\equiv Bi$	bismuth	—	$(C_6H_5)_3Bi$ triphenylbismuth
$\text{—}BiO$	bismuth oxide	—	C_6H_5BiO phenylbismuth oxide
$\text{—}BiX_2$	bismuth dihalide	—	$C_6H_5BiCl_2$ phenylbismuth dichloride
$=BiX$	bismuth halide	—	$(C_6H_5)_2BiCl$ diphenylbismuth chloride
$\equiv BiX_2$	bismuth dihalide	—	$(C_6H_5)_3BiCl_2$ triphenylbismuth dichloride
$\equiv Bi^{\oplus}$	bismuthonium	—	$(C_6H_5)_4BiCl$ tetraphenylbismuthonium chloride
$\equiv Bi=$	bismuth	—	$(C_6H_5)_5Bi$ pentaphenylbismuth

Durrant and Durrant (26), compounds of arsenic and antimony can adopt either of these configurations.

Group V elements all form quadruply-connected compounds. Antimony forms ions of the type SbX_4^- and $RSbX_3^-$, but attempts to prepare $R_2SbX_2^-$ have been unsuccessful (27). The $AsCl_4^-$ ion is also known (28), and ions of the type $ArAsCl_3^-$ may be formed in the reaction between diazonium salts and aryldichloroarsines. The structure of these ions has not been determined, but they probably can be regarded as trigonal bipyramids with the non-bonding electron pair occupying an equatorial position (29). Arsenic, antimony, and bismuth (as well as nitrogen and phosphorus) form onium ions R_4M^+, in which the four groups are tetrahedrally arranged around the central atom (sp^3). This has been shown to be the case with arsonium (30) and stibonium (31) salts by X-ray diffraction. Almost certainly the bismuthonium compounds have a similar structure. In addition to onium compounds, many other compounds containing sp^3 hybridized group V elements are known. Such compounds include arsonic (32) and arsinic acids (33) (but not stibonic or stibinic acids), and probably also the arsine oxides, arsine sulfides, and similar compounds.

Arsenic and antimony form a number of pentavalent compounds of the type MX_5, RMX_4, R_2MX_3, and R_3MX_2, where R is either an alkyl or an aryl group and X is a halogen or other negative group such as NO_3, CH_3CO_2, $\frac{1}{2}SO_4$, etc. Mixed compounds of the type $R_3As(OH)X$ and $[R_3SbX]_2O$ are also known. Bismuth forms compounds of the type R_3BiX_2 but not RMX_4 or R_2MX_3. In all of these compounds the central atom probably uses sp^3d hybrid orbitals. Where the structure of such compounds has been determined by X-ray or electron diffraction, they have all proved to be trigonal bipyramids with the more negative groups occupying the apical positions (34, 35). Polynova and Porai-Koshits (36) originally described diphenylantimony trichloride as having the configuration of a trigonal bipyramid, but in a later review paper (37) they formulated the compound as a monohydrate, $Ph_2SbCl_3 \cdot H_2O$, with octahedral geometry. Compounds of the type R_5M are also known and, with the exception of pentaphenylantimony, are trigonal bipyramids (38). Pentaphenylantimony surprisingly is a square pyramid (39, 40). Antimony also forms the inorganic ion $SbF_5^=$ (41). No organic compound related to this structure has been synthesized. Arsenic, antimony, and bismuth all form inorganic ions, AsX_6^-, SbX_6^-, and BiX_6^-, where, at least in the case of SbX_6^-, the central atom is octahedral (42); antimony forms organic ions of the type $RSbX_5^-$ and $R_2SbX_4^-$. The antimony atom in such ions is almost certainly octahedral, although no structural data are available. Apparently the corresponding bismuth compounds have not been made.

Both arsenic and antimony form compounds with As—As and Sb—Sb single bonds, but no compounds with Bi—Bi single bonds are known with certainty. No compounds with arsenic–arsenic double bonds are known; the arseno compounds, which at one time were formulated as RAs=AsR, are now known to have ring structures. Compounds involving antimony–antimony or antimony–arsenic double bonds are also unknown, although many compounds were so formulated in the older chemical literature. There is considerable evidence of p_π–d_π double bond formation between arsenic and oxygen in such compounds as triarylarsine oxides, Ar_3AsO. Probably many other compounds, such as arsonic and arsinic acids, arsine sulfides, thioarsonic acids, etc., involve some degree of p_π–d_π bonding between the arsenic and the oxygen or sulfur atom. By contrast with arsenic, the stibonic and stibinic acids are polymeric and do not possess p_π–d_π bonds between antimony and oxygen. Trialkyl- and triarylantimony sulfides, however, are monomeric (43, 44) and probably involve considerable p_π–d_π bonding between the antimony and sulfur atoms. No compounds involving p_π–p_π

Table 1-2 Mean Bond Dissociation Energy, $\bar{D}$(A – B), at 25°C for the Molecules in the Gaseous State

Bond	$\bar{D}$(A – B), kcal/mole	Compound	Reference
As—C	54.8[a]	$As(CH_3)_3$	45
As—C	63.8[a]	$As(C_6H_5)_3$	45
Sb—C	51.5[a]	$Sb(CH_3)_3$	45
Sb—C	58.3[a]	$Sb(C_6H_5)_3$	45
Bi—C	34.1[a]	$Bi(CH_3)_3$	45
Bi—C	42.2[a]	$Bi(C_6H_5)_3$	45
As—F	111.3	AsF_3	46a
Sb—F	103, 108	SbF_3	47, 48
As—Cl	68.9, 73.1	$AsCl_3$	46a, 48
Sb—Cl	74.2, 74.3	$SbCl_3$	47, 48
Bi—Cl	67.1	$BiCl_3$	[b]
As—Br	56.5, 58	$AsBr_3$	46a, 48
Sb—Br	61.0, 62	$SbBr_3$	47, 48
As—I	41.6, 42	AsI_3	46a, 48
Sb—I	45.3, 44	SbI_3	47, 48
As—O	72	As_4O_6	49
Sb—O	71	Sb_4O_6	49
As—As	38	$Me_2AsAsMe_2$	50

[a] Skinner (45) used slightly different values for the heats of sublimation of the central atom and the heats of formation of methyl and phenyl radicals, so that his values for bond energies differ slightly from those calculated by the authors who did the original work.

[b] Calculated from data given in *Selected Values of Chemical Thermodynamic Properties*, U.S. Bureau of Standards Circular 500.

bonding between arsenic, antimony, or bismuth and oxygen, sulfur, or similar atoms are known. Compounds, which in the past have been formulated with such bonding, e.g., RAs═S, RAs═O, RSb═O, are polymeric.

B. Bond energies

The mean bond dissociation energy in compounds MR_n, where R is an atom or group attached to the element M whose valency is n, is given as $\bar{D}(M - R) = \Delta H°/n$, where $\Delta H° = \Delta H_f°(M, g) + n\,\Delta H_f°(R, g) - \Delta H_f°(MR_n, g)$. The heats of combustion of trimethyl- and triphenylarsines, trimethyl- and triphenylstibines, and trimethyl- and triphenylbismuth have been determined, and from these the C—As, C—Sb, and C—Bi bond dissociation energies have been calculated. The results, which are the work of several groups of investigators, have been summarized by Skinner (45) and are given in Table 1-2. Few bond energies in organoarsenic, -antimony, and -bismuth compounds other than the trialkyl or triaryl derivatives have been determined. However, some idea of such bond energies can be seen from the values found for inorganic compounds of these elements. Some of these values for inorganic compounds are also given in Table 1-2.

The C—M bond energy values with the triaryl and trialkyl compounds of group V appear to decrease with increase in period number, whereas the M—X bond energy values (with the possible exception of Sb—F) are in the order As—X $<$ Sb—X $>$ BiX. However, as is discussed in later chapters (Sections IV-6-C, VIII-2-C and IX-1-C), there are considerable uncertainties involved in the determination of bond dissociation energies, and hence the reported data should be interpreted with caution.

C. Force constants

From infrared and Raman spectral data, the C—M force constants in trimethylarsine, -stibine, and -bismuth have been determined by several groups of workers (51–53). Siebert (53) has compared the results obtained by the three groups of workers. His results for both the stretching and bending C—M force constants are given in Table 1-3. Sheline (52) has stated that the

Table 1-3 Force Constants for P—C, As—C, Sb—C, *and* Bi—C *in the Trimethyl Compounds*

Bond	Stretching Force Constant, dynes × 10^5/cm	Bending Force Constant, dynes × 10^5/cm
P—C	2.99	0.262
As—C	2.63	0.193
Sb—C	2.18	0.145
Bi—C	1.82	0.120

Table 1-4

Bond	Ratio of Stretching to Bending Force Constant	$\chi_c - \chi_m$
P—C	12.21	0.45
As—C	12.79	0.55
Sb—C	14.41	0.75
Bi—C	13.79	0.75

ratio of stretching to bending force constants in the compounds listed in Table 1-3 should be a measure of the ionic character of the bond. Table 1-4 lists these ratios (based on Sheline's force constants) and the bond electronegativities χ_c–χ_m, which Sheline (52) calculated from the values of Gordy and which should also be a measure of ionic character.

D. Electronegativities

The electronegativities of the elements have been computed in a number of ways by a number of authors. Allred and Hensley (54) have compared and summarized the results of various authors in a review paper. These authors (who have written extensively on electronegativities) arrived at the most probable values for the group V elements by comparison of all of the various ways used for computing electronegativities. Their results are given in Table 1-5. Also in Table 1-5 are the original values of Pauling (46b) and the recent values computed by Batsanov (55). It should be noted that except for Pauling's values the electronegativities are given to three significant figures, a practice which is almost certainly not justified.

E. Bond lengths and bond angles

The bond lengths of a limited number of organoarsenic and organoantimony compounds have been determined by X-ray diffraction and in a few

Table 1-5 Comparison of the Electronegativity Values Obtained by Different Workers for Group V Elements

Element	Allred and Hensley	Pauling	Batsanov
P	2.19	2.1	2.18
As	2.18	2.0	2.19
Sb	2.06	1.9	2.06
Bi	2.16	1.9	2.14

Table 1-6 Bond Lengths and Bond Angles in Arsenic, Antimony, and Bismuth Compounds

Compound	Bond	Bond Length, A (Range)	Sum of Covalent Radii	Angle	Bond Angle°	Ref.
AsF_3	As—F	1.712 ± 0.005	1.73	F—As—F	102 ± 2	56
SbF_3	Sb—F	2.00	1.88	F—Sb—F	81.9 and 104.3	58
$AsCl_3$	As—Cl	2.161 ± 0.004 – 2.17 ± 0.02	2.20	Cl—As—Cl	98.4 ± 0.5 – 103 ± 3	56
$SbCl_3$	Sb—Cl	2.37 ± 0.02	2.38	Cl—Sb—Cl	95.2 – 99.5 ± 1.5	56, 58
$BiCl_3$	Bi—Cl	2.48 ± 0.02		Cl—Bi—Cl	100 ± 6	56
$AsBr_3$	As—Br	2.31 ± 0.03 – 2.36 ± 0.04	2.35	Br—As—Br	99.23 – 102.85	57
$SbBr_3$	Sb—Br	2.47 ± 0.04 – 2.52 ± 0.03	2.53	Br—Sb—Br	96 ± 2 – 98	56, 58
$BiBr_3$	Bi—Br	2.63 ± 0.02		Br—Bi—Br	100 ± 4	56
AsI_3	As—I	2.54 – 2.55 ± 0.03	2.54	I—As—I	98.5 – 101 ± 1.5	56
SbI_3	Sb—I	2.67 ± 0.03 – 2.75 ± 0.05	2.73	I—Sb—I	98 ± 2 – 99 ± 1	58
AsH_3	As—H	1.5192 ± 0.002 – 1.523	1.54	H—As—H	91.6 – 91.83 ± 0.33	56
SbH_3	Sb—H	1.7073 ± 0.0025	1.74	H—Sb—H	91.30 ± 0.33 – 91.5	56, 58
$(CH_3)_3As$	As—C	1.98 ± 0.02	2.00	C—As—C	96 ± 5	56
$(CF_3)_3Sb$	Sb—C	2.202	2.18	C—Sb—C	100	58
$(p\text{-}CH_3C_6H_4)_3As$	As—C	1.96 ± 0.05	2.00	C—As—C	102 ± 2	59

Table 1-6 (Continued)

Compound	Bond	Bond Length, A (Range)	Sum of Covalent Radii	Angle	Bond Angle°	Ref.
$(C_6H_5)_2AsBr$	As—C	1.99 ± 0.04	2.00	C—As—C	105 ± 2	60
	As—Br	2.40 ± 0.01	2.35	C—As—Br	95 ± 1	
CH_3AsI_2	As—C	2.07	2.00	C—As—I	90	61
	As—I	2.54 ± 0.01	2.54	I—As—I	104 ± 0.4	
$(CF_3)_3As$	As—C	2.053 ± 0.019	2.00	C—As—C	100.1 ± 3.5	56
$[(C_6H_5)_2As]_2O$	As—C	1.90	2.00	C—As—C	102 and 98	62
	As—O	1.67	1.85	As—O—As	137 ± 2	
As_4O_6	As—O	1.78 ± 0.02 – 1.80 ± 0.02	1.85	As—O—As	126 ± 3 – 128 ± 2	56
$SbCl_5$	Sb—Cl	2.29 – 2.31 (equatorial)	2.38			37
		2.34 – 2.43 (apical)				
$(CH_3)_3SbCl_2$	Sb—Cl	2.49	2.38			37
	Sb—C	2.13	2.17			
$(CH_3)_3SbBr_2$	Sb—Br	2.63	2.53			37

$(CH_3)_3SbI_2$	Sb—I	2.88	2.73			37
$(C_6H_5)_3SbCl_2$	Sb—Cl	2.48	2.38			
	Sb—C	2.15	2.17			
$(C_6H_5)_2SbCl_3$	Sb—Cl	2.43 (equatorial)	2.38			36
		2.52 (apical)				
$C_6H_5AsO_3H_2$	As—O	1.65 − 1.75	1.85	O—As—O	106.0, 106.5, and 117.0	32, 63
	As—C	1.90 − 1.97	2.00	C—As—O	104.0, 109.0, and 115	
$(CH_3)_2AsO_2H$	As—O	1.62 ± 0.03	1.85	C—As—C	109.9 ± 2.5	33
	As—C	1.91 ± 0.04	2.00	C—As—O	106.1 ± 2.3 − 114.8 ± 2.0	
$(C_6H_5)_4AsI$	As—C	1.95	2.00	C—As—C	109	56
$(C_6H_5)_3BiCl_2$	Bi—Cl	2.605 ± 0.010				64
	Bi—C	2.24 ± 0.10				
$(C_6H_5)_4SbOCH_3$	Sb—C	2.119 (equatorial)	2.18	C—Sb—C	115.8 − 123.5 (equatorial)	65
		2.199 (apical)		C—Sb—C	91.6 − 94.4 (equatorial–apical)	
	Sb—O	2.06	2.07	O—Sb—C	178.1 (apical–apical)	
				O—Sb—C	85.1 − 89.5 (apical–equatorial)	
$(C_6H_5)_5Sb$	Sb—C	2.115 ± 0.005 (axial)		C(axial)—Sb—C(basal)	96.4 − 106.0	40
	Sb—C	2.216 ± 0.007 (basal)		C(basal)—Sb—C(basal)	86.7 − 88.6	

cases by electron diffraction. There has been very little X-ray diffraction work with organobismuth compounds. Inorganic compounds of all three elements have been studied to a greater extent, and a number of bond lengths and bond angles are known. In Table 1-6 are listed a number of bond lengths and bond angles for both inorganic and organic compounds of arsenic, antimony, and bismuth. The majority of the values in Table 1-6 are taken from two sources: The Chemical Society compilation of bond lengths and angles (56), or its supplement (57); and the two recent reviews by Polynova and Porai-Koshits (37, 58) which have complete listings of all known bond lengths and angles for both organic and inorganic antimony compounds. These sources should be consulted for the original references. In Table 1-6, where several authors have determined bond lengths or bond angles, or where one author has found that the angles in a compound of the type MX_3 vary slightly, we have listed the range of values. Also included in Table 1-6 are the bond lengths calculated from the sums of the covalent radii and corrected for electronegativity differences between the atoms involved (46c).

REFERENCES

1. G. T. Morgan, *Organic Compounds of Arsenic and Antimony*, Longmans, Green and Co., London, 1918.
2. G. W. Raiziss and J. L. Gavron, *Organic Arsenical Compounds*, Chemical Catalog Co., New York, 1923.
3. W. G. Christiansen, *Organic Derivatives of Antimony*, Chemical Catalog Co., New York, 1925.
4. A. E. Goddard, "Derivatives of Arsenic," in J. N. Friend, ed., *A Text-book of Inorganic Chemistry*, Vol. XI, Part II, Griffin and Co., Ltd., London, 1930.
5. A. E. Goddard, "Derivatives of Phosphorus, Antimony, and Bismuth," in J. N. Friend, ed., *A Text-book of Inorganic Chemistry*, Vol. XI, Part III, Griffin and Co., Ltd., London, 1936.
6. E. Krause and A. von Grosse, *Die Chemie der metall-organischen Verbindungen*, Verlag Bornträger, Berlin, 1937.
7. M. Dub, *Organometallic Compounds*, Vol. III, 2nd ed., Springer-Verlag, New York, 1968.
8. F. G. Mann, *The Heterocyclic Derivatives of Phosphorus, Arsenic, Antimony, Bismuth, and Silicon*, Interscience, New York, 1950.
9. H. Gilman and H. L. Yale, *Chem. Rev.*, **30**, 281 (1942).
10. C. S. Hamilton and J. F. Morgan, "The Preparation of Aromatic Arsonic and Arsinic Acids by the Bart, Bechamp and Rosenmund Reactions," in Roger Adams, ed., *Organic Reactions*, Vol. II, Wiley, New York, 1944, pp. 415–454.
11. F. G. Mann, "The Heterocyclic Derivatives of Phosphorus, Arsenic, and Antimony," in J. W. Cook, ed., *Progress in Organic Chemistry*, Vol. IV, Butterworths, London, 1958, pp. 217–248.
12. F. G. Mann, "The Stereochemistry of the Group V Elements," in W. Klyne and P. B. D. de la Mare, eds., *Progress in Stereochemistry*, Vol. II, Butterworths, London, 1958, pp. 196–227.

13. W. Lisowski, *Wiadomosci Chemi.*, **13,** 641 (1959).
14. J. H. Harwood, *Industrial Applications of Organometallic Compounds*, Reinhold, New York, 1963.
15. H. L. Yale, "Arsenic, Antimony, and Phosphorus Compounds of Pyridine," in E. Klingsberg, ed., *Pyridine and its Derivatives*, Part 4, Interscience, New York, 1964, Chapter XVI.
16. L. Kolditz, "Halides of Phosphorus, Arsenic, Antimony, and Bismuth," in H. J. Emeléus and A. G. Sharpe, eds., *Advances in Inorganic Chemistry and Radiochemistry*, Vol. VII, Academic Press, New York, 1965, pp. 1–26.
17. C. B. Milne and A. N. Wright, "Aliphatic Organometallic and Organometalloidal Compounds," in S. Coffey, ed., *Rodd's Chemistry of Carbon Compounds*, Vol. I, Part B, 2nd ed., Elsevier, Amsterdam, 1965, pp. 218–240.
18. E. M. Marlett, *Ann. N.Y. Acad. Sci.*, **125,** Art. 1, 12 (1965).
19. W. R. Cullen, "Organoarsenic Compounds," in F. G. A. Stone and R. West, eds., *Advances in Organometallic Chemistry*, Vol. IV, Academic Press, New York, 1966, pp. 145–242.
20. A. W. Johnson, *Ylid Chemistry*, Academic Press, New York, 1966, Chapter 8.
21. G. E. Coates and K. Wade, *Organometallic Compounds, Vol. 1, The Main Group Elements*, Methuen, London, 1967.
22. G. O. Doak and L. D. Freedman, "Arsenicals, Antimonials, and Bismuthials," in A. Burger, ed., *Medicinal Chemistry*, 3rd ed., Interscience, New York, in press.
23. *C.A.*, **39,** 5939 (1945).
24. *C.A.*, **56,** 43N (1962).
25. W. C. Davies, *J. Chem. Soc.*, 462 (1935).
26. P. J. Durrant and B. Durrant, *Introduction to Advanced Inorganic Chemistry*, Longmans, London, 1962, pp. 646–7.
27. O. A. Reutov, O. A. Ptitsyna, A. N. Lovtsova, and V. F. Petrova, *Zh. Obshch. Khim.*, **29,** 3857 (1959).
28. L. Kolditz, "Halides of Arsenic and Antimony," in V. Gutmann, ed., *Halogen Chemistry*, Vol. II, Academic Press, London, 1967, p. 153.
29. R. J. Gillespie, *Angew. Chem., Int. Ed. Engl.*, **6,** 819 (1967).
30. E. Collins, D. J. Sutor, and F. G. Mann, *J. Chem. Soc.*, 4051 (1963).
31. P. J. Wheatley, *J. Chem. Soc.*, 3200 (1963).
32. A. Shimada, *Bull. Chem. Soc. Jap.*, **33,** 301 (1960).
33. J. Trotter and T. Zobel, *J. Chem. Soc.*, 4466 (1965).
34. A. F. Wells, *Z. Kristallogr.*, **99,** 367 (1938).
35. E. L. Muetterties and R. A. Schunn, *Quart. Rev.* (London), **20,** 245 (1966).
36. T. N. Polynova and M. A. Porai-Koshits, *Zh. Strukt. Khim.*, **2,** 477 (1961).
37. T. N. Polynova and M. A. Porai-Koshits, *Zh. Strukt. Khim.*, **7,** 642 (1966).
38. P. J. Wheatley, *J. Chem. Soc.*, 2206 (1964).
39. P. J. Wheatley, *J. Chem. Soc.*, 3718 (1964).
40. A. L. Beauchamp, M. J. Bennett, and F. A. Cotton, *J. Amer. Chem. Soc.*, **90,** 6675 (1968).
41. A. Byström and K-A. Wilhelmi, *Ark. Kemi*, **3,** 461 (1951).
42. G. Teufer, *Acta Crystallogr.*, **9,** 539 (1956).
43. W. J. Lile and R. C. Menzies, *J. Chem. Soc.*, 617 (1950).
44. M. Shindo, Y. Matsumara, and R. Okawara, *J. Organometal. Chem.*, **11,** 299 (1968).
45. H. A. Skinner, "The Strengths of Metal-to-Carbon Bonds," in F. G. A. Stone and R. West, eds., *Advances in Organometallic Chemistry*, Vol. II, Academic Press, New York, 1964, pp. 49–114.

46. L. Pauling, *The Nature of the Chemical Bond*, 3rd ed., Cornell Univ. Press, Ithaca, 1960. (a) p. 85, (b) p. 93, (c) p. 228.
47. R. J. Sime, *J. Phys. Chem.*, **67**, 501 (1963).
48. M. L. Huggins, *J. Amer. Chem. Soc.*, **75**, 4123 (1953).
49. F. S. Dainton, *Trans. Faraday Soc.*, **43**, 244 (1947).
50. C. T. Mortimer and H. A. Skinner, *J. Chem. Soc.*, 4331 (1952).
51. G. J. Rosenbaum, D. J. Rubin, and C. R. Sandberg, *J. Chem. Phys.*, **8**, 366 (1940).
52. R. K. Sheline, *J. Chem. Phys.*, **18**, 602 (1950).
53. H. Siebert, *Z. Anorg. Allg. Chem.*, **273**, 161 (1953).
54. A. L. Allred and A. L. Hensley, Jr., *J. Inorg. Nucl. Chem.*, **17**, 43 (1961).
55. S. S. Batsanov, *Zh. Strukt. Khim.*, **5**, 293 (1964).
56. *Tables of Interatomic Distances and Configuration in Molecules and Ions*, Special Publication No. 11, The Chemical Society, London, 1958.
57. *Tables of Interatomic Distances and Configuration in Molecules and Ions*, Supplement 1956–1959, Special Publication No. 18, The Chemical Society, London, 1965.
58. T. N. Polynova and M. A. Porai-Koshits, *Zh. Strukt. Khim.*, **7**, 146 (1966).
59. J. Trotter, *Can. J. Chem.*, **41**, 14 (1963).
60. J. Trotter, *J. Chem. Soc.*, 2567 (1962).
61. N. Camerman and J. Trotter, *Acta Crystallogr.*, **16**, 922 (1963).
62. W. R. Cullen and J. Trotter, *Can. J. Chem.*, **41**, 2983 (1963).
63. Yu. T. Struchkov, *Izv. Akad. Nauk SSSR, Otd. Khim. Nauk*, 1962 (1960).
64. D. M. Hawley, G. Ferguson, and G. S. Harris, *Chem. Commun.*, 111 (1966).
65. K. W. Shen, W. E. McEwen, S. J. La Placa, W. C. Hamilton, and A. P. Wolf, *J. Amer. Chem. Soc.* **90**, 1718 (1968).

II Arsonic and arsinic acids and their derivatives

1. PREPARATION

A. The Bart reaction

Aromatic diazo compounds react with inorganic trivalent arsenic compounds with the evolution of nitrogen and the formation of arsonic acids. When the reaction is carried out with sodium arsenite in aqueous alkaline solution, it is termed the Bart reaction (1). Although carbocyclic amines are the usual ones used, the reaction is applicable to a few amino-substituted heterocyclic compounds. Thus 3-pyridinearsonic acid (2) may be prepared from 3-aminopyridine, and 2-amino-5-pyridinearsonic acid (3) from 2,5-diaminopyridine. However, the only quinolinearsonic acids prepared by the Bart reaction are those in which the arsono group is on the benzene ring. A furanarsonic acid (4) has also been prepared by the Bart reaction.

Although the yields in the Bart reaction are generally good, in a few cases the reaction fails or gives only poor yields (5). Thus *m*-aminophenol does not give *m*-hydroxybenzenearsonic acid although the *ortho* and *para* isomers are readily prepared. Similarly, *m*-nitroaniline gives a poor yield of the arsonic acid compared with the yields of arsonic acids obtained from *o*- and *p*-nitroaniline.

The reaction is not particularly sensitive to the nature or position of the substituent group on the aromatic ring. Hamilton and Morgan (6) attempted to compare the effect of structure on the yields of arsonic acids. However, since the results compared were obtained in a number of different laboratories under widely differing reaction conditions, the conclusions reached are open to question.

Although the Bart reaction is carried out in alkaline solution, rather careful control of pH is necessary for the most satisfactory results. In strongly alkaline solution no nitrogen is evolved. For this reason sodium carbonate is frequently used as a buffer and materially increases the yields (7, 8). For example, with *o*-nitrobenzenearsonic acid (9) the yield can be increased by careful buffering between pH 8.8 and 9.2. Copper and other metallic salts have been used as catalysts, but no systematic study has been made of their effect, and in many cases the yields are not increased appreciably.

Many minor modifications of the original Bart reaction have been made which in some individual cases have resulted in higher yields. One well-known modification (10, 11) involves the addition of a dry diazonium fluoborate to an aqueous, alkaline solution of the arsenite. A major change in reaction conditions was introduced by Scheller (12). The amine is dissolved in an organic solvent and diazotized with a concentrated aqueous sodium nitrate solution in the presence of arsenic trichloride. In this case a cuprous salt is necessary to catalyze the reaction, and the mixture is usually heated. The Scheller reaction frequently gives somewhat larger yields than the Bart reaction when electron-attracting substituents are present; with electron-repelling substituents the yields are poor (13). The conditions of the Scheller reaction were further modified by Doak and Freedman (14) who demonstrated that stable, dry diazonium salts, such as the fluoborates, would react in an anhydrous solvent with arsenic trichloride to give products which were readily hydrolyzed to mixtures of arylarsonic and diarylarsinic acids.

Aromatic arsinic acids are obtained as by-products in the Bart (15) and Scheller (16) reactions. Small amounts of triarylarsine oxides have also been isolated from the Bart reaction (15). In the reaction under anhydrous conditions, studied by Doak and Freedman (14), the ratio of arsonic acid to arsinic acid varies with the solvent and catalyst used and also with the amount of water present. Strictly anhydrous conditions favor the arsonic acid while the addition of small amounts of water markedly increases the yields of arsinic acid.

It seems remarkable that 6-aminocoumarin, both by the Bart reaction and by a modified Bart reaction using arsenic trichloride, gave only tris(6-coumaryl)arsine oxide (17). Similarly, *o*-aminocinnamic acid gave tris(*o*-cinnamyl)arsine oxide although ethyl *o*-aminocinnamate gave the desired ethyl *o*-arsonocinnamate (18). No explanation for these peculiar results has been offered. The most satisfactory method for the preparation of diarylarsinic acids is the Bart or Scheller reaction in which a sodium arylarsonite or an aryldichloroarsine is used instead of sodium arsenite or arsenic trichloride. A study of the optimum reaction conditions required has been made by Takahashi (19, 20). Since aryldichloroarsines are readily prepared in a pure state by distillation, whereas the sodium arylarsonite is somewhat difficult to prepare, Takahashi prefers the Scheller conditions for preparing diarylarsinic acids. The yields by both reactions are similar; both symmetrical and unsymmetrical acids can be prepared.

Mann and Watson (21) state that the yields of diarylarsinic acids by the Bart reaction are sensitive to the pH of the reaction mixture. They give conditions for obtaining optimum yields of several arsinic acids; the variation of yield as a function of pH, however, was not given. More recently, Heaney and Millar (21a) have noted several failures of the Bart reaction when applied to the synthesis of certain halo-substituted diarylarsinic acids.

No kinetic studies of the Bart or related reactions have been made, and suggestions as to the mechanism of these reactions are based largely on analogy with similar reactions which have been followed kinetically. Waters (22) has suggested that the Bart reaction proceeds through spontaneous decomposition of a diazo hydroxide with the formation of free radicals which attack the arsenic. The presence of biphenyl derivatives (23, 24) as by-products of the reaction is cited by Waters as evidence for the free radical mechanism.

Bruker (25, 26) has suggested that a covalent diazo compound is formed as an intermediate, which rearranges by attack of the arsenic on the ring carbon attached to the nitrogen, with the concomitant evolution of nitrogen:

$$[ArN_2]X + AsX_3 \rightarrow [ArN_2][AsX_4]$$

$$[ArN_2][AsX_4] \rightarrow ArN=NAsX_4$$

$$ArN=NAsX_4 \rightarrow ArAsX_4 + N_2$$

This type of reaction sequence is unusual in the decomposition of diazo compounds and seems less probable than a free radical mechanism.

Cowdrey and Davis (27) postulate that the Bart reaction is analogous to the replacement of a diazo group by a nitro group and involves the attack of a carbonium ion on the anion of arsenous acid. Although this mechanism requires alkaline conditions, excess alkali must be avoided since this would favor formation of diazoates and isodiazoates. For this reason the Bart reaction is sensitive to pH and to the precise mode of mixing of the reactants. The reaction mechanism envisioned by Cowdrey and Davis is quite similar to the coupling of diazo compounds with phenols.

It is difficult, in the absence of kinetic data, to decide between the ionic and radical mechanisms. There can be no doubt that radicals are formed during the course of the reaction. Both Schmidt (23) and Bart (24) have shown that benzenediazonium salts yield *p*-biphenylarsonic acid and benzene as by-products; large amounts of tar are also formed. Although these products are certainly formed from free radicals, this does not demonstrate that the principal product, benzenearsonic acid, is necessarily formed in this manner.

The Scheller reaction and its modifications may proceed by a different mechanism than the Bart reaction, since the reaction conditions are quite different. It has been shown (28) that the thermal decomposition of diazonium salts in methanol proceeds by a heterolytic cleavage of the C—N bond. Unfortunately, in the case of the Scheller reaction, no kinetic studies have been made. Doak and Freedman (14) have studied the effect of different reaction conditions and different catalysts on the yields of the products. The following results were noted:

(a) The reaction between *p*-nitrobenzenediazonium fluoborate and arsenic trichloride in ethanol solution (and in the presence of a suitable catalyst) produced *p*-nitrobenzenearsonic acid, bis(*p*-nitrophenyl)arsinic acid, *p*-nitrochlorobenzene, and traces of *p*-nitrofluorobenzene. When cuprous bromide was used as the catalyst, *p*-nitrobromobenzene was also formed. No *p*-nitrophenetole could be detected.

(b) In the absence of a catalyst no arsonic or arsinic acid was formed. The principal product was *p*-nitrophenetole, with a smaller amount of *p*-nitrochlorobenzene.

(c) Cuprous and cupric salts and copper powder were the most effective catalysts. Smaller amounts of arsonic and arsinic acids were formed when ferrous chloride, potassium iodide, or hydroquinone was used, but ferric, chromous, and manganous chlorides gave no arsonic or arsinic acids. All of the effective catalysts (except cupric chloride which is probably reduced to cuprous chloride by the arsenic trichloride) were thus reducing agents whose oxidation-reduction potentials, at least in aqueous solution, lie within a relatively narrow range.

The above results are somewhat analogous to those found for the Sandmeyer reaction. The ability of copper salts and of ferrous ion, iodide ion, and hydroquinone to catalyze the Scheller reaction suggests that these catalysts serve as electron transfer agents. The mechanism of this reaction may be similar to that proposed by Waters (29) for the Sandmeyer reaction. We suggest the following reaction scheme:

$$ArN_2^+ + Cu^+ \rightarrow Ar\cdot + Cu^{2+} + N_2$$

$$Ar\cdot + AsCl_3 \rightarrow Ar\dot{A}sCl_3$$

$$Ar\dot{A}sCl_3 + ArN_2^+ \rightarrow ArAsCl_3^+ + N_2 + Ar\cdot$$

$$Ar\cdot + AsCl_3 \rightarrow ArCl + \cdot AsCl_2$$

$$Ar\cdot + Ar\overset{+}{A}sCl_3 \rightarrow Ar_2\overset{+}{A}sCl_2 + Cl\cdot$$

This accounts for the known products of the reaction, the effect of cuprous ion and other reducing agents with a similar redox potential, and the absence of aryl ethers.

B. The Béchamp reaction

The arsonation of aromatic amines, phenols, and phenyl ethers is known as the Béchamp reaction. This reaction is of considerable historical interest, since it was the investigation of the structure of arsanilic acid (formed in the Béchamp reaction) by Ehrlich and Bertheim that initiated a new era of research in the organic chemistry of arsenic.

Aniline and substituted anilines are readily arsonated by the Béchamp reaction. The yields are seldom over 25% and may be much smaller. Arsonation occurs in *para* position, but if this position is blocked some *ortho* substitution will occur. The reaction succeeds with both electron-repelling and electron-attracting substituents in the ring. Thus *o*-, *m*-, and *p*-toluidine, *o*- and *p*-chloroanilines, and *o*- and *p*-nitroanilines can be arsonated. α-Naphthylamine is arsonated in the 2-position (30), contrary to an earlier report which suggests arsonation in the 4-position. β-Naphthylamine gives only tars (31) although the desired arsonic acid can be prepared by the Bart reaction (32). There have been no reported attempts to prepare arsonic acids from N-alkyl- or N,N-dialkylanilines by the Béchamp reaction. Diphenylamine, however, gives a mixture of 4-arsono- and 4,4′-diarsonodiphenylamine (33).

The principal by-product in the Béchamp reaction with aniline is the secondary acid, $(p\text{-}H_2NC_6H_4)_2AsO_2H$. By performing the reaction at somewhat higher temperatures, this acid can be obtained as the main product of the reaction (34). Neither *o*- nor *m*-aminobenzenearsonic acid has been obtained; but since the *ortho* isomer is extremely soluble, it is quite probable that it is formed but has not been isolated. There is also considerable oxidation of the aniline as evidenced by the deep purple color of the reaction mixture.

Phenols can be arsonated by the Béchamp reaction, and the yields are somewhat better than with amines. Thus, phenol gives a 33% yield of *p*-hydroxybenzenearsonic acid (35). In addition to the *para* isomer, smaller amounts of *o*-hydroxybenzenearsonic acid, bis(*p*-hydroxyphenyl)arsinic acid, and a fourth compound which is believed to be (*o*-hydroxyphenyl)(*p*-hydroxyphenyl)arsinic acid have also been isolated from the reaction mixture (36). Both *o*- and *m*-cresol give a mixture of arsonic and arsinic acids (37). Resorcinol gives an excellent yield of 2,4-dihydroxybenzenearsonic acid, while the mono- and di-ethers of resorcinol give 2-methoxy-4-hydroxy- and 2,4-dimethoxybenzenearsonic acids, respectively (38).

No mechanistic studies of the Béchamp reaction have been reported. However, the reaction is somewhat analogous to sulfonation which has been extensively studied from the mechanistic viewpoint. Aniline can be sulfonated by two different processes: the first with oleum at 0°, the second a "baking process" in which anilinium hydrogen sulfate is heated at a high temperature. Ingold (39a) has pointed out that in the "baking process" the products tend to be the thermodynamically stable ones; their formation is largely independent of the mechanism of the reaction. Now the arsonation of aromatic amines and phenols is also a "baking process," and the same comments may apply to this reaction as to the high temperature sulfonation process. Unlike sulfonation, arsonation does not take place at low temperatures, so the mechanism is difficult to investigate. However, since the reaction only occurs with strongly electron-repelling groups and *meta*-substituted products are not found, it seems highly probable that the reaction involves the electrophilic attack of the arsenic on the activated *para* position of the ring. In the case of sulfonation, the amine is present as the anilinium ion; this gives rise to considerable metanilic acid by the oleum process at 0°. However, arsenic acid is a much weaker acid (pK_A 2.4), and a significant fraction of the amine will be present as the free base. It is not surprising, then, to find no *meta* products in the arsonation of aniline. The arsonation of phenol and of aryl ethers probably involves an electrophilic process, similar to the sulfonation of aromatic hydrocarbons.

Just as the sulfonation of phenol at high temperatures produces large amounts of bis(*p*-hydroxyphenyl)sulfone, so the arsonation of aromatic amines and phenols produces considerable amounts of the diarylarsinic acids as by-products. These can obviously be formed only by the electrophilic attack of the arsenic of the arsonic acid on a second molecule of amine or phenol.

C. The Meyer and Rosenmund reactions

Alkyl halides react with alkali metal arsenites to form alkylarsonic acids and the alkali halide. This reaction was discovered by Meyer (40) in 1883 and has been the most widely used general method for preparing alkylarsonic acids. Dehn (41, 42) thoroughly investigated the conditions for the reaction and found that it proceeds satisfactorily in the cold in an aqueous solution containing sufficient alcohol to dissolve the alkyl halide. When alcohol is used, however, there is considerable ether formation, which may be avoided by omitting the alcohol and running the reaction under heterogeneous conditions with efficient stirring (43–46). Arsenate ion is formed as a by-product. The rate of the reaction, which can be followed by titration of the unreacted arsenite with iodine, has been studied by Valeur and Delaby (47). Under satisfactory reaction conditions with suitable alkyl halides, the yields of arsonic acids are in the neighborhood of 90%.

Alkyl iodides react most rapidly, but bromides and chlorides have been used successfully. With chlorides the reaction has been conducted under pressure (48). When bromides and chlorides are used, the arsonic acids may be isolated directly by acidification; with iodides, however, acidification would reduce the arsenic and therefore the desired arsonic acids are isolated as sodium or magnesium salts. Methyl sulfate can be used instead of a methyl halide and gives an 85% yield of methanearsonic acid (49); similarly, sodium ethyl sulfate has been used for preparing ethanearsonic acid (50). Alkyl arylsulfonates have also been used for alkylating sodium arsenite (51).

Primary alkyl halides react rapidly and smoothly with alkali arsenites; in contrast secondary halides react only slowly while tertiary halides do not react. Allyl bromide gives a good yield of 2-propene-1-arsonic acid, but vinyl halides do not react (49). Ethylenechlorohydrin gives 2-hydroxyethanearsonic acid as an oil which crystallizes slowly to a hygroscopic solid (52).

An electron-attracting substituent in the alkyl chain greatly reduces the reaction rate and the yield of arsonic acid. Thus β-phenoxyethyl bromide gives only a 10% yield of 2-phenoxyethanearsonic acid, and ethylene dibromide reacts so slowly that no arsonic acid could be isolated (43). It is possible, however, to prepare di-arsonic acids (e.g., bis(β-arsonoethyl) ether and bis(β-arsonoethyl) sulfone) when the halogen atoms are sufficiently separated (49). α-Toluenearsonic acid is readily prepared from the bromide or iodide. Chloroform, bromoform, and iodoform do not react (42).

Arsinic acids are readily prepared by the Meyer reaction by substituting a sodium or potassium alkylarsonite for the alkali arsenite (43, 45, 53, 54). Mixed dialkylarsinic acids can be readily prepared (43, 53, 54). If, instead of an alkali alkylarsonite, an arylarsonite is used the alkylarylarsinic acid may be prepared in excellent yield (43, 54). Thus, ethyldichloroarsine and ethyl bromide in alkaline solution yield diethylarsinic acid; *n*-butyldichloroarsine and *n*-propyl bromide give *n*-butyl-*n*-propylarsinic acid; and phenyldichloroarsine and sodium chloroacetate give (carboxymethyl)phenylarsinic acid.

An interesting variant of the Meyer reaction was discovered by Kuskov and Vasilyev (55), who demonstrated that potassium methylarsonite and ethylene oxide give (β-hydroxyethyl)methylarsinic acid.

Aryl halides do not react readily with alkali arsenites, and earlier workers were unable to prepare arylarsonic acids by this method. In 1921, however, Rosenmund (56) prepared benzenearsonic acid from bromobenzene and potassium arsenite in a sealed tube at 180–200°. Cupric sulfate was used as a catalyst, and the yields were poor. He succeeded in preparing *o*-arsonobenzoic acid under reflux conditions. A few other investigators have used this procedure, but the yields are seldom satisfactory. *o*-Benzenediarsonic acid may be prepared from *o*-iodobenzenearsonic acid, and *p*-arsonoacetophenone from *p*-bromoacetophenone (57). Arsonic acids could not be obtained from *o*-bromo- or *o*-chlorobenzene, *p*-nitrobromobenzene, or 2,4-dinitrochlorobenzene (58). The reaction also failed with 8-chlorocaffeine,

2,4-dinitrobromobenzene, and 3 different chloroglyoxalines (59). Thus the Rosenmund reaction may be said to be of little synthetic importance.

The Meyer reaction is closely analogous to the Arbuzov rearrangement in organic phosphorus chemistry. This latter reaction, which involves the formation of dialkyl alkylphosphonate from a trialkyl phosphite and an alkyl halide, has been studied extensively from the point of view of mechanism (60). The reaction involves the nucleophilic displacement of halogen by the phosphite ester with the formation of a quasi-phosphonium ion, which is then attacked by the halide ion with elimination of RX:

$$(RO)_3P: + R'X \longrightarrow [R'P^+(OR)_3]X^- \longrightarrow (RO)_2\underset{\underset{\displaystyle O}{\|}}{P}R' + RX$$

Trialkyl arsenites do not react with alkyl halides. No reaction occurs when tributyl arsenite is heated for 8 hours with butyl iodide in a sealed tube at 210° (61). Similarly, when tributyl arsenite and benzyl chloride were heated together under reflux, the reactants were recovered unchanged (62). The arsenite ion, however, is sufficiently nucleophilic to allow the Meyer reaction to proceed, and we can envision the first step as an S_N2 reaction involving the attack of arsenite ion on the carbon, and subsequent elimination of NaX:

$$(NaO)_3As: + RX \rightarrow [RAs(ONa)_3]^+X^- \rightarrow RAsO(ONa)_2 + NaX$$

The fact that primary alkyl halides react much faster than secondary halides while tertiary halides do not react is strongly suggestive of the S_N2 mechanism.

D. The Grignard reaction

Irgolic, Zingaro, and Smith (63) have described the use of dichloro(diethylamino)arsine as an intermediate in the preparation of the C_5–C_{20} arsinic acids in yields from 50–80%. Reaction of this intermediate with Grignard reagents results in the replacement of the two chlorine atoms by alkyl groups:

$$(C_2H_5)_2NAsCl_2 + 2RMgX \rightarrow (C_2H_5)_2NAsR_2 + 2MgXCl$$

The organic product of the Grignard reaction is then hydrolyzed with dilute hydrochloric acid and finally oxidized with aqueous hydrogen peroxide:

$$(C_2H_5)_2NAsR_2 + H_3O^+ \rightarrow R_2AsOH + (C_2H_5)_2\overset{+}{N}H_2$$

$$R_2AsOH + H_2O_2 \rightarrow R_2AsO_2H + H_2O$$

More recently, bis(diethylamino)chloroarsine has been used for the synthesis of long chain arsonic acids in yields of from 50–90% (64). When the chloroarsine reacts with a Grignard reagent, a bis(N,N-diethyl) amide of an alkylarsonous acid is formed. Hydrolysis of this amide yields an alkylarsonous acid which is not isolated but oxidized with hydrogen peroxide to the arsonic acid.

E. Miscellaneous methods

(*1*) *Oxidation and hydrolysis*

All trivalent arsenic compounds containing a single organic radical attached to arsenic can be oxidized either to an arsonic acid or to a product which is readily hydrolyzed to an arsonic acid. Aliphatic arsines, arsenosoalkanes, arsenoalkanes, and esters of the type $RAs(SR')_2$ are oxidized by air to arsonic acids (65). Aromatic arsines are oxidized to a mixture of arsonic acid and the arsenoarene, but arsenosoarenes and arsenoarenes require oxidizing agents other than air. Dichloroarsines, both aliphatic and aromatic, are oxidized by chlorine to the corresponding tetrachlorides which are, in turn, readily hydrolyzed to arsonic acids. Nitric acid, hydrogen peroxide, and chloramine T are among the oxidizing agents which have been used. All trivalent arsenic compounds containing two organic radicals directly attached to arsenic are readily oxidized to arsinic acids or to products which can be hydrolyzed to arsinic acids. Tetramethyldiarsine (cacodyl) is spontaneously inflammable either in air or in an atmosphere of chlorine. Bis(dimethylarsenic) oxide (cacodyl oxide) may be oxidized with mercuric oxide or atmospheric oxygen to dimethylarsinic (cacodylic) acid (66, 67). Diphenylarsine and tetraphenyldiarsine are oxidized by air to mixtures of diphenylarsinic acid and bis(diphenylarsenic) oxide (68, 69). Diarylhaloarsines are oxidized by halogens to diarylarsenic trihalides, which in turn can be hydrolyzed to diarylarsinic acids.

(*2*) *Replacement of a sulfonic acid group*

Only one example of this reaction is known (70). When *p*-hydroxybenzenesulfonic acid was heated with arsenic acid at 135° for 4 hours, a 90% yield of *p*-hydroxybenzenearsonic acid was obtained. The apparent activation by the *p*-hydroxy group suggests that the reaction involves the electrophilic attack of arsenic on the carbon atom bearing the sulfonic acid group.

(*3*) *Preparation of arsinic acids by cyclodehydration*

Certain arsonic acids will undergo cyclodehydration to give heterocyclic arsinic acids. Thus, N-phenyl-*o*-arsanilic acid, when boiled with hydrochloric acid, gives phenazarsinic acid hydrochloride (71):

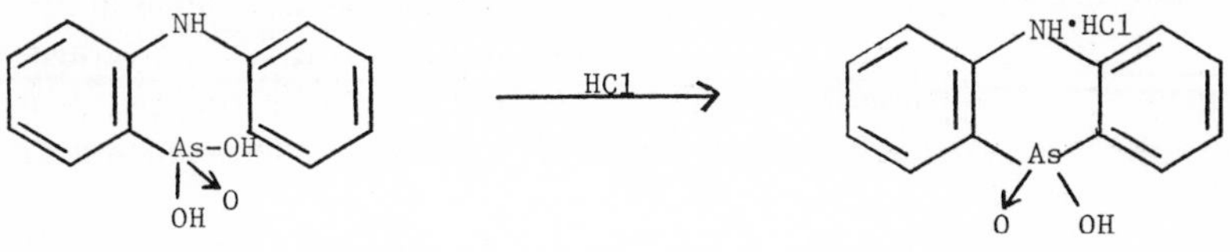

Gibson and Johnson (71) observed the formation of an oily intermediate which they postulated might be the chloride:

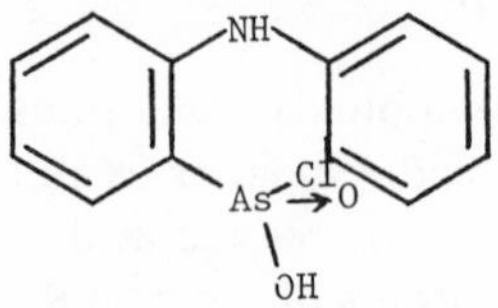

When there is a methyl group in *ortho* position to the imino group on the ring which does not contain the arsono group, cyclization occurs at a slower rate. *o*-(*m*-Chlorophenoxy)- and *o*-(*p*-chlorophenoxy)benzenearsonic acids and *o*-(*p*-toloxy)benzenearsonic acid are readily cyclized by heating with sulfuric acid (72). These reactions presumably involve protonation of the arsonic acid as the first step and subsequent electrophilic attack of the arsenic on the ring carbon atom. *o*-Phenoxybenzenearsonic acid and *o*-(*o*-chlorophenoxy)benzenearsonic acid do not undergo cyclization but are sulphonated instead (73).

(*4*) *Cleavage of tertiary arsine oxides*

The preparation of diphenylarsinic acid by the alkaline cleavage of triphenylarsine oxide is discussed in Section V-2.

2. REACTIONS

A. Acid dissociation constants

Although it has been known for a long time that arsonic acids are moderately strong acids and readily form salts, it was not until the mid-nineteen thirties that a systematic investigation of their dissociation constants was published by Backer and his co-workers (74–76). The results reported from Backer's laboratory are given in Table 2-1 as "observed pK's." The symbols pK_1 and pK_2 refer to the first and second dissociation constants of the arsono group. The acid dissociation constants of the carboxy groups present in several of the compounds were also determined and are given in the footnotes to the table. The "calculated pK's" were obtained by the present authors by using an equation derived by Branch and Calvin (77) for the dissociation constants of non-resonating acids. The level of agreement between calculated and observed pK's for the compounds in Table 2-1 is not as good as in the case of most of the examples cited by Branch and Calvin. The average discrepancy between observed and calculated pK's for the arsonic acids is 0.5 unit; the largest discrepancy is 1.1 unit. The equation of Branch and Calvin

Table 2-1 Acid Dissociation Constants of Aliphatic Arsonic Acids

$RAsO_3H_2$ R =	Observed[a]		Calculated	
	pK_1	pK_2	pK_1	pK_2
CH_3	3.61	8.24	4.4	9.3
C_2H_5	3.89	8.35	4.4	9.3
$CH_3(CH_2)_2$	4.21	9.09	4.5	9.3
$CH_3(CH_2)_3$	4.23	8.91	4.5	9.3
$CH_3(CH_2)_4$	4.14	9.07	4.5	9.3
$CH_3(CH_2)_5$	4.16	9.19	4.5	9.3
CH_2ClCH_2	3.68	8.37	4.0	8.9
$CH_3CHClCH_2$	3.76	8.39	4.1	8.9
$CH_2Cl(CH_2)_2$	3.63	8.53	4.3	9.2
$CH_3CHCl(CH_2)_2$	3.95	8.84	4.3	9.2
$CH_3CH_2CHCl(CH_2)_2$	3.71	8.77	4.3	9.2
$CH_3(CH_2)_2CHCl(CH_2)_2$	3.51	8.31	4.3	9.2
$CH_2{=}CHCH_2$	3.75	8.76	—	—
HO_2CCH_2[b]	2.94	7.68	—	—
$HO_2C(CH_2)_2$[c]	3.73	7.92	—	—
$C_2H_5CH(CO_2H)$[d]	3.01	7.64	—	—
$CH_3(CH_2)_2CH(CO_2H)$[e]	3.00	7.74	—	—
$CH_2{=}C(CO_2H)$[f]	2.91	8.60	—	—
$CH_3CH{=}C(CO_2H)$[g]	3.36	8.75	—	—
$HO_2CCH{=}C(CH_3)$[h]	2.59	8.81	—	—

[a] The dissociation constants were determined in water by potentiometric titration and are nonthermodynamic values.
[b] pK_{CO_2H}, 4.67.
[c] pK_{CO_2H}, 4.78.
[d] pK_{CO_2H}, 4.92.
[e] pK_{CO_2H}, 4.89.
[f] pK_{CO_2H}, 4.23.
[g] pK_{CO_2H}, 4.61.
[h] pK_{CO_2H}, 4.03.

predicts that the difference between pK_1 and pK_2 for the arsono group should be 4.9 units. If the carboxy-substituted compounds are excluded, the observed pK_2–pK_1 values vary from 4.46 to 5.06 (with an average value of 4.76). It is, therefore, possible to make a rough estimate of the dissociation constants of an aliphatic arsonic acid, provided that the necessary inductive constants are known. It also seems possible to calculate with fair accuracy one dissociation constant of an arsonic acid when the other dissociation constant has been experimentally determined. This last statement is *not* true for the arsonocarboxylic acids, in which the carboxy group is un-ionized during the determination of pK_1 but negatively charged during the determination of pK_2. The dissociation constants of the carboxy groups in these compounds indicate that the AsO_3H^- group has virtually no effect on the acid strength of aliphatic carboxylic acids. For example, the dissociation constant of the carboxy group in arsonoacetic acid is not significantly different from that of acetic acid.

As a class, arsonic acids are stronger acids than carboxylic acids, but weaker than sulfonic acids or phosphonic acids (78). In general, arsonic acids are weaker than arsenic acid, H_3AsO_4, itself ($pK_1 = 2.66$; $pK_2 = 5.89$) (79). An exception to this rule is trifluoromethanearsonic acid which has been

Table 2-2 Relative Strength of Acids in Acetic Acid

Compound	Relative Strength
Perchloric acid	360
Bis(trifluoromethyl)phosphinic acid	250
Hydrogen bromide	180
Sulfuric acid	32
Trifluoromethylphosphonic acid	9
Hydrogen chloride	9
Bis(trifluoromethyl)arsinic acid	3.5
Trifluoromethanearsonic acid	2.5
Trifluoroacetic acid	1
Nitric acid	1
Heptafluorobutyric acid	1

shown (80) by conductivity measurements in water to be almost completely ionized at a concentration of $0.01M$. From the conductivity data pK's of 1.11 and 5.3 were estimated. When glacial acetic acid was used as the solvent, the results summarized in Table 2-2 were obtained in which the acid strength of nitric acid, trifluoroacetic acid, and heptafluorobutyric acid were taken as unity (81). It is apparent from this table that the trifluoromethyl group has a powerful acid-strengthening effect. Some of the data, however, are difficult to understand. Thus, there seems no obvious way to explain why bis(trifluoromethyl)arsinic acid is only slightly stronger than trifluoromethanearsonic acid while bis(trifluoromethyl)phosphinic acid is almost 28 times as strong as trifluoromethylphosphonic acid.

Since the last pK's of many of the aliphatic arsonic acids are greater than 8, none of the commonly used indicators can be relied upon for determining the neutralization equivalents of these acids. The alkalimetric titration of several aliphatic diarsonic acids has been studied by Backer and Bolt (82).

They found that these acids have two relatively strongly acidic protons which can be titrated with methyl red as the indicator, a third proton which is removed at the thymolphthalein end-point (pH about 9.3), and a fourth proton which is too weakly acidic to be titrated conveniently in aqueous solution.

The acid dissociation constants of a considerable number of aromatic arsonic acids have been reported (83–94). Pressman and Brown (86) noted that the Hammett $\sigma\rho$ equation (95) applies to the pK values of the *meta*- and *para*-substituted acids and that the reaction constants (ρ) for arsonic and

Table 2-3 Reaction Constants (ρ) for the Acid Dissociation of Aromatic Acids

Reaction	ρ	s[a]	n[b]	$-\log k^{\circ}$[c]
$ArB(OH)_2 \rightleftharpoons ArBO_2H^- + H^+$	2.146[d]	0.096	14	9.700
$ArCO_2H \rightleftharpoons ArCO_2^- + H^+$	1.000[e]	—	—	4.203
$ArAsO_3H_2 \rightleftharpoons ArAsO_3H^- + H^+$	1.050	0.117	9	3.540
$ArAsO_3H^- \rightleftharpoons ArAsO_3^= + H^+$	0.874	0.102	11	8.491
$ArPO_3H_2 \rightleftharpoons ArPO_3H^- + H^+$	0.755	0.030	10	1.836
$ArPO_3H^- \rightleftharpoons ArPO_3^= + H^+$	0.949	0.058	12	6.965

[a] The standard deviation from the best straight line.
[b] The number of compounds on which the calculation of ρ is based.
[c] The calculated value of pK for $\sigma = 0$.
[d] The solvent used for the arylboric acids was 25% aqueous ethanol. Water was used for all the other series of acids.
[e] By definition.

benzoic acids have approximately the same value. Reaction constants calculated by Jaffé (96) for the dissociation of several related series of aromatic acids are given in Table 2-3. From a comparison of the reaction constants of benzoic, boric, and arsonic acids, Pressman and Brown (86) concluded that the most important resonance structures of aromatic arsonic acids do not involve formal charges on the arsenic atom, but consist of forms in which ten electrons surround the arsenic:

$$C_6H_5-As(=O)(OH)-OH \qquad C_6H_5-As(-O^-)(OH)=\overset{+}{O}-H$$

Later work (97), however, has indicated that resonance structures in which the arsenic atom has expanded its valence shell make little contribution to the ground state of aromatic arsonic acids. Therefore, an alternative interpretation of the relative values of the reaction constants was proposed by

Jaffé, Freedman, and Doak (97). They pointed out that the difference in reaction constants must depend on the effectiveness with which the electrical effects of the substituents are transmitted to the reaction site in the various compounds. The acid series under discussion differ primarily in the nature of the central atom, i.e., B, C, As, or P. It appears reasonable to assume, therefore, that the reaction constant must depend on the polarizability of the central atom. If we assume that ionic refractivities are valid measures of the polarizabilities of these atoms, it turns out that there is, in fact, a good correlation between the polarizability of the central atom and the corresponding reaction constant.

The pK of the carboxy group in *p*-carboxybenzenearsonic acid has been determined and used to calculate the substituent constant (σ) of the *p*-AsO_3H^- group. A value of -0.019 was obtained, which is not significantly different from zero. It appears, therefore, that the AsO_3H^- group has no appreciable electron-attracting or electron-repelling effect. A similar conclusion was reached earlier from a study of the acid dissociation constants of aliphatic acids.

The pK values of several *ortho*-substituted aromatic arsonic acids have been measured and are listed in Table 2-4. Since acid dissociation constants are generally insensitive to steric hindrance, the pK values of corresponding *ortho* and *para* compounds do not differ greatly except when there is intramolecular hydrogen bonding between the *ortho* substituent and the acid function. An example of such an "*ortho* effect" is provided by *o*-nitrobenzenearsonic acid which is significantly weaker than the *para*- isomer ($pK_1 = 2.90$; $pK_2 = 7.80$). The second pK is affected to a larger extent, since the proton to be removed in this dissociation is directly involved in a hydrogen bond:

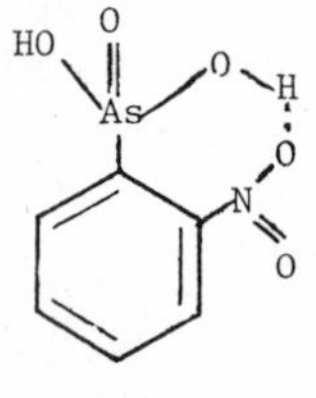

The first pK is also affected since an oxygen of the nitro group can transmit its electron-repelling effect through the hydrogen bond. Such an effect is, of course, acid-weakening (89).

Juillard (94) has determined the acid dissociation of benzenearsonic acid and a number of other organic acids in various mixtures of water and methanol. He concluded that the pK_A of an acid increases linearly with the

Table 2-4 Acid Dissociation Constants of ortho-Substituted Aromatic Arsonic Acids

Acid	$pK_{AsO_3H_2}$	$pK_{AsO_3H^-}$	Literature Reference
o-Toluenearsonic	3.82	8.85	86
o-Nitrobenzenearsonic	3.37	8.54	86
o-Aminobenzenearsonic	2.0[a]	8.66	86
o-Biphenylarsonic	4.29	9.33	89
2,4-Dichlorobenzenearsonic	3.14	—	85
1-Naphthalenearsonic	3.66	8.66	86
4-Amino-1-naphthalenearsonic	—	9.17	86

[a] Approximate value.

mole fraction of methanol in the mixture up to a mole fraction of 0.6. For benzenearsonic acid the variation in pK_A is given by

$$\Delta pK = AN$$

where A = 2.5 and N = mole fraction of methanol.

Like the aliphatic arsonic acids, the aromatic compounds are, in general, too weak to be titrated as dibasic acids in water with the aid of the usual indicators. It has been reported (98, 99), however, that neutral equivalents of aromatic arsonic acids can be obtained by first titrating with 0.1*N* sodium hydroxide to the thymolphthalein end-point, then adding an equal volume of saturated sodium chloride solution and continuing the titration of the decolorized solution until a blue color is again produced. Benzenearsonic acid itself and aromatic arsonic acids with electron-attracting substituents can be titrated accurately as *monobasic* acids to the methyl purple end-point (pH 5.4) (100). Neutral equivalents of aromatic arsonic acids can also be determined potentiometrically with an antimony electrode (101).

Comparatively little is known about the acid dissociation constants of arsinic acids. The dissociation constant of cacodylic (dimethylarsinic) acid has been reported by a number of investigators, and the compound has often been used by biochemists for buffering solutions in the pH range 5.2–7.2 (102). Using both colorimetric and electrometric methods, Kilpatrick (103) found a pK value for cacodylic acid of 6.27 at zero ionic strength. This result agrees quite well with the value of 6.2 calculated by the equation of Branch and Calvin. The pK's of several other dialkylarsinic acids have been recently reported (104); the values range from 6.51 to 7.25.

Cacodylic acid is not only an acid, but it is also a base, i.e., it can accept a proton as indicated by the following equation:

$$(CH_3)_2AsO_2H + H^+ \rightleftharpoons (CH_3)_2AsO_2H_2^+$$

Kilpatrick (103) found by catalytic measurements that the equilibrium constant of this reaction is 37.3 (at an ionic strength of 0.03). This equilibrium constant corresponds to an acid dissociation constant for the protonated species of 2.68×10^{-2} or a pK of 1.57. Other arsinic acids and even arsonic acids can also be protonated, but equilibrium constants have not been reported. The literature (105–108) contains numerous descriptions of the preparation and properties of salts of the types $[RAsO_3H_3]X$ and $[R_2AsO_2H_2]X$, where R may be aliphatic or aromatic and X is Cl, Br, NO_3, ClO_4, etc.

As indicated in Table 2-2, bis(trifluoromethyl)arsinic acid is probably the strongest acid of arsenic known. The pH of a 0.0312*M* aqueous solution is 1.67 (the pH of 0.0312*M* hydrochloric acid is 1.50). Titration of the solution with 0.1*N* sodium hydroxide yields an end-point at pH 3.7. If the titration with base is continued rapidly, a second end-point at about pH 9 is observed (109). (*Rapid* titration is necessary since the arsinic acid decomposes to fluoroform in alkaline solution.) It has been suggested that bis(trifluoromethyl)arsinic acid exists in part as $(CF_3)_2As(OH)_3$ and is therefore capable of acting as a dibasic acid. From conductivity data Hantzsch (110) had postulated that cacodylic acid may be dibasic in strongly alkaline solution. This effect is, however, very slight in the case of cacodylic acid.

The neutral equivalents of diarylarsinic acids can be determined readily by titration in aqueous alcohol with phenolphthalein as the indicator (14, 100, 111). They can also be determined by potentiometric titration using antimony (112) or glass (63) electrodes.

B. Reduction

Both aliphatic and aromatic arsonic and arsinic acids can be readily reduced; the reduction product depends both on the reducing agent and on reaction conditions. It is also possible to reduce other substituents in the molecule without reducing the arsenic atom. Thus, nitroarylarsonic acids are reduced to the corresponding aminoarylarsonic acids either with zinc in alkali (113) or catalytically with Raney nickel and hydrogen at low pressure (114). Although this latter method works well with *m*- and *p*-nitrobenzene arsonic acids, no amino compound is obtained when the nitro group is in *ortho* position to the arsonic acid (115). However, at high pressure (94–97 atmospheres) and at temperatures of 50–90°, all three isomers are readily reduced catalytically to the corresponding amino compounds (116). With the *ortho* compound some aniline is also formed. The excellent method of Mann and Wilkinson (117) which employs iron and sodium chloride can also be used to reduce *o*-nitrobenzenearsonic acid to the amino compound (118). This method is vastly superior to reduction with ferrous hydroxide which has frequently been used to reduce *o*-nitrobenzenearsonic acid to the amino-arsonic acid (119). Glucose and alkali have also been used for reducing *o*-nitro acids (120). 3-Nitro-4-aminobenzenearsonic acid has been reduced to the diamino acid with sodium hydrosulfite (121). It is surprising that the arsenic is not reduced, since this reagent is frequently used for reducing arsonic acids to the corresponding arseno compounds.

The first reduction product of the arsonic acid group is the arsenoso compound (or the dichloroarsine in hydrochloric acid solution). This reduction is of particular interest to the pharmacologist, since the therapeutic activity of arsonic acids against certain microorganisms is believed to depend upon the reduction of these acids *in vivo* to arsenoso compounds (122). Cohen, King, and Strangeways (123) studied the rate of oxidation of a series of arsenoso compounds on the assumption that this might be correlated with the rate of reduction of the corresponding arsonic acids *in vivo*, and thus with their toxicity. No such correlation was found. Baranger (124) studied the oxidation-reduction potential of the arsono–arsenoso system. The potentials registered by electrodes immersed in buffered mixtures of arylarsonic acid and its corresponding arsenoso compound were consistent with the following electrode equation:

$$2H^+ + 2e^- + ArAsO_3H_2 \rightleftharpoons ArAsO + 2H_2O$$

Measurements were made only at pH 8.9, and thus we can not be certain that the effect of pH on the electrode potential is in accord with the assumed equation. Nor was evidence presented to indicate that the potentials were independent of the total concentration of oxidant and reductant as is predicted by the above equation. Nevertheless, there seems no reason to doubt Baranger's conclusion that the arsono–arsenoso system is electromotively active and hence can be studied by potentiometric methods.

For the chemical reduction of arsonic acids to arsenoso compounds (Section III-1-B) or dihaloarsines (Section III-6-A-(1)), phosphorus trichloride or sulfur dioxide and hydriodic acid are the customary reducing agents (125, 126). With strong electron-repelling groups on the benzene ring it is often difficult or impossible to reduce the arsonic acid group without cleaving arsenic from the ring (127, 128).

Thionyl chloride has also been used as a reducing agent, but the results are unsatisfactory (129). Thus, benzenearsonic acid and thionyl chloride give only a 50% yield of phenyldichloroarsine; the remaining arsenic is cleaved from the ring with the formation of chlorobenzene and arsenic trichloride. 1-Anthraquinonearsonic acid gives only 1-chloroanthraquinone; diphenylarsinic acid gives diphenylarsenic trichloride. More recently, Kolomiets and Levskaya (130) have reported that the reaction of thionyl chloride with arsonic and arsinic acids in nonpolar solvents can be used for the preparation of compounds of the types $RAsOCl_2$ and R_2AsOCl, respectively. Unfortunately, few experimental details are given, and the new compounds are not characterized adequately. The preparation of compounds of type $RAsOF_2$ has been claimed in the patent literature (cf. Section V-5).

An interesting and unexpected reduction of arylarsonic acids has been reported by Pathak (131). When arylarsonic acids were treated with phosphorus oxychloride, chlorine was evolved and the arsenoso compound was obtained by treating the reaction mixture with alkali. The author postulates that the reaction follows the course:

$$RAsO_3H_2 \xrightarrow{POCl_3} [RAsCl_4] \longrightarrow RAsCl_2 + Cl_2$$

$$RAsCl_2 \xrightarrow{NaOH} (RAsO)_x$$

Since arylarsenic tetrachlorides are generally stable, it is difficult to understand why chlorine is evolved in the above reaction.

Arsinic acids have been reduced to the corresponding haloarsines with sulfur dioxide in hydrohalic acids (132, 133), with phosphorus trichloride (134) or with the stoichiometric amount of hypophosphorous acid in hydrochloric acid solution (135).

With stronger reducing agents both aliphatic and aromatic arsonic acids may be reduced to arseno compounds (cf. Section VI-1-A). The usual reducing agents have been hypophosphorous acid (136), phosphorous acid (137), sodium hydrosulfite (138), or stannous chloride with a trace of hydriodic acid (139). Similarly, aliphatic and aromatic arsinic acids are reduced by such reducing agents as phosphorous (140) or hypophosphorous acid (135) to bis(dialkyl)- or bis(diaryl)diarsines.

Arsonic and arsinic acids are reduced by amalgamated zinc and hydrochloric acid to the corresponding primary or secondary arsines (41, 68, 141–143). Since the arsines are all readily oxidized, these reductions are performed with the exclusion of air.

Electrolytic reduction has been extensively used for reducing arsonic acids. It is possible to obtain different compounds with different conditions of reduction. Thus with platinum, nickel, or copper electrodes, aromatic nitroarsonic acids are reduced to the corresponding aminoarsonic acids (144). With other cathodes either the arseno compound or the arsine is produced, depending on reduction conditions. With a mercury cathode in $8N$ hydrochloric acid, arsanilic acid is reduced to *p*-aminophenylarsine; but in solutions weaker than $8N$ in hydrochloric acid, the arseno compound is produced. Not only is it possible to produce the arseno compound, but the degree of polymerization (cf. Section VI-3) can be controlled by careful regulation of the current density (145). Electrolytic reduction is believed (146) to take the following course:

$$RAsO_3H_2 \rightarrow RAsO \rightarrow (RAsOH)_2 \rightarrow (RAs)_2O \rightarrow (RAs)_n \rightarrow (RAsH)_2 \rightarrow RAsH_2$$

A thorough study of electrolytic reduction of arsonic acids was made by Fichter and Elkind (147). They obtained the best results for the production of *p*-aminophenylarsine from *p*-aminobenzenearsonic acid with a cathode of amalgamated zinc and a current density of 0.08 amp/sq cm. If the reduction cell was not carefully cooled, arsenic was split from the ring with the production of aniline and arsenous acid. No reduction of the arsonic acid occurred in alkaline solution. Cacodylic acid could also be reduced electrolytically in hydrochloric acid solution with the production of cacodyl, $[(CH_3)_2As]_2$.

The reduction of arsonic acids has also been studied polarographically with the dropping mercury electrode. Breyer (85) obtained half-wave potentials for seven different arsonic acids as follows:

$C_6H_5AsO_3H_2$, −1.274; 4-$CH_3C_6H_4AsO_3H_2$, −1.290;

4-$CH_3CONHC_6H_4AsO_3H_2$, −1.300; 4-$CH_3OC_6H_4AsO_3H_2$, −1.354;

2,4-$Cl_2C_6H_3AsO_3H_2$, −1.364; 4-$HOC_6H_4AsO_3H_2$, −1.366;

4-$H_2NC_6H_4AsO_3H_2$, −1.464.

No indication of the nature of the reduction products was given. Since the majority of the work was performed in unbuffered solutions at a single pH, or in buffered solutions the strengths of which were not given, the results obtained are of doubtful significance.

Ríus and Carrancio (148) made a similar polarographic study of 3-nitro-4-hydroxy-, 3-amino-4-hydroxy-, and 3-acetamido-4-hydroxybenzenearsonic acids. The reductions were studied in buffered solutions at several different pH values. Although the nitro compound gave half-wave potentials, these could be attributed to reduction of the nitro group. They were unable to demonstrate any reduction of the amino and acetamido compounds and reached the conclusion that pentavalent (4-covalent) arsenic could not be reduced by the dropping mercury electrode. Similar results have been reported by Wallis (149).

Diphenylarsinic acid has been reduced with the dropping mercury electrode (150). A single half-wave potential at −1.2 volts at pH 3 was obtained. Maruyama and Furuya (151, 152) studied a group of bis(nitrophenyl)- and bis(aminophenyl)arsinic acids, as well as benzenearsonic acid and its three amino-substituted isomers at the dropping mercury electrode. They concluded that both the arsonic and arsinic acids were reduced to the corresponding trivalent arsenic compounds and that, in fact, only arsenicals possessing the As=O group could be reduced at the dropping mercury electrode.

C. Reactions with electrophilic reagents

(*1*) *Nitration*

The arsono and arsinico groups are strongly electron-attracting and hence are deactivating towards electrophilic reagents. Thus benzenearsonic acid is not nitrated with fuming nitric acid even at 100° but can be nitrated with anhydrous nitric acid or with mixed sulfuric–nitric acid (153, 154). The constitution of the acid was established as *m*-nitrobenzenearsonic acid by Bertheim and Benda (155). Since that time a large number of arsonic acids have been nitrated, and it has been assumed that the arsono group orients exclusively to the *meta* position. The nitration of benzenearsonic acid has recently been reinvestigated (156). By the use of the magnesium salt technique (cf. Section II-2-F), it was found that in addition to the *meta* isomer a minimum of 1.6 per cent of the *ortho* isomer was formed. This result suggests that many other nitrations of arsonic acids may give a mixture of isomers. Furthermore, since the only by-product of nitration was the *ortho* isomer, it may be concluded that the arsono group is of the —I, —M type (39b). This could occur only through double bond formation between carbon and arsenic, i.e., the expansion of the arsenic valence shell to ten electrons through the use of *d*-orbitals:

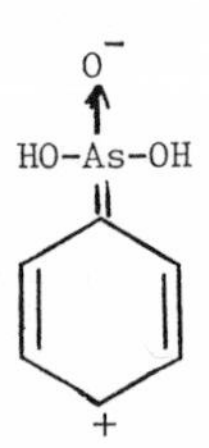

As might be expected, electron-repelling groups facilitate nitration. Thus, *p*-anisolearsonic acid gives a 95% yield of 3-nitro-4-methoxybenzenearsonic acid when nitrated with sulfuric-nitric mixture at −8° (157). Similarly, *p*-hydroxybenzenearsonic acid is readily nitrated to give 3-nitro-4-hydroxybenzenearsonic acid at 0° (158). If the nitration is carried out with sulfuric acid and fuming nitric acid (d. 1.52), 3,5-dinitro-4-hydroxybenzenearsonic acid is obtained. By contrast with electron-repelling groups, electron-attracting groups hinder nitration of aromatic arsonic acids. *m*-Arsonobenzoic acid could not be nitrated under quite drastic conditions, namely, sulfuric acid and potassium nitrate at 65° for 48 hours (159).

A large number of aromatic arsonic acids have been nitrated under a variety of conditions. The problem of orientation has frequently arisen in this regard. Often the expected compound can be made independently by the Bart or Scheller reaction. Other means for establishing orientation have also been used. Thus, 3-carboethoxyaminobenzenearsonic acid yields a mononitro derivative (160), which might be expected to be the 2-nitro, the 4-nitro or the 6-nitro compound. Orientation of this compound was established (120) as 2-nitro-3-carboethoxyaminobenzenearsonic acid by conversion of the carboethoxyamino group to a hydroxy group and reduction of the nitro to an amino group. Treatment of hydroxyaminobenzenearsonic acid with chloracetyl chloride gave 3-hydroxy-1,4-benzisoxazine-5-arsonic acid and thus established that the original compound had nitrated in the 2-position. In some cases the orientation studies have produced doubtful results and should be reexamined in the light of modern knowledge of orientation effects. Andreev (161) nitrated α-naphthalenearsonic acid and claimed that nitration occurred in the 4-position. Fusion of the nitrated compound with potassium hydroxide gave 2-naphthol, while treatment with phosphorus pentachloride gave a chloronitronaphthalene which was believed to be the 1,4-compound. Since the arsono group is strongly deactivating to electrophilic substitution, it would seem unlikely that the 4-position would be nitrated in preference to the 5- or 8-positions. The results of Andreev's work should be reinvestigated. (*o*-Arsonophenyl)phenylarsinic acid has been nitrated with a mixture of fuming nitric acid, sulfuric acid and oleum at 25° (162). A mononitro compound was obtained which presumably was (*o*-arsonophenyl)(*m*-nitrophenyl)-arsinic acid. Diphenylarsinic acid on nitration with fuming nitric acid and sulfuric acid probably gives bis(*m*-nitrophenyl)arsinic acid, although no proof of orientation has been reported. Blicke and Webster (163) have used mixed acid to nitrate (*p*-nitrophenyl)phenylarsinic acid and (*p*-hydroxyphenyl)phenylarsinic acid. The former arsinic acid was mononitrated to yield a substance which was assumed to be (*m*-nitrophenyl)(*p*-nitrophenyl)-arsinic acid. (*p*-Hydroxyphenyl)phenylarsinic acid yielded a dinitro derivative, which the authors believed to be (3-nitro-4-hydroxyphenyl)(3-nitrophenyl)-arsinic acid.

The difficulty in nitrating benzenearsonic acid, particularly in comparison with the relative ease of nitrating phenylphosphonic acid, has been considered by Freedman and Doak (156). The suggestion has been made that the difference lies in the relative base strength of benzenearsonic and phenylphosphonic acids. Since the two protonated acids have estimated pK_A values of

-1 and -3, respectively, it might be expected that benzenearsonic acid would be completely protonated in fuming nitric acid whereas phenylphosphonic acid would not. The difficulty in nitrating benzenearsonic acid then can be attributed to the strongly deactivating effect of a positively charged arsono group.

(2) *Halogenation*

Relatively few arsonic or arsinic acids containing halogen groups have been prepared by direct halogenation. Aliphatic halogen-substituted arsonic acids are invariably prepared by indirect methods such as the action of phosphorus pentachloride on β-hydroxyethyldichloroarsine and the subsequent oxidation of the resulting β-chloroethyldichloroarsine to the arsonic acid (164). Nor has it been found possible to halogenate a side chain of an aromatic arsonic acid. Thus, Hamilton and King (165) found that arsenic was cleaved from the ring when *p*-toluenearsonic acid was chlorinated at 185–190° in hexachloroethane; with hypochlorous acid 3-chloro-4-toluenearsonic acid was obtained. The direct halogenation of other aromatic arsonic acids has been achieved when electron-repelling substituents are present in the ring. At the same time, the electron-repelling group renders the carbon atom bearing the arsenic atom more susceptible to attack by electrophilic agents and thus promotes splitting of the carbon-arsenic bond. Amino- and hydroxy-substituted arsonic acids, where the substituent is *ortho* or *para* to the arsenic atom, are readily dearsonated with the formation of the corresponding trihalogenated aniline or phenol. Thus arsanilic acid, when treated with bromine either in water or dilute mineral acid, yields 2,4,6-tribromoaniline (166), and *p*-hydroxybenzenearsonic acid gives 2,4,6-tribromophenol (167). Nevertheless, it is possible to halogenate either arsanilic acid or its acetyl derivative under suitable conditions. This may be achieved by the use of a milder halogenating agent or working at lower temperatures. *p*-Acetamidobenzenearsonic acid, when chlorinated in glacial acetic acid or in aqueous acetic acid at 0° with hypochlorous acid, gives 3-chloro-4-acetamidobenzenearsonic acid. Bromination of *p*-acetamidobenzenearsonic acid could not be achieved, but 3-bromo-4-aminobenzenearsonic acid was prepared by bromination of arsanilic acid in acetic acid solution with one-half the theoretical amount of bromine. Under more drastic conditions 3,5 dichloro-, 3,5-dibromo-, and 3,5-diiodo-4-aminobenzenearsonic acids were prepared (166). In all such cases, however, some arsenic was cleaved from the ring. In marked contrast to the *para* isomer, *m*-aminobenzenearsonic acid is readily brominated in acid solution to form 2,4,6-tribromo-3-aminobenzenearsonic acid (168). The reaction of arsonic acids with sulfur tetrafluoride to form arsenic tetrafluorides and oxyfluorides is discussed in Section V-5.

(3) Coupling with diazonium salts

Evidence that diazonium salts would couple with amino- and hydroxy-benzenearsonic acids was based originally on the formation of colored products when the reactants were mixed (36, 169). In other work Benda (170) demonstrated that when benzenediazonium salts coupled with *p*-hydroxy-benzenearsonic acid, arsenic was cleaved from the ring with the formation of *p*-phenylazophenol and arsenic acid. The problem of diazo coupling with arene arsonic acids was reinvestigated by Doak, Steinman, and Eagle (171). Both *o*- and *m*-hydroxybenzenearsonic acids coupled readily with benzenediazonium chloride to give 2-hydroxy-5-phenylazo- and 5-hydroxy-2-phenylazobenzene-arsonic acids, respectively. There was no indication that any arsenic was split from the ring. With *p*-hydroxybenzenearsonic acid the reaction was somewhat more complicated. In carefully buffered solution (pH 7.3–7.4) some coupling *ortho* to the hydroxy group occurred, and 4-hydroxy-3-phenylazobenzenearsonic acid was isolated. At the same time there was an electrophilic attack of the diazonium cation on the carbon bearing the arsenic, and both *p*-phenylazophenol and 2,4-bis(phenylazo)phenol were isolated from the reaction mixture. At higher pH values, where the effect of the phenoxide ion would weaken the carbon-arsenic bond, no phenylazo-arsonic acid could be isolated, and 2,4-bis(phenylazo)phenol was the principal product. Similar results were also obtained with 4-hydroxy-1-naphthalenearsonic acid, where it was possible to isolate 4-hydroxy-3-phenylazo-1-naphthalenearsonic acid in a carefully buffered solution.

(4) Sulfonation

The only example of the sulfonation of an aromatic arsonic acid is with α-naphthalenearsonic acid which can be sulfonated with fuming sulfuric acid (172). Although the orientation of the mono-sulfonated naphthalenearsonic acid was not determined, substitution probably occurs in the 5 or 8 position, i.e., in the ring not bearing the arsono group.

(5) Miscellaneous electrophilic reactions

Sodium bisulfite solution when refluxed with 3,5-dinitro-4-hydroxybenzene-arsonic acid cleaves arsenic from the ring with the formation of 2,6-dinitro-phenol (173). Although the mechanism of this reaction is unknown, it probably involves an electrophilic attack on the carbon bearing the arsenic and subsequent separation of arsenic acid.

Both *o*- and *p*-aminobenzenearsonic acids are hydrolyzed in aqueous alkali at elevated temperatures (130–160° and 160°, respectively) (174).

D. Reactions with nucleophilic reagents

The reaction between aromatic arsonic acids and nucleophilic reagents may take place either with retention or with loss of the arsono group. In the former category are the classical reactions of organic chemistry in which a suitably activated group, usually halogen, is displaced by a nucleophilic reagent. A large number of arsonic acids have been prepared in this manner. Maclay and Hamilton (175) studied the reaction of 3-nitro-4-halogenobenzenearsonic acids with amines and phenols. In aqueous alkaline solution at 125–135°, or in amyl alcohol in the presence of anhydrous potassium carbonate, the halogen was displaced by phenoxide ion to form the 3-nitro-4-phenoxybenzenearsonic acid. Since the singly charged arsono group has a Hammett σ constant of -0.019 (96) and the σ of the doubly charged arsono group must be more negative, the arsono group in alkaline solution cannot be contributing to the activity of the aryl halide. Fourneau and Funke (176) studied the reaction of 3-nitro-4-chlorobenzenearsonic acid with ethylenediamine and piperazine and were able to obtain compounds in which either one or both amino groups reacted with the arsonic acid. A number of other workers have also studied nucleophilic reactions with 3-nitro-4-halogenobenzenearsonic acids and obtained similar results (177–179).

Although the negatively charged arsono group is not activating for nucleophilic substitution, a halogen *ortho* to the arsonic acid group may be displaced when copper is used as the catalyst. Thus Etzelmiller and Hamilton (180) found that *o*-chlorobenzenearsonic acid in amyl alcohol in the presence of potassium carbonate and copper or cuprous iodide reacted with amines, phenols, and primary aliphatic alcohols to give the corresponding *o*-amino-, *o*-phenoxy-, or *o*-alkoxybenzenearsonic acids. By contrast, *p*-bromobenzenearsonic acid could not be condensed with aniline, nor was the bromine removed by boiling with 3*N* alkali for 48 hours. It has also been reported (181) that 2-bromo-4-nitrobenzenearsonic acid, when refluxed with piperidine, gave 2-piperidino-4-nitrobenzenearsonic acid. In the course of their classical studies on phenarsazines, Gibson and co-workers (182, 183) condensed *o*-bromobenzenearsonic acid with anthranilic acid to give N-(*o*-arsonophenyl)-anthranilic acid and with benzidine to give N,N′-*p*-biphenylenedi-*o*-arsanilic acid.

The arsono group itself may be displaced by nucleophilic reagents under suitable conditions. Thus benzenearsonic acid is converted into phenol and arsenous acid by fusion with potassium hydroxide (184). Similarly *o*-nitrobenzenearsonic acid is converted to *o*-nitrophenol by boiling with aqueous alkali (57). The arsono group, when activated by the ammonio group, is readily displaced by the iodide ion. *o*-Amino- and *p*-aminobenzenearsonic acids, when boiled with potassium iodide in sulfuric acid solution, give *o*-iodo- and *p*-iodoaniline, respectively (185, 186). Obviously the iodide ion is acting as a nucleophilic reagent under these conditions. Bis(*p*-aminophenyl)arsinic acid also gives *p*-iodoaniline (and presumably potassium arsenite) under similar conditions (187).

Rupture of the arsenic–carbon bond by boiling with mineral acids occurs readily with aliphatic arsonic acids, but not with aromatic arsonic acids. The reaction was studied by Petit (188, 189) who found that heating methanearsonic or cacodylic acid with concentrated sulfuric acid gave arsenous oxide, sulfur dioxide, carbon dioxide, and water. Petit proposed that the first step in the reaction was a hydrolysis by water to form methyl alcohol and arsenous acid. In support of this idea he found that, when methanearsonic acid was heated with syrupy phosphoric acid, he could isolate methyl alcohol from the reaction mixture.

It is difficult to understand how there could be sufficient water in concentrated sulfuric acid to effect the hydrolysis of methanearsonic acid. A more likely mechanism might involve a nucleophilic attack of the bisulfate ion on the protonated acid, with the separation of arsenous acid and methyl hydrogen sulfate:

$$CH_3AsO_3H_2^+ \xrightarrow{HSO_4^-} CH_3OSO_3H + As(OH)_3$$

Oxidation of the ester by sulfuric acid might then occur to yield carbon dioxide and sulfur dioxide.

The hydrolysis of α-toluenearsonic acid by mineral acids is a more complex reaction and several different products are obtained (42). With hydrochloric acid, benzyl chloride, toluene, and arsenic trichloride are formed. Benzyl chloride is obviously formed by a nucleophilic attack of chloride ion on the methylene carbon atom. Toluene can only be formed by an attack of a proton on this same carbon atom, presumably with the separation of metaarsenic acid.

Sulfuric acid reacts with α-toluenearsonic acid to produce bibenzyl and benzaldehyde. The mechanism of this reaction is not apparent. It seems possible that benzyl radicals are formed which react with α-toluenearsonic acid to produce bibenzyl and an arsenic acid radical. The stoichiometry of the reaction is probably:

$$3C_6H_5CH_2AsO_3H_2 + H_2O \rightarrow C_6H_5CH_2CH_2C_6H_5 + C_6H_5CHO + 3As(OH)_3$$

β-Styrenearsonic acid, which is obtained in poor yield from β-bromostyrene by the Meyer reaction, is also cleaved by mineral acid with the formation of arsenous acid (190).

E. Formation of esters, anhydrides, and sulfides

(*1*) *Esters*

Until recently, few esters of arsonic acids had been prepared by direct esterification of the acids. Michaelis (154) attempted the preparation of the alkyl esters of benzenearsonic acid by two methods: (a) the reaction of sodium alkoxides with benzenearsonic dichloride ($C_6H_5AsOCl_2$) and (b) the reaction between the silver salt of the acid and alkyl chlorides. Only the latter method was successful. In this manner he prepared dimethyl and diethyl benzenearsonates. The silver salt technique has also been used by Komissarov, Maleeva, and Sorokooumov (191) for the preparation of alkyl esters of aliphatic arsonic acids.

Dialkyl alkylarsonates have also been prepared by oxidation of dialkyl alkylarsonites with selenium dioxide (191–194). Esters prepared in this manner are less pure than those obtained by means of the silver salt technique. The oxidation of alkyl dialkylarsinites with selenium dioxide (or mercuric oxide) is an unsatisfactory method for preparing the esters of dialkyl–arsinic acids (194). Thus propyl diethylarsinite gave none of the desired propyl diethylarsinate; instead the products consisted of propyl alcohol, diethylarsinic acid, and the unchanged trivalent ester. Butyl diethylarsinite gave 12 per cent butyl diethylarsinate together with diethylarsinic acid and unchanged ester.

Finally, alkyl esters of arsonic and arsinic acids have been prepared by direct esterification (61, 195–200); azeotropic distillation with benzene or toluene has usually been employed to remove the water formed during the reaction. The acids esterified by this procedure include a variety of aliphatic and aromatic arsonic acids as well as dimethyl-, phenylmethyl-, and di-phenylarsinic acids. Levskaya and Kolomiets (198) have made a study of the conditions required to obtain optimum yields of these esters.

A very interesting result was obtained when glycols, *o*-dihydric phenols, or α-hydroxy acids were used in the direct esterification reaction (61). In these cases, spirans of the following types were obtained:

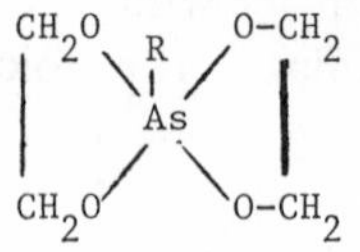

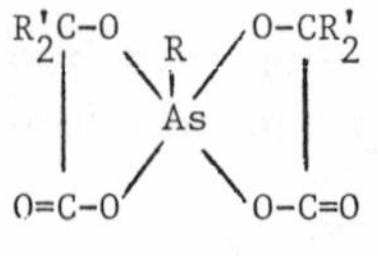

```
CH2O   R   O-CH2
 |   \ | /   |
 |    As     |
 |   /   \   |
CH2O       O-CH2
```

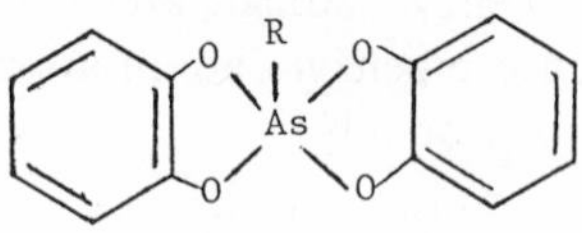

```
        O  R  O
         \ | /
 (C6H4)   As   (C6H4)
         /   \
        O     O
```

```
R'2C-O   R   O-CR'2
 |     \ | /    |
 |      As      |
 |     /   \    |
O=C-O        O-C=O
```

Two methods for synthesizing these compounds were used, namely, azeotropic distillation or heating the arsonic acid and alcohol in acetic anhydride. No analyses for the compounds were given, but the authors stated that the analytical results were in agreement with the theoretical. Somewhat similar results had been reported earlier by Englund (199, 200) who condensed arsonoacetic acid with ethylene glycol, pinacol, and pyrocatechol to give spirans. In this work, however, the products were obtained simply by heating the components together. The cryoscopic molecular weight of the spiran derived from arsonoacetic acid and ethylene glycol showed this compound to be dimeric in bromoform.

An interesting group of esters containing As—O—Si bonds has been prepared by Kary and Frisch (201, 202). Chlorosilanes were condensed with arsonic acids according to the equation:

$$2R_nSiX_{(4-n)} + R'AsO_3H_2 \longrightarrow R_n(X_{3-n})SiO\overset{\overset{\displaystyle O}{\|}}{\underset{\underset{\displaystyle R'}{|}}{As}}OSi(X_{3-n})R + 2HCl$$

Hydrolysis of the resulting products gave polymers containing Si—O—As bonds.

Esters containing the As—O—Sn bond have also been described. A number of compounds of the type $[RAs(O)(OH)O]_2Sn(n\text{-}C_4H_9)_2$ have been prepared by condensing arsonic acids with either dibutyltin diacetate or dibutyltin oxide (203). Both aliphatic and aromatic arsonic acids were used. A more extensive study of compounds containing the As—O—Sn linkage has been made by Chamberland and MacDiarmid (204). Dimethyltin dichloride reacts with both disodium methane- and benzenearsonates to form polymers of the type $[—Sn(CH_3)_2OAsR(O)O—]_x$. With the free acid, however, only one hydrogen is displaced, even in boiling benzene (with pyridine as a hydrohalide acceptor) or in the molten state; compounds of the type $[ArAs(O)(OH)O—]_2Sn(CH_3)_2$ are formed. The authors suggested that the reaction involves nucleophilic attack of the As=O group on tin; in aqueous solution ionization of the Sn—Cl bond facilitates the reaction. The silver salt of dimethylarsinic acid reacts with dimethyltin dichloride to form $[(CH_3)_2As(O)O]_2Sn(CH_3)_2$. Triphenyltin chloride and disodium benzenearsonate form $C_6H_5As(O)[OSn(C_6H_5)_3]_2$.

Henry (205) has prepared a series of organic lead arsonates and lead arsinates in which organic radicals are attached to both the lead and the arsenic atoms. Thus the reaction of sodium dimethylarsinate with triphenyllead chloride or diphenyllead dichloride leads to the compounds $(CH_3)_2As(O)OPb(C_6H_5)_3$ and $(CH_3)_2As(O)OPb(Cl)(C_6H_5)_2$, respectively. Benzylarsonic acid reacts with triphenyllead chloride in methanol solution to form $[C_6H_5CH_2As(O)(OH)O]Pb(C_6H_5)_2$, together with benzene and hydrogen chloride, whereas propanearsonic acid and triphenyllead chloride form the compound $CH_3CH_2CH_2As(O)(OH)OPb(C_6H_5)_2Cl$. The reaction of methanearsonic acid and triphenyllead chloride gives complex products of uncertain composition. The organolead arsonates are cleaved by acetic acid with the formation of organolead acetates. Compounds of the type $Ph_2AsO_2MR_3$ (where M = Ge, Sn, or Pb) are further discussed in Section VI-7.

Compounds containing the As—O—Hg linkage have been prepared by means of the reaction between an arsonic acid and an organomercury acetate (206) or hydroxide (207):

$$RAsO_3H_2 + 2CH_3CO_2HgR' \rightarrow RAsO(OHgR')_2 + 2CH_3CO_2H$$

$$RAsO_3H_2 + 2HOHgR' \rightarrow RAsO(OHgR')_2 + 2H_2O$$

The substituents (both R and R′) included a number of alkyl and aryl groups.

Aluminum, gallium, and indium derivatives of dimethylarsinic acid have been prepared by means of the following reaction (208):

$$Me_2AsO_2H + Me_3M \rightarrow Me_2AsO_2MMe_2 + MeH$$

where M = Al, Ga, or In. The yields of these compounds were in excess of 90%.

When dialkyl alkylarsonates are distilled at atmospheric pressure, considerable rupture of the arsenic–carbon bond occurs (193, 209, 210). Thus diethyl propanearsonate gave diethyl propyl arsenite together with unchanged ester; dipropyl propanearsonate gave propyl alcohol; dibutyl propanearsonate gave butyl alcohol, dibutyl propanearsonite, and dibutyl propyl arsenite. The rearrangement $RAsO(OR')_2 \rightarrow ROAs(OR')_2$ may be an intermolecular S_N2 rearrangement or intramolecular S_Ni. Unfortunately, the yields of identified products were only a small percentage of the total ester distilled, and it is difficult to arrive at any mechanism when most of the products of the reaction are not accounted for.

Chernokal'skii and co-workers (196, 211, 212) have published several detailed studies of the isomerization of alkyl esters of aliphatic arsonic acids. They found that the reaction is markedly promoted by the addition of alkyl halides. Thus, when heated alone, $CH_2{=}CHCH_2AsO(OC_2H_5)_2$, $C_2H_5AsO(OC_2H_5)_2$, and $n\text{-}C_4H_9AsO(OC_2H_5)_2$ require temperatures of 155, 227, and 240°, respectively, in order for isomerization to start. In the presence of ethyl iodide, however, all three esters begin to isomerize at 92°. The effectiveness of alkyl halides as catalysts decreases in the order RI > RBr > RCl. The authors suggested that S_N2 mechanisms are involved in the isomerization:

$$RAsO(OR'')_2 + R'X \rightleftharpoons R\overset{\oplus}{A}s(OR'')_2(OR')X^{\ominus}$$

$$\downarrow$$

$$As(OR'')_2(OR') + RX$$

The above mechanism is supported by kinetic data and electrical conductivity measurements (212).

(2) *Anhydrides*

Arsonic acids, when slowly heated, usually lose water to form anhydrides. It is possible, however, by rapid heating of the acid, or by placing it on a preheated melting point bath or block, to obtain satisfactory melting points for most arsonic acids (213). The bath or block should be only 5° lower than the melting point of the acid. Frequently, in taking melting points of arsonic acids, one observes that they soften and then resolidify to the anhydride. α-Toluenearsonic acid does not form an anhydride but on heating decomposes to form benzyl alcohol, benzaldehyde, stilbene, arsenic trioxide, and water (42).

Trifluoromethanearsonic acid, when heated in vacuum, undergoes progressive dehydration to give a pyro acid which on heating at higher temperatures gives the anhydride (80):

$$CF_3AsO_3H_2 \xrightarrow{35°/10^{-2}\,mm} CF_3\overset{\overset{O}{\uparrow}}{\underset{\underset{OH}{|}}{As}}—O—\overset{\overset{O}{\uparrow}}{\underset{\underset{OH}{|}}{As}}CF_3 \xrightarrow{73°/10^{-2}\,mm} CF_3AsO_2$$

The pyro acid shows an O—H absorption in the infrared at 4.35 μ (2299 cm^{-1}), a band which is absent in the anhydride.

The actual structure of arsonic acid anhydrides has not been determined. They are usually written as $RAsO_2$. It is more probable that they are polymeric substances formed by intermolecular loss of water. Their high melting points and insolubility in most solvents are in accord with this structure. They are rapidly hydrolyzed to the corresponding acids on treatment with water.

Mixed anhydrides of the type $RAs(O)[OP(S)(OEt)_2]_2$ have been prepared by the reaction of the disodium salt of an arsonic acid with $ClP(S)(OEt)_2$ (214). Analogous anhydrides of arsinic acids were prepared in a similar manner.

(3) *Sulfides*

There has been little modern work on sulfides of arsonic acids, and much of the older work is of questionable nature. The reaction between arsonic acids and hydrogen sulfide apparently may produce any one of three compounds: the sesquisulfide, $(RAs)_2S_3$, the disulfide, $RAsS_2$, or the thioarsenoso compound, RAsS. No systematic study of reaction conditions, however, has been made.

Benzenearsonic (215) and *p*-acetamidobenzenearsonic (216) acids when treated with hydrogen sulfide in ammoniacal solution give the corresponding sesquisulfides on acidification. *m*-Nitrobenzenearsonic acid in ammoniacal solution is reduced to *m*-thioarsenosoaniline, *m*-$H_2NC_6H_4AsS$, but in acid solution *m*-nitrophenylarsenic sesquisulfide is obtained (153). Finally, the action of hydrogen sulfide on N-(*p*-arsonophenyl)glycine in aqueous solution (216) and on 3-nitro-4-chlorobenzenearsonic acid in aqueous acetic acid (217) gives the corresponding disulfides.

Sesquisulfides have also been prepared by the reaction between arsonic acids and carbon disulfide (218–220) or potassium ethyl xanthate (220). Benzenearsonic acid with either carbon disulfide or potassium ethyl xanthate gives a mixture of $(C_6H_5As)_2S_3$, As_2S_3, and unchanged arsonic acid. If substituents are present which react with carbon disulfide or potassium ethyl xanthate, preferential reaction with the substituent may occur. Arsanilic acid reacts with potassium ethyl xanthate to give a mixture of $(p\text{-}SCNC_6H_4As)_2S_3$, $p\text{-}S{=}C(HNC_6H_4As)_2S_3$, $p\text{-}S{=}C(HNC_6H_4AsO_3H_2)_2$, and As_2S_3. The ratio between the four products is a function of the concentration of reactants used. Disulfides may also be produced in this reaction; 3-amino-4-hydroxy- and 3,4-diaminobenzenearsonic acids react with carbon disulfide to give the following compounds:

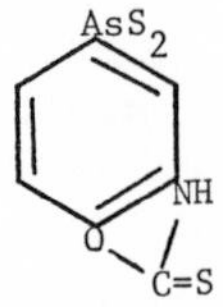

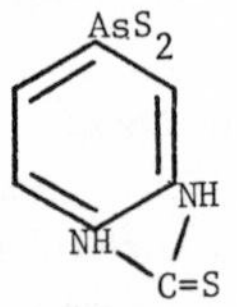

The reaction of dimethylarsinic acid with carbon disulfide gives an 11% yield of $(CH_3)_3AsS$ and a 51% yield of $[(CH_3)_2As]_2S$ (219).

When phenylarsenic sesquisulfide is treated with sodium hydrosulfide and sulfur, the sodium salt of the trithionic acid, $C_6H_5AsS_3Na_2$, is obtained (215). The free acid, however, cannot be obtained; acidification of the sodium salt regenerates the sesquisulfide.

F. Salts of arsonic and arsinic acids

Arsonic acids are dibasic and form both acid and neutral salts. The salts of alkali metals are quite soluble in water whereas the salts of heavy metals are insoluble, and the latter salts have frequently been used for isolation purposes. Of particular interest are the calcium and magnesium salts of arsonic acids which are soluble in the cold but precipitate on boiling. This property of alkaline earth salts, particularly the magnesium salt, has been known for many years, and advantage has frequently been taken of this fact to separate arsonic acids from arsenic acid, since the latter forms an insoluble magnesium salt in the cold.

A systematic investigation of the solubility properties of magnesium salts of aromatic arsonic acids has been made by Freedman and Doak (156). These authors found only one arsonic acid (N-*p*-toluylarsanilic acid) which gave an insoluble magnesium salt in the cold. *o*-Toluenearsonic and *o*-bromobenzenearsonic acids gave sparse precipitates on heating, but all other *ortho*-substituted arsonic acids investigated failed to give insoluble magnesium salts. With the exception of three acid-substituted arsonic acids, all other compounds lacking *ortho* substituents gave copious precipitates of the magnesium salts when the solutions were boiled. The behavior of arsonic acids is quite similar in this respect to the aromatic phosphonic acids (221). Advantage was taken of the differing solubility of *ortho*-substituted arsonic acids to separate the mixture of *o*- and *m*-nitrobenzenearsonic acids formed on the nitration of benzenearsonic acid.

Aliphatic arsonic acids also form insoluble magnesium salts on heating but no systematic study has been attempted. By contrast aliphatic and aromatic arsinic acids do not give insoluble magnesium salts even on boiling, but again systematic studies have not been done.

Arsonic acids have found widespread use in analytical chemistry (222). It has been found possible by varying the organic radical attached to arsenic to obtain a variety of arsonic acids which give insoluble precipitates specifically with a variety of different metals, and this has allowed the quantitative determination of such metals in mixtures. In addition a large number of arsonic acids give specific color reactions with different metals; this effect has been used for the detection of such metals and also in their determination by spectrophotometric techniques.

One of the most useful of the arsonic acids has been 2-(2-hydroxy-3,6-disulfo-1-naphthylazo)benzenearsonic acid. This reagent was originally introduced by Kuznetsov (223–224) who found it to give an insoluble red precipitate with thorium. This acid under the name Thoron or Thorin has now come into wide use for the colorimetric determination of thorium (225, 226). Interference caused by zirconium can be eliminated by the addition of *meso*-tartaric acid (227). Thoron has also been used for the determination of zirconium (228), hafnium (228), beryllium (229), lithium (230), bismuth (231), neptunium (232), and fluoride ion (233).

Closely related to Thoron is *o*-[(1,8-dihydroxy-3,6-disulfo-2-naphthyl)-azo]benzenearsonic acid (Arsenazo I or Neothorin). This compound was also introduced by Kuznetsov (223) and is of value in the colorimetric determination of aluminum (234), beryllium (235), uranium (236), thorium (237), and many other elements (238), as well as fluoride ion (239). Kuteinikov

(240) has studied the effect of pH and metal to reagent ratio on the formation of Arsenazo I complexes and has determined their visible spectra and equilibrium constants.

Several other reagents related to Thoron and Arsenazo I have been introduced by Kuznetsov (241, 242). Two of these have been named Thoron II and Arsenazo II; they are analogues of Thoron and Arsenazo I in which 3,3′-diamino-4,4′-biphenyldiarsonic acid, rather than *o*-aminobenzenearsonic acid, is coupled with the appropriate naphtholsulphonic acid. These two new reagents are said to have advantages over Thoron and Arsenazo I in the determination of some elements. An Arsenazo I derivative, [2-(2-carboxyazophenylazo)-7-(2-arsonophenylazo)-1,8-dihydroxynaphthalene-3,6-disulfonic acid, is known as Carboxyarsenazo and is a useful indicator for titrating sulfate ions (243). Organic sulfur compounds can be decomposed by the Schöniger method and then titrated with barium nitrate solution in the presence of the above indicator.

Another Arsenazo compound, Arsenazo III, was introduced by Savvin (244) in 1959. An improved method for its preparation was given by Zaikovskii and Ivanova (245) in 1963. Arsenazo III is 2,2′-[(1,8-dihydroxy-3,6-disulfo-2,7-naphthylene)bis(azo)]dibenzenearsonic acid. Although it gives color reactions with a smaller number of elements than Arsenazo I, it possesses the marked advantage that it forms stable inner complexes with certain elements so that they can be determined in strongly acid solution. It is particularly valuable for the determination of thorium, zirconium, uranium, protoactinium, scandium, and the rare earth elements. The complexes formed with Arsenzao III are 2–3 times more stable than those formed by Arsenazo I (246). The use of Arsenazo III as an analytical tool has been reviewed by Savvin (247, 248).

A number of other azo-substituted aromatic arsonic acids have been introduced as analytical reagents by various workers. Buděšínský, Haas, and Vrzalová (249) prepared a number of substituted derivatives of Arsenazo III. Arsenazo DAL (250) is similar to Arsenazo III but contains sulfonamido rather than sulfonic acid groups. It has been used for the determination of thorium. Savvin and Milyukova (251) have studied a number of Arsenazo III derivatives for the determination of plutonium. The best were Arsenazo III itself and Arsenazo-amino-ε-acid. Sulfarsazen, which contains a diazoamino group, has been introduced as a reagent for the determination of mercury (252). Palladiazo (253) is similar to Arsenazo III except that the two arsonic acid groups are *para* rather than *ortho* to the azo groups. It is a selective agent for palladium. An excellent review article by Savvin (238) lists many of the Arsenazo-type compounds that have been made, together with their chemical properties and their uses in analytical chemistry.

Pietsch and his co-workers have made a thorough investigation of a number of arsonic and arsinic acids as precipitants for metallic ions. By careful adjustment of pH many of these reagents are quite specific for individual ions. Among the acids tested in this manner were *o*-, *m*-, and *p*-nitrobenzenearsonic acids (254), *o*-, *m*-, and *p*-aminobenzenearsonic acids (255–258), *o*-, *m*-, and *p*-toluenearsonic acids (259), *o*-, *m*-, and *p*-chlorobenzenearsonic acids (260), methanearsonic acid (261), cacodylic acid (261), diethyl-, dipropyl-, and di-*n*-butylarsinic acids (262–264). Phenazarsinic acid has proved to be a valuable reagent for the detection and determination of nitric acid or nitrate ion (265, 266). Other arsonic acids which have found use in quantitative analysis include: *p*-acetamidobenzenearsonic acid for the determination of nickel (267); benzenearsonic, *p*-hydroxybenzenearsonic, propanearsonic, and *p*-(*p'*-dimethylaminophenylazo)benzenearsonic acids for the determination of zirconium (268); *p*-hydroxybenzenearsonic acid for the determination of lead (269); *m*-nitrobenzenearsonic acid for the determination of titanium (270); *o*-aminobenzenearsonic acid for the determination of gold (271); and 3-nitro-4-hydroxybenzenearsonic acid for the determination of cadmium, manganese, mercury, lead, tin, thorium, zinc, and zirconium (272, 273).

A turbimetric method for the determination of tantalum by the use of benzenearsonic acid has also been developed (274).

Pietsch (275) has studied the effect of structure and the pH of the solution on the precipitation of metallic ions. For example, 2,3- and 3,4-dimethylbenzenearsonic acids give precipitates with ions such as Zr^{4+}, Ti^{4+}, and Th^{4+}. The precipitation is a function of the pH of the solution, but this pH is the same for both acids. There is a slight difference in the pH at which precipitation occurs with the two acids and such cations as Cd^{2+}, Cr^{3+}, etc., while with cations such as Mg^{2+}, Ca^{2+}, Hg^{2+}, and Ni^{2+} precipitation occurred at pH values which differed by one pH unit. Solutions containing Sr^{2+}, Ba^{2+}, V(V), and Cr(VI) did not give precipitation with either acid at any pH. Similar studies have also been made with α- and β-naphthalene- and benzenearsonic acids (276). The effect of different ring substituents on the solubility of metallic salts of arsonic acids has also been studied (277). In general, electron-repelling groups decreased solubility, while electron-attracting groups increased solubility. A method for the determination of beryllium, tin, and bismuth by the use of benzenearsonic acid was developed on the basis of the above generalizations.

3. PHYSICAL PROPERTIES AND STRUCTURE

A. Absorption spectra

The electronic absorption spectra of a large number of aromatic arsonic and arsinic acids have been determined in water, methanol, ethanol, and

0.1*N* hydrochloric acid (88, 90–93, 238, 278–289). In the case of benzenearsonic acid, the last two solvents give almost identical results (280, 283). The absorption spectra of the benzenearsonate anions $C_6H_5AsO_3H^-$ and $C_6H_5AsO_3^=$ in aqueous solution have also been determined (280). Jaffé (280) was the first to point out that the arsono group usually causes no profound change in the characteristics of the spectrum of the parent aromatic compound. As can be seen in Table 2-5, the absorption maxima of benzenearsonic acid show bathochromic shifts of about 10 mμ, and the intensity of absorption of the secondary (benzenoid) band is increased by a factor of about 4. The

Table 2-5 Ultraviolet Absorption Maxima

Compound	λ_{max}, mμ	ε_{max}	Literature Reference
Benzene	203.5	7,400	290
	254[a]	204	
Benzenearsonic acid	214.5	7,600	283
	262[a]	740	
Diphenylarsinic acid	220	18,300	283
	263[a]	1,390	
Biphenyl	248	16,600	282
o-Biphenylarsonic acid	239	8,510	282
	276	2,890	
Arsafluorinic acid	226	25,700	282
	232.5	29,600	
	240	28,000	
	276.5	8,210	
	287.5	6,870	

[a] A number of fine structure bonds are centered around this maximum.

fine structure found in the spectrum of benzene (290) is virtually unaltered by the presence of the arsono or the arsinico (AsO_2H) group. The molar absorbance of diphenylarsinic acid is approximately twice that of benzenearsonic acid; otherwise the spectra of the two compounds are not significantly different. It seems likely that the slight bathochromic shift and the moderate hyperchromic effect noted in the spectra of these acids should be ascribed to weak resonance interaction between the "pentavalent" arsenic atom and the aromatic ring.

Substitution of an arsono group in an *ortho* position of biphenyl *does* have a marked effect on the spectrum of the parent compound (282). The intense absorption band near 250 mμ characteristic of biphenyl and its *meta* and *para* substituted derivatives is replaced by two maxima of moderate intensity. This effect has been explained by assuming that the *o*-arsono group interferes with a planar arrangement of the two phenyl rings and thus inhibits resonance interaction between them. Molecular models indicate that the two rings in *o*-biphenylarsonic acid are, in fact, almost perpendicular to one another; and the abnormally low acid dissociation constants of this compound suggest that there is a hydrogen bond between the arsono group and one of the rings (89). It seems clear that the *o*-arsono group must be very effective in restricting rotation around the carbon–carbon bond joining the two rings in the biphenyl molecule.

The ultraviolet absorption spectrum of arsafluorinic acid is characterized by a number of intense bands (282). This compound is also a biphenyl derivative in which free rotation of the rings is restricted. In this case, however, the two phenyl rings are held in a planar (or near planar) arrangement. The possibility of resonance between the rings is therefore enhanced; and, accordingly, the ultraviolet absorption of arsafluorinic acid is even more intense than that of biphenyl. The fine structure in the spectrum of this acid is a common feature of the spectra of bridged biphenyl derivatives (291).

It has often been observed that the ultraviolet absorption spectra of corresponding arsenic and phosphorus compounds are virtually identical (280, 282, 285, 286, 288, 292, 293). This generalization has proved quite useful in helping to establish the structure of certain heterocyclic derivatives of phosphorus (286, 293).

The infrared spectra of arsonic acids, arsinic acids, and related compounds containing the XO_2H group (where X is As, P, S, Se, or C) have been investigated in the laboratory of F. G. Mann (294). It was found that these compounds usually show a well-defined absorption pattern, consisting of a characteristic trio of very broad bands in three regions, viz., 2800–2400 cm^{-1}, 2350–1900 cm^{-1}, and 1720–1600 cm^{-1}. These bands are probably associated with vibrations of the OH groups under conditions of strong hydrogen bonding. Thus, they disappear when the hydroxyl groups are removed by salt formation. The band in the 2800–2400 cm^{-1} region almost certainly should be assigned to the O—H stretching vibration. The band in the 2350–1900 cm^{-1} region probably also involves an O—H vibration, but a precise assignment has not been made. The lowest frequency band was originally assigned to an O—H deformation vibration, but recent work (295) has shown that its position changes only slightly on deuteration and that the band probably should be attributed to a combination of vibrations of the XO_2H group.

Although the presence of the trio of broad bands in the infrared spectrum is a definite indication of the presence of the XO_2H group, the absence of any or even all of the bands does not necessarily rule out this group. For example, the band in the 1720–1600 cm^{-1} region is often missing or ill-defined when *two* hydroxyl groups are attached to X. This behavior is shown by some aliphatic and aromatic arsonic acids (205).

The infrared spectra of seventeen aliphatic arsonic acids have been measured in the 4000–200 cm^{-1} range by McBrearty, Irgolic, and Zingaro (64). The trio of broad bands was present in all cases, but the band near 1600 cm^{-1} was rather weak. Most of the spectra contained two very strong absorption bands at 775 and 940 cm^{-1} which were assigned to symmetric and asymmetric arsenic–oxygen stretching vibrations, respectively. The compounds were almost transparent near 600 cm^{-1}, i.e., in the region where an arsenic–carbon stretching vibration is normally observed. There were, however, two strong bands at 330 cm^{-1} and 380 cm^{-1} in all the spectra; the authors (64) suggested that these bands may arise from arsenic–carbon modes.

Bardos and co-workers (296) have recently examined the infrared spectra of a number of arsanilic acid derivatives. They observed strong, broad bands at 2800 and 2350 cm^{-1} (attributed to the AsO_3H^- group), a very strong sharp peak at 1090–1102 cm^{-1} (assigned to the C_6H_5–As bond), a strong-to-medium peak at 880–910 cm^{-1} (assigned to As=O), and a medium-to-strong peak at 760–780 cm^{-1} (As—O).

The infrared spectra of bis(trifluoromethyl)arsinic acid and its silver salt have also been determined (109). Both spectra show a band in the 1200 cm^{-1} region (assigned to C—F absorption) and two weak broad bands near 2300 cm^{-1}. It was suggested that these weak bands should be ascribed to the OH group and that they may indicate that the acid exists in part as $(CF_3)_2As(OH)_3$, which gives a silver salt $(CF_3)_2As(OH)_2(OAg)$. A similar conclusion was drawn from the results of the potentiometric titration of this arsinic acid (109).

A band near 2300 cm^{-1} has also been found (80) in the spectra of trifluoromethanearsonic acid, its mono-silver salt, and the pyro acid, $(CF_3AsO_2H)_2O$. As expected, this band is absent in the spectra of disilver trifluoromethanearsonate and the anhydride, CF_3AsO_2. All of the trifluoromethyl arsenic acids and their derivatives show an absorption band in the 830–810 cm^{-1} region; this band was ascribed to the As=O stretching vibration. The infrared spectra of di-*n*-butylarsinic and dicyclohexylarsinic acids have bands at 890 and 880 cm^{-1}, respectively, which have been assigned to As=O stretching (104).

Petit (297) and Guha (298) have examined the Raman spectra of a number of inorganic and organic arsenicals, including methanearsonic acid and cacodylic acid. The organic compounds had an intense band near 600 cm^{-1}, which was assigned to a C—As vibration. A strong band near 770 cm^{-1}, present in the oxygen-containing compounds, was attributed to the As—O bond.

Kary and Frisch (201) have reported an infrared study of some silicon-containing esters of methane- and benzenearsonic acid. A band noted at

6.20 μ (1610 cm^{-1}) was assigned to the As═O bond, while a band at 6.75 μ (1480 cm^{-1}) was assigned to C—As. It is obvious that these assignments are in marked contrast to the results of other investigators.

The infrared spectra of diphenylarsinic acid and some tin-containing esters of the type $(C_6H_5)_2AsO_2SnR_3$ have also been reported (299). Strong, complex absorption in the 850–900 cm^{-1} region was noted in all cases and was attributed to As═O stretching.

Shagidullin and co-workers (300) have recently examined the infrared spectra of a number of compounds containing the As═O group. They concluded that the stretching frequency of this group varies with the electronegativity of the substituents attached to the arsenic atom. In the case of esters of arsonic acids, the As═O asymmetric stretch was found to lie between 950 and 955 cm^{-1}.

B. Other physical measurements

The crystal structure of benzenearsonic acid has been completely determined from X-ray diffraction data (301–303). The crystal belongs to the orthorhombic system, and the unit cell has the following dimensions: $a = 10.42$, $b = 14.92$, $c = 4.70$ A. Assuming that there are four molecules in the unit cell, one can calculate the density to be 1.84 g/cc, which agrees satisfactorily with the observed value, 1.80 g/cc, reported by Schröder (304). The arrangement of the bonds around the arsenic atom is tetrahedral, with the angles between the bonds being about 110°. The three As—O distances are 1.63, 1.72, and 1.73 A, and the C—As distance is 1.90 A. This C—As bond length is very similar to other published values, but the As—O bond lengths appear to be significantly shorter than the value calculated from the single-bond covalent radii of arsenic and oxygen (305). The X-ray data indicate that each molecule of benzenearsonic acid is linked by two hydrogen bonds to other molecules so as to form endless chains. The chains in turn are held together only by van der Waals forces.

The crystal structures of arsanilic and *m*-aminobenzenearsonic acids have also been determined (306, 307). In both cases each molecular unit has four hydrogen atoms available for hydrogen bonding. The X-ray analysis shows that one of these hydrogen atoms is used to link oxygen atoms of different molecules, while the nitrogen atom forms *three* hydrogen bonds with oxygen atoms of other molecules so as to form a three dimensional network. This structural feature suggests that the formula of the aminobenzenearsonic acids is actually $^{+}NH_3C_6H_4AsO_3H^{-}$; i.e., these molecules exist as zwitterions in the solid state. This conclusion is supported by the acid dissociation constants of these compounds (86).

The crystal structure of cacodylic acid has been recently reported by Trotter and Zobel (308). As expected, the configuration around the arsenic atom is tetrahedral with bond angles in the range 106–115°. The two C—As bond lengths are equal within experimental error, and the mean value of 1.91 A agrees well with the usual C—As^V distance. Somewhat surprisingly, the As—O bond lengths also do not differ significantly, the mean value being 1.62 A. The crystal structure of cacodylic acid is built up from centrosymmetrical hydrogen-bonded dimers, which resemble the dimers characteristic of carboxylic acids. All the other intermolecular contacts correspond to normal van der Waals distances.

A crystallographic study of 3-nitro-4-hydroxybenzenearsonic acid has been reported by Goswami and Datta (309). The unit-cell dimensions were determined from Weissenberg X-ray photographs of single crystals; a is 5.54, b, 8.39, and c, 11.81 A. The calculated density of 2.03 g/cc (assuming two molecules in the unit cell) agrees favorably with the measured value of 2.05 g/cc.

Although both infrared absorption and X-ray diffraction studies show that arsonic and arsinic acids are associated in the solid state by hydrogen bond formation, these compounds give normal molecular weight values in glacial acetic acid (310) and in 95% ethanol (311). It must be concluded, therefore, that the hydrogen bonds which link the arsonic or arsinic acid molecules together in the solid state are disrupted in these solvents. Similar behavior is exhibited by carboxylic, phosphonic, and phosphinic acids. It has been recently stated (104) that molecular weight determinations of dialkylarsinic acids by a vapor pressure osmometer show that these compounds are associated. The solvent used in this study was not mentioned.

Schuster (312) has reported that the solubility of arsonic and arsinic acids in camphor is too low to allow their molecular weights to be determined by the Rast method. Thioarsenites prepared from the acids are, however, soluble in camphor and give normal molecular weights (see III-8-A).

The magnetic susceptibility of a number of arsonic and arsinic acids has been investigated by Prasad and Mulay (313–315). By using Pascal's constants for the atoms and linkages present in the organic radicals, they were able to determine that the molar susceptibilities of the arsono and arsinico groups are -55.89×10^{-6} and -49.12×10^{-6} cgs units, respectively. From these values and the constants derived for the =O and —OH groups, it was possible to calculate that the susceptibility of arsenic in these acids is -43.69×10^{-6} cgs units. This is in good agreement with the value of -43.2×10^{-6} cgs units earlier deduced by Pascal (316) for pentavalent arsenic. Prasad and Mulay also found that there is a good linear relationship between the molar susceptibility of an arsonic acid and the total number of electrons in the molecule.

The influence of substituents on the surface activity and reduction potential of aromatic arsonic acids has been studied by Breyer (85). Relationships between the reduction potential and the acidity of arsonic acids have also been investigated (84).

The atomic refraction of arsenic (for the sodium D line) has been estimated to be 8.19 in alkyl dialkylarsinates (194) and 7.78 in esters of alkylarsonic acids (317).

REFERENCES

1. H. Bart, Ger. Pat. 250,264 (Jan. 8, 1910); through *C.A.*, **6,** 3312 (1912).
2. A. Binz and C. Räth, *Justus Liebigs Ann. Chem.*, **455,** 127 (1927).
3. A. E. Tschitschibabin and A. W. Kirsanow, *Ber.*, **60,** 766 (1927).
4. W. Ried and W. Bodenstedt, *Justus Liebigs Ann. Chem.*, **679,** 77 (1964).
5. R. Nakai and Y. Yamakawa, *Bull. Inst. Chem. Res., Kyoto Univ.*, **22,** 91 (1950).
6. C. S. Hamilton and J. F. Morgan, in *Organic Reactions*, Vol. II, Wiley, New York, 1944, pp. 417–428.
7. L. Blas, *An. Soc. Espan. Fis. Quim.*, **36,** 107 (1940).
8. C. S. Palmer and R. Adams, *J. Amer. Chem. Soc.*, **44.** 1356 (1922.)
9. D. Sh. Rozina, *Zh. Prikl. Khim.*, **23,** 211 (1950); through *C.A.*, **45,** 1531 (1951).
10. A. W. Ruddy, E. B. Starkey, and W. H. Hartung, *J. Amer. Chem. Soc.*, **64,** 828 (1942).
11. R. L. Fredrickson, S. F. Bocchieri, H. J. Glenn, and L. R. Overby, *J. Assoc. Offic. Agr. Chemists*, **48,** 10 (1965).
12. E. Scheller, Ger. Pat. 522,892 (Oct. 26, 1926); *C.A.*, **25,** 3664 (1931).
13. G. O. Doak, *J. Amer. Chem. Soc.*, **62,** 167 (1940).
14. G. O. Doak and L. D. Freedman, *J. Amer. Chem. Soc.*, **73,** 5656 (1951).
15. M. Ya. Kraft and S. A. Rossina, *Dokl. Akad. Nauk SSSR*, **55,** 821 (1947); through *C.A.*, **42,** 531 (1948).
16. L. D. Freedman and G. O. Doak, *J. Amer. Chem. Soc.*, **75,** 4905 (1953).
17. M. Goswami and H. N. Das-Gupta, *J. Indian Chem. Soc.*, **8,** 417 (1931).
18. H. N. Das-Gupta, *J. Indian Chem. Soc.*, **14,** 397 (1937).
19. K. Takahashi, *Yakugaku Zasshi*, **72,** 529 (1952); *C.A.*, **46,** 9791 (1952).
20. K. Takahashi, *Yakugaku Zasshi*, **72,** 533 (1952); *C.A.*, **46,** 9791 (1952).
21. F. G. Mann and J. Watson, *J. Chem. Soc.*, 505 (1947).

21a. H. Heaney and I. T. Millar, *J. Chem. Soc.*, 5132 (1965).

22. W. A. Waters, *J. Chem. Soc.*, 2014 (1937).
23. H. Schmidt, *Justus Liebigs Ann. Chem.*, **421,** 159 (1920).
24. H. Bart, *Justus Liebigs Ann. Chem.*, **429,** 55 (1922).
25. A. B. Bruker, *Dokl. Akad. Nauk SSSR*, **58,** 803 (1947).
26. A. B. Bruker, *Zh. Obshch. Khim.*, **18,** 1297 (1948); through *C.A.*, **43,** 4647 (1949).
27. W. A. Cowdrey and D. S. Davies, *Quart. Rev.* (London), **6,** 358 (1952).
28. D. F. DeTar and M. N. Turetzky, *J. Amer. Chem. Soc.*, **77,** 1745 (1955).
29. W. A. Waters, *J. Chem. Soc.*, 266 (1942).
30. H. P. Brown and C. S. Hamilton, *J. Amer. Chem. Soc.*, **56,** 151 (1934).
31. A. A. Boon and J. Ogilvie, *Pharm. J.*, **101,** 129 (1918).
32. H. P. Brown, *Trans. Kansas Acad. Sci.*, **42,** 209 (1939).
33. H. Lieb and O. Wintersteiner, *Ber.*, **61,** 107 (1928).
34. P. A. Kober and W. S. Davis, *J. Amer. Chem. Soc.*, **41,** 451 (1919).
35. W. G. Christiansen and A. J. Norton, *Org. Syntheses*, Coll. Vol. I, 2nd ed., Wiley, New York, 1941, pp. 490–492.
36. W. A. Jacobs and M. Heidelberger, *J. Amer. Chem. Soc.*, **41,** 1440 (1919).

37. W. G. Christiansen, *J. Amer. Chem. Soc.*, **45,** 800 (1923).
38. H. Bauer, *Ber.*, **48,** 509 (1915).
39. C. K. Ingold, *Structure and Mechanism in Organic Chemistry*, Cornell University Press, Ithaca, New York, 1953: (a) p. 629; (b) pp. 261–264.
40. G. Meyer *Ber.*, **16,** 1439 (1883).
41. W. M. Dehn, *Amer. Chem. J.*, **33,** 101 (1905).
42. W. M. Dehn and S. J. McGrath, *J. Amer. Chem. Soc.*, **28,** 347 (1906).
43. A. J. Quick and R. Adams, *J. Amer. Chem. Soc.*, **44,** 805 (1922).
44. R. Pietsch, *Mikrochim. Acta*, 539 (1960).
45. R. Pietsch and E. Pichler, *Mikrochim. Acta*, 582 (1961).
46. R. Pietsch, *Monatsh. Chem.*, **96,** 138 (1965).
47. A. Valeur and R. Delaby, *Bull. Soc. Chim. Fr.*, **27,** 366 (1920).
48. G. E. Miller and S. G. Seaton (to the U.S. Dept of the Army), U.S. Pat. 2,442,372 (June 1, 1948); *C.A.*, **42,** 6842 (1948).
49. C. K. Banks, J. F. Morgan, R. L. Clark, E. B. Hatelid, F. H. Kahler, H. W. Paxton, E. J. Cragoe, R. J. Andres, B. Elpern, R. F. Coles, J. Lawhead, and C. S. Hamilton, *J. Amer. Chem. Soc.*, **69,** 927 (1947).
50. C. Ravazzoni, *Ann. Chim. Appl.*, **32,** 285 (1942).
51. M. Nagazawa, T. Hamazaki, and I. Okuda (to Ihara Agricultural Chemical Co.), Jap. Pat. 6764 (May 22, 1963); through *C.A.*, **59,** 11567 (1963).
52. R. H. Edee, *J. Amer. Chem. Soc.*, **50,** 1394 (1928).
53. R. Pietsch and P. Ludwig, *Mikrochim. Acta*, 1082 (1964).
54. R. M. Moyerman and P. J. Ehman (to Ansul Co.), U.S. Pat. 3,173,937 (March 16, 1965); through *C.A.*, **62,** 13180 (1965).
55. V. K. Kuskov and V. N. Vasilyev, *Zh. Obshch. Khim.*, **21,** 90 (1951).
56. K. W. Rosenmund, *Ber.*, **54,** 438 (1921).
57. H. J. Barber, *J. Chem. Soc.*, 2333 (1929).
58. C. S. Hamilton and C. G. Ludeman, *J. Amer. Chem. Soc.*, **52,** 3284 (1930).
59. I. E. Balaban, *J. Chem. Soc.*, 569 (1926).
60. A. Y. Garner, E. C. Chapin, and P. M. Scanlon, *J. Org. Chem.*, **24,** 532 (1959).
61. E. J. Salmi, K. Merivuori, and E. Laaksonen, *Suomen Kemistilehti*, **B, 19,** 102 (1946); *C.A.*, **41,** 5440 (1947).
62. S. O. Lawesson, *Ark. Kemi*, **10,** 167 (1956).
63. K. Irgolic, R. A. Zingaro, and M. R. Smith, *J. Organometal. Chem.*, **6,** 17 (1966).
64. C. F. McBrearty, Jr., K. Irgolic, and R. A. Zingaro, *J. Organometal. Chem.*, **12,** 377 (1968).
65. G. Kamai and N. A. Chadaeva, *Izv. Kazansk. Filiala Akad. Nauk SSSR, Ser. Khim. Nauk*, 69 (1957); through *C.A.*, **54,** 6521 (1960).
66. R. Bunsen, *Justus Liebigs Ann. Chem.*, **46,** 1 (1843).
67. M. Fioretti and M. Portelli, *Ann. Chim.* (Rome), **53,** 1869 (1963).
68. W. M. Dehn and B. B. Wilcox, *Amer. Chem. J.*, **35,** 1 (1906).
69. C. W. Porter and P. Borgstrom, *J. Amer. Chem. Soc.*, **41,** 2048 (1919).
70. T. Ueda (to Dai Nippon Drug Manufg. Co.), Jap. Pat. 3973 (July 24, 1951); through *C.A.*, **47,** 8092 (1953).
71. C. S. Gibson and J. D. A. Johnson, *J. Chem. Soc.*, 2499 (1927).
72. M. S. Leslie and E. E. Turner, *J. Chem. Soc.*, 1170 (1934).
73. E. Roberts and E. E. Turner, *J. Chem. Soc.*, **127,** 2004 (1925).
74. H. J. Backer and C. C. Bolt, *Rec. Trav. Chim. Pays-Bas*, **54,** 186 (1935).
75. H. J. Backer and C. H. K. Mulder, *Rec. Trav. Chim. Pays-Bas*, **55,** 357 (1936).
76. H. J. Backer and R. P. van Oosten, *Rec. Trav. Chim. Pays-Bas*, **59,** 41 (1940).

77. G. E. K. Branch and M. Calvin, *The Theory of Organic Chemistry*, Prentice-Hall. Inc., New York, 1941, p. 204.
78. L. D. Freedman and G. O. Doak, *Chem. Rev.*, **57**, 479 (1957).
79. V. G. Chukhlantsev, *Zh. Fiz. Khim.*, **33**, 3 (1959); through *C.A.*, **53**, 21067 (1959).
80. H. J. Emeléus, R. N. Haszeldine, and R. C. Paul, *J. Chem. Soc.*, 881 (1954).
81. H. J. Emeléus, R. N. Haszeldine, and R. C. Paul, *J. Chem. Soc.*, 563 (1955).
82. H. J. Backer and C. C. Bolt, *Rec. Trav. Chim. Pays-Bas*, **54**, 47 (1935).
83. J. Prat, *Bull. Soc. Chim. Fr.*, **53**, 1475 (1933).
84. H. Erlenmeyer and E. Willi, *Helv. Chim. Acta*, **18**, 733 (1935).
85. B. Breyer, *Ber.*, **71B**, 163 (1938).
86. D. Pressman and D. H. Brown, *J. Amer. Chem. Soc.*, **65**, 540 (1943).
87. A. I. Portnov, *Zh. Obshch. Khim.*, **18**, 594 (1948); through *C.A.*, **43**, 57 (1949).
88. D. W. Margerum, C. H. Byrd, S. A. Reed, and C. V. Banks, *Anal. Chem.*, **25**, 1219 (1953).
89. H. H. Jaffé, L. D. Freedman, and G. O. Doak, *J. Amer. Chem. Soc.*, **76**, 1548 (1954).
90. C. H. Byrd, *Iowa State Coll. J. Sci.*, **29**, 389 (1954–55).
91. A. E. Klygin and V. K. Pavlova, *Zh. Anal. Khim.*, **14**, 167 (1959).
92. V. A. Mikhailov, *Zh. Anal. Khim.*, **16**, 141 (1961).
93. B. Buděšínský, *Collect. Czech. Chem. Commun.*, **28**, 2902 (1963).
94. J. Juillard, *Bull. Soc. Chim. Fr.*, 3069 (1964).
95. L. P. Hammett, *Physical Organic Chemistry*, McGraw-Hill Book Co., Inc., New York, 1940, Chapter VII.
96. H. H. Jaffé, *Chem. Rev.*, **53**, 191 (1953).
97. H. H. Jaffé, L. D. Freedman, and G. O. Doak, *J. Amer. Chem. Soc.*, **75**, 2209 (1953).
98. H. King and G. V. Rutterford, *J. Chem. Soc.*, 2138 (1930).
99. A. Cohen, H. King, and W. I. Strangeways, *J. Chem. Soc.*, 3236 (1931).
100. L. D. Freedman and G. O. Doak, *J. Amer. Chem. Soc.*, **77**, 6374 (1955).
101. T. Ueda and K. Takahashi, *Yakugaku Zasshi*, **71**, 974 (1951); *C.A.*, **46**, 1714 (1952).
102. M. Plumel, *Bull. Soc. Chim. Biol.*, **30**, 129 (1948).
103. M. L. Kilpatrick, *J. Amer. Chem. Soc.*, **71**, 2607 (1949).
104. A. Merijanian and R. A. Zingaro, *Inorg. Chem.*, **5**, 187 (1966).
105. C. P. A. Kappelmeier, *Rec. Trav. Chim. Pays-Bas*, **49**, 57 (1930).
106. J. Prat, *C. R. Acad. Sci., Paris*, **198**, 583 (1934).
107. G. J. Burrows, *J. Proc. Roy. Soc. N.S. Wales*, **68**, 72 (1935).
108. M. Sartori and E. Recchi, *Ann. Chim. Appl.*, **29**, 128 (1939).
109. H. J. Emeléus, R. N. Haszeldine, and E. G. Walaschewski, *J. Chem. Soc.*, 1552 (1953).
110. A. Hantzsch, *Ber.*, **37**, 1076 (1904).
111. R. J. Garascia and I. V. Mattei, *J. Amer. Chem. Soc.*, **75**, 4589 (1955).
112. K. Takahashi, *Yakugaku Zasshi*, **72**, 523 (1952); *C.A.*, **46**, 9790 (1952).
113. C. Oechslin (to Établissements Poulenc Freres), U.S. Pat. 1,798,030 (March 24, 1931); through *C.A.*, **25**, 2738 (1931).
114. M. R. Stevinson and C. S. Hamilton, *J. Amer. Chem. Soc.*, **57**, 1298 (1935).
115. G. O. Doak, unpublished results.
116. M. Inoue and S. Kimura, *Yakugaku Zasshi*, **53**, 1105 (1933); through *C.A.*, **28**, 3391 (1934).
117. F. G. Mann and A. J. Wilkinson, *J. Chem. Soc.*, 3336 (1957).
118. G. O. Doak and L. D. Freedman, unpublished results.
119. W. A. Jacobs, M. Heidelberger, and I. P. Rolf, *J. Amer. Chem. Soc.*, **40**, 1580 (1918).
120. G. Newbery, M. A. Phillips, and R. W. E. Stickings, *J. Chem. Soc.*, 3051 (1928).

121. A. Bertheim, *Ber.*, **44,** 3092 (1911).
122. C. Voegtlin and H. W. E. Smith, *J. Pharmacol. Exp. Therap.*, **15,** 475 (1920).
123. A. Cohen, H. King, and W. I. Strangeways, *J. Chem. Soc.*, 2866 (1932).
124. P. M. Baranger, *Bull. Soc. Chim. Fr.*, **51,** 203 (1932).
125. P. Ehrlich and A. Bertheim, *Ber.*, **43,** 917 (1910).
126. T. Kanehori, *Takamine Kenkyusho Nempo*, **10,** 39 (1958); *C.A.*, **55,** 1488 (1958).
127. C. K. Banks and C. S. Hamilton, *J. Amer. Chem. Soc.*, **62,** 3142 (1940).
128. A. Cohen, H. King, and W. I. Strangeways, *J. Chem. Soc.*, 2505 (1932).
129. W. Steinkopf and S. Schmidt, *Ber.*, **61,** 675 (1928).
130. A. F. Kolomiets and G. S. Levskaya, *Zh. Obshch. Khim.*, **36,** 2024 (1966).
131. B. Pathak, *Sci. Cult.* (Calcutta), **16,** 331 (1951); *C.A.*, **46,** 1482 (1952).
132. J. F. Norris, *Ind. Eng. Chem.*, **11,** 817 (1919).
133. G. P. van der Kelen, *Bull. Soc. Chim. Belges*, **65,** 343 (1956).
134. W. Steinkopf and W. Mieg, *Ber.*, **53,** 1013 (1920).
135. V. Auger, *C. R. Acad. Sci., Paris*, **142,** 1151 (1906).
136. V. Auger, *C. R. Acad. Sci, Paris*, **138,** 1705 (1904).
137. A. Binz, H. Bauer, and A. Hallstein, *Ber.*, **53,** 416 (1920).
138. A. Michaelis and C. Schulte, *Ber.*, **15,** 1953 (1882).
139. P. Ehrlich and A. Bertheim, *Ber.*, **44,** 1260 (1911).
140. A. Michaelis, *Justus Liebigs Ann. Chem.*, **321,** 141 (1902).
141. A. W. Palmer and W. M. Dehn, *Ber.*, **34,** 3594 (1901).
142. E. J. Cragoe, Jr., R. J. Andres, R. F. Coles, B. Elpern, J. F. Morgan, and C. S. Hamilton, *J. Amer. Chem. Soc.*, **69,** 925 (1947).
143. R. C. Cookson and F. G. Mann, *J. Chem. Soc.*, 618 (1947).
144. K. Matsuyima and H. Nakata, *Mem. Coll. Sci. Kyoto Univ.*, **A12,** 63 (1929); through *C.A.*, **23,** 4939 (1929).
145. M. Ya. Kraft, O. I. Korzina, and A. S. Morozova, *Sb. Statei Obshch. Khim.*, **2,** 1356 (1953); through *C.A.*, **49,** 5347 (1955).
146. S. V. Vasil'ev and G. D. Vovchenko, *Vestn. Mosk. Univ.*, **5,** No. 3, *Ser. Fiz.-Mat. i Estest. Nauk*, No. **2,** 73 (1950); through *C.A.*, **45,** 6594 (1951).
147. F. Fichter and E. Elkind, *Ber.*, **49,** 239 (1916).
148. A. Ríus and H. Carrancio, *An. Real Soc. Espan. Fis. Quim., Ser. B.*, **47,** 767 (1951); *C.A.*, **46,** 10010 (1952).
149. C. P. Wallis, *J. Electroanal. Chem.*, **1,** 307 (1960).
150. R. Brdička *Collect. Czech. Chem. Commun.*, **7,** 457 (1955).
151. M. Maruyama and T. Furuya, *Bull. Chem. Soc. Jap.*, **30,** 650 (1957).
152. M. Maruyama and T. Furuya, *Bull. Chem. Soc. Jap.*, **30,** 657 (1957).
153. A. Michaelis and H. Loesner, *Ber.*, **27,** 263 (1894).
154. A. Michaelis, *Justus Liebigs Ann. Chem.*, **320,** 271 (1902).
155. A. Bertheim and L. Benda, *Ber.*, **44,** 3297 (1911).
156. L. D. Freedman and G. O. Doak, *J. Org. Chem.*, **24,** 1590 (1959).
157. R. G. Fargher, *J. Chem. Soc.*, **117,** 865 (1920).
158. L. Benda and A. Bertheim, *Ber.*, **44,** 3445 (1911).
159. G. O. Doak, H. G. Steinman, and H. Eagle, *J. Amer. Chem. Soc.*, **67,** 719 (1945).
160. Farbw. v. M. L. & B., Ger. Pat. 256, 343 (Dec. 22, 1911); through *C.A.*, **7,** 2097 (1913).
161. N. Andreev, *Zh. Russ. Fiz.-Khim. Obshchest.*, **45,** 1980 (1913); through *C.A.*, **8,** 1422 (1914).
162. H. Wieland and W. Rheinneimer, *Justus Liebigs Ann. Chem.*, **423,** 1 (1921).
163. F. F. Blicke and G. L. Webster, *J. Amer. Chem. Soc.*, **59,** 534 (1937).
164. S. M. Sherlin and G. I. Epstein, *Zh. Russ. Fiz.-Khim. Obshchest.*, **60,** 1487 (1928); through *C.A.*, **23,** 2158 (1929).

165. C. S. Hamilton and W. N. King, *J. Amer. Chem. Soc.*, **55,** 1689 (1933).
166. A. Bertheim, *Ber.*, **43,** 529 (1910).
167. A. Bertheim, *Ber.*, **41,** 1853 (1908).
168. D. R. Nijk, *Rec. Trav. Chim. Pays-Bas*, **41,** 461 (1922).
169. A. Bertheim, *Ber.*, **41,** 1655 (1908).
170. L. Benda, *Ber.*, **44,** 3449 (1911).
171. G. O. Doak, H. G. Steinman, and H. Eagle, *J. Amer. Chem. Soc.*, **66,** 197 (1944).
172. A. E. Hill and A. K. Balls, *J. Amer. Chem. Soc.*, **44,** 2051 (1922).
173. G. Newbery and M. A. Phillips, *J. Chem. Soc.*, 3046 (1928).
174. S. Orlić, *Arhiv. Hem. i Tehnol.*, **12,** 153 (1938); through *C.A.*, **34,** 6407 (1940).
175. W. D. Maclay and C. S. Hamilton, *J. Amer. Chem. Soc.*, **54,** 3310 (1932).
176. E. Fourneau and A. Funke, *Bull. Soc. Chim. Fr.*, **43,** 889 (1928).
177. H. King, *J. Chem. Soc.*, 1049 (1927).
178. H. J. Barber, *J. Chem. Soc.*, 471 (1929).
179. R. L. McGeachin, *J. Amer. Chem. Soc.*, **71,** 3755 (1949).
180. R. E. Etzelmiller and C. S. Hamilton, *J. Amer. Chem. Soc.*, **53,** 3085 (1931).
181. A. A. Abou-Ouf and Y. M. Abou-Zeid, *Egyptian Pharm. Bull.*, **40,** 53 (1958); through *C.A.*, **54,** 19670 (1960).
182. H. Burton and C. S. Gibson, *J. Chem. Soc.*, 247 (1927).
183. C. S. Gibson and J. D. A. Johnson, *J. Chem. Soc.*, 2204 (1928).
184. W. La Coste, *Justus Liebigs Ann. Chem.*, **208,** 1 (1881).
185. P. Ehrlich and A. Bertheim, *Ber.*, **40,** 3292 (1907).
186. L. Benda, *Ber.*, **44,** 3304 (1911).
187. F. L. Pyman and W. C. Reynolds, *J. Chem. Soc.*, **93,** 1180 (1908).
188. G. Petit, *C. R. Acad. Sci., Paris*, **205,** 322 (1937).
189. G. Petit, *Ann. Chim.* (Paris), **16,** 5 (1941).
190. H. N. Das-Gupta, *J. Indian Chem. Soc.*, **14,** 400 (1937).
191. Ya. F. Komissarov, A. Ya. Maleeva, and A. S. Sorokooumov, *Dokl. Akad. Nauk SSSR*, **55,** 719 (1947); *C.A.*, **42,** 3721 (1948).
192. L. M. Werbel, T. P. Dawson, J. R. Hooton, and T. E. Dalbey, *J. Org. Chem.*, **22,** 452 (1957).
193. G. Kamai and B. D. Chernokal'skii, *Zh. Obshch. Khim.*, **30,** 1176 (1960).
194. G. Kamai and B. D. Chernokal'skii, *Zh. Obshch. Khim.*, **30,** 1536 (1960); through *C.A.*, **55,** 1415 (1961).
195. B. D. Chernokal'skii, V. S. Gamayurova, and G. Kh. Kamai, *Izv. Vyssh. Ucheb. Zaved., Khim. Khim. Tekhnol.*, **8,** 959 (1965); through *C.A.*, **64,** 17630 (1966).
196. B. D. Chernokal'skii, V. S. Gamayurova, and G. Kh. Kamai, *Zh. Obshch. Khim.*, **36,** 1673 (1966).
197. G. S. Levskaya and A. F. Kolomiets, *Zh. Obshch. Khim.*, **36,** 2024 (1966).
198. G. S. Levskaya and A. F. Kolomiets, *Zh. Obshch. Khim.*, **37,** 905 (1967).
199. B. Englund, *Ber.*, **59,** 2669 (1926).
200. B. Englund, *J. Prakt. Chem.*, **120,** 179 (1929).
201. R. M. Kary and K. C. Frisch, *J. Amer. Chem. Soc.*, **79,** 2140 (1957).
202. R. M. Kary and K. C. Frisch (to American Smelting and Refining Co. and E. F. Houghton and Co.), U.S. Pat. 2,863,893 (Dec. 9, 1958); through *C.A.*, **53,** 9148 (1959).
203. A. W. Walde, H. E. Van Essen, and T. W. Zbornik (to Dr. Salsbury's Laboratories), U.S. Pat. 2,762,821 (Sept. 11, 1956); *C.A.*, **51,** 4424 (1957).
204. B. L. Chamberland and A. G. MacDiarmid, *J. Chem. Soc.*, 445 (1961).
205. M. C. Henry, *Inorg. Chem.*, **1,** 917 (1962).
206. M. Nagasawa and F. Yamamoto (to Ihara Agricultural Chemical Co.), Jap. Pat. 6200 (May 16, 1963); through *C.A.*, **60,** 4185 (1964).

207. Y. Hirota and H. Oda (to Sankyo Co., Ltd.), Jap. Pat. 10,912 (June 17, 1964); through *C.A.*, **61,** 12035 (1964).
208. H. Schmidbaur and G. Kammel, *J. Organometal. Chem.*, **14,** P28 (1968).
209. Ya. F. Komissarov, A. S. Sorokooumov, and A. Ya. Maleeva, *Dokl. Akad. Nauk SSSR*, **56,** 53 (1947); through *C.A.*, **42,** 520 (1948).
210. G. Kamai and B. D. Chernokal'skii, *Dokl. Akad. Nauk SSSR*, **128,** 299 (1959).
211. B. D. Chernokal'skii, V. S. Gamayurova, and G. Kh. Kamai, *Dokl. Akad. Nauk SSSR*, **166,** 144 (1966).
212. B. D. Chernokal'skii, V. S. Gamayurova, and G. Kh. Kamai, *Dokl. Akad. Nauk SSSR*, **166,** 384 (1966).
213. J. F. Morgan and C. S. Hamilton, *J. Amer. Chem. Soc.*, **66,** 874 (1944).
214. T. Sato, M. Nagasawa, M. Kado, and R. Kubota (to Ihara Agricultural Chemical Co.), Jap. Pat. 5562 (July 26, 1957); through *C.A.*, **52,** 11119 (1958).
215. C. Schulte, *Ber.*, **15,** 1955 (1882).
216. Farbwerke vorm. Meister Lucius & Brüning, Höchst a. M., Ger. Pat. 205,617 (Nov. 23, 1907); through *Chem. Zentr.*, **80, I,** 807 (1909).
217. K. Burschkies and M. Rothermundt, *Ber.*, **69B,** 2721 (1936).
218. J. G. Everett, *J. Chem. Soc.*, 670 (1929).
219. W. T. Reichle, *Inorg. Chem.*, **1,** 650 (1962).
220. J. G. Everett, *J. Chem. Soc.*, 1691 (1930).
221. L. D. Freedman and G. O. Doak, *J. Amer. Chem. Soc.*, **77,** 6221 (1955).
222. F. J. Welcher, *Organic Analytical Reagents*, Vol. IV, Van Nostrand, New York, 1948, pp. 49–76.
223. V. I. Kuznetsov, *Dokl. Akad. Nauk SSSR*, **31,** 898 (1941); through *C.A.*, **37,** 845 (1943).
224. V. I. Kuznetsov, *Zh. Obshch. Khim.*, **14,** 914 (1944); through *C.A.*, **39,** 4561 (1945).
225. P. F. Thomason, M. A. Perry, and W. M. Byerly, *Anal. Chem.*, **21,** 1239 (1949).
226. C. V. Banks and C. H. Byrd, *Anal. Chem.*, **25,** 416 (1953).
227. F. S. Grimaldi and M. H. Fletcher, *Anal. Chem.*, **28,** 812 (1956).
228. A. D. Horton, *Anal. Chem.*, **25,** 1331 (1953).
229. L. P. Adamovich and A. P. Mirnaya, *Sovremen. Metody Analiza v Met. Sbornik*, 172 (1955); through *C.A.*, **52,** 2653 (1958).
230. P. F. Thomason, *Anal. Chem.*, **28,** 1527 (1956).
231. H. A. Mottola, *Anal. Chim. Acta*, **27,** 136 (1962).
232. R. D. Britt, Jr., *Anal. Chem.*, **33,** 969 (1961).
233. A. D. Horton, P. F. Thomason, and F. J. Miller, *Anal. Chem.*, **24,** 548 (1952).
234. V. I. Kuznetsov and R. B. Golubtsova, *Zavod. Lab.*, **21,** 1422 (1955); through *C.A.*, **50,** 7004 (1956).
235. V. I. Kuznetsov, *Zh. Anal. Khim.*, **10,** 276 (1955).
236. V. I. Kuznetsov, *Zh. Anal. Khim.*, **13,** 220 (1958); through *C.A.*, **52,** 12673 (1958).
237. J. S. Fritz and E. C. Bradford, *Anal. Chem.*, **30,** 1021 (1958).
238. S. B. Savvin, *Usp. Khim.*, **32,** 195 (1963).
239. K. Emi and T. Hayami, *Nippon Kagaku Zasshi*, **76,** 1291 (1955); through *C.A.*, **51,** 12747 (1957).
240. A. F. Kuteinikov, *Zavod. Lab.*, **24,** 1050 (1958); through *C.A.*, **54,** 15069 (1960).
241. V. I. Kuznetsov, *Zh. Anal. Khim.*, **14,** 7 (1959).
242. V. I. Kuznetsov and S. B. Savvin, *Zh. Prikl. Khim.*, **32,** 2329 (1959).
243. K. F. Novikova, N. N. Basargin and M. F. Tsyganova, *Zh. Anal. Khim.*, **16,** 348 (1961).
244. S. B. Savvin, *Dokl. Akad. Nauk SSSR*, **127,** 1231 (1959).
245. F. V. Zaikovskii and V. N. Ivanova, *Zh. Anal. Khim.*, **18,** 1030 (1963).

246. S. B. Savvin and M. P. Volynets, *Zh. Neorg. Khim.*, **8,** 2470 (1963).
247. S. B. Savvin, *Talanta*, **8,** 673 (1961).
248. S. B. Savvin, *Zh. Anal. Khim.*, **17,** 788 (1962).
249. B. Buděšínský, K. Haas, and D. Vrzalová, *Collect. Czech. Chem. Commun.*, **30,** 2373 (1965).
250. B. Buděšínský and B. Menclová, *Talanta*, **14,** 523 (1967).
251. S. B. Savvin and M. S. Milyukova, *Zh. Anal. Khim.*, **21,** 1075 (1966).
252. A. M. Lukin, K. A. Smirnova, and G. S. Petrova, *Tr., Vses. Nauch.-Issled. Inst. Khim. Reaktivov Osobo Chist. Khim. Veshchestv.*, **29,** 290 (1966); through *C.A.*, **68,** 1737 (1968).
253. J. A. Pérez-Bustamante and F. Burriel-Martí, *Anal. Chim. Acta*, **37,** 49 (1967).
254. R. Pietsch, *Mikrochim. Acta*, 1019 (1955).
255. A. Musil and R. Pietsch, *Z. Anal. Chem.*, **140,** 421 (1953).
256. A. Musil and R. Pietsch, *Z. Anal. Chem.*, **142,** 81 (1954).
257. R. Pietsch, *Mikrochim. Acta*, 954 (1955).
258. R. Pietsch, *Z. Anal. Chem.*, **152,** 168 (1956).
259. A. Musil and R. Pietsch, *Mikrochim. Acta*, 796 (1957).
260. R. Pietsch, *Mikrochim. Acta*, 854 (1959).
261. R. Pietsch, *Mikrochim. Acta*, 220 (1958).
262. R. Pietsch and E. Pichler, *Mikrochim. Acta*, 954 (1962).
263. R. Pietsch and E. Pichler, *Mikrochim. Acta*, 914 (1961).
264. R. Pietsch, *Mikrochim. Acta*, 37 (1962).
265. R. Pietsch, *Mikrochim. Acta*, 1490 (1956).
266. R. Pietsch, *Mikrochim. Acta*, 1672 (1956).
267. A. I. Portnov, *Zh. Obshch. Khim.*, **18,** 601 (1948); through *C.A.*, **43,** 57 (1949).
268. G. Charlot, *Anal. Chim. Acta*, **1,** 218 (1947).
269. A. I. Portnov, *Zh. Obshch. Khim.*, **18,** 605 (1948); through *C.A.*, **43,** 57 (1949).
270. E. Vandalen and R. P. Graham, *Anal. Chim. Acta*, **12,** 489 (1955).
271. T. C. Chen and S. K. Yeh, *Hua Hsueh Hsueh Pao*, **23,** 474 (1957); through *C.A.*, **52,** 16982 (1958).
272. F. Szabadváry, J. Takács, and L. Erdey, *Z. Anal. Chem.*, **182,** 88 (1961).
273. W. Hubicki and R. Cienciala, *Ann. Univ. Mariae Curie-Sklodowska, Lublin-Polonia, Sect. AA*, **8,** 77 (1953); through *C.A.*, **51,** 6426 (1957).
274. R. Saint-James and T. Lecomte, *Anal. Chim. Acta* **24,** 155 (1961).
275. R. Pietsch, *Mikrochim. Acta*, 705 (1957).
276. R. Pietsch, *Mikrochim. Acta*, 699 (1957).
277. A. I. Portnov and A. I. Vasyutinskiĭ, *Tr. Komissii Anal. Khim., Akad. Nauk SSSR, Inst. Geokhim. Anal. Khim.*, **11,** 192 (1960); through *C.A.*, **55,** 12142 (1961).
278. K. Brand and E. Rosenkranz, *Pharm. Zentralh. Deut.*, **79,** 489 (1938).
279. V. Hamon, *Ann. Chim.* (Paris), **2,** 233 (1947).
280. H. H. Jaffé, *J. Chem. Phys.*, **22,** 1430 (1954).
281. R. B. Angier, A. L. Gazzola, J. Semb, S. M. Gadekar, and J. H. Williams, *J. Amer. Chem. Soc.*, **76,** 902 (1954).
282. L. D. Freedman, *J. Amer. Chem. Soc.*, **77,** 6223 (1955).
283. A. G. Maddock and N. Sutin, *Trans. Faraday Soc.*, **51,** 184 (1955).
284. A. L. Gazzola and R. B. Angier (to American Cyanamid Co.), U.S. Pat. 2,760,960 (Aug. 28, 1956); *C.A.*, **51,** 3676 (1957).
285. L. D. Freedman and G. O. Doak, *J. Org. Chem.*, **24,** 638 (1959).
286. L. D. Freedman, G. O. Doak, and J. R. Edmisten, *J. Org. Chem.*, **26,** 284 (1961).
287. F. Korte and F. Wüsten, *Chem. Ber.*, **96,** 2841 (1963).

288. G. O. Doak, L. D. Freedman, and J. B. Levy, *J. Org. Chem.*, **29,** 2382 (1964).
289. M. I. Ermakova, L. N. Vorontsova, and N. I. Latosh, *Zh. Obshch. Khim.*, **37,** 649 (1967).
290. L. Doub and J. M. Vandenbelt, *J. Amer. Chem. Soc.*, **69,** 2714 (1947).
291. R. N. Jones, *J. Amer. Chem. Soc.*, **67,** 2127 (1945).
292. F. G. Mann, I. T. Millar, and B. B. Smith, *J. Chem. Soc.*, 1130 (1953).
293. L. D. Freedman and G. O. Doak, *J. Org. Chem.*, **21,** 238 (1956).
294. J. T. Braunholtz, G. E. Hall, F. G. Mann, and N. Sheppard, *J. Chem. Soc.*, 868 (1959).
295. L. C. Thomas and R. A. Chittenden, *Spectrochim. Acta*, **20,** 489 (1964).
296. T. J. Bardos, N. Datta-Gupta, and P. Hebborn, *J. Med. Chem.*, **9,** 221 (1966).
297. G. Petit, *C. R. Acad. Sci., Paris*, **218,** 414 (1944).
298. M. P. Guha, *Sci. Cult.* (Calcutta), **7,** 315 (1941); *C.A.*, **36,** 3730 (1942).
299. I. G. M. Campbell, G. W. A. Fowles, and L. A. Nixon, *J. Chem. Soc.*, 3026 (1964).
300. R. R. Shagidullin, I. A. Lamanova, and A. K. Urazgil'deeva, *Dokl. Akad. Nauk SSSR*, **174,** 1359 (1967).
301. J. H. Bryden, *Acta Crystallogr.*, **12,** 558 (1959).
302. A. Shimada, *Bull. Chem. Soc. Jap.*, **32,** 309 (1959).
303. A. Shimada, *Bull. Chem. Soc. Jap.*, **33,** 301 (1960).
304. H. Schröder, *Ber.*, **12,** 561 (1879).
305. L. Pauling, *The Nature of the Chemical Bond*, 3rd Ed., Cornell University Press, Ithaca, N.Y., 1960, Chapter 7.
306. A. Shimada, *Bull. Chem. Soc. Jap.*, **34,** 639 (1961).
307. A. Shimada, *Bull. Chem. Soc. Jap.*, **35,** 1600 (1962).
308. J. Trotter and T. Zobel, *J. Chem. Soc.*, 4466 (1965).
309. K. N. Goswami and S. K. Datta, *Indian J. Phys.*, **37,** 651 (1963).
310. L. D. Freedman and G. O. Doak, *J. Org. Chem.*, **21,** 1533 (1956).
311. B. R. Ezzell, M. S. Thesis, North Carolina State University, Raleigh, North Carolina, 1966.
312. G. Schuster, *J. Pharm. Chim.*, **19,** 497 (1934).
313. M. Prasad and L. N. Mulay, *J. Chem. Phys.*, **19,** 1051 (1951).
314. L. N. Mulay, *Proc. Indian Acad. Sci.*, **A, 34,** 245 (1951).
315. M. Prasad and L. N. Mulay, *J. Chem. Phys.*, **20,** 201 (1952).
316. P. Pascal, *C. R. Acad. Sci., Paris*, **174,** 1698 (1922).
317. E. J. Salmi and E. Laaksonen, *Suomen Kemistilehti*, **B, 19,** 108 (1946).

III Arsonous and arsinous acids, $RAs(OH)_2$ and R_2AsOH, and their derivatives

1. PREPARATION OF ARSONOUS AND ARSINOUS ACIDS

A. By hydrolysis

The hydrolysis of compounds of the type $RAsX_2$, where X is a group such as halo, alkoxy, alkylmercapto, dialkylamino, or cyano, gives rise to either the arsonous acid or to the corresponding anhydride (arsenoso compound). Whether the acid or the anhydride is formed depends primarily on the nature of the R group. When R is aliphatic the arsenoso compound is the only product. When R is aromatic and the ring is unsubstituted or contains electron-repelling substituents, the arsenoso compound again is formed. An arsenoso compound is also formed by the hydrolysis of ferrocenyldichloroarsine (1). When the aromatic ring contains electron-attracting substituents such as nitro or carboxy groups, the arsonous acid is usually formed (2); in certain cases, however, depending upon the conditions of isolation, either the arsenoso compound or the arsonous acid can be obtained (3, 4, 5). It would seem probable that an equilibrium exists between the two forms

$$(RAsO)_n + nH_2O \rightleftharpoons nRAs(OH)_2$$

which is shifted to the right by electron-attracting substituents, to the left by electron-repelling substituents.

The hydrolysis of *o*-hydroxyphenyldichloroarsine does not give the expected arsenoso compound but the following anhydride:

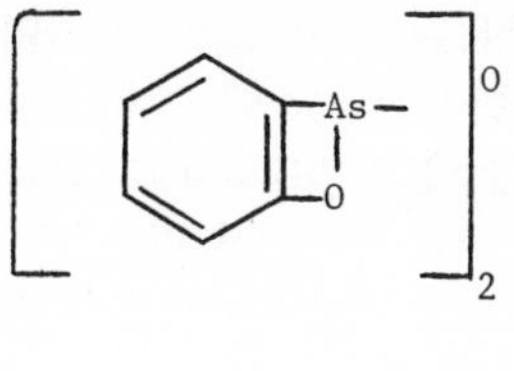

This compound was first prepared by Kalb (6), who diazotized *o*-aminobenzenearsonic acid and treated the solution with sulfurous acid. The anhydride separated from solution and was recrystallized from acetic acid. Blicke and Webster (7) prepared the same compound by the conventional hydrolysis of *o*-hydroxyphenyldichloroarsine with sodium carbonate. Molecular weight determinations were in agreement with the proposed structure. The formation of the stable 4-membered ring is surprising, and Doak, Eagle, and Steinman (3) isolated only *o*-arsenosophenol by the hydrolysis of the dichloroarsine with aqueous ammonia.

The hydrolysis of compounds of the type R_2AsX usually gives either arsinous acids, R_2AsOH, or their anhydrides, $[R_2As]_2O$. In the majority of cases studied X has been a halogen. A number of diarylarsinous acids have been prepared by Takahashi (8) by the hydrolysis of diarylchloroarsines. With diphenylchloroarsine, bis(diphenylarsenic) oxide was obtained, but with substituted diarylchloroarsines, regardless of whether the substituent was electron-attracting or electron-repelling, the arsinous acid was obtained. These results are in contrast to the findings of other workers (9, 10) who obtained only the anhydrides when electron-repelling groups were present.

The hydrolysis of 4-acetamido-4′-nitrodiphenylchloroarsine gives a sticky arsenic oxide which crystallizes from hot benzene with two molecules of benzene of crystallization (11). By heating *in vacuo* at 98° for 2 hours one molecule of benzene was removed, but the second molecule of benzene could only be removed at 180° at which temperature decomposition occurred. This surprising result has no obvious explanation. It is possible that the π-electron system of the benzene ring is forming a stable bond with the *d*-orbitals of an arsenic atom. Other bis(diarylarsenic) oxides containing electron-attracting groups on the benzene ring should be studied for their ability to coordinate with unsaturated molecules.

The ease of hydrolysis of the As—X bond in the dichloroarsines and chloroarsines is a function of the nature of both the X and the R groups. In general, where X is R_2N, RO, or CN, the compounds are readily hydrolyzed by water. Where X is $CH_3\overset{\overset{O}{\|}}{C}—O—$, hydrolysis is effected only on boiling with water for several days (12). Where X is halogen an equilibrium is established:

$$RAsX_2 + H_2O \rightleftharpoons RAs(OH)_2 + 2HX$$

Many dihaloarsines are soluble in water without appreciable hydrolysis, which can be brought about only by the use of strong alkali. The difluoroarsines are possible exceptions to this generalization. They are readily hydrolyzed by water; and, in general, the difluoroarsines cannot be regenerated by the addition of 40% hydrofluoric acid (13). Sporzyński (14) and Spada (15), however, were able to prepare methyl-α-naphthylfluoroarsine by treating bis(methyl-α-naphthylarsenic) oxide with 75% hydrofluoric acid.* In a similar manner 10-fluoro-5,10-dihydrophenarsazine was obtained by treating the corresponding oxide with hydrofluoric acid (16). There is also some evidence that the diiodoarsines are less susceptible to hydrolysis than the dichloroarsines (17). Thus, the ease of hydrolysis is probably in the order $F > Cl > Br > I$. No quantitative data, however, are available.

When the R group is considered, it is found that electron-attracting groups decrease the ease of hydrolysis and electron-repelling groups increase it. Thus CF_3AsI_2, CF_3AsCl_2, $(CF_3)_2AsI$, and $(CF_3)_2AsCl$ are stable to water at room temperature (18). By contrast many amino- and hydroxy-substituted dihaloarsines are readily hydrolyzed in aqueous solution (17). The solubility of the product also is important, and the hydrolysis of haloarsines may be carried to completion by precipitation of the hydrolysis product.

The mechanism of hydrolysis of the haloarsines has not been studied quantitatively. The reaction may involve the nucleophilic attack of a water molecule or alkoxide ion on the arsenic atom, followed by a reversible elimination of HX:

$$RAsX_2 + OH_2 \rightleftharpoons RAs(X)(X)(OH_2) \rightleftharpoons RAs(X)—OH + HX$$

Such a mechanism has been suggested for the hydrolysis of inorganic covalent halides of the nonmetallic type (19). Alternately the hydrolysis may involve an S_N1 type reaction:

$$RAsX_2 \rightleftharpoons R\overset{+}{As}X + X^-$$

$$R\overset{+}{As}X + H_2O \rightleftharpoons RAs(X)\text{-}\overset{+}{O}H_2$$

$$RAs(X)\text{-}\overset{+}{O}H_2 + X^- \longrightarrow RAs(X)\text{-}OH + HX$$

Waters and Williams (20) have discussed the hydrolysis of β-chlorovinyldichloroarsine (Lewisite I) and phenyldichloroarsine. They suggest the following mechanism:

$$ClCH{=}CHAsCl_2 + 2H_2O \overset{\text{fast}}{\rightleftharpoons} ClCH{=}CHAs(OH)_2 + 2HCl$$

$$ClCH{=}CHAs(OH)_2 \underset{\text{slow}}{\rightleftharpoons} H_2O + ClCH{=}CHAsO \underset{\text{slow}}{\rightleftharpoons} (ClCH{=}CHAsO)_n$$

* The work described in refs. 14 and 15 is identical.

The evidence for the above scheme is entirely qualitative and is not convincing. In particular the formation of the monomer, ClCH=CHAsO, is open to question (cf. III-8-A). Until the kinetics of this hydrolysis have been studied, the mechanism must remain in doubt.

The intermediate $R\overset{\overset{\displaystyle X}{|}}{As}\text{—}OH$ has not been isolated, although Newbery and Phillips (17) suggest that in the hydrolysis of certain highly substituted dihaloarsines there is evidence for its formation. However, the reaction of phenyldichloroarsine with sodium ethoxide gives the intermediate, $C_6H_5As(OC_2H_5)Cl$, which can be isolated and purified (21). There can be little doubt that the hydrolysis of dihaloarsines is a two-step reaction.

In certain cases attempted hydrolysis of the dihaloarsines results in cleavage of the arsenic–carbon bond. An interesting example is benzyldichloroarsine, which is hydrolyzed by either water or alkali to arsenous acid and either benzaldehyde or benzoic acid (22). It is difficult to write an acceptable mechanism for the formation of these products, and the reaction should be reinvestigated. In other cases some disproportionation, obviously catalyzed by hydroxide ion, occurs. Thus, diethylchloroarsine when treated with alkali gives a mixture of triethylarsine, arsenosoethane, and bis(diethylarsenic) oxide (23). When 2-hydroxyethylmethylchloroarsine is treated with excess alkali, ethylene is eliminated with the formation of disodium methanearsonate (24). The mechanism of this interesting reaction has not been investigated.

The reduction of 2-hydroxy-4-aminobenzenearsonic acid with sulfur dioxide in hydrohalic acid solution, as well as with hypophosphorous acid, hydrogen in the presence of Raney nickel, phosphorus trichloride, or sodium hydrosulfite, causes cleavage of the carbon–arsenic bond, but the reduction can be achieved in 2*N* sulfuric acid solution with sulfur dioxide and hydriodic acid (25). The products of the cleavage reaction are not stated but are presumably *m*-aminophenol and arsenous acid. It is difficult to see why sulfuric acid does not cleave the bond as does hydrochloric acid.

Emeléus and co-workers (18, 26) have studied the hydrolysis of a number of trifluoromethyl-substituted arsenicals. Compounds of the types $As(CF_3)_3$, $(CF_3)_2AsX$, CF_3AsX_2, $(CF_3)_3AsX_2$, and $[(CF_3)_2As]_2O$ (where X may be a halogen and in certain cases a cyano or thiocyano group) liberate fluoroform rapidly and quantitatively on treatment with aqueous base. The authors (18, 26) postulate that the strong electron-attracting trifluoromethyl group facilitates nucleophilic attack of hydroxide ion on the arsenic atom, with rupture of the carbon–arsenic bond. Bis(trifluoromethyl)arsinous acid apparently is decomposed in acid, neutral, or basic solution with the liberation of fluoroform, and all attempts to isolate this compound have met with failure.

Hartmann and Nowak (27) have reported that compounds of the type $R_2AsC{\equiv}CAsR_2$ can be hydrolyzed to arsinous acids:

$$R_2AsC{\equiv}CAsR_2 + 2H_2O \rightarrow 2R_2AsOH + H{-}C{\equiv}C{-}H$$

Where R is ethyl, the reaction occurs slowly with cold water. The corresponding cyclohexyl and phenyl derivatives are cleaved only by alkali, while the α-naphthyl compound requires heating with a silver nitrate solution.

B. By reduction

The reduction of arsonic acids to arsenoso compounds is usually performed in a hydrohalic acid with sulfur dioxide and with iodide ion as a catalyst; i.e., it involves the formation of an intermediate dihaloarsine. In some cases, however, it is possible to reduce arsonic acids directly to arsenoso compounds (28, 29). Both arsenosomethane (28) and arsenosoethane (29) have been prepared in this manner. Auger (28) noted that $CH_3AsO_3Na_2$ and CH_3AsO_3Ca could be reduced with sulfur dioxide alone, whereas the free acid was reduced only if iodide ion was present. Benzenearsonic acid can be reduced with sodium bisulfite (29, 30).

The reduction of arsonic acids with phenylhydrazine is of special interest. Ehrlich and Bertheim (31) reduced arsanilic acid with phenylhydrazine in methanol solution and obtained a 43.4% yield of *p*-arsenosoaniline. No mention of other products was made. This reaction, however, is apparently not a general one, and different products can be obtained depending upon reaction conditions. Wieland (32) obtained only triphenylarsine by the reduction of either benzenearsonic or diphenylarsinic acid with phenylhydrazine.

Seide and co-workers (33) found that arsenic acid reacted with phenylhydrazine in the presence of copper or copper salts to give a mixture of arsenosobenzene, diphenylarsinous acid, and triphenylarsine. By the use of copper bronze as a catalyst, these same authors (34) have recently prepared *p*-chloro- and *p*-bromophenyldichloroarsine and bis(*p*-chlorophenyl)- and bis(*p*-bromophenyl)chloroarsines by the reaction of arsenic acid and *p*-chloro- or *p*-bromophenylhydrazine. *p*-Fluorophenylhydrazine, however, gave only bis(*p*-fluorophenyl)chloroarsine.

Other investigators (35) have also obtained a mixture of phenyldichloro- and diphenylchloroarsines and triphenylarsine from arsenic acid and phenylhydrazine. Copper compounds were used as catalysts, the most effective of which was cupric oxide or hydroxide. Bruker and Maklyaev (36) obtained a mixture of bis(diphenylarsenic) oxide and triphenylarsine (the maximum As—C bond formation was 40%) by the reaction of phenylhydrazine and benzenearsonic acid without the use of copper. With arsenic acid, however, copper was necessary for the formation of a carbon–arsenic bond. These authors (36) postulate that copper serves as an electron transfer agent in the arylation of arsenic acid. In the case of benzenearsonic acid, however, they state that copper is not necessary since electron transfer is facilitated by resonance of the benzene ring.

Bruker (37) has suggested that the first step in the reaction of arsenic acid with phenylhydrazine is the oxidation of hydrazine to benzenediazo hydroxide which then couples with arsenic acid to form a covalent arsenic compound. This is followed by migration of the phenyl group on the arsenic with the concomitant evolution of nitrogen. There is no evidence for this mechanism

$$ArN{=}N{-}As(OH)_4 \rightarrow ArAsO_3H_2 + N_2 + H_2O$$

and the intramolecular nucleophilic attack seems unlikely.

C. By oxidation

Arsines and arseno compounds can be oxidized to arsenoso compounds or arsonous acids under carefully controlled conditions. The method is not a preparative procedure and the arsenoso compound is usually further oxidized to the arsonic acid (38). Methylarsine can be oxidized to arsenosomethane with dry oxygen in the presence of mercury; diphenylarsine, when oxidized in the air, gives 88% diphenylarsinic acid and 12% bis(diphenylarsenic) oxide (39).

Although these oxidations have not been studied from the mechanistic viewpoint, the oxidation of other more reactive organometallic compounds has been investigated. The oxidation of Grignard (40) and dialkylzinc reagents (41) with oxygen gives compounds of the type ROMgX and $(RO)_2Zn$. It has been shown that the mechanism involves the formation of peroxides, by attack of the oxygen molecule on the metal, followed by reduction of the peroxide by unreacted Grignard reagent or dialkylzinc. Abraham (41) postulates that the autooxidation of many other organometallic compounds proceeds in this manner. The oxidation of arsines may proceed by a similar type of mechanism:

$$RAsH_2 + O_2 \longrightarrow RAs^{\oplus}(H)_2(O{-}O^{\ominus}) \longrightarrow RAsH(O{-}OH)$$

$$RAsH(O{-}OH) + RAsH_2 \longrightarrow 2RAs(OH)_2$$

In many cases arseno compounds can be isolated as intermediates in the oxidation of arsines. These must arise by condensation of arsenoso compounds with unreacted arsines (cf. Section VI-1-C).

Arseno compounds themselves, particularly aromatic arseno compounds containing electron-repelling substituents, can be oxidized by air to arsenoso compounds. There is considerable evidence that the therapeutic activity of arseno compounds (e.g., arsphenamine) is dependent on this oxidation, and that the arseno compounds, as such, are therapeutically inactive (42). Oxidation of an aqueous solution of 3,3′-diamino-4,4′-dihydroxyarsenobenzene (arsphenamine) by air at 20° gives 3-amino-4-hydroxyarsenosobenzene as the sole product (43). In spite of the importance of the oxidation of arseno compounds in relation to their therapeutic activity, no studies on the mechanism of this oxidation have been reported.

D. The reaction between thioesters and arsonous acids

Compounds of the type $RAs(SR')_2$ will react under appropriate conditions with arsonous acids with interchange of the alkylmercapto and hydroxy groups:

$$RAs(SR')_2 + R''As(OH)_2 \rightleftharpoons R''As(SR')_2 + RAs(OH)_2$$

This reaction has been used to prepare arsonous acids or arsenoso compounds otherwise difficult to prepare. Thus, 4-acetamido-2-hydroxyarsenosobenzene is prepared only in small yield by reduction of the corresponding arsonic acid, but may be readily prepared by the following exchange reaction (44):

$$\text{2-HO-4-}CH_3CONHC_6H_3As(SCH_2CO_2H)_2 + p\text{-}HO_2CC_6H_4As(OH)_2 \rightarrow$$

$$\text{2-HO-4-}CH_3CONHC_6H_3AsO + p\text{-}HO_2CC_6H_4As(SCH_2CO_2H)_2 + H_2O$$

E. The reaction of arsenic trioxide with Grignard reagents

Arsinous acids or bis(diarylarsenic) oxides (but not arsonous acids or arsenoso compounds) can be formed by the reaction between Grignard reagents and arsenic trioxide (45, 46). Arsenic trioxide when added to an ether solution of the Grignard reagent dissolves. The compound $R_2AsOMgBr$ then precipitates from solution. This is then hydrolyzed to the secondary oxide. Some tertiary arsine and some arsinic (or arsinous) acid are usually formed; and if the reaction is continued by heating for long periods on the water bath, the tertiary arsine is usually the sole product.

A number of bis(diarylarsenic) oxides as well as bis(dibenzylarsenic) oxide have been prepared by this reaction. It is also possible to prepare these oxides by the reaction between aromatic arsenoso compounds and Grignard reagents (47, 48). Mixed bis(diarylarsenic) oxides can be prepared in this manner.

The fact that the reaction between Grignard reagents and arsenic trioxide gives principally the diaryl compounds, whereas arsenic trichloride and Grignard reagents almost always give the tertiary arsine, may be due in part to the greater reactivity of the halides. In addition, there must be some steric hindrance for attack of the carbanion on the bis(diarylarsenic) oxide. Thus, α-naphthylmagnesium bromide and arsenic trioxide give only the secondary oxide, nor does this react further on heating with the Grignard reagent (46). By contrast, methylmagnesium bromide reacts with bis(di-α-naphthylarsenic) oxide to give di-α-naphthylmethylarsine.

F. The Cadet reaction

When an intimate mixture of arsenic trioxide and potassium acetate is heated together and the products distilled, the distillate consists of approximately 56% cacodyl, $[(CH_3)_2As]_2$, and 40% cacodyl oxide, $[(CH_3)_2As]_2O$ (49). This is known as the Cadet reaction. By using propionic acid in place of acetic acid, a mixture of ethyl cacodyl and ethyl cacodyl oxide is obtained (50). The Cadet reaction is of considerable historical importance, since it constituted the first preparation of an organic arsenic compound (51). Such eminent chemists as Berzelius, Laurent, Dumas, Gerhardt, Frankland, and Kolbe studied the reaction (52). The most modern method for preparing cacodyl and cacodyl oxide consists in passing a mixture of arsenic trioxide and acetic acid over an alkali metal acetate catalyst at 300–400° (50).

The mechanism of the reaction has not been definitely established. Dehn and Wilcox (39) were able to demonstrate that, in addition to cacodyl and cacodyl oxide, polymeric methyl arsenic compounds were formed. In an extensive study of the reaction products Valeur and Gailliot (49, 53) demonstrated that trimethylarsine was present in small amounts (2.6%). They also investigated the nature of the polymeric methyl arsenic derivatives. They obtained a liquid which was successfully fractionated into two components which were formulated as $(CH_3)_7As_3$ and $(CH_3)_5As_3$. A brick red solid, which separated from Cadet's liquid on standing, was formulated as $(CH_3As)_5$. These authors suggest that the first product of the Cadet reaction consists of polymeric methyl arsenic compounds and that cacodyl and trimethylarsine are formed by disproportionation of these polymers through pyrolysis.

The mechanism of the reaction has also been investigated by Titov and Levin (54, 55). These authors also suggest, in agreement with Valeur and Gailliot, that the principal products of the Cadet reaction arise from pyrolysis of polymeric methyl arsenic derivatives. They suggest the following reaction sequence for the formation of the polymeric arsenic compounds:

$$CH_3CO_2K + As_2O_3 \rightleftharpoons O{=}AsOCOCH_3 + KAsO_2$$

$$CH_3CO_2^- + CH_3{-}\overset{\overset{\displaystyle O}{\|}}{C}{-}OAs{=}O \rightleftharpoons CH_2{=}\overset{\overset{\displaystyle O^-}{|}}{C}{-}OAs{=}O + CH_3CO_2H$$

$$O{=}As{-}O{-}As{=}O + CH_2{=}\overset{\overset{\displaystyle O^-}{|}}{C}{-}O{-}As{=}O \longrightarrow AsO_2^- + O{=}AsCH_2CO_2As{=}O$$

$$O{=}AsCH_2COOAs{=}O + CH_3CO_2^- \rightleftharpoons O{=}AsCH_2COO^- + CH_3COOAs{=}O$$

$$O{=}AsCH_2COO^- \longrightarrow O{=}AsCH_2^- + CO_2$$

$$O{=}AsCH_2^- + As_2O_3 \longrightarrow O{=}AsCH_2As{=}O + AsO_2^-$$

The $O{=}AsCH_2As{=}O$ will then react further with the acetate ion, leading to the eventual formation of polymers of the type:

$$\begin{array}{c} \quad\quad\quad\quad\quad\quad | \\ {-}CH_2{-}As{-}CH_2{-}As{-}CH_2{-} \\ | \quad\quad\quad\quad \\ CH_2{-}As{-}CH_2 \\ | \\ CH_2{-} \end{array}$$

Evidence for this mechanism is based partially on the isolation of diarsenosomethane which upon heating with potassium acetate gave a product similar to Cadet's liquid.

2. REACTIONS OF ARSONOUS AND ARSINOUS ACIDS

A. Acid dissociation constants

The arsonous acids are weak acids which do not dissolve in aqueous ammonia or alkali bicarbonate solutions unless other acid substituents are present. They do dissolve, however, in alkali hydroxide solutions. Arsenoso compounds similarly dissolve in alkali hydroxide solutions and thus behave as pseudo acids. Although the literature (28) states that salts of the type $RAs(OM)_2$ are formed, no analyses of such salts have been reported. The pK_A's of arsonous acids have not been determined accurately. However, Eagle and co-workers (56) have titrated arsenosobenzene and four acid substituted derivatives (m-CO_2Na, p-OCH_2CO_2Na, p-SO_3Na, and p-CO_2Na) potentiometrically. The pK_A's of all five compounds were quite similar, approximately 11. The pK_A's of arsinous acids (or of their anhydrides) have not been determined.

There is some evidence that the oxygen of the arsenoso group can be protonated. Thus Petit (57) isolated deliquescent solids with the formulas $CH_3AsO{\cdot}H_2SO_4$ and $C_2H_5AsO{\cdot}H_2SO_4$ by the reaction between arsenosomethane or arsenosoethane and concentrated sulfuric acid. Although the structure of these compounds has not been determined, they may be formed by protonation of the oxygen atom attached to the arsenic.

B. Reactions with electrophilic reagents

(*1*) *Mercuration*

Arsonous acids and arsenoso compounds react with most electrophilic reagents with oxidation of the arsenic. With some reagents, e.g., mercuric oxide, there may be cleavage of the carbon–arsenic bond (58, 59). Thus, arsenosobenzene, when boiled with mercuric chloride in alcoholic sodium hydroxide, gives a 25% yield of diphenylmercury together with metallic mercury and mercuric oxide. Phenylmercuric hydroxide and arsenosobenzene in alkaline solution also give diphenylmercury. Bis(diphenylarsenic) oxide, however, is oxidized by mercuric oxide to diphenylarsinic acid.

The mercuration of aromatic arsenoso compounds by means of mercuric acetate in acetic acid has been thoroughly investigated by Hiratsuka (60–63). He found that the arsenoso group (like the amino and hydroxy groups) strongly activated the aromatic ring toward mercuration. The mercury normally entered *para* to the arsenoso group; if the *para* position was already occupied, the mercury took an *ortho* position. *Ortho*- and *p*-arsenosotoluene and α-arsenosonaphthalene were found to react rapidly with mercuric acetate at room temperature, but the expected products could not be isolated. Instead, *o*- and *p*-tolyl- and α-naphthylmercury acetates were obtained. Hiratsuka has suggested the following mechanism for the splitting of arsenic from the ring:

$$CH_3\text{-}C_6H_4\text{-}AsO + Hg(OAc)_2 \longrightarrow \left[CH_3\text{-}C_6H_3(AsO)(HgOAc) \right]$$

$$\longrightarrow CH_3\text{-}C_6H_4\text{-}HgOAc + As_2O_3$$

Arsenosobenzene itself was readily converted at room temperature to *p*-arsenosophenylmercury acetate. The latter compound on heating in acetic acid at 40–50° yielded *p*-arsenosodiphenylmercury. At 100° phenylmercury acetate and a small amount of diphenylmercury were obtained.

Mercuric acetate at room temperature could also be used to mercurate the following arsenoso compounds: 2,4-dimethylphenyl-, *p*-bromophenyl-, *o*- and *p*-chlorophenyl-, and 2,4-dichlorophenyl-. On prolonged heating in acetic acid, arsenic trioxide was cleaved and the mercury moved to the position originally occupied by the arsenoso group.

Arsenoso compounds containing nitro or carboxy substituents could be mercurated only by heating with mercuric acetate. In general, these mercurated derivatives were very stable and could not be hydrolyzed even by heating in acetic acid solution. 3-Nitro-4-arsenosophenylmercury acetate, however, was relatively unstable and could be easily degraded to *o*-nitrophenylmercury acetate:

$$\text{(}NO_2\text{)}C_6H_4\text{–AsO} + Hg(OAc)_2 \longrightarrow \text{AcOHg–}C_6H_3\text{(}NO_2\text{)–AsO} \xrightarrow{H_2O} \text{(}NO_2\text{)}C_6H_4\text{–HgOAc}$$

Most of the arsenosoarylmercury acetates could be converted to the corresponding arsenosoarylmercury chlorides by treatment with aqueous sodium chloride. *p*-Arsenosophenylmercury acetate itself, however, was degraded by sodium chloride to phenylmercury chloride. Hiratsuka has suggested that *p*-arsenosophenylmercury chloride was first formed as follows:

$$\text{AcOHg–}C_6H_4\text{–AsO} + Cl^- \longrightarrow \text{ClHg–}C_6H_4\text{–AsO} + OAc^-$$

The product, however, was unstable in the solution made alkaline by the acetate ion formed in the reaction and split off arsenic with the concomitant migration of the chloromercuri group to the position originally occupied by the arsenic. 3-Carboxy-4-arsenosophenylmercury acetate also lost arsenic upon treatment with sodium chloride and yielded *o*-carboxyphenylmercury chloride:

$$\text{AcOHg–}C_6H_3\text{(COOH)–AsO} \xrightarrow{NaCl} \left[\text{ClHg–}C_6H_3\text{(COOH)–AsO}\right] \longrightarrow \text{(COOH)}C_6H_4\text{–HgCl}$$

In general, treatment of an arsenosoarylmercury acetate with hydrochloric acid yielded the corresponding dichloroarsinoarylmercury chloride. Some of the compounds, however, lost *mercury* during this process, and the arsenic migrated to the position originally occupied by the mercury. Thus *p*-arsenosophenylmercury acetate was converted by hydrochloric acid to phenyldichloroarsine. 2-Arsenoso-5-bromophenylmercury acetate was converted to *m*-bromophenyldichloroarsine:

$$2\text{-AcOHg-}5\text{-Br-}C_6H_3AsO \xrightarrow{HCl} [2\text{-ClHg-}5\text{-Br-}C_6H_3AsCl_2] \rightarrow m\text{-BrC}_6H_4AsCl_2$$

Similarly, 4-arsenoso-2-bromophenylmercury acetate was converted to *o*-bromophenyldichloroarsine:

$$4\text{-OAs-}2\text{-Br-}C_6H_3HgOAc \xrightarrow{HCl} [4\text{-Cl}_2As\text{-}2\text{-Br-}C_6H_3HgCl] \rightarrow o\text{-BrC}_6H_4AsCl_2$$

These remarkable reactions thus provide a method for converting *p*-bromophenyl arsenic compounds to *m*-bromophenyl arsenic compounds which in turn can be converted to *o*-bromophenyl arsenicals. Interestingly enough, 2-arsenoso-5-chlorophenylmercury acetate did not undergo this rearrangement but reacted normally:

$$2\text{-AcOHg-}4\text{-Cl-}C_6H_3AsO + 3HCl \rightarrow 2\text{-ClHg-}4\text{-Cl-}C_6H_3AsCl_2$$

3-Carboxy-4-arsenosophenylmercury acetate (which is split by sodium chloride) reacted normally with hydrochloric acid:

$$4\text{-AcOHg-}2\text{-HOOC-}C_6H_3AsO + 3HCl \rightarrow 4\text{-ClHg-}2\text{-HOOC-}C_6H_3AsCl_2$$

Hiratsuka has also studied the reaction of aromatic thioarsenoso compounds with mercuric acetate in acetic acid. When a large excess of thioarsenosobenzene was used, he obtained a crystalline yellow solid the analysis of which corresponded to $(C_6H_5AsS)_2(p\text{-AcOHgC}_6H_4AsS)$. Treatment of this substance with additional mercuric acetate yielded first a crystalline red solid of composition $(C_6H_5AsS)(p\text{-AcOHgC}_6H_4AsS)$ and finally an acetic acid soluble material which Hiratsuka suggested was (p-$AcOHgC_6H_4AsS$)-($AcOH$). When the acetic acid solution was allowed to stand, white leaflets of p-thioarsenosophenylmercury acetate were obtained. This compound was quite stable and did not lose mercury even on heating. When it was treated with alkali, a black compound was formed to which was assigned the following structure:

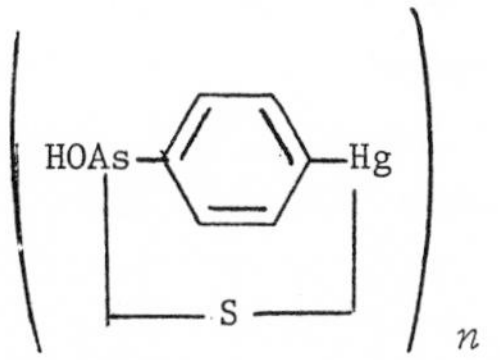

HOAs—C_6H_4—Hg, with As and Hg bridged by S; $(\ldots)_n$

This black compound could be reconverted to p-thioarsenosophenylmercury acetate by reaction with acetic acid.

Hiratsuka found that bis(diphenylarsenic) oxide could be mercurated to bis[bis(p-acetoxymercuriphenyl)arsenic] oxide by prolonged heating with mercuric acetate. Reaction of the acetoxymercuri compound with hydrochloric acid yielded bis(p-chloromercuriphenyl)chloroarsine. Upon treatment with aqueous sodium cyanide, the acetoxymercuri compound underwent a curious reaction in which mercury was split from the ring and diphenylarsinic acid was formed. Sodium cyanide also removed the mercury from p-arsenosophenylmercury acetate, but in this case the arsenic was not oxidized concomitantly:

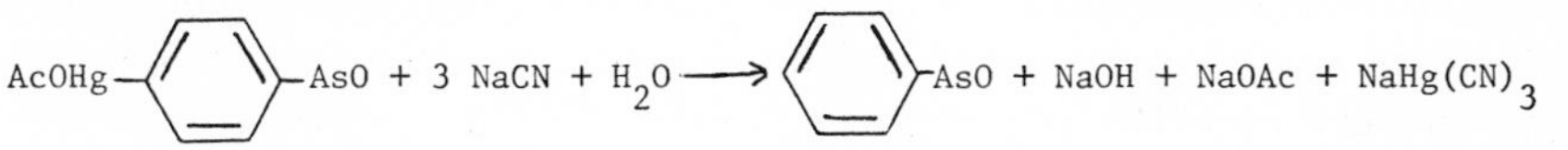

$$AcOHg\text{—}C_6H_4\text{—}AsO + 3\ NaCN + H_2O \longrightarrow C_6H_5\text{—}AsO + NaOH + NaOAc + NaHg(CN)_3$$

(2) Halogenation

It is not clear what products are obtained in the reaction between arsenoso compounds or the secondary arsenic oxides and halogens. Earlier workers (64, 65) all mention the formation of "oxyhalides" according to the reactions:

$$RAsO + X_2 \rightarrow RAsOX_2$$

$$(R_2As)_2O + X_2 \rightarrow (R_2AsX_2)_2O$$

Only in the case of $C_6H_5As(O)Cl_2$ was a satisfactory analysis obtained, and only chlorine was determined. Most other compounds failed to give satisfactory analytical values and the suggested formulas were based on the amount of halogen absorbed.

These difficulties have not been solved by later work. By the action of chlorine on bis(diphenylarsenic) oxide in aqueous solution, Kappelmeier (66) obtained principally $[(C_6H_5)_2AsCl_2]_2O$; from the mother liquors smaller amounts of $[(C_6H_5)_2As(OH)_2]Cl$ were isolated. In contrast Burrows and Lench (67), by the same procedure, obtained only $[(C_6H_5)_2As(OH)_2]Cl$. At the present time it is impossible to reach any definite conclusions as to the nature of the products formed by the action of halogens on arsenoso compounds or secondary arsenic oxides.

Michaelis (68) attempted the halogenation of dimethoxyphenylarsine and obtained a compound to which he assigned the structure $C_6H_5As(OCH_3)_2Cl_2$. Only analytical values for chlorine were reported.

The reaction of diarylarsinous acids with aqueous iodine gives diarylarsinic acids (69). It has been made the basis for the potentiometric determination of arsinous acids; an antimony electrode was used in the titrations.

C. Reactions with nucleophilic reagents

The sodium salts of aromatic arsonous acids react with alkyl iodides to form sodium arylalkylarsinates and sodium iodide (70):

$$ArAs(ONa)_2 + RI \rightarrow ArRAsO_2Na + NaI$$

This reaction is similar in nature to the Meyer reaction and must involve the nucleophilic attack of an alkyl group on the arsenic. The reaction of disodium phenylarsonites with alkyl iodides is in contrast to the failure of esters of the type $RAs(OR')_2$ and R_2AsOR' to react with alkyl iodides in an Arbuzov-type rearrangement (71).

Arsenosomethane reacts with sulfuric acid at elevated temperatures with the production of arsenous acid, carbon dixoide, and sulfur dioxide (57). The first step in the reaction is undoubtedly oxidation to the arsonic acid followed by a nucleophilic attack of the bisulfate ion (cf. Section II-2-D).

D. Other reactions

Arsenoso compounds are readily reduced to arseno compounds and oxidized to arsonic acids. Hydrogen peroxide is frequently used as the oxidizing agent. Cystine has also been used as the oxidizing agent, and the rate constants for the following reaction, where X is either an electron-attracting or electron-repelling group, have been obtained (72):

$$XC_6H_4AsO + (SCH_2CH(NH_2)COOH)_2 + 2H_2O \rightleftharpoons XC_6H_4AsO_3H_2 + 2HSCH_2CH(NH_2)COOH$$

When heated above their melting points, arylarsonous acids and arsenoso compounds rearrange to triarylarsines and arsenic trioxide (73):

$$3(ArAsO)_n \rightarrow nAr_3As + nAs_2O_3$$

Although the mechanism of this rearrangement has not been studied with arsenic compounds, a kinetic study of the rearrangement of arylstiboso compounds has been made (cf. Section VIII-4). It is probable that both types of compounds react by a similar mechanism.

o-Nitroarsenosobenzene undergoes a curious reaction (74). When dissolved in alcoholic hydrogen chloride and exposed to sunlight, brown crystals separate from solution. These crystals dissolve in alkali carbonate or bicarbonate solutions, and also form a magnesium salt with magnesia mixture. Karrer (74) postulates that these crystals are the anhydride of *o*-nitrosobenzenearsonic acid, *o*-$NOC_6H_4AsO_2$, formed by migration of an oxygen from the nitro group to the arsenic. The compound, however, does not give the usual nitroso reactions, and its structure is uncertain.

Dessy and co-workers (75) have found that a solution of bis(diphenylarsenic) oxide undergoes a two-electron electrolytic reduction to give the yellow diphenylarsenide ion and diphenylarsenite:

$$Ph_2AsOAsPh_2 + 2e^- \rightarrow Ph_2As^- + Ph_2AsO^-$$

Upon addition of two equivalent amounts of diphenylbromoarsine to the yellow solution, the color disappears, and tetraphenyldiarsine and bis(diphenylarsenic) oxide are formed:

$$Ph_2As^- + Ph_2AsO^- + 2Ph_2AsBr \rightarrow Ph_2AsAsPh_2 + Ph_2AsOAsPh_2 + 2Br^-$$

3. ESTERS

Esters of arsonous and arsinous acids can be prepared by several different methods. Esters of arsonous acids were first prepared by Michaelis (68) by the reaction between aryldichloroarsines and sodium alkoxides or phenoxides. Esters of alkylarsonous acids can be prepared by the same reaction (76, 77). Similarly, esters of arsinous acids can be prepared from diarylhaloarsines and sodium alkoxides (71, 76, 78). If a base such as pyridine is used, alcohols will react with dichloroarsines directly without the necessity of preparing the alkoxides (79–83). In some cases, monoesters of the type RAs(Cl)OR′ are formed (82).

A number of cyclic esters of arsonous acids of the type

```
        O—CH—R'
       /   |
   RAs     |
       \   |
        O—CH—R'
```

have been prepared by the reaction between glycols or *o*-dihydric phenols and dichloroarsines (84,85). The 2-chloroethyl ester of benzenearsonous acid may be prepared by the reaction of ethylene oxide and phenyldichloroarsine (86):

$$C_6H_5AsCl_2 + \underset{\diagdown \, O \, \diagup}{CH_2CH_2} \longrightarrow (ClCH_2CH_2O)_2AsC_6H_5$$

With substituted ethylene oxides, however, only one halogen of the dihaloarsine is replaced. Chlorodiphenylarsine does not react with ethylene oxide or substituted ethylene oxides.

Although the mechanism of the reaction of alkoxides or ethylene oxides with dihaloarsines has not been studied, it may involve nucleophilic attack of oxygen on arsenic followed by elimination of halide ion.

Esters of arsonous acids can be prepared by direct esterification of arsenoso compounds. This reaction was first investigated by Chwalinski (87) who prepared several esters of benzenearsonous acid by refluxing arsenosobenzene and the appropriate alcohol with calcium chloride. In this manner, he prepared the dimethyl, diethyl, diphenyl, di-β-chloroethyl, and diallyl esters of benzenearsonous acid. Esters of arsinous acids have been prepared from the secondary arsenic oxides and alcohols (78, 88–90). Kamai and Khisamova (78) have prepared the compound $(C_3H_7)_2As—OCH_2CH_2O—As(C_3H_7)_2$ from bis(dipropylarsenic) oxide and ethylene glycol. Bromination of the glycol ester gave the tetrabromide $(C_3H_7)_2As(Br)_2OCH_2CH_2OAs(Br)_2(C_3H_7)_2$. Treatment of esters of dipropylarsinous acid with acetic anhydride yielded the interesting acetoxy compound, $(C_3H_7)_2AsOCOCH_3$.

Esters of alkylarsonous acids have also been prepared by ester exchange reactions. Thus, allyl alcohol reacts with diethyl ethanearsonite to give the diallyl ester (91). It is interesting to note that these allyl esters were not polymerized by means of benzoyl peroxide; this fact indicates that the trivalent arsenic interfered with the free radical reaction.

Closely related to the ester exchange reaction is the preparation of diethyl benzenearsonite from dichlorophenylarsine and triethyl antimonite (92):

$$C_6H_5AsCl_2 + (EtO)_3Sb \rightarrow C_6H_5As(OEt)_2 + EtOSbCl_2$$

Another method for preparing esters of arsonous acids involves heating an alcohol or phenol with a bis(dialkylamino)alkylarsine or a bis(dialkylamino)-arylarsine (93).

Esters of arsonous and arsinous acids are rather unstable substances. They are rapidly hydrolyzed by moisture and oxidized by air. They can be distilled *in vacuo* without decomposition, but when an attempt was made to distill them at atmospheric pressure, oxidation and rearrangement occurred (94):

$$RAs(OR)_2 \xrightarrow{O_2} As(OR)_3$$

They form addition compounds with mercuric chloride and with cuprous salts (85, 88).

In addition to esters containing the As—O—C bond, a few esters have been prepared containing As—O—Si bonds. Chamberland and MacDiarmid (95) condensed phenyldiiodoarsine with the sodium salt of triphenylsilanol and obtained the compound $C_6H_5As[OSi(C_6H_5)_3]_2$. Reaction of the same diiodoarsine with diphenylsilanediol in the presence of ammonia gave the dimer:

$$C_6H_5As \begin{matrix} \diagup OSi(C_6H_5)_2O \diagdown \\ \diagdown OSi(C_6H_5)_2O \diagup \end{matrix} AsC_6H_5$$

Distillation of this dimer gave a glassy material of higher molecular weight which the authors believe to be largely a trimer of the composition $[OSi(C_6H_5)_2OAsC_6H_5]_3$.

The interesting compound $(CH_3)_2AsOSi(CH_3)_3$ has been prepared by the reaction between dimethylchloroarsine and the lithium or sodium salt of trimethylsilanol (96, 97):

$$(CH_3)_2AsCl + LiOSi(CH_3)_3 \rightarrow LiCl + (CH_3)_2AsOSi(CH_3)_3$$

When this ester is treated with thionyl chloride, dimethylchloroarsine is regenerated. Trimethylchlorosilane and sulfur dioxide are the other products of the reaction. The ester is oxidized explosively when heated to 60° in the air.

Kamai and Miftakhova (98) have prepared a series of esters of the types $RAs(ON{=}CMe_2)_2$ and $R_2AsON{=}CMe_2$ by the reaction of dichloroarsines and chloroarsines with acetoxime.

4. SULFIDES AND SELENIDES

Thioarsenoso compounds can be prepared by treating an alcoholic solution of the arsenoso compound or the dichloroarsine with hydrogen sulfide or with an alkali hydrosulfide (54, 99–103). Some splitting of the carbon–arsenic bond occurs concurrently and results in the formation of arsenic trisulfide and the hydrocarbon. Some arsonic acids are reduced by hydrogen sulfide in acid solution with formation of thioarsenoso compounds (cf. Section II-2-E). Primary arsines can be converted to thioarsenoso compounds by reaction with sulfur (104), thionyl chloride (105), or sulfinylaniline (105). Treatment of one mole of 3,3′-diamino-4,4′-dihydroxyarsenobenzene with sulfur dioxide yields $\frac{2}{3}$ of a mole of the corresponding thioarsenoso compound and $\frac{4}{3}$ of a mole of the arsenoso compound (106).

Diphenylchloroarsine reacts with hydrogen sulfide to give bis(diphenylarsenic) sulfide, $[(C_6H_5)_2As]_2S$ (100). Bis(diarylarsenic) sulfides are also formed by the reaction of Grignard reagents and arsenic trisulfide. Thus, *p*-tolylmagnesium bromide gives a mixture of $(p\text{-}CH_3C_6H_4)_3AsS$, $[(p\text{-}CH_3C_6H_4)_2As]_2S$, and $(p\text{-}CH_3C_6H_4)_3As$ (107). Bis(dialkylarsenic) sulfides can be prepared by the treatment of secondary arsines with sulfur (39, 108). The reaction of dimethylarsine with sulfur yields not only bis(dimethylarsenic) sulfide but also "cacodyl disulfide," $(Me_2As)_2S_2$. The structure of this compound will be discussed in III-8-C. The preparation of bis(dimethylarsenic) sulfide by the treatment of dimethylarsinic acid with carbon disulfide has been described in II-2-E-(3).

Mercaptans (or their salts) react with a variety of trivalent and pentavalent arsenicals to yield compounds of the types $RAs(SR')_2$ and $R_2As(SR')$. These substances have often been called "thioarsenites." The arsenicals used in this reaction include arsonic acids (109–111), arsinic acids (112–114), arsonous acids (20, 103), arsenoso compounds (115–117), dihaloarsines (118–122), haloarsines (76, 123–125), dialkyl arsonites (126), alkyl arsinites (127), diarylarsinous acids (113), bis(diarylarsenic) oxides (113, 128), bis(dialkylamino)alkylarsines, and bis(dialkylamino)arylarsines (93).

Cullen and co-workers (129, 130) have developed several other methods for the preparation of compounds of the type R_2AsSR'. One method used was the reaction between a tetraalkyldiarsine and a disulfide:

$$R_2AsAsR_2 + R'SSR' \rightarrow 2R_2AsSR'$$

This method was investigated with diarsines where R was CH_3 or CF_3 and with a variety of disulfides. Yields varying between 70 and 100% were obtained, and the method seems to be a general one. Two other methods for preparing these compounds involve the reaction of a sulfenyl chloride with either a diarsine, R_2AsAsR_2, or a secondary arsine, R_2AsH. Another method used for preparing this type of compound was the cleavage of a disulfide by a secondary arsine:

$$R_2AsH + R'SSR' \rightarrow R_2AsSR' + R'SH$$

Finally, Cullen (108) has described the preparation of $(CF_3)_2AsSCF_3$ by means of the following reaction:

$$(CF_3)_2AsSAs(CF_3)_2 + CF_3I \rightarrow (CF_3)_2AsSCF_3 + (CF_3)_2AsI$$

The thioarsenites are of considerable importance in pharmacology since it is believed that organic arsenicals owe their parasiticidal activity to their ability to combine with sulfhydryl groups in certain essential enzymes. A large amount of work on the following reaction has been summarized in a review by Eagle and Doak (131):

$$RAsO + 2R'SH \rightleftharpoons RAs(SR')_2 + H_2O$$

The chemistry of the thioarsenites has been quite extensively studied, particularly by Cohen, King, and Strangeways (115). The thioarsenites are stable in acid or slightly alkaline solution but are readily hydrolyzed in sodium hydroxide solution. The hydrolysis has been followed both polarimetrically and by means of the nitroprusside test.

The thioarsenites are well-defined crystalline solids with reproducible melting points. Thioglycolamide has been recommended (109) as a reagent for the characterization of arsonic acids; a number of thioarsenites have been prepared with this reagent, and their melting points have been determined.

In addition to the thioarsenites a number of other compounds containing the As—S bond are known. These compounds may be considered mixed anhydrides of arsonous (or arsinous) acids and thio acids such as thioic acids (132, 133), dithiocarbamic acids (133–136), xanthic acids (133, 137, 138), phosphorothioic acids (139), phosphorodithioic acids (133, 137, 140), and thiosulfonic acids (133). They are prepared by the reaction of an arsenoso compound, a dihaloarsine, or a haloarsine with a salt of the appropriate thio acid.

The reaction between 10-chlorophenoxarsine and phenothiazine has been reported (141) to yield an unusual compound with an As—S linkage:

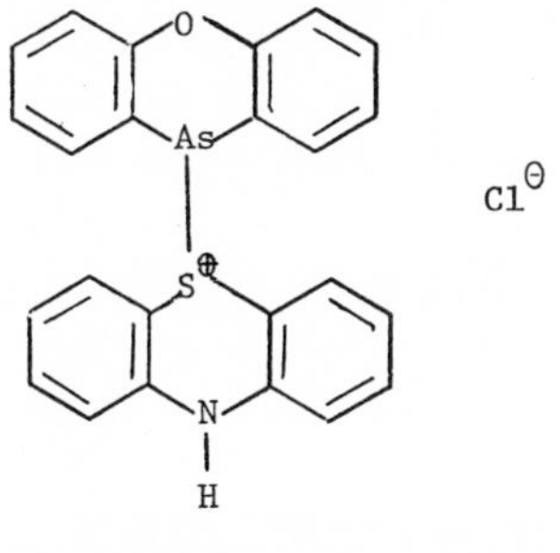

O
As
S⊕
N
H
Cl⊖

Emeléus and co-workers (142) have prepared two compounds containing the As—Se bond by the reaction of bis(trifluoromethyl)iodoarsine and $Hg(SeCF_3)_2$ or $Hg(SeCF_2CF_2CF_3)_2$.

5. AMIDES

Both dihaloarsines and monohaloarsines react with ammonia and with amines. The reaction of dihaloarsines is complex, and a number of products have been isolated from the reaction. Michaelis (68) obtained $C_6H_5As(Cl)NHC_4H_9$ and $C_6H_5As(Cl)N(C_4H_9)_2$ by the reaction between dichlorophenylarsine and butylamine or dibutylamine, respectively. With triethylamine, however, he obtained a product to which he assigned a structure with pentacovalent nitrogen and arsenic, namely, $Cl_2(C_6H_5)As{=}N(C_2H_5)_3$.

With ammonia and phenyldichloroarsine he obtained a product to which he assigned the imide structure, $C_6H_5As{=}NH$. Diphenylchloroarsine and ammonia give the amide, $(C_6H_5)_2AsNH_2$. Methyldichloroarsine and ammonia, however, give the imide, $CH_3As{=}NH$, which then polymerizes to $(CH_3AsNH)_n$ (143). Isolation of a monomeric imide with a nitrogen-arsenic double bond is surprising; without molecular weight data, this result should be viewed with caution.

The reaction between arsonous halides and amines was later investigated by Doak (144). He found that the nature of the product often varied with the ratio of the reactants and the order of addition. Thus, ethyldichloroarsine and piperidine gave $C_2H_5As(NC_5H_{10}{\cdot}HCl)_2$ or $C_2H_5As(Cl)NC_5H_{10}$, but dimethylchloroarsine and piperidine gave only $(CH_3)_2AsNC_5H_{10}$. Other workers (145–148) have prepared a number of compounds of the types $RAs(NR_2')_2$ and R_2AsNR_2' by the reaction of dihaloarsines or haloarsines with secondary amines.

A study of the reaction between ammonia or amines and bis(trifluoromethyl)chloroarsine or trifluoromethyldichloroarsine has been made by Cullen and Emeléus (146). Bis(trifluoromethyl)chloroarsine and liquid ammonia gave $[(CF_3)_2As]_2NH$, but with gaseous ammonia $(CF_3)_2AsNH_2$ was also formed. With primary or secondary amines, compounds of the types $(CF_3)_2AsNHR$ and $(CF_3)_2AsNR_2$, respectively, were obtained. By contrast trifluoromethyldichloroarsine reacted with ammonia to give principally fluoroform. No products containing the As—N linkage were isolated.

Scherer and Schmidt (149) have found that reaction of haloarsines with primary amines results in the formation of two As—N bonds:

$$2(CH_3)_2AsCl + 3CH_3NH_2 \longrightarrow (CH_3)_2As{-}\underset{\displaystyle |}{\overset{\displaystyle CH_3}{N}}{-}As(CH_3)_2 + 2CH_3NH_3Cl$$

Banks and co-workers (150) reported that the imide $C_2H_5As{=}NC_2H_5$ is the product of the reaction between ethyldichloroarsine and ethylamine. No molecular weight data were given, but the relatively high boiling point (165–175° at 3.5 torr) suggests that the imide is polymeric.

Amides of dialkylarsinous acids have also been made by the reaction of Grignard or lithium reagents with dialkylaminodichloroarsines (151, 152):

$$2RMgX + (R_2'N)AsCl_2 \rightarrow R_2AsNR_2' + 2MgXCl$$

This type of reaction has been previously discussed in connection with the synthesis of arsinic acids (cf. Section II-1-D). The reaction of methyllithium with bis(dimethylamino)chloroarsine has been found to lead to the formation of a mixture of Me_2AsNMe_2 and $MeAs(NMe_2)_2$ (153).

Silicon-containing amino derivatives of arsenic, antimony, and bismuth have been recently prepared by means of the following type of reaction (154, 155):

$$n(CH_3)_3Si{-}\underset{}{\overset{CH_3}{\overset{|}{N}}}{-}Li + (CH_3)_{3-n}MX_n \longrightarrow [(CH_3)_3Si{-}\overset{CH_3}{\overset{|}{N}}]_nM(CH_3)_{3-n} + nLiX$$

where M=As, Sb, or Bi, X = Cl or Br, and $n = 1, 2$, or 3. The compounds are colorless or light yellow liquids which can be readily distilled *in vacuo*. Sensitivity to oxygen and moisture decreases in going from the monoamino to the trisamino compounds. The proton NMR spectra of the compounds clearly distinguish three kinds of methyl groups, i.e., those attached to silicon, nitrogen, and the metal.

Sisler and Stratton (156) have studied the reactions of diphenylchloroarsine with chloramine and with mixtures of chloramine and ammonia. At −78° chloramine reacts with an ethereal solution of the chloroarsine to yield a white solid which has been formulated as $[Ph_2As(NH_2)Cl]Cl$. Heating this solid in dimethylformamide (DMF) solution produces a mixture of two compounds. The major component has the following structure:

$$[Ph_2\overset{Cl}{\overset{|}{As}}\text{=-=}N\text{=-=}\overset{Cl}{\overset{|}{As}}Ph_2]Cl$$

This substance has a very strong band at 955 cm^{-1} which was assigned to the As=-=N=-=As group. The structure of the minor component was not established. The reaction of diphenylchloroarsine with a mixture of chloramine and ammonia at ordinary temperatures and pressures yields (after recrystallization from chloroform) a solid formulated as:

$$[Ph_2\overset{NH_2}{\overset{|}{As}}\text{=-=}N\text{=-=}\overset{NH_2}{\overset{|}{As}}Ph_2]Cl\cdot CHCl_3$$

Heating this substance in DMF solution yields another solid, $(Ph_2AsN)_3\cdot DMF$; the structure of this compound is presumably analogous to that of the phosphazenes. Reichle (157) has prepared a somewhat similar compound, $(Ph_2AsN)_4$, by the thermal decomposition of the azide Ph_2AsN_3 (cf. Section III-7).

The amides of arsonous and arsinous acids are generally liquids which can be distilled at low pressure without decomposition. The hydrohalides of the type $RAs(NRR'\cdot HX)_2$ and $R_2As(NRR'\cdot HX)$ are high melting solids which can be separated from the accompanying ammonium halides by vacuum sublimation (144). The arsenic–nitrogen bond is readily hydrolyzed by water to give the corresponding arsenoso compound or the bis(dialkylarsenic) or bis(diarylarsenic) oxides. $(CF_3)_2AsN(CH_3)_2$, however, reacts with water to give fluoroform, dimethylamine, and arsenious acid.

$CH_3As[N(CH_3)_2]_2$ reacts with alcohols or thiols with rupture of the arsenic–nitrogen bond and the formation of esters or thioesters (93). With 2-ethylmercaptoethanol, the compound $CH_3As(OCH_2CH_2SC_2H_5)_2$ is obtained. Other compounds prepared by this reaction include $CH_3As(SCH_2CH_2SC_2H_5)_2$, $CH_3As(p\text{-}SC_6H_4SCH_3)_2$, and $CH_3As(OC_2H_5)_2$.

In connection with a study of the relative stability of the As—N and P—N bonds, Singh and Burg (158) have prepared $(CF_3)_2AsNHP(CF_3)_2$, $(CF_3)_2AsN(CH_3)P(CF_3)_2$, and $(CF_3)_2AsN[P(CF_3)_2]_2$. These compounds are easily cleaved by hydrogen chloride at the As—N bond to give quantitative yields of bis(trifluoromethyl)chloroarsine. Reaction with boron trichloride or with ammonia also results in cleavage of the As—N bond.

6. DIHALO- AND MONOHALOARSINES

A. Preparation

(*1*) *Reduction*

The reduction of arsonic and arsinic acids has been discussed in Section II-2-B. The chloro- and bromoarsines are prepared by reduction with sulfur dioxide in hydrochloric or hydrobromic acid. The reduction of *o*-benzenediarsonic acid ordinarily does not give the expected *o*-phenylenebis(dichloroarsine) but instead the cyclic structure (6, 159):

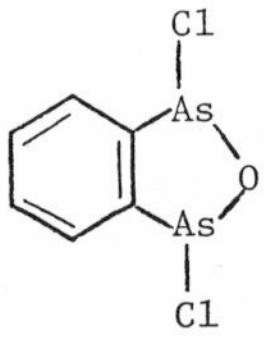

Chatt and Mann (160) were able to convert this compound with thionyl chloride to the desired bis(dichloroarsine). More recent work (161, 162), however, has shown that under slightly different reduction conditions the bis(dichloroarsine) may be prepared directly from the diarsonic acid.

The iodoarsines are prepared by reduction with hydriodic acid (163–165). Even for the preparation of dichloro- and dibromoarsines by sulfur dioxide reduction, catalytic amounts of hydriodic acid are usually added. The actual reduction is thus achieved by hydriodic acid, and the resulting iodine in turn is reduced by the sulfur dioxide.

Hamer and Leckey (166) have reinvestigated the reduction of arsonoacetic acid to dichloroarsinoacetic acid with phosphorus trichloride. They found that, if the reaction conditions are carefully controlled, the dichloroarsine can be prepared in satisfactory yields. By essentially the same procedure they prepared two new compounds, dibromoarsinoacetic acid and 3-dichloroarsinopropanoic acid.

(*2*) *Action of hydrohalic acids on arsonous and arsinous acids*

As stated in Section III-1-A, dihaloarsines (and monohaloarsines) are in equilibrium with the corresponding arsonous (or arsinous) acids in aqueous solution:

$$RAsX_2 + 2H_2O \rightleftharpoons RAs(OH)_2 + 2HX$$

Accordingly, it is possible to prepare haloarsines by adding an excess of the hydrohalic acid to the arsonous or arsinous acids or to their anhydrides.

(*3*) *Disproportionation*

The following reactions occur, usually at elevated temperatures:

$$(1)\quad 2RAsCl_2 \rightleftharpoons R_2AsCl + AsCl_3$$

$$(2)\quad 3RAsCl_2 \rightleftharpoons R_3As + 2AsCl_3$$

$$(3)\quad 2R_2AsCl \rightleftharpoons R_3As + RAsCl_2$$

This was one of the earliest known methods for preparing haloarsines. The trialkyl- and triarylarsines could be readily obtained by other methods, and redistribution was a route to compounds containing one or two organic groups. The reaction is still used to obtain compounds difficult to prepare by other methods. Thus, vinyldibromo- and divinylbromoarsines are readily prepared from trivinylarsine and arsenic tribromide (167). The reaction is unsuccessful, however, with trimethylarsine and arsenic trichloride, or with trimethylarsine and methyldichloroarsine (168). Instead, unstable addition compounds of the type $(CH_3)_3As{\cdot}AsCl_3$ and $(CH_3)_3As{\cdot}As(CH_3)Cl_2$ are formed. These were spontaneously inflammable when heated in air.

The mechanism of the disproportionation reaction has been studied by several groups of workers (169–172). Most of the studies have used the unsubstituted phenyl compounds (169–171). No redistribution takes place in this case below 200°. Reaction (1) is the first to occur as the temperature is raised above 200°. The apparent energy of activation is 35 ± 2 kcal. The $\Delta H°$ is 4.8 kcal/mole and the $\Delta S°$ is 3.6 cal/degree. Reactions (2) and (3) occur at higher temperatures. $\Delta H°$ for reaction (2) is 20.1 kcal/mole with

$\Delta S° = 20.3$ cal/degree; $\Delta H°$ for reaction (3) is 10.5 kcal/mole with $\Delta S° =$ 13.2 cal/degree. These results suggest that there are considerable differences in the bond strengths of the carbon–arsenic bonds in the three types of compounds (171).

(*4*) *From organometallic compounds*

The formation of carbon–arsenic bonds from organic mercurials and arsenic trihalides was one of the earliest known methods, and it is still used to a limited extent (64, 173, 174). The use of organic mercurials is of particular value for establishing the carbon–arsenic bond in compounds which are sensitive to more drastic conditions. β-Pyridyldichloroarsine hydrochloride is readily prepared from β-chloromercuripyridine and arsenic trichloride; the free base, however, cannot be obtained (175). Similarly α-chloromercurifuran and arsenic trichloride give a mixture of α-furyldichloro-, di-α-furylchloro-, and tri-α-furylarsines (176, 177). Mercuric chloride has been found to catalyze the disproportionation of the chloro- and dichloroarsines to the tri-α-furylarsine (178). Thiophene arsenicals can also be prepared from bis(2-thienyl)mercury or 2-thienylmercurichloride (161, 179–181).

Aliphatic dihaloarsines can be prepared from the corresponding dialkylmercurials (182), but better results are often obtained with tetraalkyllead compounds. Thus tetraethyllead has been used for the preparation of both ethyldichloro- and diethylchloroarsines (183) while vinyldichloroarsine has been prepared from tetravinyllead (184). Several haloarsines have been made by the interaction of a dihaloarsine and a tetraalkyllead compound. For example, Kamai and Gatilov (185) have prepared ethylphenylchloroarsine in 90% yield by the reaction of phenyldichloroarsine with tetraethyllead. Similarly methylvinylbromoarsine and methylethylbromoarsine have been prepared from methyldibromoarsine and the appropriate lead compound (186).

Divinylbromo- and vinyldibromoarsines have been prepared from mixed aliphatic tin compounds (167):

$$(n\text{-}C_4H_9)_2Sn(CH{=}CH_2)_2 + 2AsBr_3 \rightarrow (n\text{-}C_4H_9)_2SnBr_2 + 2CH_2{=}CHAsBr_2$$

$$(n\text{-}C_4H_9)_2Sn(CH{=}CH_2)_2 + AsBr_3 \rightarrow (n\text{-}C_4H_9)_2SnBr_2 + (CH_2{=}CH)_2AsBr$$

Vinyldibromoarsine was prepared in 39.4% yield, divinylbromoarsine in 81% yield. It is interesting to note that these vinyl compounds may have been contaminated with the corresponding ethyl derivatives. Kaesz and Stone (187) noted that the vinyl bromide which was obtainable commercially at the time of this work contained 20% ethyl bromide. Since vinyl bromide was used to prepare the tin compounds, the resulting arsines may well have been contaminated with the ethyl derivatives (188).

In recent years trialkylaluminum compounds have found considerable use. The reaction of triisobutylaluminum and arsenic trichloride in ether solution gives a mixture of isobutyldichloroarsine and diisobutylchloroarsine (189). It has been demonstrated (190) that gas-liquid chromatography can be used for the separation of tertiary arsines and haloarsines. This procedure may prove to be useful for separating mixtures of arsines and haloarsines obtained by the reaction of arsenic halides and organometallic reagents.

The reaction of Grignard reagents with arsenic trichloride usually yields tertiary arsines (cf. Section IV-4-A). In a few cases, however, the Grignard reaction has been used for preparing haloarsines or dihaloarsines. Thus, Hartmann and Nowak (27) prepared di-α-naphthylchloroarsine in 50% yield by the reaction between α-naphthylmagnesium bromide and arsenic trichloride. Similarly, Issleib and Tzschach (190a) obtained a good yield of dicyclohexylchloroarsine by the interaction of two moles of cyclohexylmagnesium chloride and arsenic trichloride at −20°. More recently, Green and Kirkpatrick (190b, 190c) prepared pentafluorophenyldichloroarsine by the reaction of the corresponding Grignard reagent with a four-fold excess of arsenic trichloride. Bis(pentafluorophenyl)chloroarsine was obtained from the dichloroarsine:

$$C_6F_5AsCl_2 + C_6F_5MgBr \rightarrow (C_6F_5)_2AsCl$$

They also found that the chlorines in (dimethylamino)dichloroarsine could be replaced stepwise:

$$C_6F_5MgBr + Me_2NAsCl_2 \rightarrow C_6F_5As(Cl)NMe_2$$

$$C_6F_5As(Cl)NMe_2 + C_6F_5MgBr \rightarrow (C_6F_5)_2AsNMe_2$$

The dimethylaminobis(pentafluorophenyl)arsine thus prepared could be converted to the chloroarsine by treatment with dry hydrogen chloride:

$$(C_6F_5)_2AsNMe_2 + HCl \rightarrow (C_6F_5)_2AsCl$$

(*5*) *Arsonation with arsenic trihalides*

Trivalent arsenic compounds are generally poor electrophiles. Arsenic trichloride, however, will react with benzene to form dichlorophenylarsine (64, 191). The reaction is extremely slow and requires a high temperature to obtain a satisfactory yield. The product is contaminated with large amounts of diphenyl, which is extremely difficult to separate from the dichloroarsine. However, if the ring is sufficiently activated with electron-repelling groups, a carbon–arsenic bond can be readily established. N,N-Dialkylanilines have been investigated for this purpose. The reaction was discovered by Michaelis and Rabinerson (192) who obtained a mixture of *p*-dimethylaminoarsenosobenzene and tris(*p*-dimethylaminophenyl)arsine by the reaction of arsenic trichloride with dimethylaniline. A number of other workers in the field of arsenic chemistry have also reported on this reaction. Mroczkowski (193) extended the reaction to a number of ring-substituted dimethylanilines and also prepared *p*-arsenoso-N-methylaniline from arsenic trichloride and N-methylaniline. A group of Indian workers (194) have obtained *p*-(N,N-ethylmethylamino)phenyl-, 2-methyl-4-(N,N-dimethylamino)phenyl-, and 4-(N,N-dimethylamino)naphthyldichloroarsines, together with the corresponding tertiary arsines, from N,N-ethylmethylaniline, N,N-dimethyl-*m*-toluidine, and N,N-dimethyl-α-naphthylamine, respectively. None of the above authors has mentioned the presence of diarylchloroarsines, although they are probably formed in the reaction.

More recently, the reaction of N,N′-dimethyl-N,N′-diphenylethylenediamine and arsenic trichloride has been found (195) to give N,N′-dimethyl-N,N′-bis(*p*-arsenosophenyl)ethylenediamine.

Compounds other than amines can also be arsonated with arsenic trichloride. Thus, dibenzofuran has been converted in this manner to the 2-dichloroarsino derivative (196). Oxidation of the latter compound yielded an arsonic acid which was found to be identical with the acid prepared from 2-aminodibenzofuran by the Bart reaction. The reaction of ferrocene with arsenic trichloride under Friedel-Crafts conditions has been shown (1) to give only monoferrocenyl derivatives of arsenic.

β-Chlorovinyldichloroarsine reacts readily with benzene in the presence of aluminum chloride to give a mixture of β-chlorovinylphenylchloroarsine and β-chlorovinyldiphenylarsine (197).

(*6*) *Halogen exchange*

Although bromo- and chloroarsines can readily be prepared by a number of methods, the fluoro- and iodoarsines are usually prepared from the chloroarsines. When aqueous solutions of dichloroarsines are treated with potassium iodide or hydriodic acid, the sparingly soluble diiodoarsines separate from solution as deep yellow crystals (17). Similarly, a considerable number of iodoarsines and diiodoarsines have been prepared by the reaction of the chloro compounds with sodium iodide in acetone (14, 15, 198–201). For the preparation of the fluoroarsines from other haloarsines, anhydrous ammonium fluoride, silver fluoride, or sodium fluoride in acetone has been used (13, 18, 147, 202–204). Silver chloride and silver bromide have been used for converting iodoarsines to chloroarsines and bromoarsines, respectively (147, 203, 205).

(*7*) *The addition of arsenic halides to unsaturated compounds*

In contrast to phosphorus compounds, the reaction of arsenicals with unsaturated linkages has not been investigated to an appreciable extent. The exception to this is the catalyzed addition of arsenic halides to acetylene. The products of the reaction were first elucidated by Green and Price (206). These authors (206) used anhydrous aluminum chloride and arsenic trichloride and correctly assigned the structures $ClCH{=}CHAsCl_2$, $(ClCH{=}CH)_2AsCl$, and $(ClCH{=}CH)_3As$ to the products of the reaction. These compounds, particularly the dichloroarsine, are among the most powerful vesicants ever prepared. They are known by the common name Lewisites, after W. L. Lewis, who also made a comprehensive study of the reaction. The reaction is extremely slow unless a catalyst is present. Aluminum trichloride, however, promotes the formation of the secondary and tertiary compounds at the expense of the desired primary compound. Mercuric chloride and cuprous chloride are preferred catalysts for the formation of the primary product (207, 208).

Phenyldichloroarsine will also react with acetylene (catalyzed by aluminum trichloride) to give a mixture of products which include the primary, secondary, and tertiary Lewisites, β-chlorovinylphenylchloroarsine, bis(β-chlorovinyl)phenylarsine, β-chlorovinyldiphenylarsine, phenyldichloroarsine, and triphenylarsine (209, 210).

The mechanism of addition of arsenic halides to acetylene has been studied by several workers. Prat (211) has suggested that the reaction involves the electrophilic attack of arsenic on the carbon atom with the formation of an arsonium salt, followed by attack of chloride ion on one carbon atom with rupture of the arsenic–carbon bond:

$$AsCl_3 + C_2H_2 \longrightarrow \left[Cl_2\overset{+}{As} \begin{matrix} CH \\ \| \\ CH \end{matrix} \right] Cl^- \longrightarrow ClCH{=}CHAsCl_2$$

The resulting β-chlorovinyldichloroarsine then reacts further with acetylene by the same mechanism to form bis(β-chlorovinyl)chloroarsine and eventually tris(β-chlorovinyl)arsine. This mechanism ignores the catalytic effect of aluminum trichloride. It seems more likely that the arsenic trichloride combines with the aluminum trichloride and that the electrophile is a species such as $[\overset{+}{As}Cl_2][AlCl_4]^-$. Lewis and Perkins (212) have also considered the mechanism of this addition reaction. The mercuric chloride catalyzed addition involves first the formation of β-chlorovinylmercurichloride by the electrophilic attack of mercury on the unsaturated linkage, followed by reaction of the organic mercurial with arsenic trichloride (207).

The Lewisites are capable of existing in two geometric forms. The first published investigation of the isomerism was by Donohue, Humphrey, and

Schomaker (213) who quote, however, an earlier British report on the separation of primary Lewisite, β-chlorovinyldichloroarsine, into two isomers to which *cis* and *trans* configurations had been assigned. They (213) studied the structure of the two (impure) isomers and on the basis of the two non-bonded As—Cl distances (3.30 and 4.45 A) assigned the *cis* and *trans* structures to the two isomers. The authors pointed out, however, that the results obtained do not exclude the possibility that the two isomers are position rather than geometric isomers, i.e., that one compound is α-chlorovinyldichloroarsine.

Subsequent to the above work a complete study of the two isomeric primary Lewisites was published in England. In the first paper of the series Hewett (214) gives the chemistry of the two isomers and methods for their separation. When mercuric or cuprous chloride is used as the catalyst, the resulting primary Lewisite is less vesicant than the primary compound obtained from the reaction using aluminum trichloride. All three products give the same arsonic acid on oxidation, but from the mother liquors of the acid obtained by the use of mercuric or cuprous chloride, a second, very soluble acid was obtained in impure form. All attempts to obtain two pure isomeric Lewisites by fractional distillation were unsuccessful. Finally, by fractional distillation of the Lewisites and oxidation of the fractions, two arsonic acids were obtained which were further purified by fractional crystallization. Reduction of these acids gave two pure primary Lewisites. The higher boiling (presumably *trans*) isomer evolved acetylene on treatment with sodium hydroxide; the lower boiling (presumably *cis*) isomer did not. The latter could be partially isomerized by ultraviolet irradiation, but the reaction was accompanied by considerable disproportionation.

In the second paper (215) the dipole moments of the two isomers were reported. The lower boiling isomer gave a value of 2.20 D; the higher boiling isomer gave a value of 2.61 D. These results confirm the identity of these isomers as the *cis* and *trans* forms, respectively. The two dipole moments agreed quite well with the calculated values, but differed significantly from the value calculated for α-chlorovinyldichloroarsine, thereby ruling out this compound as one of the two species.

In the final paper of the series Whiting (216) determined a number of physical constants for the two isomers. The *cis* isomer melted at $-44.7°$; it boiled at 169.8°/760 mm; the d_4^{25} was 1.8598; and n_D^{25} was 1.5859. The corresponding values for the *trans* isomer were $-1.2°$, 196.6°/760 mm, $d_4^{25} =$ 1.8793, and $n_D^{25} = 1.6067$.

Arsenic trichloride adds to 1-octyne with the formation of 2-chlorooctenyl-1-dichloroarsine (217). It is not clear from the abstract whether a catalyst was used. The reaction of phenyldichloroarsine with phenylacetylene at 140–150° has been found (218) to yield 3-chloro-1-phenylarsindole.

(*8*) *The reaction of arsenic halides with diazoalkanes*

Arsenic trichloride in ether solution reacts vigorously with diazoalkanes, with the evolution of nitrogen and the formation of arsenicals containing a chloromethyl group. The reaction was first investigated by Braz and Yakubovich (219) who obtained chloromethyldichloroarsine from diazomethane and arsenic trichloride. Diazoethane produced only bis(α-chloroethyl)-chloroarsine, unless an excess of diazoethane was used in which case some tris(α-chloroethyl)arsine was also formed (220, 221). α-Chloroethyldichloroarsine could not be isolated from the reaction mixture.

The reaction of diphenylchloroarsine and diazomethane gave the unexpected product methylenebis(diphenylarsine) oxide, $(C_6H_5)_2AsCH_2As(O)(C_6H_5)_2$ (222). The authors suggested that traces of water in the ether hydrolyzed the chloroarsine to bis(diphenylarsenic) oxide. This was followed by a nucleophilic attack of arsenic on diazomethane, followed by loss of nitrogen, ionization, and recombination of the ions to form the product:

$$[(C_6H_5)_2As]_2O + \overset{+}{C}H_2N{=}\overset{-}{N} \longrightarrow (C_6H_5)_2\underset{\substack{|\\ CH_2-N=\overset{-}{N}}}{\overset{+}{A}s}{-}O{-}As(C_6H_5)_2 \longrightarrow$$

$$N_2 + (C_6H_5)_2As\overset{-}{C}H_2 + (C_6H_5)_2\overset{+}{A}sO \longrightarrow (C_6H_5)_2AsCH_2As(O)(C_6H_5)_2$$

Seyferth (223) has pointed out that the above mechanism could be tested by treating the readily available bis(diphenylarsenic) oxide with diazomethane.

The mechanism of methylenation of metal and metalloid halides has been considered by several workers. Their ideas have been summarized by Seyferth (223) in a review article. Yakubovich and Ginsburg (224) consider that a methylene radical formed by the cleavage of diazomethane attacks the metal or metalloid halide. Other workers (225, 226) have postulated that the reaction involves the nucleophilic attack of diazomethane (which exists as a resonance hybrid of a number of charge-separated forms, several of which involve a carbanionic form) on the metal. Seyferth (223) considers that the experimental evidence favors the polar mechanism. In particular he points out that all of the elemental halides which react with diazoalkanes possess vacant *d*-orbitals which could be readily attacked by the anionic carbon. No kinetic studies of the mechanism have been undertaken.

(*9*) *Pyrolysis of arsenic dihalides*

The pyrolysis of arsenic dihalides with the elimination of RX has been used for the preparation of diarylhaloarsines (27, 227). In practice, a tertiary arsine (usually containing at least one aliphatic group) is halogenated to the arsenic dihalide. Pyrolysis then results in the elimination of the alkyl halide and the formation of the chloroarsine. For example, dimethyl-α-naphthylarsine can be readily brominated in carbon tetrachloride to dimethyl-α-naphthylarsenic dibromide. When this is pyrolyzed, methyl bromide is evolved and methyl-α-naphthylbromoarsine remains (14, 15):

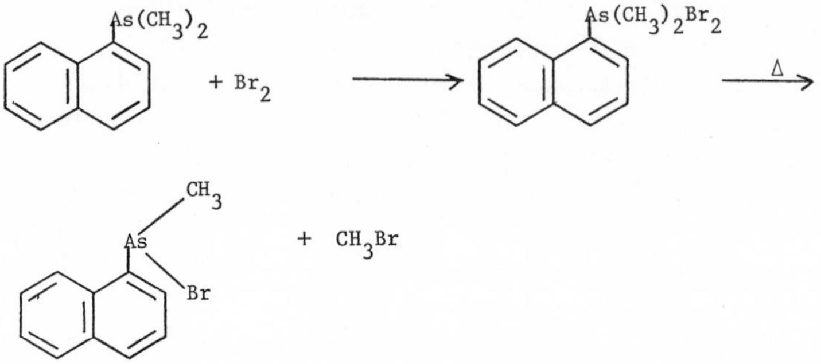

$$As(CH_3)_2 \quad + Br_2 \longrightarrow As(CH_3)_2Br_2 \xrightarrow{\Delta}$$

$$CH_3 \quad As \quad Br \quad + CH_3Br$$

Trivinylarsenic diiodide is pyrolyzed to divinyliodoarsine and vinyl iodide (167, 228).

The elimination reaction has been used successfully for the preparation of heterocyclic arsenicals. Thus 1-methylpentamethylenearsine, when chlorinated and then pyrolyzed, eliminated methyl bromide with the formation of 1-chloropentamethylenearsine (229).

In the preparation of 1-methylarsindoline, Turner and Bury (230) converted dimethyl-2-phenylethylarsine to the corresponding dichloride, which lost methyl chloride on heating to give methyl-2-phenylethylchloroarsine. Cyclization of this chloroarsine with elimination of hydrogen chloride gave the desired heterocyclic compound.

Cullen (231) has considered the ease of elimination of different groups. Diphenyltrifluoromethylarsenic dibromide when heated to 120° lost trifluoromethyl bromide, whereas methylphenyltrifluoromethylarsenic dibromide lost principally methyl bromide together with smaller amounts of trifluoromethyl bromide. The ease of elimination was thus in the order $CH_3 > CF_3 > C_6H_5$. The actual mechanism of the reaction has not been studied.

A reaction which may go by a somewhat similar pathway has been used for the preparation of methyldichloroarsine (232). Cacodylic acid was treated with hydrogen chloride until saturated, and the resulting product distilled. Methyl chloride was eliminated with the formation of methyldichloroarsine:

$$(CH_3)_2AsO_2H + HCl \rightarrow [(CH_3)_2As(OH)_2]Cl$$

$$[(CH_3)_2As(OH)_2]Cl + 2HCl \rightarrow CH_3AsCl_2 + CH_3Cl + 2H_2O$$

Dialkyl- and diarylarsenic trihalides undergo a similar elimination reaction when heated. Dicyclohexylarsenic trichloride gives cyclohexyldichloroarsine at 80–90° (233); and diphenylarsenic trichloride heated in a stream of carbon dioxide gives phenyldichloroarsine (64).

(10) From β-chlorovinylarsines

A relatively new method, developed by Knunyants and Pil'skaya (234), depends upon the evolution of acetylene from β-chlorovinylarsines. For example, bis(β-chlorovinyl)chloroarsine is converted by means of a Grignard reagent to a tertiary arsine. This is then treated with alcoholic potassium hydroxide whereupon acetylene is evolved; treatment of the product with hydrogen chloride then forms the dichloroarsine. Similarly, β-chlorovinyldichloroarsine can be used for the preparation of dialkyl- or diarylchloroarsines:

$$ClCH{=}CHAsCl_2 + 2RMgX \rightarrow ClCH{=}CHAsR_2 \xrightarrow[(2)\ HCl]{(1)\ KOH} R_2AsCl + C_2H_2$$

The mechanism of the reaction has not been studied.

(11) The reaction between aryldichloroarsines and arsenoso compounds

Aryldichloroarsines react with arsenoso compounds according to the following equation:

$$ArAsCl_2 + 3ArAsO \rightarrow 2Ar_2AsCl + As_2O_3$$

The reaction was the subject of an isolated observation by Sir W. J. Pope and E. E. Turner during the first World War and is often referred to as the Pope-Turner reaction. The reaction was thoroughly investigated by Barker and co-workers in 1949 (30, 235, 236). As the reaction is ordinarily performed, the product is contaminated with phenyldichloroarsine and triphenylarsine, both of which are formed by disproportionation of the diphenylchloroarsine. The latter authors found that the condensation is catalyzed by a number of substances: iron, ferric chloride, zinc chloride, and zinc oxide. Zinc oxide proved to be the best catalyst, and with its use it is possible to prepare diphenylchloroarsine in 91.4% yield by heating the reactants in stoichiometric amounts at 120° for 2.5 hours. The product, however, contains 4.6% phenyldichloroarsine.

Heaney and Millar (237) have used the Pope-Turner reaction to prepare bis(*o*-bromophenyl)chloroarsine after they were unsuccessful in obtaining this compound via the Bart reaction.

(12) The cleavage of 10-alkyl-5,10-dihydrophenarsazines

When a 10-alkyl-5,10-dihydrophenarsazine is heated with hydrochloric acid, two arsenic–carbon bonds are ruptured with the formation of diphenylamine hydrochloride and an alkyldichloroarsine. The 10-alkylphenarsazine is readily prepared by the action of the appropriate Grignard reagent on 10-chloro-5,10-dihydrophenarsazine:

$$\text{(10-chloro-5,10-dihydrophenarsazine)} + \text{RMgX} \longrightarrow \text{(10-R-5,10-dihydrophenarsazine)}$$

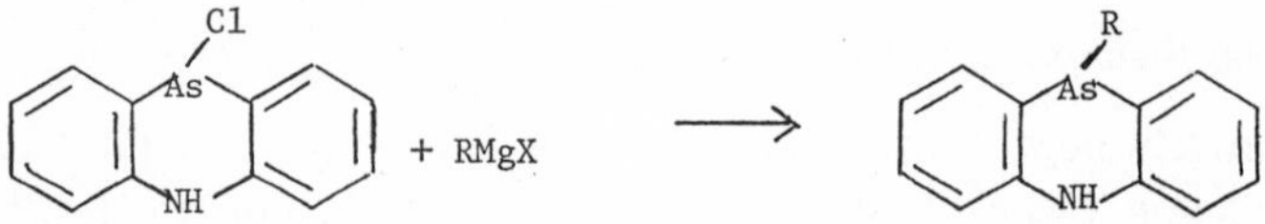

This method has been used by Gibson and Johnson (238) for preparing a number of alkyldichloroarsines in a high state of purity. Phenyldichloroarsine can also be prepared in this manner (239). The reaction has also been used to prepare perfluorovinyldichloroarsine, $CF_2{=}CFAsCl_2$, in pure form (240, 241).

The reaction mechanism may involve a stepwise electrophilic attack of protons on the ring carbons attached to arsenic, followed by a rupture of carbon–arsenic bonds.

(*13*) *The reaction of diazocarboxylates with arsenic trichloride*

Although diazonium salts react with trivalent arsenicals to produce arylarsonic and diarylarsinic acids, Reutov and Bundel (242) have demonstrated that diazocarboxylates undergo an entirely different type of reaction with arsenic trichloride, forming aryldichloro- and diarylchloroarsines. In practice these were not isolated but hydrolyzed to the arsenoso compounds or bis(diarylarsenic) oxides (or arsinous acids). The authors suggested the following reaction sequence:

$$ArN{=}NCO_2K + AsCl_3 \rightarrow$$
$$[ArN{=}NCO_2AsCl_2] + KCl \rightarrow ArAsCl_2 + N_2 + CO_2$$

Presumably the unstable intermediate decomposes with the formation of an aryl carbanion and $AsCl_2^+$, followed by direct combination of the carbanion and the arsenic fragment. Further reaction of $ArAsCl_2$ with the diazocarboxylate would then lead to the formation of the diarylchloroarsine.

The authors (242) also demonstrated that a second reaction occurs which they formulated as:

$$2ArN_2CO_2K + 4AsCl_3 + O_2 \rightarrow$$
$$2[Ar\overset{+}{N}_2][AsCl_4]^- + 2AsOCl + 2KCl + 2CO_2$$

Thus, the addition of zinc dust to the reaction mixture resulted in a vigorous evolution of nitrogen. If the original reaction was carried out in the absence of oxygen, no evolution of nitrogen occurred on the addition of zinc dust.

An attempt was made to isolate a diazonium chloroarsenite by the reaction of $[C_6H_5\overset{+}{N}_2][FeCl_4]^-$ and arsenic trichloride. An unstable solid was formed, the nitrogen analysis of which suggested a mixture of $[C_6H_5AsN_2][AsCl_4]$ and $[C_6H_5AsN_2]_2[AsCl_5]$.

(14) The reaction of free radicals with arsenic and with arsenic halides

In extensive studies of the reaction of aryl radicals with metals, Hanby and Waters (243) prepared phenyldichloro- and diphenylchloroarsines, together with triphenylarsine, by the reaction between benzenediazonium chlorozincate, arsenic trichloride, and zinc dust in acetone. The method is not of preparative value for the haloarsines; its use for the preparation of triarylarsines as well as the mechanism of the reaction is considered in Section IV-4-F.

The reaction has also been studied by Reutov and Bundel (244) who obtained mixtures of diarylchloroarsines, aryldichloroarsines, triarylarsines, and triarylarsine dihalides by the action of powdered iron on diazonium chlorozincates or chloroferrates and arsenic trichloride (or dichlorophenylarsine) in acetone solution. The ratio between the diazonium salt and the arsenic trichloride had considerable effect on the yields of products formed, and by using an excess of arsenic trichloride, it was possible to obtain the dichloroarsine as the principal product.

The free radical method has proved to be of value for the preparation of trifluoromethyl arsenicals. When powdered arsenic metal and trifluoroiodomethane are heated in a Carius tube or steel autoclave to 220°, a mixture of trifluoromethyldiiodo-, bis(trifluoromethyl)iodoarsines, tris(trifluoromethyl)arsine, and arsenic triiodide is formed (26). The mechanism of the reaction has been considered by Walaschewski (203), who suggests two possible schemes. One mechanism involves the homolytic cleavage of trifluoroiodomethane into trifluoromethyl and iodine radicals which attack the arsenic to form the four arsenic-containing products. The second scheme involves the formation of an activated but undissociated trifluoroiodomethane molecule, which attacks the arsenic with the formation of trifluoromethyldiiodoarsine, which then disproportionates to form the products:

$$As + CF_3I^* \rightarrow CF_3AsI\cdot$$

$$CF_3AsI\cdot + CF_3I^* \rightarrow CF_3AsI_2 + CF_3\cdot$$

A somewhat similar reaction has been used to prepare vinyldihalo- and divinylhaloarsines by passing vinyl bromide or vinyl chloride over a mixture of arsenic, copper, and zinc heated to 450° (245).

Another reaction, which must involve free radicals, has been described by Henglein (246). When phosphorus trichloride, phosphorus oxychloride, arsenic trichloride, or silicon tetrachloride was irradiated in the presence of certain hydrocarbons, organic chlorine-containing compounds of these elements and hydrogen chloride were obtained. Thus arsenic trichloride and cyclohexane, when irradiated with electrons in the beam of a 3.3 MeV Van de Graaff generator, gave cyclohexyldichloroarsine in 25% yield. The author (246) proposed a free radical mechanism for the reaction in which the electron beam excites the hydrocarbon and metalloid halide to excited or ionized molecules which then dissociate into free radicals:

$$RH, MCl_3 \rightarrow RH^*, MCl_3^* \rightarrow R\cdot + H\cdot, MCl_2\cdot + Cl\cdot$$

Reaction of these radicals will produce the various products found in the reaction. Obviously, unless the hydrocarbon contains only one type of carbon atom (primary, secondary, or tertiary) a mixture of products will result, since hydrogen may be eliminated from all types of carbon atoms. In spite of this limitation, the reaction may well be of value in obtaining organometallic compounds which are otherwise difficult to synthesize.

(15) The preparation of heterocyclic haloarsines by cyclodehydrohalogenation

Dihaloarsines containing the dihaloarsino group in a suitable position for cyclization readily lose hydrogen chloride under mild conditions with the formation of the heterocyclic chloroarsine:

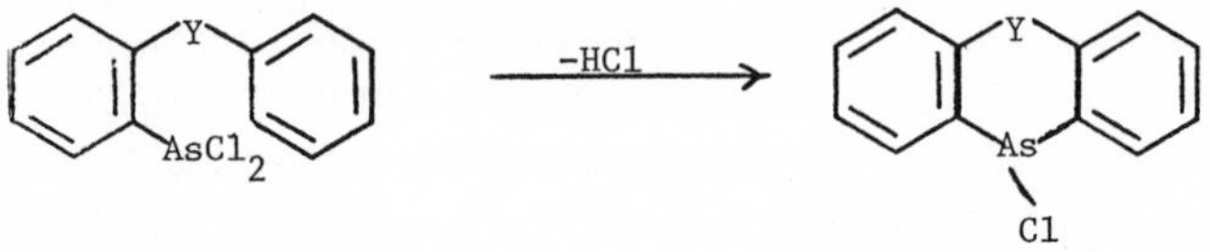

Y $AsCl_2$ —HCl→ Y As Cl

Y may be oxygen, nitrogen, sulfur, or a similar group. The reaction often succeeds where the cyclodehydration of arsonic acids is unsuccessful. The preparation and properties of heterocyclic derivatives of arsenic were reviewed by Mann (247) in 1950. More recently, Campbell and Poller (248) have used cyclodehydrohalogenation for the preparation of 5-chloro-3-methoxydibenzoarsole:

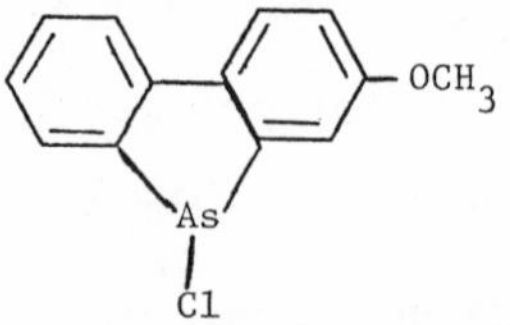

OCH_3 As Cl

The mechanism of the cyclization has been studied by Turner and coworkers (249, 250). These authors studied the rates of cyclization of a number of substituted *o*-dichloroarsinodiphenyl ethers, in which a substituent was on the ring not containing the dichloroarsino group. The rate of cyclization was increased markedly by electron-repelling groups and retarded by electron-attracting groups. The authors concluded that the reaction involves an electrophilic attack of arsenic on the ring carbon atom.

(*16*) *Miscellaneous methods*

Dihalo- and haloarsines have been prepared by a number of other methods, none of which appears to be of general preparative importance. One of these methods involves the reaction of a trialkyl- or triarylarsine with a halogen or a halogen compound. The halogen compounds used include thallic chloride, cobaltic fluoride, and hydrogen chloride. Thus, Goddard (251) obtained 2,4- and 2,5-dimethylphenyldichloroarsines by treating the corresponding trixylylarsines with thallic chloride. Similarly, bis(trifluoromethyl)fluoroarsine has been prepared by the reaction of tris(trifluoromethyl)arsine with cobaltic fluoride (18). The use of hydrogen chloride for cleaving 10-alkyl-5,10-dihydrophenarsazines has been discussed in III-3-A-(12). Both bromine and iodine have been used for converting tris(trifluoromethyl)arsine to the corresponding haloarsines and dihaloarsines (18).

A few primary and secondary arsines have been found to react with halogens to form dihalo- and haloarsines. Thus phenylarsine reacts instantly with two moles of iodine to give a quantitative yield of phenyldiiodoarsine (104, 252). Two diarsines, $C_6H_5As(H)(CH_2)_nAs(H)C_6H_5$) where $n = 3$ or 6, have been converted to the dibromides, $C_6H_5As(Br)(CH_2)_nAs(Br)C_6H_5$, by treatment with bromine (253). Similarly, $CHF_2CF_2AsH_2$ and $(CHF_2CF_2)_2AsH$ on chlorination yield the corresponding dichloro- and chloroarsines, respectively (254). Reagents other than the halogens have also been used for converting primary and secondary arsines to dihalo- and haloarsines. Thus, phenylarsine reacts with thionyl chloride in benzene solution to yield phenyldichloroarsine (as well as thioarsenosobenzene) (105), while dimethylarsine has been found to react with acetyl halides to give dimethylhaloarsines and several other products (255). The reaction of secondary arsines with a sulfenyl chloride or with any of a number of cyclic vinyl chlorides produces haloarsines as by-products (130, 256).

The cleavage of amides of arsonous and arsinous acids by hydrogen halides has been used for preparing dihalo- and haloarsines (145, 152). Both symmetrical and unsymmetrical haloarsines have been made by this procedure.

Phenyldichloro- and diphenylchloroarsines, together with triphenylarsine, are formed when benzene and arsenic trichloride are irradiated with neutrons (257). The action of alkyl halides on tetraalkyldiarsines gives a mixture of trialkylarsine and dialkylhaloarsine:

$$R_2As{-}AsR_2 + R'X \rightarrow R_2R'As + R_2AsX$$

Tetramethyldiarsine, for example, reacts with trifluoromethyliodomethane to form dimethyltrifluoromethylarsine and dimethyliodoarsine (258). Tetraalkylarsines are also cleaved by sulfenyl chlorides to give chloroarsines (130).

Arsenic trioxide reacts with acyl chlorides in the presence of aluminum chloride to give methylenebis(dichloroarsines) (259):

$$As_2O_3 + 5RCH_2COCl \xrightarrow{AlCl_3} RCH(AsCl_2)_2 + 2(RCH_2CO)_2O + CO_2 + HCl$$

The mechanism of this reaction has been elucidated by Gutbier and Plust (260), who made use of the mass spectrometer to identify the intermediate products. They showed that arsenic trioxide dissolves in acetyl chloride at 100° with the formation of arsenic trichloride, arsenic triacetate, and acetic anhydride, together with small amounts of acetoxydichloroarsine, $CH_3CO_2AsCl_2$. The reaction takes the following course:

$$As_2O_3 + 6CH_3COCl \rightleftharpoons 2AsCl_3 + 3(CH_3CO)_2O$$

$$AsCl_3 + 3(CH_3CO)_2O \rightleftharpoons As(CH_3CO_2)_3 + CH_3COCl$$

$$2AsCl_3 + As(CH_3CO_2)_3 \rightleftharpoons 3(CH_3CO_2)AsCl_2$$

At temperatures of 90° or above, arsenic triacetate pyrolyzes as follows:

$$As(CH_3CO_2)_3 \rightarrow CH_3CO_2AsO + (CH_3CO)_2O$$

$$CH_3CO_2AsO \rightarrow CH_3AsO + CO_2$$

The arsenosomethane thus formed reacts with acetyl chloride to give methyldichloroarsine, a reaction previously reported in the chemical literature. For the formation of methylenebis(dichloroarsine), the authors proposed the following reaction:

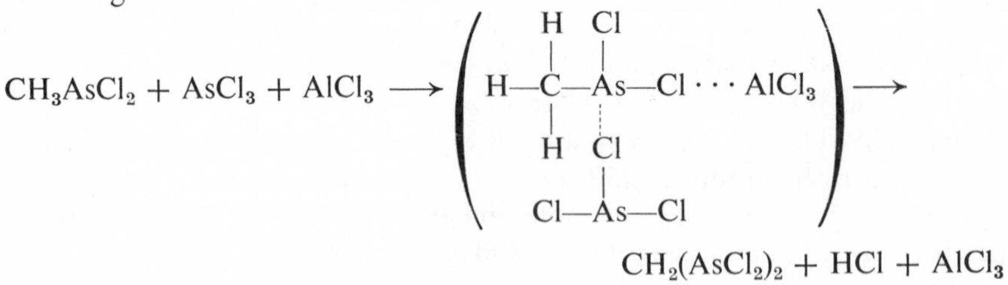

$$CH_3AsCl_2 + AsCl_3 + AlCl_3 \longrightarrow \left(\begin{array}{c} \mathrm{H}\quad \mathrm{Cl} \\ |\quad\; | \\ \mathrm{H{-}C{-}As{-}Cl \cdots AlCl_3} \\ |\quad\; \vdots \\ \mathrm{H}\quad \mathrm{Cl} \\ | \\ \mathrm{Cl{-}As{-}Cl} \end{array} \right) \longrightarrow$$

$$CH_2(AsCl_2)_2 + HCl + AlCl_3$$

Some proof of this mechanism was obtained by showing that when arsenic triacetate was heated with arsenic trichloride in the presence of aluminum chloride, methylenebis(dichloroarsine) was formed.

B. Reactions

Dihalo- and haloarsines are reactive compounds. They are readily oxidized by halogens to compounds of the type $RAsX_4$ or R_2AsX_3. They are reduced to arsines or arseno compounds and are hydrolyzed to arsonous and arsinous acids or their anhydrides. In addition, they react with Grignard or lithium reagents to give tertiary arsines (IV-4-A), with alcohols or alkoxides to give esters (III-3), with amines to give amides or imides (III-5), with phosphines to form phosphonium salts (VI-7), with hydrogen sulfide, mercaptans or thio acids to give sulfides (III-4), and with selenols to give selenides (III-4). They also undergo metathetical reactions with metallic halides (III-6-A-(6)), pseudohalides (III-7), and carboxylates (III-7). They undergo disproportionation (III-6-A-(3)), cyclodehydrohalogenation (III-6-A-(15) and IV-4-G), addition to unsaturated linkages (III-6-A-(7)), and the Friedel-Crafts reaction (III-6-A-(5)). The dihaloarsines react with arsenoso compounds to yield haloarsines (III-6-A-(11)). Halo- and dihaloarsines also undergo a number of miscellaneous reactions, some of which are discussed below.

Aryldichloro-, diarylchloro-, and diarylcyanoarsines react with acyl chlorides in carbon disulfide solution in the presence of aluminum chloride (261). The carbon–arsenic bond is ruptured with the formation of the corresponding ketone:

$$RAsCl_2 + R'COCl \rightleftharpoons R'COR + AsCl_3$$

The reaction is useful in the determination of the point of attachment of the arsenic to the aromatic ring.

Phenyldichloro- and diphenylchloroarsines react with anhydrous aluminum chloride to form the addition compounds $(C_6H_5AsCl_2)_3AlCl_3$ and $(C_6H_5)_2AsCl{\cdot}AlCl_3$, respectively (262). Both of these compounds are viscous syrups which decompose on heating to give tetraphenylarsonium chloride and metallic arsenic.

Ethyldichloroarsine reacts with *p*-$PhC_6H_4N{=}NC(S)NHNHC_6H_4Ph$-*p* to give the following compound (263):

$$\begin{array}{l} p\text{-}C_6H_5C_6H_4N{=}N{-}\overset{}{C}{=}N{-}N{-}C_6H_4C_6H_5\text{-}p \\ \qquad\qquad\qquad\qquad\quad | \qquad\quad | \\ \qquad\qquad\qquad\qquad\quad S{-\!\!-\!\!-}As \\ \qquad\qquad\qquad\qquad\qquad\qquad | \\ \qquad\qquad\qquad\qquad\qquad\quad C_2H_5 \end{array}$$

A similar type of condensation is also given by arsenobenzene and by arsenic trichloride.

The reaction of alkyl iodides with iodo- or diiodoarsines in the presence of mercury generally gives arsonium triiodomercurates (264):

$$RAsI_2 + Hg + R'I \rightarrow [RR_3'As][HgI_3]$$

However, the reaction of dimethyliodoarsine with ethyl iodide and mercury yields the arsine complex $(CH_3)_2AsC_2H_5 \cdot HgI_2$. Trifluoromethyl iodide usually reacts with iodo- or diiodoarsines to give the expected tertiary arsines (231, 258):

$$RAsI_2 + Hg + CF_3I \rightarrow RAs(CF_3)_2$$

The reaction of trifluoromethyl iodide with diethyliodoarsine and mercury, however, yields mainly ethylbis(trifluoromethyl)arsine; and, from the reaction with di-*n*-butyliodoarsine, the only tertiary arsine isolated was *n*-butylbis(trifluoromethyl)arsine (201). The reaction of trifluoromethyl iodide with bis(trifluoromethyl)iodoarsine and mercury gives the expected tertiary arsine in 62% yield as well as tetrakis(trifluoromethyl)diarsine (265). Trifluoromethyl iodide reacts with iodo- and diiodoarsines in the absence of mercury at 170° to yield a complex mixture of products (266).

The electrochemical behavior of several organic derivatives of group V elements has been studied by Dessy and co-workers (75). The electrolytic reduction of diphenylbromoarsine was found to yield the diphenylarsenide ion. Two one-electron steps were identified:

$$Ph_2AsBr + e^- \rightarrow \tfrac{1}{2}Ph_2AsAsPh_2 + Br^-$$

$$\tfrac{1}{2}Ph_2AsAsPh_2 + e^- \rightarrow Ph_2As^-$$

The first step probably proceeds via coupling of diphenylarsenic radicals. The identity of the tetraphenyldiarsine formed in this step was established by polarographic comparison of the solution after the first-wave electrolysis of diphenylbromoarsine with a solution of authentic tetraphenyldiarsine. Addition of diphenylbromoarsine to the solution of diphenylarsenide formed in the second step was found to give tetraphenyldiarsine:

$$Ph_2AsBr + Ph_2As^- \rightarrow Ph_2AsAsPh_2 + Br^-$$

The electrolytic reduction of di-α-naphthylbromoarsine also proceeds through two one-electron steps to yield an arsenide ion. This latter species shows a third polarographic wave, which probably indicates transfer of an electron into a π^* orbital of the naphthalene ring system. The reduction of phenyldichloroarsine was found to involve a two-electron step, but no details of this reaction have been published.

7. MISCELLANEOUS COMPOUNDS OF THE TYPE $RAsX_2$ AND R_2AsX

In addition to the dihalo- and haloarsines, a number of compounds of the type $RAsX_2$ and R_2AsX are known where X is a group such as cyano, thiocyano, cyanato, or azido. Such compounds are usually formed by the metathetical reaction between the chloroarsines and silver, sodium, or lithium salts (e.g., AgCN, NaSCN, and LiN_3) (147, 150, 157, 161, 182, 203, 229). Compounds of the type $RAs(CN)_2$ have been studied by Grichkiewitch-Trohimowski, Mateyak, and Zablotski (267). Phenyl- and α-naphthyldicyanoarsines were readily prepared from the corresponding dichloroarsines and silver cyanide. Methyldicyanoarsine was prepared from dimethylcyanoarsine and cyanogen bromide. In the latter case an unstable intermediate, $(CH_3)_2As(CN)_2Br$, presumably was formed and lost methyl bromide to form the desired methyldicyanoarsine. The dicyanoarsines are crystalline solids, readily hydrolyzed by water to the corresponding arsenoso compounds.

Diphenylcyanoarsine may be prepared by the action of hydrogen cyanide on bis(diphenylarsenic) oxide (34) or ethyl diphenylarsinite (21). Methyl-α-naphthylcyanoarsine has been prepared by the reaction of dimethyl-α-naphthylarsine with cyanogen bromide and the subsequent elimination of methyl bromide from the pentavalent arsenical (14, 15). Similarly, cyclohexylmethylphenyl- and cyclohexyldipropylarsines, when treated with cyanogen bromide and the product heated, eliminate methyl or propyl bromide to give cyclohexylphenyl- and cyclohexylpropylcyanoarsines (233). When the product formed by the addition of cyanogen bromide to methyl-α-naphthylphenylarsine is heated, both methyl bromide and acetonitrile are eliminated, and a mixture of α-naphthylphenylcyanoarsine and α-naphthylphenylbromoarsine is formed (268).

A number of compounds of the types $RAs(O_2CR')_2$ and R_2AsO_2CR' have been prepared. The methods used include: (1) the reaction of halo- or dihaloarsines with metallic carboxylates (147, 150, 269); (2) the reaction of esters of arsonous or arsinous acids with acid anhydrides (12, 78, 270); (3) the reaction of arsenoso compounds or bis(diarylarsenic) oxides with acid anhydrides (102); (4) the reaction of bis(diarylarsenic) oxides with carboxylic acids in methanol solution (271); (5) the reaction of amides of arsonous acids with acid anhydrides (272); and (6) the reaction of thioarsenites with acid anhydrides (119).

Acetyldiphenylarsine has been prepared by the reaction between diphenylarsine and acetyl chloride (76). The compound when oxidized in aqueous solution yields a mixture of diphenylarsinic and acetic acids. Acetyldimethylarsine can similarly be prepared from dimethylarsine and acetyl chloride, or from dimethylarsine and $CH_3CONH_2{\cdot}BF_3$ (255). Acetyl bromide or ketene reacts with dimethylarsine to give acetyldimethylarsine in a less pure form.

Trichloroacetyl chloride and dimethylarsine give dimethyltrichloroacetylarsine. The preparation of keto-substituted arsines via magnesium derivatives of primary and secondary arsines is discussed in Section IV-4-C.

The preparation of carboxydiphenyl- and carbamyldiphenylarsines has been claimed in the patent literature (273, 274). The latter was prepared by the oxidation of cyanodiphenylarsine with hydrogen peroxide. In view of the ease of hydrolysis of the As—CN linkage, the existence of these compounds is doubtful.

Dessy and co-workers (75) have prepared a solution of diphenylarsenic perchlorate, $Ph_2As^+ClO_4^-$, by the addition of an equimolar amount of silver perchlorate to a "glyme" solution of a diphenylhaloarsine. Electrolytic reduction of $Ph_2As^+ClO_4^-$ yields a yellow solution containing the diphenylarsenide ion. Addition of silver perchlorate to this yellow solution gives a clear, golden brown solution which apparently contains the covalent compound, (diphenylarsenic)silver. It is stated that no silver ion is present in this solution. Electrochemical degradation of the solution regenerates the yellow diphenylarsenide anion and deposits metallic silver.

8. PHYSICAL PROPERTIES AND STRUCTURE

A. Molecular weights

Cryoscopic and ebullioscopic molecular weights have been reported for a considerable number of organic arsenicals dissolved in nonpolar solvents. Normal molecular weights were found for all compounds whose conventional structural formulas indicate that the arsenic atom is linked to other atoms only by *single* bonds. This group of compounds includes halo- and dihaloarsines (1, 101, 166), arsinous acids and their anhydrides (101, 275), amides of arsonous and arsinous acids (145, 153, 277), thioarsenites (134, 276), and anhydrides of thioarsinous acids (101). On the other hand, compounds whose conventional structural formulas show an arsenic atom linked by a *double* bond to some other atom are invariably associated. Thus arsenoso compounds give molecular weight values which are two, three, or four times the formula weights (1, 101, 183); and thioarsenoso compounds exhibit degrees of polymerization ranging from two to six (100, 101, 105, 278). Similarly, imides of the empirical formula, RAsNH, have been shown to be associated in benzene solution (143); the phenyl derivative is tetrameric, and the methyl compound is hexameric. Substituted imides are also associated. The N-*tert*-butyl imide of methylarsonous acid is dimeric, while the N-phenyl imide has a molecular weight corresponding to the formula, $(CH_3AsNC_6H_5)_{2.36}$ (277).

It has been suggested (279) that arsenosobenzene exists in a monomeric as well as a polymeric form, but supporting molecular weight data have not been reported. Titov and Levin (54) have isolated two forms of diarsenosomethane, only one of which is soluble in benzene. They suggested that one of these substances may be the dimer of the other. Sollott and Peterson (1) have shown by means of vapor pressure osmometry in pyridine that arsenosoferrocene can be isolated in a dimeric form. A pyridine-soluble polymeric form of arsenosoferrocene was also reported, although the molecular weight of this substance was apparently not determined. In an earlier paper, Sollott and co-workers (280) had concluded that the polymeric material was bis(ferrocenylhydroxyarsenic) oxide. Kharasch and co-workers (183) have noted that the boiling point of arsenosoethane (158° at 10 torr) is much higher than the boiling point of ethyldichloroarsine (74° at 50 torr). This observation is consistent with the fact that the arsenoso compound is trimeric in benzene, while the dihaloarsine has a normal molecular weight. Normal molecular weights in the vapor phase have been observed for bis(trifluoromethyl)chloro- and bis(trifluoromethyl)fluoroarsine, and methyl and *tert*-butyl bis(trifluoromethyl)arsinite (281). The vapor-phase molecular weight of the hydrazine derivative, $(CF_3)_2AsNHN(CH_3)_2$, is also normal (282).

B. Absorption spectra

The literature contains relatively few absorption spectra of compounds belonging to the classes discussed in this chapter. Furthermore, the data that are available must be interpreted with caution since these compounds are often rather reactive and unstable. It has been noted (283), for example, that hexane solutions of diphenylchloroarsine become cloudy on exposure to light and that the ultraviolet absorption spectra of the resulting solutions are affected by illumination.

During the nineteen thirties, Mohler and his co-workers (283–285) made an extensive investigation of the ultraviolet absorption characteristics of a variety of chemical warfare agents. A number of trivalent arsenicals were included in this study; some of the results are given in Table 3-1. It is somewhat surprising that the saturated compounds, arsenic trichloride and ethyldichloroarsine, have ultraviolet absorption bands of considerable intensity. The spectra of the α,β-unsaturated chloroarsines are more complex and presumably reflect conjugation of the nonbonded electrons on the arsenic atom with the π-electron system of the chlorovinyl group. The spectra of trifluoromethyldiiodoarsine and bis(trifluoromethyl)iodoarsine have also been reported (18) and appear to resemble the spectrum of arsenic triiodide. A strong hyperchromic effect found in the spectrum of diphenylcyanoarsine is probably associated with a significant amount of resonance between the arsenic atom and the benzene rings.

Table 3-1 Ultraviolet Absortpion Maxima[a]

Compound	λ_{max}, mμ	log ε_{max}	Literature Reference
Arsenic trichloride	205[b]	3.7	283
	240[b]	2.7	
Ethyldichloroarsine	241	3.25	283
2-Chlorovinyldichloroarsine	214	4.00	283
	230[b]	3.8	
	240[b]	3.3	
Bis(2-chlorovinyl)chloroarsine	209	4.15	283
	230[b]	3.8	
	240[b]	3.3	
Diphenylcyanoarsine	200[b]	5.2	283
	227	4.40	
	242	4.0	
	247[b]	3.7	
	254	3.4	
	260[b]	3.1	
	270	3.0	
Bis(diphenylarsenic) oxide	222	4.50	286
	261[c]	3.87	

[a] The spectrum of bis(diphenylarsenic) oxide was determined in 95% ethyl alcohol. Hexane was used as the solvent for the other compounds.

[b] Shoulder.

[c] A number of fine structure bands are centered around this maximum.

The ultraviolet spectrum of bis(diphenylarsenic) oxide is particularly difficult to interpret—the position of the bands suggests only *weak* resonance interaction between the arsenic atoms and the rings, but the high intensity of the 261 mμ band does not appear consistent with this idea. Jaffé (287) has determined the spectrum of arsenosobenzene and has concluded that the arsenoso group causes considerable perturbation of the benzene spectrum; his reported results for this compound are not in good agreement, however, with those from another laboratory (286). The spectrum of 2-amino-4-arsenosophenol has been determined in the 250–400 mμ region and appears to resemble the spectrum of *o*-aminophenol (288).

The Raman spectrum of arsenosomethane has been examined by Petit (289). A strong band was observed at 585 cm^{-1} and was assigned to carbon–arsenic stretching. In contrast to the spectra of pentavalent arsenic acids which contain a strong band near 770 cm^{-1} attributed to an arsenic–oxygen vibration, the spectrum of arsenosomethane contains two weak bands at 745 cm^{-1} and 826 cm^{-1}.

Van der Kelen and Herman (290) have carefully studied the Raman spectra of methyldichloro-, methyldibromo-, and methyldiiodoarsine, as well as the corresponding dimethylhaloarsines. For methyldichloro- and dimethylchloroarsine, depolarization factor measurements were made. The results of this spectroscopic investigation showed that the molecules of all of the compounds are pyramidal and belong to the symmetry class $C_{1h} = C_s$. The C—As stretching vibrations are found in the narrow range of 565 to 582.5 cm^{-1}. The As—Cl stretching frequencies lie between 360 and 388 cm^{-1}, while the corresponding As—Br bands are between 263 and 278 cm^{-1}. Unfortunately, the As—I frequencies are difficult to pick out unambiguously in these spectra. Claeys and van der Kelen (291) have used the Raman data for calculating the stretching and bending force constants of the CH_3—As, As—Cl, and As—Br bonds in the halo- and dihaloarsines.

Shagidullin and Pavlova (292–294) have recently reported an infrared and Raman investigation of several trivalent arsenicals, including phenyldichloroarsine, diphenylchloroarsine, and the diethyl esters of ethanearsonous and benzenearsonous acids. A strong band near 650 cm^{-1} found in the infrared spectra of the esters was assigned to the As—O stretch; the Raman spectra of these compounds also showed absorption in this region. The infrared and Raman spectra of ethyl ethanearsonite had a strong band at 580 cm^{-1} which was assigned to As—C stretching. The aromatic As—C stretch of ethyl benzenearsonite was located in the Raman spectrum at 672 cm^{-1}. The phenyl–arsenic grouping was also characterized by intense infrared absorption at 1435 and 1080 cm^{-1}. The chloro compounds gave a band in the infrared at about 390 cm^{-1} which was ascribed to the As—Cl bond.

The infrared spectrum of bis(trifluoromethyl)fluoroarsine has been reported in considerable detail (281). Bands observed at 328 and 346 cm^{-1} were assigned to As—CF_3 stretching. Similar bands were also found in the spectra of methyl and *tert*-butyl bis(trifluoromethyl)arsinite. A band at 692 cm^{-1} in the spectrum of the fluoroarsine was assigned to the As—F stretching mode. The As—O stretching frequencies in the methyl and *tert*-butyl esters were located at 634 and 653 cm^{-1}, respectively. Infrared spectra of other trifluoromethyl-substituted trivalent arsenicals, e.g., haloarsines (18, 266, 295), dihaloarsines (18, 296), cyanoarsines (18), thiocyanoarsines (18), amides (146, 158), sulfides (108, 142), and selenides (142), have also been reported.

The infrared spectra of thioarsenosomethane and thioarsenosobenzene have been examined by Zingaro and co-workers (278). A strong absorption at 565 cm^{-1} in the spectrum of the methane derivative was assigned to the As—S stretch. It is not obvious why this vibration occurs at a much higher frequency than the 465 cm^{-1} band observed for thioarsenosobenzene or the 472.8 cm^{-1} band observed for trimethylarsine sulfide.

Sollott and Peterson (1) have recorded the infrared spectra of arsenosobenzene, bis(diphenylarsenic) oxide, arsenosoferrocene, and oxybis(ferrocenylchloroarsine). In every case, strong bands in the 700–750 cm^{-1} region were observed which were attributed to stretching modes of the As—O—As group. Bardos and co-workers (297) have independently concluded that intense absorption in the 680–780 cm^{-1} region is characteristic of this linkage. They have also described the spectra of several thioarsenites of the following type:

```
         S —CH2
        /     |
  ArAs        |
        \     |
         S — CH2
```

The infrared and proton NMR spectra of two esters of dimethylthioarsinous acid have been determined by Cullen and co-workers (298), while Nyquist and co-workers (133) have investigated the infrared sepectra of a large number of 10-substituted phenoxarsine derivatives in which the substituent was RCO_2, RC(O)S, $ROCS_2$, RSO_2S, $(RO)_2PS_2$, or $(RO)_2P(O)$. The latter workers also examined the infrared spectrum of 10,10′-oxybis(phenoxarsine); a strong band was found at 688 cm^{-1} and was assigned to the asymmetric As—O—As stretching vibration.

In addition to the studies reviewed above, the literature contains descriptions of the infrared spectra of numerous esters (95, 97, 98), amides (145, 147), and sulfides (108) of arsonous and arsinous acids. The infrared spectrum of an impure sample of diphenylarsinous azide, $(C_6H_5)_2AsN_3$, has also been reported (157). This spectrum exhibits the bands typical of a monosubstituted phenyl group as well as a sharp, intense band at 2075 cm^{-1}, presumably due to the azide linkage.

Peterson and Thé (282) have described the proton and ^{19}F NMR spectra of the hydrazine derivative, $(CF_3)_2AsNHN(CH_3)_2$. The proton spectrum consists of two absorptions in approximately a 6:1 ratio, corresponding to six equivalent protons on the two methyl groups and one proton on the nitrogen atom. The ^{19}F spectrum contains a single line which remains sharp from $+30°$ to $-50°$; this result is consistent with rapid quadrupolar relaxation of the ^{75}As nucleus and with free rotation about the As—N bond. Thus, the ^{19}F data provide no evidence for any significant $N(p_\pi)$—$As(d_\pi)$ dative bonding in this compound.

Bardos and co-workers (297) have reported the proton NMR spectra of a number of aromatic arsenoso and dithiarsolane compounds.

C. Molecular structure

Trotter and his co-workers (299–308) have used X-ray diffraction measurements for the determination of the crystal and molecular structures of bis(diphenylarsenic) oxide and of a number of haloarsines and related compounds. It was found that crystals of the oxide (305) are monoclinic, with four molecules in the unit cell. The mean As—O length in this compound is 1.67 A, close to the corresponding distance in benzenearsonic acid but significantly shorter than the As—O length of 1.78 A found in As_4O_6. The As—O—As angle (137°) is considerably larger than the usual valency angle at oxygen, but is similar to the corresponding angle in As_4O_6. The large angle is probably the result of the formation of p_π-d_π bonds involving donation of oxygen lone-pair electrons to vacant 4*d* orbitals of the arsenic atoms.

Crystals of chloro-, bromo-, and iododiphenylarsine were found to be essentially isomorphous (299, 300, 303). The bond lengths in the three compounds agree well with the values calculated from covalent radii except for As—Br, which is slightly longer than the corresponding bond in other compounds. The orientation of the two benzene rings relative to the arsenic atom indicates that one ring is in a position where interaction with the arsenic lone pair must be negligible while the other ring may be able to interact to a considerable extent.

The existence of several crystalline forms of 10-chloro-5,10-dihydrophenarsazine and the possibility of geometrical isomerism have prompted the X-ray investigation of this chloroarsine (308). It was found that there is only one stable form; the other forms contain solvent of crystallization which is readily lost. Detailed analysis of the solvent-free crystals showed that the molecule is slightly folded about the As · · · N axis, the angle between the benzene rings being 169° 20′ and the chlorine atom being outside this angle. Since each benzene ring is thus displaced by only about 5° from a completely planar arrangement, the isolation of stable geometrical isomers seems unlikely. The C—As and the C—N bonds are slightly but significantly shorter than normal single-bond distances, a fact which suggests interaction of the arsenic and nitrogen lone pair electrons with the π-electron systems of the benzene rings. In addition there is the possibility of d_π-p_π bonding between the π-electrons and the vacant 4*d* orbitals of the arsenic atom. The C—As—C angle is somewhat smaller than the usual values found in arsenic–phenyl compounds, while the C—N—C angle is considerably larger than normal. Both of these facts indicate the existence of strain in the central ring.

Methyldiiodoarsine is the simplest dihaloarsine which is a solid at room temperature. X-ray diffraction has shown (306) that the arsenic atom in this compound has the usual pyramidal configuration found in trivalent derivatives of arsenic and that the bond lengths agree well with those calculated from covalent radii. The I—As—I angle (104°) is a little larger than the angles in AsI_3, but the angles involving the carbon atom do not differ significantly from normal values.

The X-ray structure of dimethylcyanoarsine has also been determined in Trotter's laboratory (304). The compound exhibits no unusual bond distances or angles, but the As · · · N *intermolecular* distance is considerably less than the sum of the van der Waals radii. It was suggested that the short intermolecular contact indicates a charge-transfer bond involving donation of nitrogen lone pair electrons to the vacant arsenic 4*d* orbitals. This intermolecular bonding may explain why dimethylcyanoarsine is a solid, mp 30°, while the dimethylhaloarsines are liquids at room temperature. It has been estimated that the energy of the charge-transfer bond is of the order of one or two kilocalories per mole.

A number of interesting features have been revealed by the X-ray analysis of 1,3-dichloro-1,3-dihydro-2,1,3-benzoxadiarsole (301):

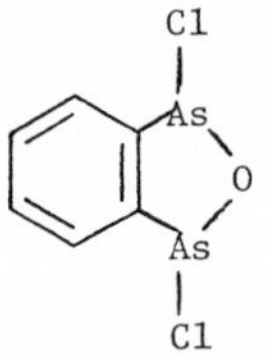

Cl
As
O
As
Cl

The molecule is completely planar except for the chlorine atoms, which lie one on either side of the plane of the other atoms. The O—As—C angles within the five-membered ring are only 77°, instead of the 95–105° usually found for trivalent arsenic compounds. In addition, the oxygen valency angle (151°) is even larger than the 128° observed in As_4O_6. In spite of these abnormal valency angles, the five-membered ring is unusually stable and, in fact, is unattacked by hot concentrated hydrochloric acid. Cullen and Trotter (301) have proposed that the stability, planarity, and unusual valency angles indicate that the five-membered ring may have some degree of aromatic character. The large oxygen valency angle suggests that the oxygen is approaching a state of digonal hybridization and that the two lone pairs on the oxygen are available for p_π-bonding. Accordingly, d_π-p_π bonding involving vacant 4*d* orbitals of the arsenic atoms should be favored and may account for the fact that the As—O distance is significantly shorter than the corresponding distance in either As_4O_6 or in the arsenate ion. It is also possible that d_π-p_π bonding may occur between the arsenic atoms and the π-electron system of the benzene ring. No shortening of the C—As bonds, however, was observed, but these distances were determined with less precision than the As—O length.

"Cacodyl disulfide" was first prepared by Bunsen over one hundred years ago and has generally been considered to have a true disulfide structure:

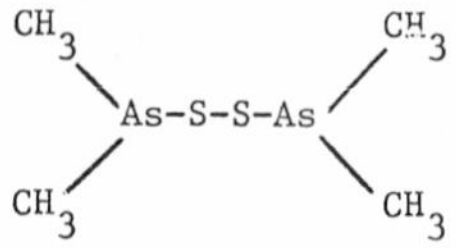

$$(CH_3)_2As{-}S{-}S{-}As(CH_3)_2$$

Chemical evidence suggested, however, that the compound contains pentavalent arsenic. X-ray analysis has recently demonstrated (307) that the compound has *both* trivalent and pentavalent arsenic in the molecule:

$$(CH_3)_2As(=S)-S-As(CH_3)_2$$

The trivalent arsenic atom has a pyramidal configuration, with angles 96–99°; the pentavalent arsenic is tetrahedral (angles 101–116°). The As—S distances are 2.28 A for As^{III}—S and 2.21 for As^{V}—S. The difference between these values is significant and suggests that pentavalent arsenic has a slightly smaller covalent radius. The As^{V}=S distance is 2.07 A and indicates that this is a true double bond, since the difference between single- and double-bond radii for sulfur is believed to be about 0.10 A. The other bond distances and angles observed in this molecule require no special comment, but the $As^{III} \cdots As^{III}$ intermolecular distance is abnormally short, i.e., less than the sum of the van der Waals radii. It has been suggested that this close contact is the result of charge-transfer bonding involving donation of lone-pair electrons of a trivalent arsenic atom to vacant 4*d* orbitals of a trivalent arsenic atom in another molecule. Similar interactions have been noted in other structures (e.g., dimethylcyanoarsine) in which nitrogen acts as the electron-donor and arsenic the electron-acceptor. In the present case trivalent arsenic appears to act as both electron-donor and electron-acceptor.

The structures of several haloarsines have been studied by electron diffraction measurements. By assuming that the C—As distance and the C—As—C angle in dimethylchloro-, dimethylbromo-, and dimethyliodoarsines are identical with those in trimethylarsine, Skinner and Sutton (309) were able to derive values for the As—X distance and the X—As—X angle (where X was Cl, Br, or I). The angles were found to be in the 92–102° range, and the bond lengths agreed well with those calculated from the sums of the normal single-bond covalent radii. Electron diffraction has also been used to investigate the *cis* and *trans* isomers of 2-chlorovinyldichloroarsine (Lewisite) (213). The As—Cl distance in these compounds was found to be normal, but the C—As distance, 1.90 A, was significantly less than 1.98 A, the distance found in trimethylarsine. The observed shortening of the C—As bond may be the result of conjugation of the non-bonded arsenic electrons with the olefinic double bond. The C—Cl distance in the Lewisites also appears to be less than the sum of the covalent radii, but it is about the same as the values found in the chloroethylenes.

Microwave spectroscopy has been used by Nugent and Cornwell (310) to investigate the structure of methyldifluoroarsine. Assuming that the C—H bond length in this compound is 1.095 A, these authors were able to derive the C—As and As—F distances as well as the various bond angles. The values obtained are in substantial agreement with those reported for other arsenic compounds.

D. Other physical measurements

Baxter and co-workers (311) were apparently the first to obtain vapor pressure data for chloro- and dichloroarsines. Somewhat later Gibson and Johnson (238) reported vapor pressure and density data for eight alkyldichloroarsines. Still more recently, the vapor pressures, densities, and refractive indices of seven carefully purified dihaloarsines were measured by Redemann and co-workers (312). Volatilities and Trouton constants for a number of trifluoromethyl-substituted arsenicals are also available (18, 281). In addition, the vapor pressure, density, refractive index, freezing-point, viscosity, and dipole moment of *cis*- and *trans*-2-chlorovinyldichloroarsine have been reported (215, 216). Dipole moments of several other halo- and dihaloarsines have been measured (313–316), and the bond moments of the arsenic–halogen and arsenic–methyl bonds have been calculated (316).

Kamai and co-workers (81, 82, 317, 318) have made an extensive study of the atomic refractivity of arsenic in organoarsenic compounds. Using the sodium D-line, they found an average value of 11.75 for the atomic refractivity of arsenic in bis(alkylphenylarsenic) oxides, a range of 11.27–11.98 in alkyl alkylphenylarsinites, a value of 10.46 in di-*n*-butyl benzenearsonite, and a value of about 10.25 in esters and ester chlorides of alkylarsonous acids. The value for $C_6H_5As(OR)Cl$ was 11.0. Values for the atomic refractivity of arsenic in haloarsines, cyanoarsines, and arsenoso compounds are available from the work of earlier investigators (319).

Mortimer and Skinner (320) have reported a thermochemical investigation of the formation of dimethyliodoarsine by the reaction between tetramethyldiarsine and iodine. From their data and some general considerations concerning the strength of the As—As bond, they concluded that the dissociation energy of the As—I bond in the iodoarsine probably lies within the range 47–50 kcal/mole. This value appears to be significantly greater than the dissociation energy of 42.6 kcal/mole reported for the As—I bond in arsenic triiodide. On the other hand, the dissociation energy of this bond in phenyldiiodoarsine appears to be normal (252).

Buu-Hoi and co-workers (321) have investigated the mass spectrometry of 10-chloro-5,10-dihydrophenarsazine (adamsite). The molecular peaks (m/e = 277 and m/e = 279) are practically nonexistent and represent only 0.2% of the base peak. At m/e = 242, there is an important peak which corresponds to the loss of a chlorine atom. The most intense peak (which was therefore taken as the base peak) is at m/e = 241 and corresponds to phenarsazine:

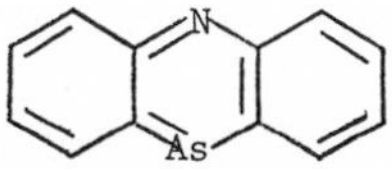

The ease of formation of this species suggests that the arsenic atom may be a member of a truly aromatic ring system. Loss of arsenic from the phenarsazine gives a peak at m/e = 167 which corresponds to the formation of carbazole. In a later paper Buu-Hoi and co-workers (322) reported similar results with several analogs of adamsite.

The mass spectra of trifluoromethyldichloroarsine as well as trifluoromethylarsine, bis(trifluoromethyl)arsine, tris(trifluoromethyl)arsine, and tetrakis(trifluoromethyl)diarsine have been recently examined (323). In all cases, rearranged fluorocarbon fragments (e.g., CF_2H^+, CFH_2^+, and CF_2Cl^+) and species containing As—F bonds (e.g., AsF^+, AsF_2^+, and As_2F^+) were observed. It was shown that the rearrangements did not occur in the heated inlet system of the mass spectrometer, and it was suggested that the rearranged ions were formed by intramolecular transfer of fluorine. The use of electron impact mass spectrometry for the determination of the first ionization potentials of several halo- and dihaloarsines is discussed in Section IV-6-C.

The diamagnetic susceptibility of diphenylchloroarsine and phenyl diphenylarsinite has been reported by Pascal (324).

REFERENCES

1. G. P. Sollott and W. R. Peterson, Jr., *J. Org. Chem.*, **30,** 389 (1965).
2. A. Michaelis and H. Loesner, *Ber.*, **27,** 263 (1894).
3. G. O. Doak, H. Eagle, and H. G. Steinman, *J. Amer. Chem. Soc.*, **62,** 168 (1940).
4. J. F. Oneto and E. L. Way, *J. Amer. Chem. Soc.*, **63,** 762 (1941).
5. E. L. Way and J. F. Oneto, *J. Amer. Chem. Soc.*, **64,** 1287 (1942).
6. L. Kalb, *Justus Liebigs Ann. Chem.*, **423,** 39 (1921).
7. F. F. Blicke and G. L. Webster, *J. Amer. Chem. Soc.*, **59,** 534 (1937).
8. K. Takahashi, *Yakugaku Zasshi*, **72,** 1144 (1952).
9. A. Michaelis and L. Weitz, *Ber.*, **20,** 48 (1887).
10. H. Wieland and W. Rheinheimer, *Justus Liebigs Ann. Chem.*, **423,** 1 (1921).
11. H. Bauer, *J. Amer. Chem. Soc.*, **67,** 591 (1945).
12. G. Kamai and N. A. Chadaeva, *Dokl. Akad. Nauk SSSR*, **109,** 309 (1956); through *C.A.*, **51,** 1876 (1957).
13. L. H. Long, H. J. Emeléus, and H. V. A. Briscoe, *J. Chem. Soc.*, 1123 (1946).
14. A. Sporzyński, *Rocz. Chem.*, **14,** 1293 (1934).
15. A. Spada, *Atti Soc. Nat. Mat. Modena*, **72,** 34 (1941).
16. I. Kageyama and S. Nakanishi (to Osoka Kinzoku Kogyo Co., Ltd., and Kansai Paint Co., Ltd.), Brit. Pat. 861,500 (Feb. 22, 1961); through *C.A.*, **55,** 21152 (1961).
17. G. Newbery and M. A. Phillips, *J. Chem. Soc.*, 2375 (1928).
18. H. J. Emeléus, R. N. Haszeldine, and E. G. Walaschewski, *J. Chem. Soc.*, 1552 (1953).

19. H. J. Emeléus and J. S. Anderson, *Modern Aspects of Inorganic Chemistry*, D. Van Nostrand Co., Princeton, 3rd Edition, 1960, p. 244.
20. W. A. Waters and J. H. Williams, *J. Chem. Soc.*, 18 (1950).
21. A. McKenzie and J. K. Wood, *J. Chem. Soc.*, **117**, 406 (1920).
22. A. Michaelis and U. Paetow, *Justus Liebigs Ann. Chem.*, **233**, 60 (1886).
23. E. Gryszkiewicz-Trochimowski, M. Buczwinski, and J. Kwapiszewski, *Rocz. Chem.*, **8**, 423 (1928); through *C.A.*, **23**, 1614 (1929).
24. V. K. Kuskov and V. N. Vasilyev, *Zh. Obshch. Khim.*, **21**, 90 (1951).
25. C. K. Banks and C. S. Hamilton, *J. Amer. Chem. Soc.*, **62**, 3142 (1940).
26. G. R. A. Brandt, H. J. Emeléus, and R. N. Haszeldine, *J. Chem. Soc.*, 2552 (1952).
27. H. Hartmann and G. Nowak, *Z. Anorg. Allg. Chem.*, **290**, 348 (1957).
28. V. Auger, *C. R. Acad. Sci., Paris*, **137**, 925 (1903).
29. J. F. Norris, *Ind. Eng. Chem.*, **11**, 817 (1919).
30. R. L. Barker, E. Booth, W. E. Jones, A. F. Millidge, and F. N. Woodward, *J. Soc. Chem. Ind.*, **68**, 289 (1949).
31. P. Ehrlich and A. Bertheim, *Ber.*, **43**, 917 (1910).
32. H. Wieland, *Justus Liebigs Ann. Chem.*, **431**, 30 (1923).
33. O. A. Seide, S. M. Scherlin, and G. J. Bras, *J. Prakt. Chem.*, **138**, 225 (1933).
34. O. A. Zeide, S. M. Sherlin, and A. B. Bruker, *Zh. Obshch. Khim.*, **28**, 2404 (1958).
35. R. L. Barker, E. Booth, W. E. Jones, and F. N. Woodward, *J. Soc. Chem. Ind.*, **68**, 277 (1949).
36. A. B. Bruker and F. L. Maklyaev, *Dokl. Akad. Nauk SSSR*, **63**, 271 (1948); through *C.A.*, **43**, 2592 (1949).
37. A. B. Bruker, *Zh. Obshch. Khim.*, **18**, 1297 (1948); through *C.A.*, **43**, 4647 (1949).
38. W. M. Dehn, *Amer. Chem. J.*, **33**, 101 (1905).
39. W. M. Dehn and B. B. Wilcox, *Amer. Chem. J.*, **35**, 1 (1906).
40. C. Walling and S. A. Buckler, *J. Amer. Chem. Soc.*, **77**, 6032 (1955).
41. M. H. Abraham, *J. Chem. Soc.*, 4130 (1960).
42. H. Eagle, *J. Pharmacol. Exp. Ther.*, **66**, 423 (1939).
43. M. Ya. Kraft and V. V. Katyshkina, *Dokl. Akad. Nauk SSSR*, **99**, 89 (1954); through *C.A.*, **49**, 4233 (1955).
44. M. A. Phillips, *J. Chem. Soc.*, 192 (1941).
45. F. Sachs and H. Kantorowicz, *Ber.*, **41**, 2767 (1908).
46. K. Matsumiya and M. Nakai, *Mem. Coll. Sci. Kyoto Imp. Univ.*, **8A**, 309 (1925); through *C.A.*, **19**, 3086 (1925).
47. F. F. Blicke and F. D. Smith, *J. Amer. Chem. Soc.*, **51**, 1558 (1929).
48. F. F. Blicke and F. D. Smith, *J. Amer. Chem. Soc.*, **51**, 3479 (1929).
49. A. Valeur and P. Gailliot, *C. R. Acad. Sci., Paris*, **185**, 956 (1927).
50. R. C. Fuson and W. Shive, *J. Amer. Chem. Soc.*, **69**, 559 (1947).
51. Cadet, *Mém. Math. Phys.*, **3**, 623 (1760).
52. J. S. Thayer, *J. Chem. Educ.*, **43**, 594 (1966).
53. A. Valeur and P. Gailliot, *C. R. Acad. Sci., Paris*, **185**, 779 (1927).
54. A. I. Titov and B. B. Levin, *Sb. Statei Obshch. Khim.*, **2**, 1469 (1953); *C.A.*, **49**, 4503 (1955).
55. A. I. Titov and B. B. Levin, *Sb. Statei Obshch. Khim.*, **2**, 1473 (1953); *C.A.*, **49**, 4504 (1955).
56. H. Eagle, R. B. Hogan, G. O. Doak, and H. G. Steinman, *J. Pharmacol. Exp. Ther.*, **70**, 221 (1940).
57. G. Petit, *Ann. Chim.* (Paris), **16**, 5 (1941).
58. A. N. Nesmejanow and K. A. Kozeschkow, *Ber.*, **67**, 317 (1934).

59. K. A. Kocheshkov and A. N. Nesmeyanov, *Zh. Obshch. Khim.*, **4,** 1102 (1934); *C.A.*, **29,** 3993 (1935).
60. K. Hiratsuka, *Kogyo Kagaku Zasshi*, **58,** 935 (1937).
61. K. Hiratsuka, *Kogyo Kagaku Zasshi*, **58,** 1051 (1937).
62. K. Hiratsuka, *Kogyo Kagaku Zasshi*, **58,** 1060 (1937).
63. K. Hiratsuka, *Kogyo Kagaku Zasshi*, **58,** 1163 (1937).
64. W. La Coste and A. Michaelis, *Justus Liebigs Ann. Chem.*, **201,** 184 (1880).
65. A. Michaelis, *Justus Leibigs Ann. Chem.*, **321,** 141 (1902).
66. C. P. A. Kappelmeier, *Rec. Trav. Chim. Pays-Bas*, **49,** 57 (1930).
67. G. J. Burrows and A. Lench, *J. Proc. Roy. Soc. N.S. Wales*, **70,** 300 (1937).
68. A. Michaelis, *Justus Liebigs Ann. Chem.*, **320,** 271 (1902).
69. K. Takahashi, *Yakugaku Zasshi*, **72,** 523 (1952).
70. A. Bertheim, *Ber.*, **48,** 350 (1915).
71. G. Kamai and V. M. Zoroastrova, *Zh. Obshch. Khim.*, **10,** 921 (1940); through *C.A.*, **35,** 3241 (1941).
72. A. Cohen, H. King, and W. I. Strangeways, *J. Chem. Soc.*, 2866 (1932).
73. W. La Coste and A. Michaelis, *Ber.*, **11,** 1887 (1878).
74. P. Karrer, *Ber.*, **47,** 1783 (1914).
75. R. E. Dessy, T. Chivers, and W. Kitching, *J. Amer. Chem. Soc.*, **88,** 467 (1966).
76. W. Steinkopf, I. Schubart, and S. Schmidt, *Ber.*, **61,** 678 (1928).
77. Ya. F. Komissarov, A. Ya. Maleeva, and A. S. Sorokooumov, *Dokl. Akad. Nauk SSSR*, **55,** 719 (1947); *C.A.*, **42,** 3721 (1948).
78. G. Kamai and Z. L. Khisamova, *Zh. Obshch. Khim.*, **30,** 3611 (1960).
79. L. M. Werbel, T. P. Dawson, J. R. Hooton, and T. E. Dalbey, *J. Org. Chem.*, **22,** 452 (1957).
80. G. Kamai and R. K. Zaripov, *Izv. Vyssh. Ucheb. Zaved., Khim. Khim. Tekhnol.*, **5,** 938 (1962); *C.A.*, **59,** 5194 (1963).
81. G. Kamai and R. K. Zaripov, *Tr. Kazansk. Khim.-Tekhnol. Inst.*, No. 30, 77 (1962); *C.A.*, **60,** 4181 (1964).
82. G. Kamai and R. G. Miftakhova, *Zh. Obshch. Khim.*, **33,** 2904 (1963).
83. G. Kamai, I. N. Azerbaev, and P. K. Zaripov, *Izv. Akad. Nauk Kaz. SSR, Ser. Khim.*, **16,** 85 (1966); *C.A.*, **65,** 13752 (1966).
84. G. Kamai and N. A. Chadaeva, *Dokl. Akad. Nauk SSSR*, **86,** 71 (1952); through *C.A.*, **47,** 6365 (1953).
85. G. Kamai, *Uchenye Zapiski Kazan. Gosudarst. Univ.*, **115,** 43 (1955); through *C.A.*, **53,** 1205 (1959).
86. M. S. Malinovskii, *Zh. Obshch. Khim.*, **10,** 1918 (1940); through *C.A.*, **35,** 4736 (1941).
87. S. Chwalinski, *Rocz. Chem.*, **18,** 443 (1938).
88. I. M. Starshov and G. Kamai, *Zh. Obshch. Khim.*, **24,** 2044 (1954).
89. G. Kamai and B. D. Chernokal'skii, *Zh. Obshch. Khim.*, **30,** 1536 (1960); through *C.A.*, **55,** 1415 (1961).
90. G. Kamai and Yu. F. Gatilov, *Zh. Obshch. Khim.*, **35,** 1239 (1965).
91. G. Kamai and N. A. Chadaeva, *Izv. Vyssh. Ucheb. Zaved., Khim. Khim. Tekhnol.*, **2,** 601 (1959); through *C.A.*, **54,** 7606 (1960).
92. G. Kamai and N. A. Chadaeva, *Izv. Akad. Nauk SSSR, Otd. Khim. Nauk*, 585 (1957); through *C.A.*, **51,** 14585 (1957).
93. H. G. Schick and G. Schrader (to Farbenfabriken Bayer Akt.-Ges.), Ger. Pat. 1,094,746 (Dec. 15, 1960); through *C.A.*, **55,** 25754 (1961).
94. B. A. Arbuzov and M. K. Saikina, *Zh. Fiz. Khim.*, **34,** 2344 (1960); through *C.A.*, **55,** 13034 (1961).

95. B. L. Chamberland and A. G. MacDiarmid, *J. Amer. Chem. Soc.*, **83,** 549 (1961).
96. H. Schmidbaur and M. Schmidt, *Angew. Chem.*, **73,** 655 (1961).
97. H. Schmidbaur, H. S. Arnold, and E. Beinhofer, *Chem. Ber.*, **97,** 449 (1964).
98. G. Kamai and R. G. Miftakhova, *Zh. Obshch. Khim.*, **35,** 2001 (1965).
99. C. Schulte, *Ber.*, **15,** 1955 (1882).
100. A. E. Kretov and A. Ya. Berlin, *Zh. Obshch. Khim.*, **1,** 411 (1931); through *C.A.*, **26,** 2415 (1932).
101. F. F. Blicke and F. D. Smith, *J. Amer. Chem. Soc.*, **52,** 2946 (1930).
102. A. Étienne, *Bull. Soc. Chim. Fr.*, 47 (1947).
103. C. K. Banks, J. Controulis, D. F. Walker, E. W. Tillitson, L. A. Sweet, and O. M. Gruhzit, *J. Amer. Chem. Soc.*, **70,** 1762 (1948).
104. F. F. Blicke and L. D. Powers, *J. Amer. Chem. Soc.*, **55,** 1161 (1933).
105. L. Anschütz and H. Wirth, *Chem. Ber.*, **89,** 1530 (1956).
106. M. Ya. Kraft and I. A. Batshchouk, *Dokl. Akad. Nauk SSSR*, **55,** 723 (1947); *C.A.*, **42,** 3742 (1948).
107. K. Matsumiya and M. Nakai, *Mem. Coll. Sci. Kyoto Imp. Univ.*, **10,** 57 (1926); through *C.A.*, **21,** 904 (1927).
108. W. R. Cullen, *Can. J. Chem.*, **41,** 2424 (1963).
109. H. J. Barber, *J. Chem. Soc.*, 1024 (1929).
110. E. J. Cragoe, Jr., and C. S. Hamilton, *J. Amer. Chem. Soc.*, **67,** 536 (1945).
111. I. H. Witt and C. S. Hamilton, *J. Amer. Chem. Soc.*, **68,** 1078 (1945).
112. H. King and R. J. Ludford, *J. Chem. Soc.*, 2086 (1950).
113. K. Takahashi, *Yakugaku Zasshi*, **72,** 1148 (1952).
114. T. Ueda (to Dai Nippon Drug Manufg. Co.), Jap. Pat. 1073 (Feb. 19, 1955); through *C.A.*, **51,** 2859 (1957).
115. A. Cohen, H. King, and W. I. Strangeways, *J. Chem. Soc.*, 3043 (1931).
116. T. H. Maren, *J. Amer. Chem. Soc.*, **68,** 1864 (1946).
117. A. D. Ainley and D. G. Davey, *Brit. J. Pharmacol. Chemother.*, **13,** 244 (1958).
118. G. Schuster, *J. Pharm. Chim.*, **17,** 331 (1933).
119. G. Kamai and N. A. Chadaeva, *Dokl. Akad. Nauk SSSR*, **115,** 305 (1957); *C.A.*, **52,** 6161 (1958).
120. G. Kamai and N. A. Chadaeva, *Izv. Kazansk. Filiala Akad. Nauk SSSR, Ser. Khim. Nauk*, No. 4, 69 (1957); through *C.A.*, **54,** 6521 (1960).
121. G. Kamai and N. A. Chadaeva, *Zh. Obshch. Khim.*, **31,** 3554 (1961).
122. H. Yamashina, Y. Nagae, and S. Sasaki (to Toa Agricultural Chemical Co., Ltd), Jap. Pat. 21,072 (Oct. 10, 1963); through *C.A.*, **60,** 3015 (1964).
123. W. R. Cullen, *Can. J. Chem.*, **40,** 575 (1962).
124. N. A. Chadaeva, G. Kamai, and K. A. Mamakov, *Dokl. Akad. Nauk SSSR*, **157,** 371 (1964).
125. M. P. Osipova, G. Kh. Kamai, and N. A. Chadaeva, *Zh. Obshch. Khim.*, **37,** 1660 (1967).
126. N. A. Chadaeva, G. Kh. Kamai, and G. M. Usacheva, *Zh. Obshch. Khim.*, **36,** 704 (1966).
127. N. A. Chadaeva, G. Kh. Kamai, and K. A. Mamakov, *Zh. Obshch. Khim.*, **36,** 916 (1966).
128. K. Takahashi and T. Ueda (to Dai Nippon Pharmaceutical Co., Ltd.,) U.S. Pat. 2,701,812 (Feb. 8, 1955); through *C.A.*, **50,** 1907 (1956).
129. W. R. Cullen, P. S. Dhaliwal, and W. B. Fox, *Inorg. Chem.*, **3,** 1332 (1964).
130. W. R. Cullen and P. S. Dhaliwal, *Can. J. Chem.*, **45,** 379 (1967).
131. H. Eagle and G. O. Doak, *Pharmacol. Rev.*, **3,** 107 (1951).

132. E. Urbschat and P. E. Frohberger (to Farbenfabriken Bayer A.-G.), U.S. Pat. 2,767,114 (Oct. 16, 1956); through *C.A.*, **51,** 5354 (1957).
133. R. A. Nyquist, H. J. Sloane, J. E. Dunbar, and S. J. Strycker, *Appl. Spectrosc.*, **20,** 90 (1966).
134. F. F. Blicke and U. O. Oakdale, *J. Amer. Chem. Soc.*, **54,** 2993 (1932).
135. E. R. H. Jones and F. G. Mann, *J. Chem. Soc.*, 401 (1955).
136. E. Hayashi and M. Kado (to Ihara Agricultural Chemicals Co., Ltd.), Jap. Pat. 17,027 (Aug. 3, 1965); through *C.A.*, **63,** 18121 (1965).
137. C. A. Peri, *Gazz. Chim. Ital.*, **89,** 1315 (1959).
138. H. Oda, H. Sumi, and Y. Tanaka, *Takamine Kenkyusho Nempo*, **11,** 193 (1959); *C.A.*, **55,** 6768 (1961).
139. M. Nagasawa and Y. Imamiya (to Ihara Agricultural Chemical Co.), Jap. Pat. 8116 (June 5, 1963); through *C.A.*, **59,** 11567 (1963).
140. K. D. Shvetsova-Shilovskaya, N. N. Mel'nikov, E. N. Andreeva, L. P. Bocharova, and Yu. N. Sapozhkov, *Zh. Obshch. Khim.*, **31,** 845 (1961).
141. S. J. Strycker (to Dow Chemical Co.), U.S. Pat. 3,117,123 (Jan. 7, 1964); *C.A.*, **61,** 667 (1964).
142. H. J. Emeléus, K. J. Packer, and N. Welcman, *J. Chem. Soc.*, 2529 (1962).
143. W. Ipatiew, G. Rasuwajew, and W. Stromski, *Ber.*, **62,** 598 (1929).
144. G. O. Doak, *J. Amer. Pharm. Assoc.*, **24,** 453 (1935).
145. K. Mödritzer, *Chem. Ber.*, **92,** 2637 (1959).
146. W. R. Cullen and H. J. Emeléus, *J. Chem. Soc.*, 372 (1959).
147. W. R. Cullen and L. G. Walker, *Can. J. Chem.*, **38,** 472 (1960).
148. G. Kamai and Z. L. Khisamova, *Dokl. Akad. Nauk SSSR*, **156,** 365 (1964).
149. O. J. Scherer and M. Schmidt, *Angew. Chem.*, **76,** 787 (1964).
150. C. K. Banks, J. F. Morgan, R. L. Clark, E. B. Hatlelid, F. H. Kahler, H. W. Paxton, E. J. Cragoe, R. J. Andres, B. Elpern, R. F. Coles, J. Lawhead, and C. S. Hamilton, *J. Amer. Chem. Soc.*, **69,** 927 (1947).
151. A. Tzschach and W. Lange, *Z. Anorg. Allg. Chem.*, **326,** 280 (1964).
152. N. K. Bliznyuk, G. S. Levskaya, and E. N. Matyukhina, *Zh. Obshch. Khim.*, **35,** 1247 (1965).
153. H. J. Vetter and H. Nöth, *Z. Anorg. Allg. Chem.*, **330,** 233 (1964).
154. O. J. Scherer and M. Schmidt, *Angew. Chem.*, **76,** 144 (1964).
155. O. J. Scherer, P. Hornig, and M. Schmidt, *J. Organometal. Chem.*, **6,** 259 (1966).
156. H. H. Sisler and C. Stratton, *Inorg. Chem.*, **5,** 2003 (1966).
157. W. T. Reichle, *Tetrahedron Lett.*, 51 (1962).
158. J. Singh and A. B. Burg, *J. Amer. Chem. Soc.*, **88,** 718 (1966).
159. C. S. Hamilton and C. G. Ludeman, *J. Amer. Chem. Soc.*, **52,** 3284 (1930).
160. J. Chatt and F. G. Mann, *J. Chem. Soc.*, 610 (1939).
161. L. J. Goldsworthy, W. H. Hook, J. A. John, S. G. P. Plant, J. Rushton, and L. M. Smith, *J. Chem. Soc.*, 2208 (1948).
162. K. C. Eberly and G. E. P. Smith, Jr., *J. Org. Chem.*, **22,** 1710 (1957).
163. J. F. Oneto, *J. Amer. Chem. Soc.*, **60,** 2058 (1938).
164. J. F. Oneto and E. L. Way, *J. Amer. Chem. Soc.*, **62,** 2157 (1940).
165. R. J. Garascia, G. W. Batzis, and J. O. Kroeger, *J. Org. Chem.*, **25,** 1271 (1960).
166. D. Hamer and R. G. Leckey, *J. Chem. Soc.*, 1398 (1961).
167. L. Maier, D. Seyferth, F. G. A. Stone, and E. G. Rochow, *J. Amer. Chem. Soc.*, **79,** 5884 (1957).
168. A. Valeur and P. Gaillot, *Bull. Soc. Chim. Fr.*, **41,** 1318 (1927).
169. G. D. Parkes, R. J. Clarke, and B. H. Thewlis, *J. Chem. Soc.*, 429 (1947).

170. A. G. Evans and E. Warhurst, *Trans. Faraday Soc.*, **44,** 189 (1948).
171. H. D. N. Fitzpatrick, S. R. C. Hughes, and E. A. Moelwyn-Hughes, *J. Chem. Soc.*, 3542 (1950).
172. W. R. Cullen, *Can. J. Chem.*, **41,** 317 (1963).
173. C. D. Nenitzescu, D. A. Isacescu, and C. Gruescu, *Bull. Soc. Chim. Romania*, **A, 20,** 135 (1938); through *C.A.*, **34,** 1979 (1940).
174. G. Drefahl and G. Stange, *J. Prakt. Chem.*, **10,** 257 (1960).
175. N. P. McCleland and R. H. Wilson, *J. Chem. Soc.*, 1497 (1932).
176. W. G. Lowe and C. S. Hamilton, *J. Amer. Chem. Soc.*, **57,** 1081 (1935).
177. W. G. Lowe and C. S. Hamilton, *J. Amer. Chem. Soc.*, **57,** 2314 (1935).
178. J. F. Morgan, E. J. Cragoe, B. Elpern, and C. S. Hamilton, *J. Amer. Chem. Soc.*, **69,** 932 (1947).
179. C. Finzi, *Gazz. Chim. Ital.*, **45,** II, 280 (1915).
180. W. Steinkopf, *Justus Liebigs Ann. Chem.*, **413,** 310 (1917).
181. C. Finzi, *Gazz. Chim. Ital.*, **55,** 824 (1925).
182. W. Steinkopf and W. Mieg. *Ber.*, **53,** 1013 (1920).
183. M. S. Kharasch, E. V. Jensen, and S. Weinhouse, *J. Org. Chem.*, **14,** 429 (1949).
184. L. Maier, *Tetrahedron Lett.*, No. 6, 1 (1959).
185. G. Kamai and Yu. F. Gatilov, *Zh. Obshch. Khim.*, **31,** 1844 (1961).
186. L. Maier, *J. Inorg. Nucl. Chem.*, **24,** 1073 (1962).
187. H. D. Kaesz and F. G. A. Stone, *Spectrochim. Acta*, **15,** 360 (1959).
188. D. Seyferth, Personal communication.
189. L. I. Zakharkin and O. Yu. Okhlobystin, *Dokl. Akad. Nauk SSSR*, **116,** 236 (1957).
190. B. J. Gudzinowicz and H. F. Martin, *Anal. Chem.*, **34,** 648 (1962).
190a. K. Issleib and A. Tzschach, *Angew. Chem.*, **73,** 26 (1961).
190b. M. Green and D. Kirkpatrick, *Chem. Commun.*, 57 (1967).
190c. M. Green and D. Kirkpatrick, *J. Chem. Soc.*, **A,** 483 (1968).
191. W. La Coste and A. Michaelis, *Ber.*, **11,** 1883 (1878).
192. A. Michaelis and J. Rabinerson, *Justus Liebigs Ann. Chem.*, **270,** 139 (1892).
193. S. Mroczkowski, *Inaug. Dissertation*, Rostock, 1910.
194. P. S. Varma, K. S. V. Raman, and K. M. Yashoda, *J. Indian Chem. Soc.*, **16,** 515 (1939).
195. Société des usines chimique Rhône-Poulenc, Fr. Pat., 1, 159,169 (June 24, 1958); through *C.A.*, **54,** 18437 (1960).
196. W. C. Davies and C. W. Othen, *J. Chem. Soc.*, 1236 (1936).
197. H. N. Das-Gupta, *J. Indian Chem. Soc.*, **14,** 231 (1937).
198. W. Steinkopf and G. Schwen, *Ber.*, **54,** 1437 (1921).
199. F. F. Blicke and S. R. Safir, *J. Amer. Chem. Soc.*, **63,** 575 (1941).
200. A. N. Nesmeyanov, O. A. Reutov, Yu. G. Bundel, and I. P. Beletskaya, *Izv. Akad. Nauk SSSR, Otd. Khim. Nauk*, 929 (1957); *C.A.*, **52,** 4533 (1958).
201. W. R. Cullen, *Can. J. Chem.*, **39,** 2486 (1961).
202. M. Sartori and E. Recchi, *Ann. Chim. Appl.*, **29,** 128 (1939).
203. E. G. Walaschewski, *Chem. Ber.*, **86,** 272 (1953).
204. G. Hayashi and M. Kado (to Ihara Agricultural Chemicals Co., Ltd.), Jap. Pat. 23,793 (Oct. 19, 1965); through *C.A.*, **64,** 3602 (1966).
205. H. J. Emeléus *J. Chem. Soc.*, 2979 (1954).
206. S. J. Green and T. S. Price, *J. Chem. Soc.*, **119,** 448 (1921).
207. W. E. Jones, R. J. Rosser, and F. N. Woodward, *J. Soc. Chem. Ind.*, **68,** 258 (1949).
208. C. L. Hewett, W. E. Jones, H. W. Vallender, and F. N. Woodward, *J. Soc. Chem. Ind.*, **68,** 263 (1949).

209. H. N. Das-Gupta, *J. Indian Chem. Soc.*, **14,** 349 (1937).
210. C. K. Banks, F. H. Kahler, and C. S. Hamilton, *J. Amer. Chem. Soc.*, **69,** 933 (1947).
211. J. Prat, *Mem. Serv. Chim. Etat* (Paris), **33,** 395 (1947); through *C.A.*, **44,** 3435 (1950).
212. W. L. Lewis and G. A. Perkins, *Ind. Eng. Chem.*, **15,** 290 (1923).
213. J. Donohue, G. Humphrey, and V. Schomaker, *J. Amer. Chem. Soc.*, **69,** 1713 (1947).
214. C. L. Hewett, *J. Chem. Soc.*, 1203 (1948).
215. C. A. McDowell, H. G. Emblem, and E. A. Moelwyn-Hughes, *J. Chem. Soc.*, 1206 (1948).
216. G. H. Whiting, *J. Chem. Soc.*, 1209 (1948).
217. R. Fusco and T. Cottignoli, *Farm. Ital.*, **11,** 89 (1943); through *C.A.*, **38,** 6054 (1944).
218. H. N. Das-Gupta, *J. Indian Chem. Soc.*, **15,** 495 (1938).
219. G. I. Braz and A. Ya. Yakubovich, *Zh. Obshch. Khim.*, **11,** 41 (1941); through *C.A.*, **35,** 5459 (1941).
220. A. Ya. Yakubovich, V. A. Ginsburg, and S. P. Makarov, *Dokl. Akad. Nauk SSSR*, **71,** 303 (1950); through *C.A.* **44,** 8320 (1950).
221. A. Ya. Yakubovich and S. P. Makarov, *Zh. Obshch. Khim.*, **22,** 1528 (1952).
222. G. H. Cookson and F. G. Mann, *J. Chem. Soc.*, 2895 (1949).
223. D. Seyferth, *Chem. Rev.*, **55,** 1155 (1955).
224. A. Ya. Yakubovich and V. A. Ginsburg, *Zh. Obshch. Khim.*, **22,** 1783 (1952).
225. L. Hellerman and M. D. Newman, *J. Amer. Chem. Soc.*, **54,** 2859 (1932).
226. R. Huisgen, *Oesterr. Chem.-Ztg.*, **55,** 237 (1954).
227. K. Mislow, A. Zimmerman, and J. T. Melillo, *J. Amer. Chem. Soc.*, **85,** 594 (1963).
228. L. Maier, D. Seyferth, F. G. A. Stone, and E. G. Rochow, *Z. Naturforsch.*, **B, 12,** 263 (1957).
229. W. Steinkopf, I. Schubart, and J. Roch, *Ber.*, **65,** 409 (1932).
230. E. E. Turner and F. W. Bury, *J. Chem. Soc.*, **123,** 2489 (1923).
231. W. R. Cullen, *Can. J. Chem.*, **38,** 445 (1960).
232. E. V. Zappi, *An. Soc. Quim. Arg.*, **3,** 447 (1915); through *C.A.*, **10,** 1523 (1916).
233. W. Steinkopf, H. Dudek, and S. Schmidt, *Ber.*, **61,** 1911 (1928).
234. I. L. Knunyants and V. Ya. Pil'skaya, *Izv. Akad. Nauk SSSR, Otd. Khim. Nauk*, 472 (1955); through *C.A.*, **50,** 6298 (1956).
235. R. L. Barker, E. Booth, W. E. Jones, A. F. Millidge, and F. N. Woodward, *J. Soc. Chem. Ind.*, **68,** 285 (1949).
236. R. L. Barker, E. Booth, W. E. Jones, and F. N. Woodward, *J. Soc. Chem. Ind.*, **68,** 295 (1949).
237. H. Heaney and I. T. Millar *J. Chem. Soc.* 5132 (1965).
238. C. S. Gibson and J. D. A. Johnson, *J. Chem. Soc.*, 2518 (1931).
239. O. Seide and J. Gorski, *Ber.*, **62,** 2186 (1929).
240. R. N. Sterlin, L. N. Pinkina, R. D. Yatsenko, and I. L. Knunyants, *Khim. Nauka i Promy.*, **4,** 800 (1959); through *C.A.*, **54,** 14103 (1960).
241. R. N. Sterlin, R. D. Yatsenko, L. N. Pinkina, and I. L. Knunyants, *Izv. Akad. Nauk SSSR, Otd. Khim. Nauk*, 1991 (1960).
242. O. A. Reutov, and Yu. G. Bundel, *Izv. Akad. Mauk SSSR, Otd. Khim. Nauk*, 1041 (1952).
243. W. E. Hanby and W. A. Waters *J. Chem. Soc.*, 1029 (1946).
244. O. A. Reutov and Yu. G. Bundel, *Zh. Obshch. Khim.*, **25,** 2324 (1955).
245. L. Maier, E. G. Rochow, and W. C. Fernelius, *J. Inorg. Nucl. Chem.*, **16,** 213 (1961).
246. A. Henglein, *Int. J. Appl. Radiat. Isotopes*, **8,** 156 (1960).
247. F. G. Mann, *The Heterocyclic Derivatives of Phosphorus, Arsenic, Antimony, Bismuth, and Silicon*, Interscience Publishers, Inc., New York, 1950.

248. I. G. M. Campbell and R. C. Poller, *J. Chem. Soc.*, 1195 (1956).
249. E. Roberts and E. E. Turner, *J. Chem. Soc.* **127,** 2004 (1925).
250. J. D. C. Mole and E. E. Turner, *J. Chem. Soc.*, 1720 (1939).
251. A. E. Goddard, *J. Chem. Soc.*, **123,** 1161 (1923).
252. C. T. Mortimer and H. A. Skinner, *J. Chem. Soc.*, 3189 (1953).
253. A. Tzschach and G. Pacholke, *Chem. Ber.*, **97,** 419 (1964).
254. Kh. R. Rauer, A. B. Bruker, and L. Z. Soborovskii, *Zh. Obshch. Khim.*, **35,** 1162 (1965).
255. H. Albers, W. Künzel, and W. Schuler, *Chem. Ber.*, **85,** 239 (1952).
256. W. R. Cullen and P. S. Dhaliwal, *Can. J. Chem.*, **45,** 719 (1967).
257. A. N. Nesmeyanov and E. G. Ippolitov, *Vestn. Mosk. Univ.*, **10,** No. 10, *Ser. Fiz. Mat. i Estestv. Nauk, No.* **7,** 87 (1955); through *C.A.*, **50,** 9906 (1956).
258. W. R. Cullen, *Can. J. Chem.*, **38,** 439 (1960).
259. F. Popp, *Chem. Ber.*, **82,** 152 (1949).
260. H. Gutbier and H. G. Plust, *Chem. Ber.*, **88,** 1777 (1955).
261. M. S. Malinovskii, *Zh. Obshch. Khim.*, **19,** 130 (1949).
262. D. R. Lyon and F. G. Mann, *J. Chem. Soc.*, 666 (1942).
263. D. S. Tarbell and J. F. Bunnett, *J. Amer. Chem. Soc.*, **69,** 263 (1947).
264. M. M. Baig and W. R. Cullen, *Can. J. Chem.*, **39,** 420 (1961).
265. W. R. Cullen, *Can. J. Chem.*, **41,** 322 (1963).
266. W. R. Cullen, *Can. J. Chem.*, **40,** 426 (1962).
267. E. Grichkiewitch-Trohimowski, L. Mateyak, and Zablotski, *Bull. Soc. Chim. Fr.* **41,** 1323 (1927).
268. J. Klippel, *Rocz. Chem.*, **10,** 777 (1930); through *C.A.*, **25,** 1516 (1931).
269. E. J. Cragoe, Jr., R. J. Andres, R. F. Coles, B. Elpern, J. F. Morgan, and C. S. Hamilton, *J. Amer. Chem. Soc.*, **69,** 925 (1947).
270. G. Kamai and N. A. Chadaeva, *Izv. Akad. Nauk SSSR, Otd. Khim Nauk*, 1779 (1960).
271. M. M. Koton and F. S. Florinskii, *Dokl. Akad. Nauk SSSR*, **137,** 1368 (1961).
272. H. G. Schicke and G. Schrader (to Farbenfabriken Bayer A.-G.), Ger. Pat. 1,112,739 (Nov. 25, 1959); through *C.A.*, **56,** 4798 (1962).
273. A. Job and H. Guinot, Fr. Pat. 521,119 (July 7, 1921); through *Chem. Zentr.*, **92,** IV, 870 (1921).
274. A. Job and H. Guinot, Fr. Pat. 521,469 (July 15, 1921); through *Chem. Zentr.*, **92,** IV, 870 (1921).
275. F. F. Blicke and J. F. Oneto, *J. Amer. Chem. Soc.*, **57,** 749 (1935).
276. G. Schuster, *J. Pharm. Chim.*, **19,** 497 (1934).
277. H. J. Vetter and H. Nöth, *Z. Naturforsch.*, **B, 19,** 167 (1964).
278. R. A. Zingaro, R. E. McGlothin, and R. M. Hedges, *Trans. Faraday Soc.*, **59,** 798 (1963).
279. W. Steinkopf, S. Schmidt, and H. Penz, *J. Prakt. Chem.*, **141,** 301 (1934).
280. G. P. Sollott, J. L. Snead, S. Portnoy, W. R. Peterson, Jr., and H. E. Mertwoy, *U.S. Dept. Com., Office Tech. Serv., A. D. 611869*, Vol. II, pp. 441–452 (1965); *C.A.*, **63,** 18147 (1965).
281. A. B. Burg and J. Singh, *J. Amer. Chem. Soc.*, **87,** 1213 (1965).
282. L. K. Peterson and K. I. Thé, *Chem. Commun.*, 1056 (1967).
283. H. Mohler and J. Sorge, *Helv. Chim. Acta*, **22,** 235 (1939).
284. H. Mohler and J. Pólya, *Helv. Chim. Acta*, **19,** 1222 (1936).
285. H. Mohler and J. Pólya, *Helv. Chim. Acta*, **19,** 1239 (1936).
286. A. G. Maddock and N. Sutin, *Trans. Faraday Soc.*, **51,** 184 (1955).

287. H. H. Jaffé, *J. Chem. Phys.*, **22,** 1430 (1954).
288. J. Eisenbrand, *Arch. Pharm.*, **269,** 683 (1931).
289. G. Petit, *C. R. Acad. Sci., Paris,* **218,** 414 (1944).
290. G. P. van der Kelen and M. A. Herman, *Bull. Soc. Chim. Belges,* **65,** 350 (1956).
291. E. G. Claeys and G. P. van der Kelen, *Spectrochim. Acta,* **22,** 2103 (1966).
292. R. R. Shagidullin and T. E. Pavlova, *Izv. Akad. Nauk SSSR, Ser. Khim.*, 2117 (1963).
293. R. R. Shagidullin and T. E. Pavolva, *Izv. Akad. Nauk SSSR, Ser. Khim.*, 995 (1965).
294. R. R. Shagidullin and T. E. Pavlova, *Izv. Akad. Nauk SSSR, Ser. Khim.*, 2091 (1966).
295. W. R. Cullen and N. K. Hota, *Can. J. Chem.*, **42,** 1123 (1964).
296. H. C. Clark and C. J. Willis, *Proc. Chem. Soc.*, 282 (1960).
297. T. J. Bardos, N. Datta-Gupta, and P. Hebborn, *J. Med. Chem.*, **9,** 221 (1966).
298. W. R. Cullen, P. S. Dhaliwal, and W. B. Fox, *Inorg. Chem.*, **3,** 1332 (1964).
299. W. R. Cullen and J. Trotter, *Can. J. Chem.*, **39,** 2602 (1961).
300. J. Trotter, *J. Chem. Soc.*, 2567 (1962).
301. W. R. Cullen and J. Trotter, *Can. J. Chem.*, **40,** 1113 (1962).
302. J. Trotter, *Can. J. Chem.*, **40,** 1590 (1962).
303. J. Trotter, *Can. J. Chem.*, **41,** 191 (1963).
304. N. Camerman and J. Trotter, *Can. J. Chem.*, **41,** 460 (1963).
305. W. R. Cullen and J. Trotter, *Can. J. Chem.*, **41,** 2983 (1963).
306. N. Camerman and J. Trotter, *Acta Crystallogr.*, **16,** 922 (1963).
307. N. Camerman and J. Trotter, *J. Chem. Soc.*, 219 (1964).
308. A. Camerman and J. Trotter, *J. Chem. Soc.*, 730 (1965).
309. H. A. Skinner and L. E. Sutton, *Trans. Faraday Soc.*, **40,** 164 (1944).
310. L. J. Nugent and C. D. Cornwell, *J. Chem. Phys.*, **37,** 523 (1962).
311. G. P. Baxter, F. K. Bezzenberger, and C. H. Wilson, *J. Amer. Chem. Soc.*, **42,** 1386 (1920).
312. C. E. Redemann, S. W. Chaikin, R. B. Fearing, and D. Benedict, *J. Amer. Chem. Soc.*, **70,** 637 (1948).
313. H. Mohler, *Helv. Chim. Acta,* **21,** 784 (1938).
314. H. Mohler, *Helv. Chim. Acta,* **21,** 789 (1938).
315. E. G. Claeys, G. P. van der Kelen, and Z. Eeckhaut, *Bull. Soc. Chim. Belges,* **70,** 462 (1961).
316. E. G. Claeys, *J. Organometal. Chem.* **5,** 446 (1966).
317. G. Kamai and I. M. Starshov, *Dokl. Akad. Nauk SSSR,* **96,** 995 (1954); through *C.A.*, **48,** 13318 (1954).
318. G. Kamai and B. D. Chernokal'skii, *Dokl. Akad. Nauk SSSR,* **149,** 850 (1963).
319. E. Gryszkiewicz-Trochimowski and S. F. Sikorski, *Bull. Chim. Soc. Fr.*, **41,** 1570 (1927).
320. C. T. Mortimer and H. A. Skinner *J. Chem. Soc.*, 4331 (1952).
321. N. P. Buu-Hoi, M. Mangane, and P. Jacquignon, *J. Heterocycl. Chem.*, **3,** 149 (1966).
322. N. P. Buu-Hoi, M. Mangane, and P. Jacquignon, *J. Heterocycl. Chem.*, **3,** 374 (1966).
323. R. C. Dobbie and R. G. Cavell, *Inorg. Chem.*, **6,** 1450 (1967).
324. P. Pascal, *C. R. Acad. Sci., Paris,* **218,** 57 (1944).

IV Arsines

1. PREPARATION OF PRIMARY AND SECONDARY ARSINES

A. By reduction

Primary and secondary arsines, $RAsH_2$ and R_2AsH, are commonly prepared by reduction. Any arsenical containing a single R group bonded to arsenic will be reduced to the corresponding primary arsine if a suitable reducing agent is used. Similarly, an arsenical containing two R groups will be reduced to the secondary arsine. One of the most satisfactory reducing agents is zinc dust and hydrochloric acid (1–4); amalgamated zinc (5–8), zinc–copper couple (9), and platinized zinc (10) with hydrochloric acid have also been used. Both arsonic and arsinic acids, or their salts, are readily reduced (2, 4–6, 11); other arsenicals, such as halo- or dihaloarsines (3, 9, 10), arseno compounds (1), or diaryl- or dialkylarsenic oxides (7), can be reduced equally well. Methylphenylchloroarsine has been reduced to the secondary arsine with zinc amalgam in methanol (12).

Lithium aluminum hydride has also been used in the preparation of primary and secondary arsines, but the results have not always been entirely satisfactory. Several dialkylchloroarsines have been successfully reduced with lithium aluminum hydride to the corresponding dialkylarsines (13, 14), but the reduction of dimethylchloroarsine was unsuccessful except in a single experiment (15). Both trifluoromethyldiiodo- and bis(trifluoromethyl)iodoarsines have been reduced with lithium aluminum hydride, but in both cases better results were obtained when the reduction was carried out with hydrochloric acid and zinc coated with copper (16).

Wiberg and Mödritzer (17, 18) obtained excellent results in the preparation of phenylarsine and diphenylarsine by the use of lithium borohydride. With phenylarsenic tetrachloride and lithium borohydride, an 81% yield of phenylarsine was obtained. Phenyldichloroarsine gave only a 70% yield of phenylarsine when reduced under the same conditions. Lithium aluminum hydride was not as satisfactory as the borohydride. The authors attributed this to a reaction between the arsine and aluminum hydride:

$$PhAsH_2 + AlH_3 \rightarrow PhAsH_2{\cdot}AlH_3 \rightarrow PhAs(H)AlH_2 + H_2$$

Diphenylarsenic trichloride and diphenylchloroarsine were similarly reduced with either lithium borohydride or lithium aluminum hydride to diphenylarsine in excellent yield. The reduction of phenylarsenic tetrachloride with either lithium borohydride or lithium aluminum hydride also gave phenylarsine.

Bis(trifluoromethyl)chloroarsine has been reduced to bis(trifluoromethyl)arsine by hydrogen chloride in the presence of metallic mercury (19). Tetrakis(trifluoromethyl)diarsine is obtained as a by-product. The author suggested that the diarsine is first formed by the following reaction:

$$2(CF_3)_2AsCl + Hg \rightarrow (CF_3)_2As{-}As(CF_3)_2 + HgCl_2$$

Cleavage of the diarsine with hydrogen chloride then gives the arsine and starting material:

$$(CF_3)_2As{-}As(CF_3)_2 + HCl \rightarrow (CF_3)_2AsH + (CF_3)_2AsCl$$

A convenient method for preparing bis(trifluoromethyl)arsine by the reduction of bis(trifluoromethyl)iodoarsine or tetrakis(trifluoromethyl)diarsine with mercury and hydrogen iodide has been described by Cavell and Dobbie (20). Reduction of trifluoromethyldiiodoarsine, or of a mixture of trifluoromethylarsenic tetramer and pentamer, with the same reducing agent gives trifluoromethylarsine.

Reduction to arsines can also be achieved electrolytically (21–23). The conditions for electrolytic reduction have been described in III-2-B.

B. From metallic arsenides and alkyl or aryl halides

Arsine, when passed into a solution of metallic sodium or potassium in liquid ammonia, gives a solution of the metal arsenide, $MAsH_2$. The arsenide will then react with an alkyl halide to give the primary alkyl arsine (24). The reaction can also be used for the preparation of pure secondary arsines. Thus phenylarsine reacts with sodium or potassium to give C_6H_5AsHNa, or C_6H_5AsHK, which will react with a variety of alkyl halides to give mixed alkylphenylarsines (8, 25, 26). Calcium metal in liquid ammonia reacts with arsine to form the diarsenide, $Ca(AsH_2)_2$, which with methyl chloride forms a mixture of methyl-, dimethyl-, and trimethylarsine (15).

When triarylarsines or alkyldiarylarsines are heated with potassium metal in dioxane, one aryl group is eliminated, and a potassium diaryl- or arylalkylarsenide is obtained. These compounds react with water to give excellent yields of secondary arsines (13, 27). Similarly, the Grignard compound $(C_6H_5)_2AsMgBr$ reacts with water to give diphenylarsine (28).

C. By the Friedel-Crafts reaction

When a mixture of arsenic trichloride, anhydrous aluminum chloride, benzene, and isopentane is stirred together at 35° in a nitrogen atmosphere, two layers are formed. The upper layer is separated, dried, and fractionated to yield a mixture of phenylarsine, diphenylarsine, and phenylpentanes (29). It has been suggested that the reaction involves first the phenylation of arsenic trichloride to form phenyldichloroarsine. This then reacts with the isopentane to form phenylarsine and two moles of a chloropentane. The phenylpentanes are then formed by a Friedel-Crafts reaction with benzene:

$$C_6H_6 + AsCl_3 \xrightarrow{AlCl_3} C_6H_5AsCl_2 + HCl$$

$$C_6H_5AsCl_2 + 2(CH_3)_2CHCH_2CH_3 \longrightarrow C_6H_5AsH_2 + 2C_5H_{11}Cl$$

$$C_5H_{11}Cl + C_6H_6 \xrightarrow{AlCl_3} C_6H_5C_5H_{11} + HCl$$

These results are very surprising. La Coste and Michaelis (30) had shown, many years previously, that arsenic trichloride will react with benzene in the presence of aluminum trichloride, but the reaction is extremely slow, and high temperatures and repeated recycling are required to give a satisfactory yield of phenyldichloroarsine. Further, the reaction of phenyldichloroarsine with the saturated hydrocarbon is amazing, and if it occurs as suggested, it must be catalyzed by aluminum chloride or some other catalyst, since dichloroarsines do not ordinarily react with saturated hydrocarbons. The mild conditions of the reaction, together with the unexpected products, suggest that this interesting reaction should be further investigated.

D. By the addition of arsines to carbonyl compounds

The rather unstable primary arsine $(CF_3)_2C(OH)AsH_2$ has been prepared by the addition of arsine to hexafluoroacetone (31). The preparation of $(CF_3)_2C(OH)As(H)CH_3$ by the addition of methylarsine to hexafluoroacetone is discussed in connection with the reactions of primary and secondary arsines (Section IV-2).

2. REACTIONS OF PRIMARY AND SECONDARY ARSINES

Both primary and secondary arsines, aliphatic and aromatic, are extremely reactive compounds. Dimethyl- (7) and diethylarsines (32), as well as methylphenylarsine (12), are all spontaneously inflammable in air. Even arsines which are not spontaneously inflammable are rapidly oxidized in the air and must be handled in an inert atmosphere. It has been stated (16) that bis-(trifluoromethyl)arsine slowly decomposes at room temperature and that the decomposition is greatly accelerated by traces of impurities. Cullen (33), however, has reported that this arsine showed no signs of decomposition after three months at 20° in a sealed tube.

The oxidation of primary arsines may yield arseno compounds (6), arsenoso compounds (18), or arsonic acids (11). Similar products are also formed by the oxidation of secondary arsines (18, 34). Methylarsine, however, when oxidized with nitric acid, yields arsenic pentoxide, formic acid, and nitrogen dioxide (6).

The arsines are strong reducing agents. Phenylarsine reacts with phenylmercuric chloride or diphenylmercury to give mercury, benzene, and arsenobenzene (35). Methylarsine and aqueous silver nitrate give metallic silver and methanearsonic acid (6). Phenylarsine reacts with iodine to give phenyldiiodoarsine, arsenobenzene, or 1,2-diiodo-1,2-diphenyldiarsine (36). The specific product obtained in the reaction with iodine is a function of the ratio of the reactants. The reaction of diphenylarsine with bromine in ethereal solution yields diphenylarsenic tribromide (18). Diphenylarsine reacts with nitrosobenzene to form arsenobenzene and azobenzene (37). It was suggested, without proof, that a compound containing the As=N linkage is first formed, which subsequently disproportionates to the final products.

Diphenylarsine reacts with ethylene oxide in a sealed tube in a carbon dioxide atmosphere to form 1,2-bis(diphenylarsino)ethane (38). Phenylarsine, under similar reaction conditions, gives an undistillable product which may be 1,2-bis(hydroxyphenylarsino)ethane. Tetraphenyldiarsine is formed when diphenylarsine reacts with 1,2-diiodo-1,2-diphenyldiarsine (39). Arsenobenzene is also formed in this reaction. Diarylarsines react with bis(diarylarsenic) oxides to form symmetrical tetraaryldiarsines (2, 40):

$$2Ar_2AsH + (Ar_2'As)_2O \rightarrow (Ar_2As)_2 + (Ar_2'As)_2 + H_2O$$

Unsymmetrical tetraaryldiarsines, $Ar_2AsAsAr_2'$, are not formed in the reaction.

The addition of primary and secondary arsines to unsaturated compounds has been but little investigated. Mann and co-workers (8, 11, 41, 42) found that primary and secondary arsines add to the double bond of acrylonitrile to form compounds of the types $RAs(CH_2CH_2CN)_2$ and $R_2AsCH_2CH_2CN$, respectively. The reaction is base catalyzed, and the resulting tertiary arsines are remarkably stable towards oxidizing agents. The addition of phenylarsine to ethyl acrylate in the presence of acetic acid has also been studied (43). More recently, Uhlig and Maaser (44) have prepared 2-(2-diphenylarsinoethyl)pyridine by the reaction between diphenylarsine and 2-vinylpyridine. Cullen and co-workers (45) have found that the addition of dimethylarsine to hexafluoropropene at 100° yields (1,1,2,3,3,3-hexafluoropropyl)dimethylarsine. By contrast, secondary arsines do not add to hexafluorocyclobutene but displace one of the vinyl fluorine atoms to form tertiary arsines.

Adams and Palmer (46, 47) have found that primary aromatic arsines react with aldehydes in three different ways:

$$ArAsH_2 + 2RCHO \longrightarrow ArAs(CHOHR)_2$$

$$2ArAsH_2 + 4RCHO \longrightarrow ArAs\begin{matrix} O-CHR \\ RCH-O \end{matrix}AsAr + 2RCH_2OH$$

$$nArAsH_2 + 2nRCHO \longrightarrow (C_6H_5As)_n + 2nRCH_2OH$$

The first reaction takes place readily when the arsine, the aldehyde, and a little concentrated hydrochloric acid are stirred vigorously together in the cold. The heterocyclic compounds are formed by allowing a mixture of arsine and aldehyde to remain for a day or two in the presence of anhydrous hydrogen chloride at ordinary temperature. The third reaction takes place at a higher temperature without a catalyst or in the presence of acetic acid. More recently, Cullen and Styan (48) have reported that hexafluoroacetone reacts with dimethylarsine and methylarsine to give $(CH_3)_2AsC(CF_3)_2OH$ and $CH_3As(H)C(CF_3)_2OH$, respectively. The latter compound reacts slowly with excess hexafluoroacetone to give an unidentified solid and 1,1,1,3,3,3-hexafluoro-2-propanol.

Tzschach and Schwarzer (49) have prepared compounds of the types $R_2As(CONHC_6H_5)$ and $RAs(CONHC_6H_5)_2$ by the addition of primary or secondary arsines to phenylisocyanate in the presence of dibutyltin diacetate:

$$R_nAsH_{3-n} + (3\text{-}n)C_6H_5NCO \rightarrow R_nAs(CONHC_6H_5)_{3-n}$$

Cyclohexylisocyanate did not react under the conditions employed, but the carbamoyl compounds could be readily obtained by the reaction of lithium diphenylarsenide (or dilithium phenylarsenide) with cyclohexylisocyanate and treatment of the reaction product with alcohol:

$$Li_{3-n}AsR_n + (3\text{-}n)C_6H_{11}NCO \rightarrow R_nAs(CONLiC_6H_{11})_{3-n}$$

$$R_nAs(CONLiC_6H_{11})_{3-n} + C_2H_5OH \rightarrow R_nAs(CONHC_6H_{11})_{3-n}$$

The authors studied a number of reactions of the carbamoylarsines and also described their infrared spectra and other physical properties.

Cullen and co-workers (50) have investigated the addition of secondary arsines to hexafluorobut-2-yne and 3,3,3-trifluoropropyne. The reaction with hexafluorobut-2-yne yields predominantly the *trans* isomer of the expected butene derivative:

$$R_2AsH + CF_3C{\equiv}CCF_3 \rightarrow R_2AsC(CF_3){=}CHCF_3$$

The trifluoropropyne and dimethylarsine give three products (in the ratio of 3:3:2): *cis*- and *trans*-$Me_2AsCH{=}CHCF_3$ and $Me_2AsC(CF_3){=}CH_2$.

Dimethylarsine reacts with acetyl bromide in ether in a nitrogen atmosphere to give dimethylbromoarsine, arsenomethane, and acetyldimethylarsine (10). In addition, a substance formulated as $[(CH_3)_2(CH_3CO)_3As_2]Br{\cdot}2(C_2H_5)_2O$ is obtained. Acetyldimethylarsine could not be obtained in a pure state in the above reaction, but was prepared by the reaction between dimethylarsine and $CH_3CONH_2{\cdot}BF_3$. The same tertiary arsine can also be obtained from dimethylarsine and ketene. Dimethylarsine and trichloroacetyl chloride give dimethylchloroarsine and trichloroacetyldimethylarsine. Dimethylarsine reacts with trifluoroacetyl chloride in a similar manner (51).

Diphenylarsine reacts with chloroacetyl chloride, chloroacetyl bromide, phosgene, phosphorus trichloride, or benzenesulfonyl chloride to give diphenylchloro- or diphenylbromoarsine (52). Phenylarsine and thionyl chloride give a mixture of thioarsenosobenzene and phenyldichloroarsine (53). Sulfuryl chloride and phenylarsine give arsenobenzene, which reacts further with sulfuryl chloride to give phenyldichloroarsine. Sulfinylaniline, C_6H_5NSO, reacts with phenylarsine to give C_6H_5AsSO and aniline. The latter arsenic derivative is also obtained by reaction of the Grignard reagent, $C_6H_5As(MgBr)_2$, with thionyl chloride. Dimethylarsine reacts with carbon disulfide to give $[(CH_3)_2As]_2CS$ (10).

Cullen and co-workers (45, 54) have found that fluorine or chlorine can be displaced from certain halo-substituted cycloolefins by secondary arsines:

$$R_2AsH + \overline{XC{=}CY(CF_2)_nCF_2} \longrightarrow R_2As\overline{C{=}CY(CF_2)_nCF_2} + HX$$

where X = F or Cl; Y = F, Cl, H, C_2H_5, or $As(CH_3)_2$; and n = 1, 2, or 3. Dimethylchloroarsine is formed as a by-product when R = CH_3 and X = Cl. The reaction of alkyl halides with secondary arsines to form tertiary arsines has been noted in the older literature (7).

The reaction of dialkyl- or diarylarsines with dialkyl- or diarylchloroarsines has been used for preparing symmetrical tetraalkyl- or tetraaryldiarsines (7, 33, 55). The reaction of secondary arsines with dihaloarsines is discussed in Section VI-4-C.

Cullen (33) has found that the reaction of dimethylarsine with hydrogen chloride yields a solid which apparently has the formula $(CH_3)_2AsH_2Cl$. This substance has no measurable vapor pressure at $-130°$ and is completely dissociated into the starting materials at 20°. In contrast, the reaction of dimethylarsine with hydrogen bromide or hydrogen iodide appears to form an onium compound which decomposes to hydrogen and the haloarsine (7):

$$Me_2AsH + HX \rightarrow Me_2AsH_2X \rightarrow Me_2AsX + H_2$$

The acidities of phenylarsine and of a number of secondary arsines in tetrahydrofuran have been compared with the acidities of other very weak acids, including several primary and secondary phosphines (55a). The arsines were found to be significantly stronger than the corresponding phosphines. The pK_A values of the arsines ranged from 20.3 for diphenylarsine to 32.0 for di-*tert*-butylarsine.

The preparation of organometallic derivatives of primary and secondary arsines is discussed in Section VI-7. The conversion of primary and secondary arsines to sulfur derivatives is discussed in Section III-4. The reaction of primary and secondary arsines with diborane is discussed in Section VI-7. Primary and secondary arsines also react with alkyl halides to form quaternary arsonium salts (Section V-9-A).

3. PHYSICAL PROPERTIES OF PRIMARY AND SECONDARY ARSINES

Possibly because of their toxicity and ease of oxidation, there have been relatively few investigations of the physical properties of arsines. Methylarsine has been found to be monomolecular with a heat of vaporization of 5.39 kcal/mole (56). The melting point of this arsine is $-143°$, and it boils at 2°. Dimethylarsine melts at $-136.1°$, boils at 37.1°, and has a Trouton's constant of 21.3 (15). The heat of the following reaction has been studied by Mortimer and Skinner (57):

$$C_6H_5AsH_2 + 2I_2 \rightarrow C_6H_5AsI_2 + 2HI$$

The results obtained in this study lead to the conclusion that the bond dissociation energy of the As—H bond in phenylarsine is larger by 17.0 ± 2.3 kcal/mole than the dissociation energy of the As—I bond in phenyldiiodoarsine.

The infrared and Raman spectra of methylarsine and its deuterated analogs CH_3AsD_2, CD_3AsH_2, and CD_3AsD_2 have been recently investigated (58). Complete vibrational assignments were made, and a normal coordinate calculation was carried out. The most intense bands in the spectra of these compounds occur at about 2100 cm^{-1} for the two $—AsH_2$ compounds and at about 1500 cm^{-1} for the two $—AsD_2$ compounds; these peaks were assigned to the As—H and As—D stretching vibrations, respectively. Moderately strong bands at 563, 585, 518, and 535 cm^{-1} in CH_3AsH_2, CH_3AsD_2, CD_3AsH_2, and CD_3AsD_2, respectively, were assigned to the C—As stretching vibrations.

The proton NMR spectra of arsine, diphenylarsine, and a number of related silanes, germanes, stannanes, amines, and phosphines have been investigated by Ryan and Lehn (59). They found that protons attached to the group IV or group V elements absorb at lower fields whenever phenyl groups are also attached to these elements. The magnitude of the displacement caused by a phenyl group appears to be dependent on the electronegativity of the element to which it is bonded.

The mass spectra of trifluoromethylarsine and bis(trifluoromethyl)arsine are discussed in Section III-8-D; the use of electron impact mass spectrometry for the determination of the ionization potentials of several primary and secondary arsines is discussed in Section IV-6-C.

4. PREPARATION OF TERTIARY ARSINES

A. From organometallic compounds

The reaction between a Grignard reagent and arsenic halides is a well-known method for the preparation of both trialkyl- and triarylarsines. The yields with aliphatic Grignard reagents are often poor, but the method offers the advantage that the products are not contaminated with haloarsines. A very pure sample of trimethylarsine has been prepared from methylmagnesium iodide and arsenic trichloride in dibutyl ether (60). Triethyl-, tri-*n*-butyl-, triisobutyl-, tri-*n*-pentyl-, and triisohexylarsines (but not tri-*tert*-butylarsine) have all been prepared from the corresponding Grignard reagents. The yields were not large, and some metallic arsenic was always formed in the reaction (61). Tri-*n*-butylarsine has been obtained in 50% yield by performing the reaction in di-*n*-butyl ether and distilling the products (62). The reaction is quite successful with long chain alkyl groups such as dodecyl and tetradecyl (63). Other aliphatic arsines readily prepared from the Grignard reaction include triallylarsine (64, 65), tris(perfluoroethyl)arsine (66), and trivinylarsine (67, 68). Tricyclohexylarsine is also readily prepared from the corresponding Grignard reagent (69). Benzylmagnesium bromide and arsenic trichloride give tribenzylarsine; the by-products are tribenzylarsine oxide and dibenzylarsinic acid (70). Although no explanation for the formation of these by-products has been offered, it has been demonstrated (71) that tribenzylarsine is readily oxidized in the air, in the presence of catalytic impurities, to a mixture of dibenzylarsinic acid, benzaldehyde, and arsenic trioxide. The Grignard reagent prepared from bromomethyltrimethylsilane reacts with arsenic trichloride to give $(Me_3SiCH_2)_3As$ (72).

Triarylarsines are also readily prepared from the Grignard reagent and arsenic trichloride. Thus, *p*-bromophenylmagnesium bromide gives tris(*p*-bromophenyl)arsine (73). Similarly, tris(pentafluorophenyl)arsine has been prepared from the corresponding Grignard reagent (74); and tri-α-pyridylarsine can be obtained from α-pyridylmagnesium bromide (75). The use of tetrahydrofuran rather than ethyl ether in the Grignard reaction may lead to good yields of tertiary arsines (76). Thus, tri-β-styrylarsine, tri-α-furylarsine, and triphenylarsine have been prepared in excellent yields by this procedure.

Mixed trialkyl- or triarylarsines can be obtained by using dichloro- or monochloroarsines rather than arsenic trihalides. Thus, bis(β-chlorovinyl)-chloroarsine and propylmagnesium iodide give bis(β-chlorovinyl)propylarsine (77); and *p*-bromophenyldichloroarsine and either methyl- or propylmagnesium bromides give *p*-bromophenyldimethyl- and *p*-bromophenyldipropylarsines (78). The widely-used chelatiug agent, *o*-phenylenebis(dimethylarsine), *o*-$[As(CH_3)_2]_2C_6H_4$, can be prepared by the action of methylmagnesium bromide on *o*-bis(dichloroarsino)benzene (79).

Interesting results have been obtained with di-Grignard reagents. When *o*-diiodobenzene is treated with magnesium and the resulting solution treated with dimethyliodoarsine, α-biphenylyldimethylarsine is formed (80). The following mechanism has been suggested:

2 (o-$C_6H_4I_2$) + Mg ⟶ 2,2'-diiodobiphenyl + Mg ⟶

2,2'-bis(MgI)biphenyl + $(CH_3)_2AsI$ ⟶ 2-IMg-2'-$As(CH_3)_2$-biphenyl

$\xrightarrow{H_2O}$ 2-$As(CH_3)_2$-biphenyl

The Grignard reagent formed from *o*-bromoiodobenzene and three equivalents of magnesium reacts with dimethyliodoarsine to give a mixture of (*o*-bromophenyl)dimethylarsine, phenyldimethylarsine, 2-biphenylyldimethylarsine,

2,2′-biphenylylenebis(dimethylarsine), and *o*-terphenyl-2,2″-ylenebis(dimethylarsine) (81). In order to explain these results, Heaney, Mann, and Millar (81) proposed that benzyne is formed as a reactive intermediate. Di-Grignard reagents have been used for the preparation of heterocyclic tertiary arsines. Thus, the Grignard reagents obtained from 1,5-dibromopentane (82) and 1,4-dibromobutane (83) have been condensed with methyldichloroarsine and phenyldichloroarsine, respectively, to yield the following compounds:

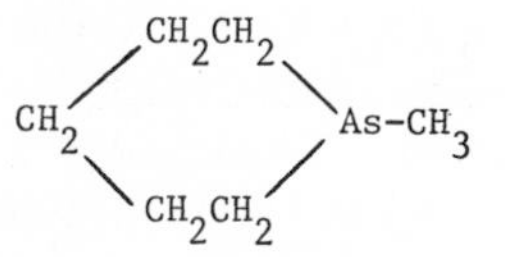

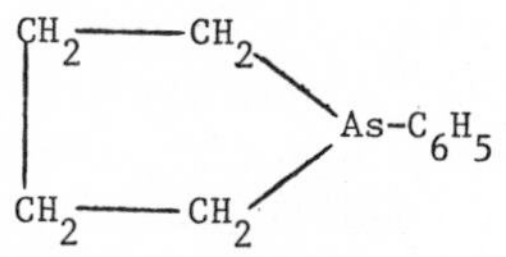

$$CH_2\begin{matrix} CH_2CH_2 \\ CH_2CH_2 \end{matrix}As{-}CH_3 \qquad \begin{matrix} CH_2{-}CH_2 \\ CH_2{-}CH_2 \end{matrix}As{-}C_6H_5$$

A number of tertiary arsines containing the acetylenic linkage have been prepared from organometallic reagents. Thus, the di-Grignard reagent from acetylene reacts with diethylbromoarsine (84), with dicyclohexyl-, diphenyl-, and di-α-naphthylchloroarsines (84), and a number of iodoarsines (85) to form di-tertiary arsines. In other papers a number of unsymmetrical acetylene arsenicals have been reported (86–89). Thus from diphenylchloroarsine and either $C_6H_5C{\equiv}CMgBr$ or $C_6H_5C{\equiv}CNa$, $C_6H_5C{\equiv}CAs(C_6H_5)_2$ was obtained. Other acetylenic tertiary arsines of this type prepared were $(\alpha\text{-}C_{10}H_7)_2AsC{\equiv}CC_6H_5$ and $(\alpha\text{-}C_{10}H_7)_2AsC{\equiv}CH$. The reaction of arsenic trichloride and $C_6H_5C{\equiv}CMgBr$ gave $(C_6H_5C{\equiv}C)_3As$ (87).

Although arsenic halides are usually used in the preparation of triarylarsines, the Grignard reagent will react with other arsenicals. Phenylmagnesium bromide reacts with finely powdered arsenic to produce a 5.9% yield of triphenylarsine (90). Benzylmagnesium chloride reacts with tributyl arsenite to give a 7% yield of tribenzylarsine (91). Dibenzylarsinic acid and bibenzyl are also formed in this reaction. Phenylmagnesium bromide reacts with tetraphenyldiarsine to give triphenylarsine (92). Grignard reagents also react with arsenic trioxide to give tertiary arsines, but the reaction often gives mixtures of arsenicals. Tributyl-, triamyl-, trihexyl-, triheptyl-, and trioctylarsines have all been prepared in 60 to 80% yield from the corresponding Grignard reagents and arsenic trioxide (93). Phenylmagnesium bromide and arsenic trioxide give a mixture of bis(diphenylarsenic) oxide, diphenylarsinic acid, and triphenylarsine (94, 95). α-Naphthylmagnesium bromide gives only bis(di-α-naphthylarsenic) oxide, regardless of the reaction conditions. Grignard reagents also react with arsenic trisulfide to produce a mixture of arsenicals, which includes the tertiary arsine (96). Thus, *p*-tolylmagnesium bromide and arsenic trisulfide give a mixture of tri-*p*-tolylarsine sulfide, bis(di-*p*-tolylarsenic) sulfide, and tri-*p*-tolylarsine.

In recent years lithium reagents have been used extensively in place of Grignard reagents, particularly in cases in which the Grignard reagent is difficult or impossible to obtain. Vinyllithium and arsenic trichloride give a 51% yield of trivinylarsine (97). α-Pyridyllithium and arsenic trichloride react to form tri-α-pyridylarsine trihydrochloride (98). Treatment of the hydrochloride with aqueous ammonia gives the tertiary arsine as a dihydrate. The anhydrous arsine has also been prepared from the lithium reagent (98, 99). Both *cis*- and *trans*-tripropenylarsine, as well as triisopropenylarsine, have been prepared from the corresponding lithium reagents and arsenic trichloride (100).

The lithium reagent from α-bromopicoline reacts with dimethyliodoarsine to form α-picolyldimethylarsine (101). The latter arsine reacts with moist air with rupture of a carbon–arsenic bond and the formation of α-picoline and cacodylic acid. 2,2′-Dilithiobibenzyl reacts readily with phenyldichloroarsine to give the following heterocycle (102):

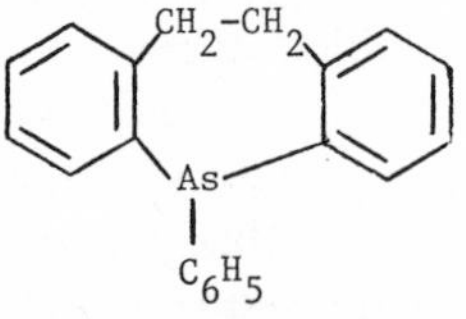

CH_2-CH_2

As

C_6H_5

2,2′-Dilithiobiphenyl reacts with dimethyliodoarsine to give 2,2′-biphenylenebis(dimethylarsine) (103). An interesting type of heterocyclic arsenical has been prepared by the reaction of 1,4-dilithio-1,2,3,4-tetraphenylbutadiene and aryldichloroarsines (104, 105). Thus when phenyldichloroarsine is used, the following compound is formed:

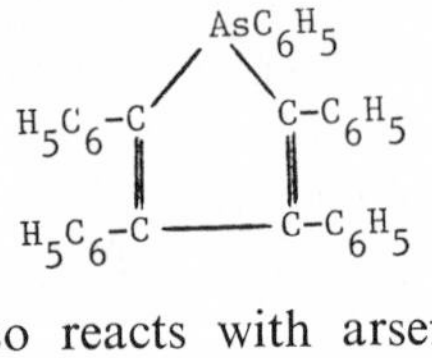

AsC_6H_5

H_5C_6-C $C-C_6H_5$

H_5C_6-C —— $C-C_6H_5$

The dilithio compound also reacts with arsenic trichloride to form the following chloroarsine (106):

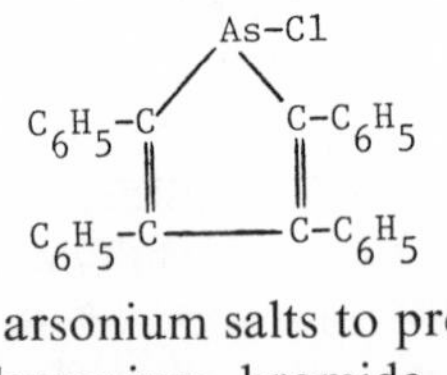

$As-Cl$

C_6H_5-C $C-C_6H_5$

C_6H_5-C —— $C-C_6H_5$

Lithium alkyls react with arsonium salts to produce trialkylarsines. Thus, butyllithium and tetraethylarsonium bromide give triethylarsine together with a mixture of hydrocarbons (107). Ethyllithium and triethylbutylarsonium bromide give butyldiethylarsine and a mixture of saturated and unsaturated hydrocarbons. Phenyllithium reacts with finely powdered arsenic to give a 27% yield of triphenylarsine (90).

A series of reactions employing both the Grignard and lithium reagents has been employed by Jones and Mann (108) for the preparation of phenylenediarsines. First an alkyl Grignard reagent reacts with *o*-bromophenyldichloroarsine to produce an *o*-bromophenyldialkylarsine. This, with *n*-butyllithium, then gives an *o*-lithiophenyldialkylarsine, which can then react further with a dialkylhaloarsine to form a mixed phenylenediarsine:

$$o\text{-}BrC_6H_4AsCl_2 + 2RMgX \longrightarrow o\text{-}BrC_6H_4AsR_2 + n\text{-}C_4H_9Li \longrightarrow$$

$$o\text{-}LiC_6H_4AsR_2 + R'_2AsI \longrightarrow o\text{-}C_6H_4(AsR_2)(AsR'_2)$$

When dialkylchlorophosphines are used in the last stage of the reaction, dialkyl(*o*-dialkylphosphinophenyl)arsines are formed. The latter compounds as well as the diarsines have been used as chelating agents.

In recent years organoaluminum compounds have been used to prepare tertiary arsines. For example, triethylarsine can be readily prepared from triethylaluminum and arsenic trichloride in 80% yield (109, 110). The same arsine can also be prepared by the reaction between ethylaluminum sesquichloride (111) or lithium tetraethylaluminate (112) and arsenic trichloride. A number of trialkylarsines have been obtained by the alkylation of arsenic trioxide with aluminum alkyls (113).

Organic mercurials have also been used for the preparation of tertiary arsines. Since the mercurials are less reactive than Grignard, lithium, or aluminum reagents, the resulting tertiary arsines are usually mixed with halo- and dihaloarsines. Nevertheless, the use of organic mercurials has proved useful for preparing sensitive tertiary arsines that cannot be prepared by other methods. Thus tri-α-thienylarsine, admixed with the chloro- and dichloroarsines, has been obtained from di-α-thienylmercury and arsenic-trichloride (114). Tri-α-furylarsine, admixed with the chloro- and dichloroarsines, has been prepared from α-chloromercurifuran and arsenic trichloride (115). It was later found (116) that mercury catalyzed the reaction of the tertiary arsine with arsenic trichloride; if the mercury was removed from the reaction as it formed, tri-α-furylarsine could be obtained in a pure state. Tri-β-furylarsine has been obtained from β-chloromercurifuran and arsenic trichloride (117). Recently, several tertiary arsines have been prepared by the reaction of organomercurials with thioarsenosobenzene (118).

Organozinc compounds were used for the preparation of tertiary arsines before the discovery of the Grignard reagents (119). Very pure trimethylarsine can be prepared from dimethylzinc and arsenic trichloride (60).

Trimethylarsine has been prepared from dimethylcadmium and arsenic trichloride. The arsine was not isolated but was converted to tetramethylarsonium iodide, which was isolated in an overall yield of 55% (120). More recently, diethyl(*p*-nitrophenyl)arsine has been prepared by the interaction of the ethylcadmium reagent and *p*-nitrophenyldichloroarsine (121).

Organotin compounds can also be converted to tertiary arsines. For example, the reaction of tetraphenyltin with finely-powdered arsenic at 320° for 18 hours gives a 40% yield of triphenylarsine (122). Two diarsines containing acetylenic linkages have been prepared by the displacement of the trimethyltin group (123):

$$(CH_3)_3SnC{\equiv}C{-}C{\equiv}CSn(CH_3)_3 + 2Ar_2AsCl \rightarrow Ar_2AsC{\equiv}C{-}C{\equiv}CAsAr_2 + 2(CH_3)_3SnCl$$

Similarly, phenyldichloroarsine has been found to react with $Ph_2SnCH_2CO_2$-CH_3 to yield $PhAs(CH_2CO_2CH_3)_2$ (118).

B. By a "Wurtz-Fittig" type reaction

The preparation of a tertiary arsine by the interaction of an aryl or alkyl halide, an arsenic halide, and sodium is sometimes considered as a type of Wurtz-Fittig reaction (124). Although this method has been largely supplanted by the use of Grignard or lithium reagents, it may occasionally be the method of choice. Thus, triphenylarsine can be readily obtained in good yields by the reaction of either bromobenzene (125) or chlorobenzene (126, 127) with arsenic trichloride and sodium. Other triarylarsines prepared by this procedure include tri-*o*-biphenylylarsine (128), tri-α-thienylarsine (129), and the *ortho*, *meta* (124), and *para* (130) isomers of tris(methoxyphenyl)arsine. However, an attempt to prepare trifluoromethyl-substituted phenylarsines by the Wurtz-Fittig procedure was unsuccessful (124). Das-Gupta (131) was unable to prepare tri-β-styrylarsine by the use of di-β-styrylmercury, but he did obtain the desired arsine from β-bromostyrene, arsenic trichloride, and sodium. The tertiary arsine, however, was admixed with β-styryldichloro- and di-β-styrylchloroarsines. The Wurtz-Fittig reaction has also been employed for converting benzyl chloride to tribenzylarsine (71). This reaction must be conducted in an inert atmosphere, or the arsine is oxidized to dibenzylarsinic acid, tribenzylarsine oxide, and arsenic trioxide. The Wurtz-Fittig reaction has also been used with chloro- or dichloroarsines in place of arsenic trichloride. Thus, α,α′-dibromo-*o*-xylene reacts with aryldichloroarsines in the presence of sodium to give isoarsindolines (132, 133):

CH_2 AsAr CH_2

Similarly, diphenylchloroarsine and 4-bromo-2,6-di-*tert*-butylphenol react with excess sodium in tetrahydrofuran to yield the sodium salt of diphenyl-(4-hydroxy-3,5-di-*tert*-butylphenyl)arsine (134). Kamai (135) has used metals other than sodium to condense dialkyliodoarsines with alkyl halides. Ethylmethyliodoarsine and ethyl bromoacetate in ether solution condense in the presence of zinc metal to give ethyl methylethylarsinoacetate. Methylphenyliodoarsine and ethyl bromoacetate were condensed by means of magnesium metal to give ethyl methylphenylarsinoacetate. In another experiment Kamai (135) first allowed an ethereal solution of ethylphenyliodoarsine to react with magnesium and then added ethyl β-iodopropionate to yield ethyl β-ethylphenylarsinopropionate. This procedure may have involved the intermediate formation of $(C_6H_5)C_2H_5AsMgBr$. The preparation of tertiary arsines from metallic derivatives of primary and secondary arsines is described in the following section (IV-4-C).

The preparation of trifluoromethyl-substituted tertiary arsines by the reaction of trifluoroiodomethane and an iodoarsine in the presence of mercury has been discussed in Section III-6-B.

C. The reaction of halides, ethers, and sulfides with metallic derivatives of primary and secondary arsines

In the nineteen thirties Blicke and co-workers (92, 136) obtained triphenylarsine by the reaction of arsenosobenzene, bis(diphenylarsenic) oxide, or tetraphenyldiarsine with sodium amalgam and bromobenzene. The following mechanism was suggested in the case of arsenosobenzene:

$$(C_6H_5AsO)_n + 2n\text{Na} \longrightarrow nC_6H_5As\begin{cases}\text{ONa}\\ \text{Na}\end{cases}$$

$$C_6H_5As\begin{cases}\text{ONa}\\ \text{Na}\end{cases} + C_6H_5Br \longrightarrow (C_6H_5)_2AsONa + NaBr$$

$$2(C_6H_5)_2AsONa \longrightarrow [(C_6H_5)_2As]_2O + Na_2O$$

$$[(C_6H_5)_2As]_2O + 2Na \longrightarrow (C_6H_5)_2AsNa + (C_6H_5)_2AsONa$$

$$(C_6H_5)_2AsNa + C_6H_5Br \longrightarrow (C_6H_5)_3As + NaBr$$

A similar mechanism involving sodium diphenylarsenide was suggested for the oxide and the diarsine. In recent years the formation of tertiary arsines by the reaction of organic halides with metallic derivatives of primary and secondary arsines has become an important synthetic procedure. Some examples are given below.

Alkali dialkyl- and diarylarsenides readily displace halogen from a variety of alkyl, aryl, and vinyl halides. Thus, Tzschach and Lange (27) reported that potassium di-α-naphthylarsenide reacts with ethyl bromide in dioxane to yield ethyldi-α-naphthylarsine. Similarly, Phillips and Vis (137) found that bromobenzene reacts with sodium dimethylarsenide in tetrahydrofuran to give the expected dimethylphenylarsine. They also showed that *m*-bromotoluene reacts under the same conditions to form only dimethyl-*m*-tolylarsine; this result suggests that a benzyne intermediate is not involved, since such a mechanism would be expected to lead to a mixture of products. In agreement with this conclusion, Sindellari and Centurioni (138) demonstrated that *o*-, *m*-, and *p*-bromoanisole react with sodium dimethylarsenide to yield the corresponding *o*-, *m*-, and *p*-methoxyphenyldimethylarsines. Aguiar and coworkers (139, 140) have found that the reaction of vinyl halides with lithium diphenylarsenide in tetrahydrofuran occurs with retention of configuration. Thus, *cis*- and *trans*-β-bromostyrene yield *cis*- and *trans*-β-styryldiphenylarsine, respectively (139), while *trans*-α-bromostilbene yields *trans*-(1,2-diphenylvinyl)diphenylarsine (140).

Di-, tri-, and tetra-tertiary arsines can also be prepared via alkali derivatives of secondary arsines. Issleib and Tzschach (13) found that 1,4-dichlorobutane reacts with two moles of potassium diphenylarsenide or dicyclohexylarsenide to yield $(C_6H_5)_2As(CH_2)_4As(C_6H_5)_2$ or $(C_6H_{11})_2As(CH_2)_4As(C_6H_{11})_2$. In later work (141), a similar procedure was used to prepare arsines of the type $(C_6H_5)_2As(CH_2)_nAs(C_6H_5)_2$, where $n = 3$, 4, 5, or 6. Feltham and co-workers (4, 142) found that two moles of sodium dimethylarsenide react vigorously with *o*-dichlorobenzene to yield *o*-phenylenebis(dimethylarsine). Analysis of the crude reaction product by gas-liquid chromatography failed to reveal any (*o*-chlorophenyl)dimethylarsine. Similarly, Zorn and co-workers (143) prepared *p*-phenylenebis(diphenylarsine) from potassium diphenylarsenide and *p*-dibromobenzene, and Müller and co-workers (134) obtained diphenyl(4-hydroxy-3,5-di-*tert*-butylphenyl)arsine from 4-iodo-2,6-di-*tert*-butylphenol. *Cis*- and *trans*-1,2-dichloroethene react with sodium dimethylarsenide in tetrahydrofuran, but the structure of the resulting di-tertiary arsines has not been established (137). 1,1,1-Tris(dimethylarsinomethyl)ethane has been prepared by the reaction of sodium dimethylarsenide in tetrahydrofuran with either 1,1,1-tris(chloromethyl)ethane (137) or 1,1,1-tris(bromomethyl)ethane (4). The reaction of sodium diphenylarsenide in liquid ammonia with 1,3-dibromo-2,2-bis(bromomethyl)propane has been used to synthesize the new tetra-tertiary arsine, tetrakis(diphenylarsinomethyl)methane (144).

The reaction of a dichloroalkane with an equimolar amount of an alkali dialkyl- or diarylarsenide may lead to the formation of a chloroalkyl-substituted tertiary arsine. Thus, Tzschach and Lange (141) found that potassium diphenylarsenide reacts at room temperature with several dichlorides according to the following scheme (where $n = 2, 4, 5$, or 6):

$$KAs(C_6H_5)_2 + Cl(CH_2)_nCl \rightarrow KCl + Cl(CH_2)_nAs(C_6H_5)_2$$

Under similar conditions, the analogous dibromoalkanes yield di-tertiary arsines and arsonium salts. The following heterocyclic arsonium salt was isolated from the reaction of potassium diphenylarsenide and 1,5-dibromopentane:

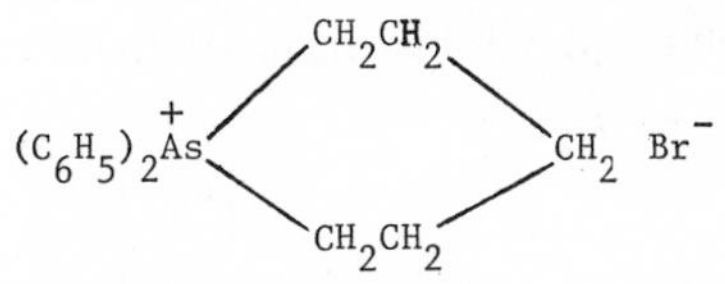

The reaction with 1,2-dibromoethane does not yield a tertiary arsine, but rather ethylene and tetraphenyldiarsine. In a later paper, Tzschach and Fischer (145) reported that lithium dicyclohexylarsenide reacts at −10° in dioxane with a number of dichlorides (where $n = 3, 4, 5$, or 6):

$$LiAs(C_6H_{11})_2 + Cl(CH_2)_nCl \rightarrow Cl(CH_2)_nAs(C_6H_{11})_2$$

With the exception of 4-chlorobutyldicyclohexylarsine, the compounds were stable and could be distilled *in vacuo*. The 4-chlorobutyl compound cyclized on standing at room temperature for several weeks or on being heated to 170° to yield the following heterocyclic arsonium salt:

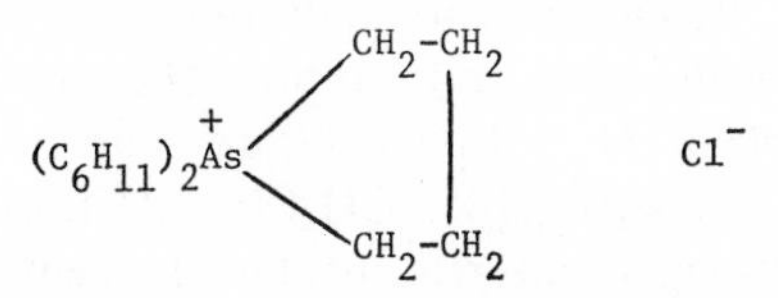

A number of tertiary arsines have been prepared by the reaction of organic halides with magnesium derivatives of primary arsines. As early as 1924, Job and co-workers (146) succeeded in condensing phenylarsylenebis(magnesium bromide), $C_6H_5As(MgBr)_2$, with bis(2-chloroethyl) sulfide to yield 4-phenyl-1,4-thiarsenane:

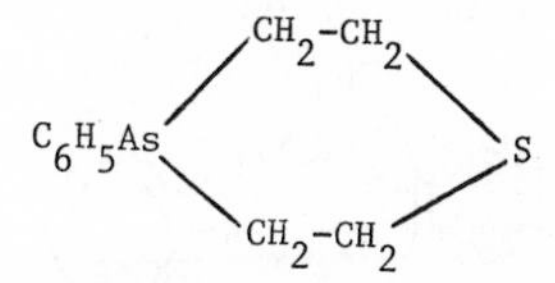

Mann and his co-workers (147–149) have extended this type of reaction to the preparation of other heterocyclic tertiary arsines. For example, from o,α,α'-dibromoxylene and phenylarsylenebis(magnesium bromide), 2-phenylisoarsindoline (**1**) was prepared in 80% yield (147).

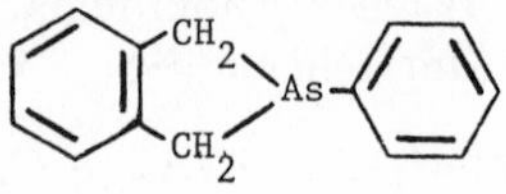

1

With the corresponding dichloride, however, the yield of 2-phenylisoarsindoline (**1**) was only 40%; phenyldichloroarsine and *sym*-tribenzcyclododecatriene were obtained as by-products. The condensation of phenylarsylenebis(magnesium bromide) and 1,8-bis(bromomethyl)naphthalene yielded the heterocyclic compound **2**.

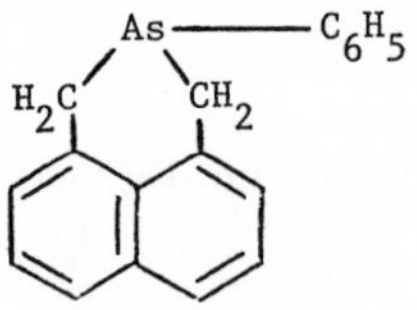

2

Cookson and Mann (150) have utilized the magnesium reagent for the preparation of 5-phenyl-5,6-dihydroarsanthridine (**3**). The following synthesis was employed: (See top of page 137 opposite.)

The magnesium reagent has certain limitations as a synthetic tool. Thus, when it reacted with bromoacetonitrile or with bromoacetone diethylketal, the sole products isolated were arsenobenzene and phenyldibromoarsine (147). Similarly, Job, Reich, and Vergnaud (146) obtained only arsenobenzene from the reagent and phosgene or 1,2-dibromoethane. Ethyl bromoacetate not only gave arsenobenzene and phenyldibromoarsine, but also gave a 5% yield of the desired arsine, $C_6H_5As(CH_2CO_2C_2H_5)_2$ (147). Ethylene oxide also reacted with the magnesium reagent, but no identifiable products could be obtained.

Thornton (151) has found that heterocyclic arsines can be prepared by the use of dilithium phenylarsenide, which is much easier to prepare than phenylarsylenebis(magnesium bromide). Thus, the reaction of the dilithium compound with α-bromo-o-(2-bromoethyl)toluene gives a 58% yield of pure 2-phenyl-1,2,3,4-tetrahydroisoarsinoline:

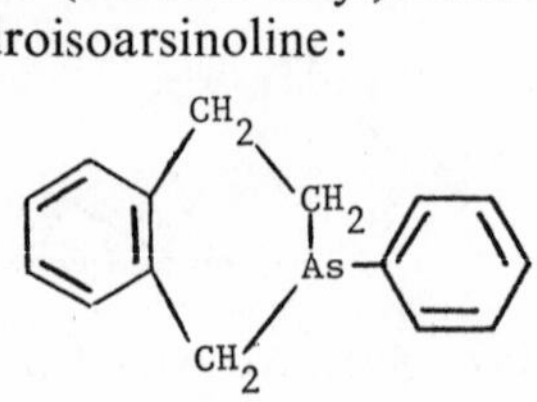

$$2\ \text{2-}(BrCH_2)C_6H_4C_6H_5 + C_6H_5As(MgBr)_2 \longrightarrow (2\text{-}C_6H_5C_6H_4CH_2)_2AsC_6H_5$$

$$+\ Cl_2 \longrightarrow (2\text{-}C_6H_5C_6H_4CH_2)_2AsCl_2C_6H_5 \xrightarrow{\Delta} 2\text{-}C_6H_5C_6H_4CH_2As(Cl)C_6H_5$$

$$\xrightarrow[CS_2]{AlCl_3} \text{(cyclic) } CH_2\text{—}As\text{-}C_6H_5$$

3

Beeby, Cookson, and Mann (147) had earlier made the same heterocycle from the magnesium reagent. Thornton also found that dilithium phenylarsenide reacts with ethyl bromide to give a low yield of diethylphenylarsine. No tertiary arsine was isolated from the reaction of the dilithium compound with 1,2-dibromoethane; the only crystalline substance obtained was arsenobenzene. This result is similar to an earlier observation of Job and co-workers (146), who isolated only arsenobenzene from the reaction of phenylarsylenebis(magnesium bromide) and 1,2-dibromoethane.

Compounds of the types $RAs(COR)_2$, $RAs(CO_2R)_2$, and $R_2As(COR)$ have been prepared from magnesium derivatives of primary and secondary arsines. Thus, Job and co-workers (146, 152) prepared $C_6H_5As(COCH_3)_2$ and $C_6H_5As(CO_2C_2H_5)_2$ by the reaction of phenylarsylenebis(magnesium bromide) with acetyl chloride or ethyl chloroformate, respectively. Similarly, Albers and co-workers (10) obtained $(CH_3)_2AsCOCH_3$ by the interaction of dimethylarsinomagnesium bromide, $(CH_3)_2AsMgBr$, and acetyl chloride. The preparation of keto-substituted arsines by other methods is discussed in Section III-7.

In addition to their reaction with organic halides, alkali dialkyl- and diarylarsenides also react with some ethers and sulfides with cleavage of the C—O or C—S bond and formation of tertiary arsines. Thus, Mann and Pragnell (153) prepared methyldiphenylarsine from lithium diphenylarsenide and anisole, and benzyldiphenylarsine from benzyl phenyl sulfide. Tzschach and Deylig (154) have shown that lithium and potassium derivatives of secondary arsines add exothermally to epoxides. Hydrolysis of the addition product yields a 2-hydroxy-substituted tertiary arsine. Thus, potassium diphenylarsenide and propylene oxide yield diphenyl(2-hydroxypropyl)arsine; and, similarly, cyclohexene oxide and lithium dicyclohexylarsenide give dicyclohexyl(2-hydroxycyclohexyl)arsine. The lithium and potassium derivatives add to trimethylene oxide and tetrahydrofuran under more drastic conditions. After hydrolysis of the reaction mixtures, 3-hydroxy- and 4-hydroxyalkyl-substituted tertiary arsines are obtained.

D. By redistribution

At temperatures above 200°, the following reactions take place with the formation of tertiary arsines:

$$3RAsCl_2 \rightleftharpoons R_3As + 2AsCl_3$$

$$2R_2AsCl \rightleftharpoons R_3As + RAsCl_2$$

However, since the tertiary arsines are readily prepared by other methods, the reaction is more often used for the preparation of dichloro- and monochloroarsines rather than the preparation of tertiary arsines. The mechanism of the reaction has been discussed in Section III-6-A-(3).

The redistribution of diarylchloroarsines has occasionally been used for preparative purposes. Bis(*p*-aminophenyl)chloroarsine dihydrochloride, when heated in aqueous solution for 5 minutes, gives tris(*p*-aminophenyl)-arsine (155). Diethylchloroarsine rearranges in alkaline solution to give triethylarsine, bis(diethylarsenic) oxide, and arsenosoethane (156). In addition to haloarsines, arsenoso compounds also rearrange under the influence of acid to tertiary arsines. Thus α-*p*-Arsenosoanilinoacetamide, when warmed in 3% acetic acid, rearranges to tris(*p*-carbamylmethylaminophenyl)-arsine according to the equation (157):

$$3p\text{-}OAsC_6H_4NHCH_2CONH_2 \rightarrow (p\text{-}H_2NCOCH_2NHC_6H_4)_3As + As_2O_3$$

Similarly, bis(*p*-aminophenyl)arsinous acid rearranges in acid solution to give tris(*p*-aminophenyl)arsine (155).

E. The pyrolysis of quaternary arsonium salts

When quaternary arsonium salts containing a suitable leaving group are pyrolyzed, a tertiary arsine is formed according to the following equation:

$$[R_3R'As]^+X^- \xrightarrow{\Delta} R_3As + R'X$$

The reaction has been used extensively by Mann and co-workers for the preparation of a variety of heterocyclic arsenicals.

o-Phenylenebis(dimethylarsine) can be diquaternized with 1,2-dibromoethane. The resulting salt, when heated *in vacuo*, gives a mixture of the two compounds **4** and **5** (158). **4** can be diquaternized again with 1,2-dibromoethane. The resulting salt when heated in *vacuo* loses methyl bromide to give the heterocycle **6**. Ethylenebis(methylphenylarsine) can be diquaternized with 1,2-dibromoethane to give the heterocyclic compound **7**, which loses methyl bromide on heating to give 1,4-diphenyl-1,4-diarsenane (**8**) (159).

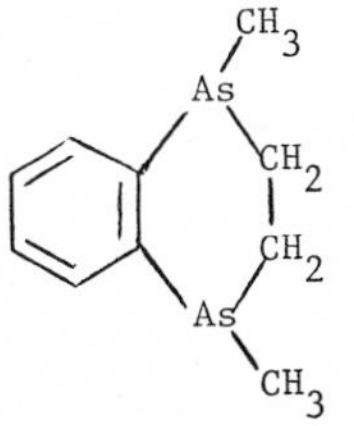

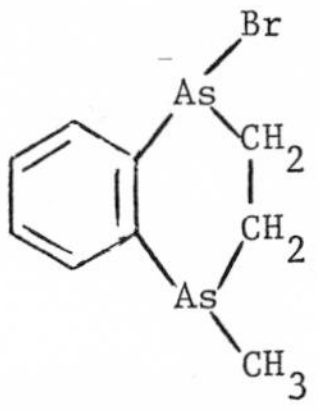

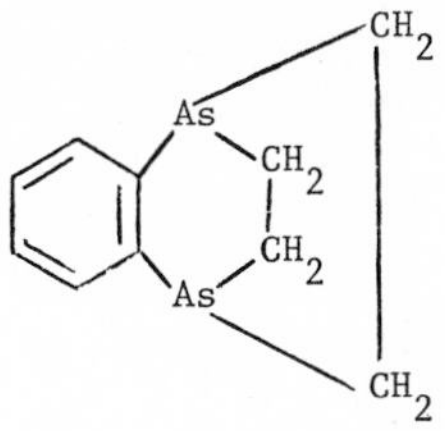

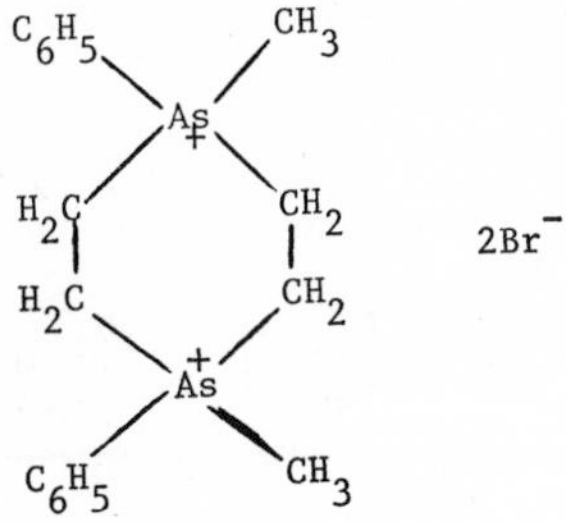

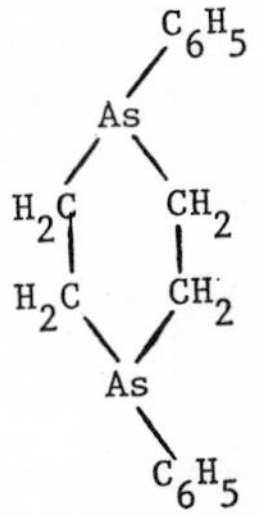

o-Phenylenebis(dimethylarsine) reacts with *o*,α,α′-dibromoxylene to give the heterocycle **9**. This compound on heating undergoes a remarkable transformation, which leads to the formation of the new heterocycle **10** (160).

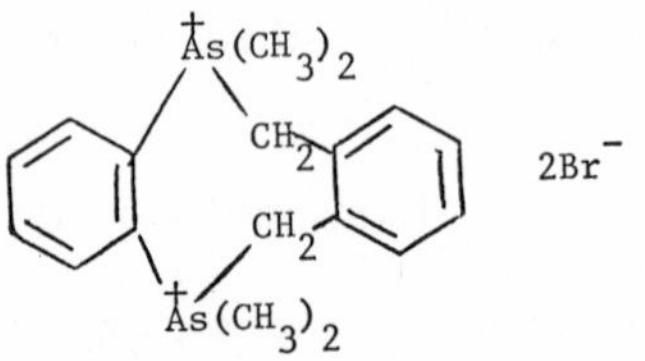

9

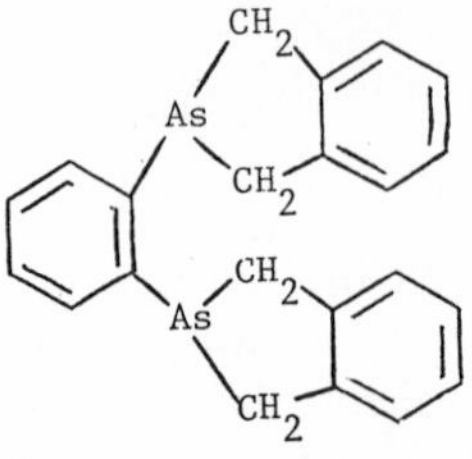

10

Several interesting heterocycles have been prepared from 5,10-dihydro-5,10-dimethylarsanthrene (**11**) (161). When **11** is treated with 1,2-dibromoethane, the expected salt does not form. Instead a compound is obtained to

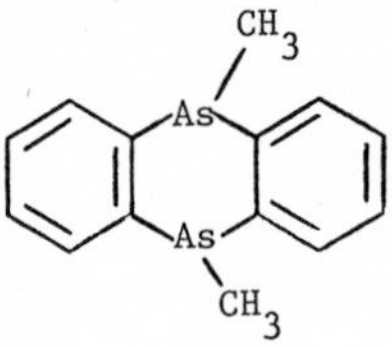

11

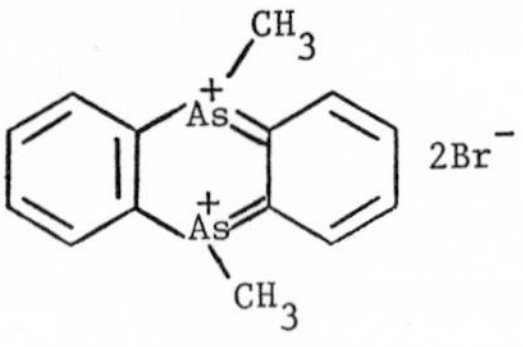

12

which Jones and Mann (161) assigned the structure **12**. This compound on pyrolysis loses one mole of methyl bromide to give **13**. The authors speculated that this compound is formed via the intermediate **14**. When the quaternization of **11** is carried out in methanol solution, either **15** or **16** is formed, depending upon the relative concentrations of the reactants; **16**, on heating, regenerates the starting arsanthren **11**.

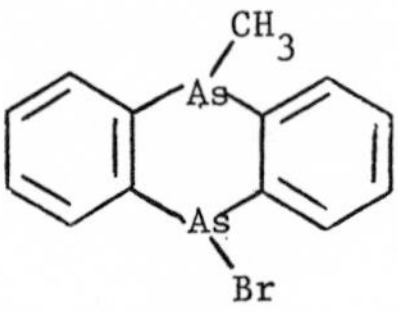

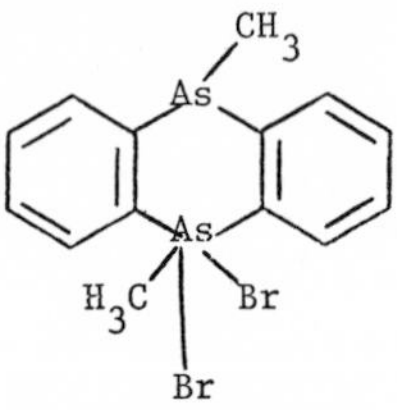

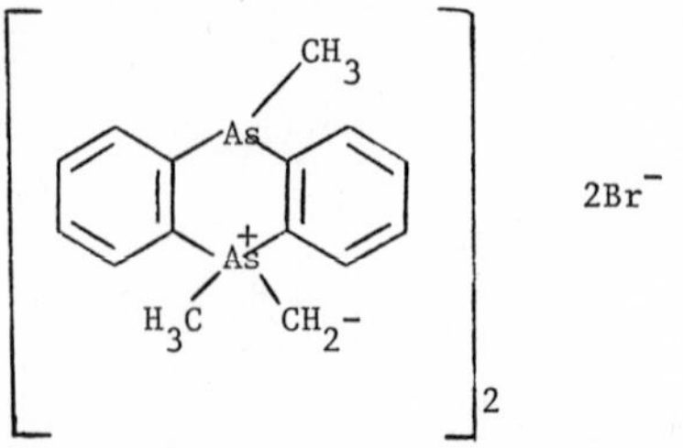

15

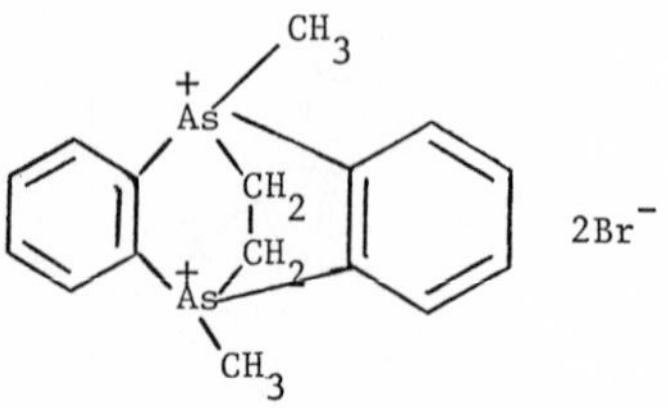

16

2,2′-Biphenylenebis(dimethylarsine) undergoes diquaternization with dibromomethane, 1,2-dibromoethane, 1,3-dibromopropane, and *o*,α,α′-dibromoxylene (103). Loss of methyl bromide from the quaternary salts obtained from dibromomethane and *o*,α,α′-dibromoxylene gives the expected heterocyclic arsines **17** and **18**, respectively. The quaternary salts ob-

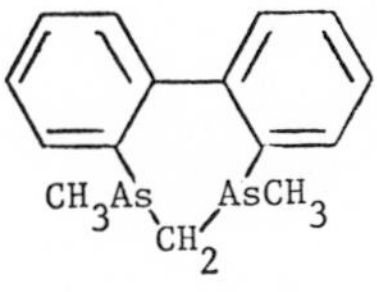

17

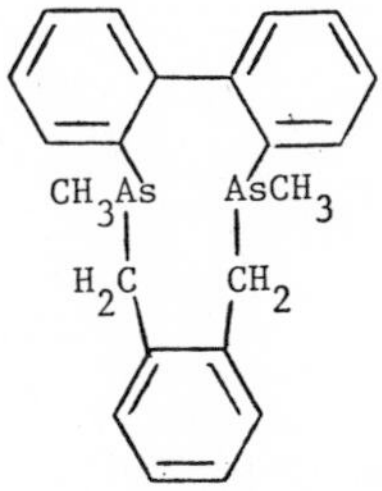

18

tained from 1,2-dibromoethane and 1,3-dibromopropane, however, undergo ring contraction on heating with the formation of 5-methyldibenzarsole (**19**).

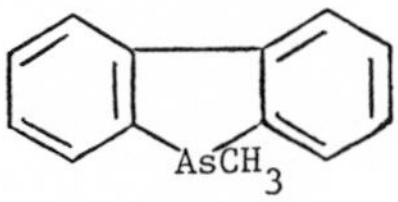

19

The decomposition of a quaternary arsonium salt has also been used to prepare the interesting spirocyclic arsine, hexahydro-1*H*-arsolo[1,2-*a*]-arsole (**20**) (162). The arsonium salt $[Me_3AsCH(CH_2CH_2CH_2OEt)_2]Br$

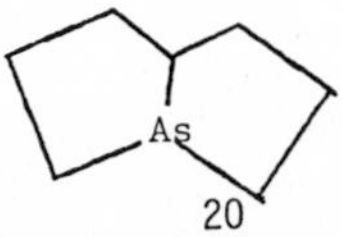

20

was first prepared and then treated with hydrobromic acid to yield $[Me_3AsCH(CH_2CH_2CH_2Br)_2]Br$. This bromo compound was not isolated, but on heating to 200° it lost three moles of methyl bromide to form the spirocyclic arsine **20**.

Mann and co-workers (163) have recently investigated the pyrolysis of arsonium hydroxides as a means of preparing heterocyclic tertiary arsines. The quaternary arsonium hydroxide **21** readily lost methanol on heating to give **17**, identical with the compound obtained by pyrolysis of the dibromide. Pyrolysis of the dihydroxide **22**, however, gave **23**.

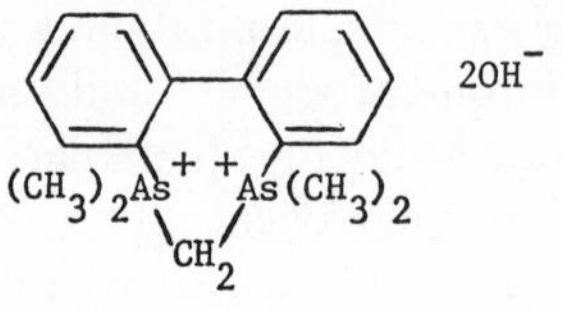

21

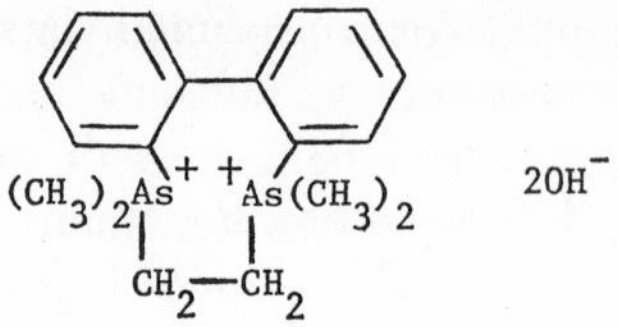

22

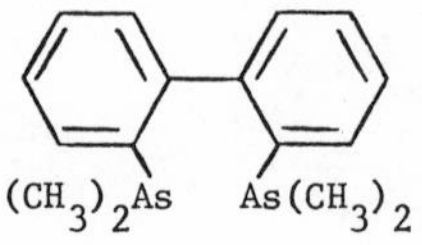

23

Jones and Mann (108) have investigated the pyrolysis of mixed phosphonium arsonium halides. (*o*-Diethylphosphinophenyl)diethylarsine, when diquaternized with ethylene, trimethylene, or *o*-xylene dibromides, gives the interesting heterocyclic salts **24**, **25**, and **26**, respectively. Pyrolysis of **24** and

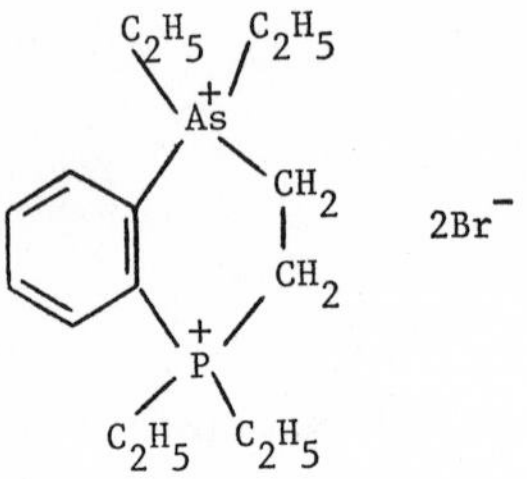

24

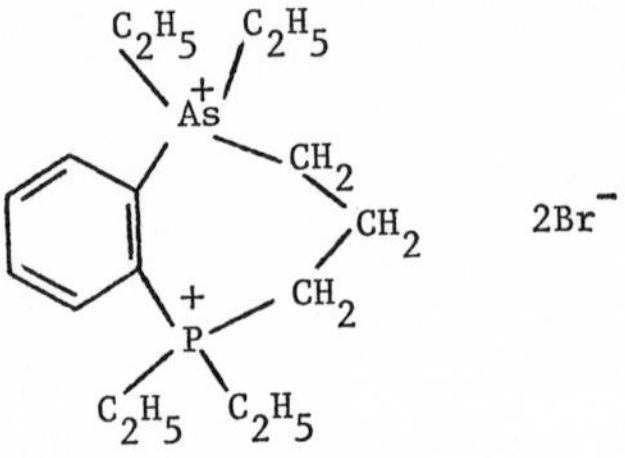

25

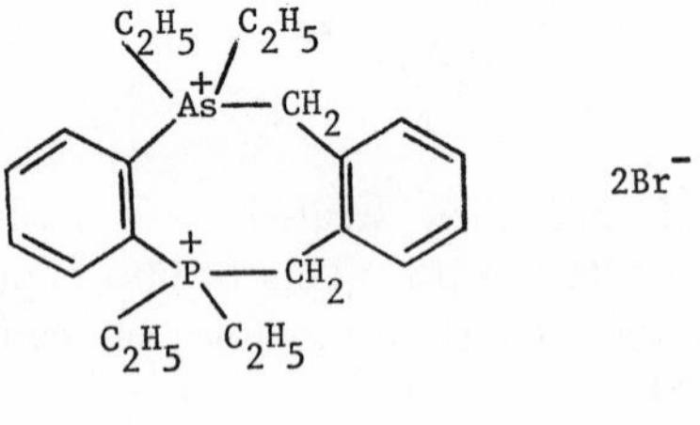

26

26 resulted in loss of one molecule of ethyl bromide from the arsenic atom with the formation of the corresponding arsine. All attempts to bring about loss of ethyl bromide from the phosphonium salt by further heating resulted in extensive decomposition of the molecule.

Thornton (164) has also prepared heterocyclic arsines by the thermal decomposition of arsonium salts.

F. By the Diazo reaction

The formation of a carbon–metal bond by the decomposition of a diazonium "double salt" in the presence of a metal was first described by Nesmeyanov and is frequently referred to as the Nesmeyanov reaction. The reaction was first used for the preparation of organic mercurials by decomposing a diazonium halomercurate in acetone with copper powder (165, 166). The reaction was later extended to the preparation of organometallic compounds of germanium, tin, lead, bismuth, thallium, and arsenic. The results of Nesmeyanov and of other workers have been summarized by Nesmeyanov, Makarova, and Tolstaya (167). Reutov and Bundel (168, 169) decomposed $[ArN_2]$ $[FeCl_4]$ and $[ArN_2]_2$ $[ZnCl_4]$ with powdered iron in the presence of either arsenic trichloride or phenyldichloroarsine and obtained mixtures of compounds of the type $ArAsX_2$, Ar_2AsX, Ar_3As, and Ar_3AsX_2. If zinc metal was used for the decomposition, no Ar_3AsX_2 was formed.

In 1937 Waters (170) published the first of a series of papers dealing with the decomposition of diazo compounds suspended in an organic solvent and kept neutral by the addition of calcium carbonate. Waters concluded that the products of the decomposition were best explained by homolytic decomposition of the diazo compound. In the presence of a metal, organometallic compounds are formed. With mercury, phenylmercuric chloride is obtained; with antimony, triphenylantimony dichloride is obtained (171). When powdered arsenic is used, compounds of the type $Ar_3As(OH)OAr$ are produced (172–174).

The method was later modified by Hanby and Waters (175). By the addition of powdered zinc to a mixture of aryldiazonium chlorozincate and arsenic trichloride in acetone, they were able to prepare triarylarsines. If phenyldichloroarsine was used rather than arsenic trichloride, mixed triarylarsines could be prepared. Hanby and Waters believed that the zinc reduces the arsenic trichloride to free arsenic which is then attacked by aryl free radicals. Waters (176) also suggested that the Nesmeyanov reaction in general involves reduction of a metal halide to a free metal.

There has been considerable controversy concerning the mechanism of the Nesmeyanov reaction and of Waters' modification of this reaction. There can be but little doubt that the original conditions used by Waters, wherein diazo compounds decompose in neutral solution with attack on powdered metals, involve free radicals. The evidence for this mechanism is extremely convincing. This does not mean, however, that the Nesmeyanov reaction and the modification used by Hanby and Waters for the production of triarylarsines necessarily involves homolytic decomposition of a diazo compound.

Nesmeyanov and co-workers (167) have suggested that the homolytic decomposition of diazonium chlorides and similar salts brought about by the action of a powdered metal transfers a free radical directly from the diazonium cation to the metal or metalloid of the anion. When a diazonium fluoborate or iodonium fluoborate is used, however, a fluorine atom would be formed by homolytic decomposition, a result which Nesmeyanov and co-workers (167, 177, 178) believe to be improbable. In support of these contentions, they demonstrated that the products of the decomposition of diazonium chlorides and diazonium fluoborates (or the corresponding iodonium salts) in organic solvents were quite different. Thus, the decomposition of benzeneiodonium chloride gives phenylmercuric chloride in the presence of mercury, but pyridine is not attacked. By contrast, when benzeneiodonium fluoborate decomposes in the presence of pyridine, N-phenylpyridinium fluoborate is obtained.

Bruker (179, 180) considers that the first step in the reaction is the transfer of the anion of the diazonium salt to the metal or metalloid halide; this new salt then rearranges to a covalent diazo compound. There follows an intramolecular electrophilic attack of the metal on the phenyl group with the liberation of nitrogen. Bruker's mechanism is summarized in the following equations:

$$[ArN_2]^+X^- + MX_3 \rightarrow [ArN_2]^+[MX_4]^- \rightarrow$$

$$[ArN{=}N{-}MX_3]^+X^- \rightarrow ArMX_4 + N_2$$

In the absence of a free metal, Bruker believes that the reaction stops with the $ArMX_4$ product. When a metal is present, however, $ArMX_4$ is reduced to $ArMX_2$, which can then react further with the diazonium salt. Thus, the final product of the reaction in the presence of a metal is Ar_3M. It should be pointed out, however, that diarylarsinic acids and triarylarsine oxides are formed, even in the absence of a free metal (181, 182). No experimental evidence for the suggested mechanism is offered by Bruker, and the intramolecular electrophilic reaction seems unlikely.

The best experimental study of the Nesmeyanov-type reaction has been done by Reutov (183). He pointed out that the mechanism suggested by Waters is improbable for several reasons. Thus, mercuric chloride will form phenylmercuric chloride just as well when silver is used as when copper is used, but silver cannot reduce mercuric chloride to metallic mercury. Nor can copper reduce tin or lead tetrachlorides to the metal, even though organometallic compounds are formed from these metallic halides. Reutov further pointed out that the reaction proceeds with aryldichlorostibines as well as with antimony trichloride and that the aryldichlorostibine could not be reduced to metallic antimony. It might also be mentioned that Hanby and Waters (175) obtained mixed triarylarsines from an aryldichloroarsine, a diazonium chlorozincate, and zinc metal. Further, the present authors (184) have been unable to reduce arsenic trichloride to metallic arsenic with zinc dust in acetone or ethyl acetate. Reutov also made a kinetic study of the following reaction:

$$[XC_6H_4SbCl_4][YC_6H_4N_2Cl] + Fe \rightarrow XC_6H_4(YC_6H_4)SbCl_3 + N_2 + FeCl_2$$

He demonstrated that, if Y is H and X is varied, the rate of the reaction increases in the following order:

$$NO_2 < Cl < H < CH_3 < C_2H_5O$$

If X is H and Y is varied, the reaction rate increases in the order $C_2H_5O < CH_3 < H$. Where Y is NO_2 or $COOC_2H_5$, the double salt was largely dissociated, and the products of the reaction were in part those of the decomposition of the diazonium halide.

Reutov pointed out that the results of this study can only be interpreted in terms of a heterolytic decomposition of the diazonium double salt. He concluded further that the Hanby-Waters reaction proceeds by means of a heterolytic decomposition rather than by a free radical mechanism.

G. By the cyclization of chloroarsines

Just as certain dichloroarsines can be cyclized to heterocyclic chloroarsines by cyclodehydrohalogenation (cf. Section III-6-A-(15)), so can secondary chloroarsines be cyclized to heterocyclic tertiary arsines. The reaction in this case, however, is not as successful and has not been used to any extent. Turner and Bury (185) have cyclized methyl(2-phenylethyl)chloroarsine by heating with aluminum trichloride in carbon disulfide solution. Similarly, methyl-(3-phenylpropyl)chloroarsine can be cyclized to give 1-methyl-1,2,3,4-tetrahydroarsinoline (186). Cookson and Mann (150) have cyclized (2-biphenylylmethyl)phenylchloroarsine to give 5-phenyl-5,6-dihydroarsanthridine (**3**.)

Although the above cyclizations were successful, methyl(*o*-phenoxyphenyl)-chloroarsine could not be cyclized by this procedure (187). Jones and Mann

$As-C_6H_5$

27

(188) readily prepared 1-phenylarsindoline (**27**) by cyclodehydration of phenylethylarsinic acid and reduction of the resulting arsine oxide with sulfur dioxide. They showed that, if the phenylethylarsinic acid is first reduced to the chloroarsine and then cyclized by use of aluminum chloride, the yield is cut by one-half.

H. By miscellaneous methods

Tertiary arsines, usually admixed with halo- and dihaloarsines, have been obtained by the direct arsenation of suitably activated aromatic compounds, by the addition of arsenic halides to unsaturated compounds, by the reaction of arsenic halides with diazoalkanes, and by the reaction of arsenic metal with free radicals. All of these methods have been described in the section dealing with the dihalo- and monohaloarsines (cf. Section III-6-A-(5), (7), (8), and (14)). Tertiary arsines are also obtained as a by-product of the Cadet reaction (Section III-1-F) and by the reaction between phenylhydrazine and benzenearsonic acid, diphenylarsinic acid, or arsenic acid (Section III-1-B). A number of tertiary arsines have been prepared from primary and secondary arsines—either by addition to unsaturated compounds or by the displacement of halogen. These reactions are discussed in section IV-2. The reduction of quaternary arsonium salts to tertiary arsines is mentioned in Section V-9-B. Arseno compounds and diarsines can also be converted to tertiary arsines; these reactions are discussed in Sections VI-2 and VI-5, respectively. Tertiary arsines have been prepared by the reaction of alkyl halides with sodium arsenide, but this procedure has been little used (189).

Dimethyl(trifluoromethyl)arsine can be prepared by the reaction between trimethylarsine and trifluoroiodomethane at room temperature according to the equation (190):

$$2(CH_3)_3As + CF_3I \rightleftharpoons CF_3As(CH_3)_2 + [(CH_3)_4As]^+I^-$$

It is necessary to remove the arsonium salt in order to drive the reaction to the right, and only one methyl group can be replaced by this reaction. At higher temperatures methyl iodide reacts with trifluoromethyl-substituted tertiary arsines to replace a trifluoromethyl group (191):

$$(CF_3)_2AsCH_3 + CH_3I \rightarrow CF_3As(CH_3)_2 + CF_3I$$

(Perfluoroalkyl)dimethylarsines have also been prepared by the reaction of perfluoroalkyl iodides with cacodyl disulfide or (ethylthio)dimethylarsine (192). Tris(trifluoromethyl)arsine has been obtained by the pyrolysis of (methylamino)bis(trifluoromethyl)arsine (193) or bis[bis(trifluoromethyl)arsenic] sulfide (192).

It has long been known that molds, growing on the paste used to fasten wallpaper, may produce toxic gaseous arsenicals (Gosio-gas) when the wallpaper contains arsenic compounds. Earlier workers were unable to identify the volatile arsenicals produced, although diethylarsine was suspected. In 1933 Challenger and co-workers launched an investigation of this problem. The earlier papers (194, 195) in the series contain excellent reviews of the work of previous investigators. In their earlier work Challenger and co-workers (194) used *Penicillium brevicaule* (*Scopulariopsis brevicaulis*) cultivated on bread crumbs that contained arsenic trioxide. The gas which was produced was aspirated through mercuric chloride solution to give two compounds, $(CH_3)_3As{\cdot}2HgCl_2$ and $(CH_3)_3As{\cdot}HgCl_2$. When the gas was aspirated through nitric acid solution, the hydroxynitrate $(CH_3)_3As(OH)NO_3$ was produced. The action of the mold on bread that contained disodium methanearsonate or sodium cacodylate also produced trimethylarsine. When disodium ethanearsonate was used, ethyldimethylarsine was produced. With sodium selenate or sodium selenite, dimethyl selenide was produced, but no gaseous products were produced when sulfur, phosphorus, bismuth, lead, or mercury compounds were used (195). Later work (196, 197, 198), however, demonstrated that when sulfur-containing amino acids of the type $RSCH_2CH(NH_2)COOH$ were used, mercaptans and sulfides were produced. When potassium tellurate was used, dimethyl telluride was produced (199).

Finally, radioactive tracers were used to investigate the biological mechanism of the methylation of the arsenic (200). A number of different substances, all labelled with carbon-14, were added to the culture medium. Only with labelled methionine was a labelled product produced. It was concluded that the mold contains an enzyme that transfers the methyl group directly from methionine to the arsenic, selenium, or sulfur.

Kamai and Chernokal'skiĭ (201) have prepared triethyl-, tripropyl-, and butyldimethylarsines by the pyrolysis of the corresponding arsine oxides. Ethyl and propyl alcohols were obtained as by-products of the first two reactions; butyl butylmethylarsinite was obtained as a by-product of the last reaction. The mechanism of the reaction is not known.

Auger (202) obtained trimethylarsine together with metallic arsenic by the pyrolysis of arsenomethane. In a similar manner the pyrolysis of cacodyl gives trimethylarsine (203). Nesmeyanov and Ippolitov (204) have obtained triphenylarsine, admixed with dichlorophenyl- and chlorodiphenylarsines, by the irradiation of a mixture of arsenic trichloride and benzene with neutrons.

Arsine oxides can be readily reduced in acid solution to tertiary arsines in a manner similar to the reduction of arsonic and arsinic acids. Thus, Jones and Mann (205) reduced 5-phenyl-10(5*H*)-acridarsinone 5-oxide (**28**) to the

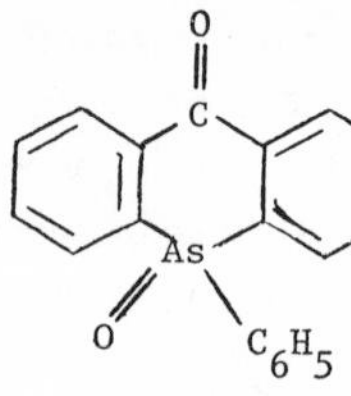

28

corresponding tertiary arsine with sulfur dioxide in hydrochloric acid solution. Methylbis(3-nitro-4-bromophenyl)arsine oxide and tris(4-bromophenyl)-arsine oxide have been reduced to tertiary arsines with hypophosphorous acid and hydriodic acid in acetic acid solution (206). In one case (73), an arsine oxide has been reduced to a tertiary arsine by means of Raney nickel and hydrogen at 1000 psi. Since arsine oxides are usually prepared from the arsines, conversion of oxides to tertiary arsines is not of great preparative importance.

The interesting heterocyclic tertiary arsine 2,3,5,6,7,8-hexakis(trifluoromethyl)-1,4-diarsabicyclo[2,2,2]octa-2,5,7-triene (**29**) has been prepared by Krespan (207) by the reaction of metallic arsenic with 2,3-diiodohexa-

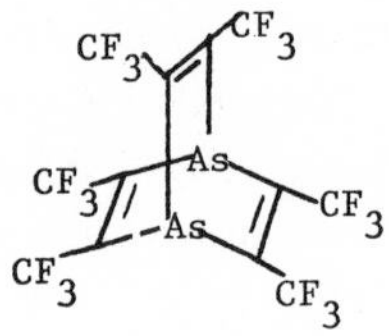

29

fluoro-2-butene at 200° for 10 hours. The phosphorus analog is formed when red phosphorus is heated with hexafluoro-2-butyne and a catalytic amount of iodine. Evidence for the bicyclo structure for these arsenic and phosphorus compounds is based on analytical and spectral data. Molecular weights, however, have not been determined.

Dimethyl(trifluoromethyl)arsine has been prepared in 23% yield by the decarboxylation of dimethyl(trifluoroacetoxy)arsine at 205° (208). The acetoxy compound was prepared by the reaction of chlorodimethylarsine and silver trifluoroacetate according to the reaction:

$$(CH_3)_2AsCl + AgO_2CCF_3 \rightarrow (CH_3)_2AsO_2CCF_3 + AgCl$$

The reaction of ketene with the amide $(CH_3CH_2CH_2)_2AsN(C_2H_5)_2$ has been found to yield $(CH_3CH_2CH_2)_2AsCH_2CON(C_2H_5)_2$ (118). Arsine adds to formaldehyde in aqueous solution in the presence of a catalyst to give tris(hydroxymethyl)arsine (209).

5. REACTIONS OF TERTIARY ARSINES

A. With electrophilic reagents

Although tertiary arsines are much weaker bases than tertiary amines (210), they do demonstrate nucleophilic properties in many of their reactions. Gillespie and Robinson (211) could not detect $(C_6H_5)_3As^+H$ when triphenylarsine was dissolved in sulfuric acid, but Peach and Waddington (212) were able to detect this ion in liquid hydrogen chloride by means of conductivity measurements. Their results indicate that triphenylarsine behaves as a strong base when dissolved in liquid hydrogen chloride:

$$(C_6H_5)_3As + 2HCl \rightarrow (C_6H_5)_3AsH^+ + HCl_2^-$$

Conductimetric titration of this solution with boron trichloride gave a sharp break at a 1:1 molar ratio. The following reaction was inferred:

$$(C_6H_5)_3AsH^+ + BCl_3 + HCl_2^- \rightarrow (C_6H_5)_3AsH^+BCl_4^- + HCl$$

The tetrachloroborate formed in this way was found to lose hydrogen chloride at about −25° to yield the stable adduct $(C_6H_5)_3As{\cdot}BCl_3$. Peach and Waddington also found that triphenylarsine reacts with methanesulfonic acid in liquid hydrogen chloride to give a solid product, the infrared spectrum of which indicated that the methanesulfonate ion was present.

Triphenylarsine reacts with sulfur trioxide to form an adduct, $(C_6H_5)_3As{\cdot}SO_3$ (213). At room temperature this substance decomposes in about twelve hours to triphenylarsine oxide and sulfur dioxide. By contrast tricyclohexylarsine is oxidized by sulfur trioxide to form the sulfate $(C_6H_{11})_3AsSO_4$. The same compound can be obtained by the addition of sulfur trioxide to tricyclohexylarsine oxide. Hydrolysis of the sulfate yields the hydroxy bisulfate $(C_6H_{11})_3As(OH)(OSO_3H)$.

Trichloroiodomethane reacts with trimethylarsine and with triphenylarsine to give low-melting adducts of the type $R_3As{\cdot}CCl_3I$ (214). These form nonconducting solutions in nitrobenzene and probably contain an arsenic–iodine bond, since carbon tetrachloride fails to form an analogous adduct with tertiary arsines.

Aromatic tertiary arsines can be oxidized to triarylarsine oxides with suitable oxidizing agents. Thus, triphenylarsine and selenium dioxide give triphenylarsine oxide, but the product is contaminated with triphenylarsine selenide (215). Better results are obtained if potassium permanganate (136, 206), hydrogen peroxide (127, 149), or phenylbenzenethiolsulfinate (216) is used. The oxidation of triarylarsines to free radicals has been studied by several groups of investigators (134, 217).

The oxidation of tertiary alkyl arsines is more complicated than the oxidation of aromatic arsines. Early workers in the field claimed to have obtained trialkylarsine oxides by the air oxidation of trialkylarsines (189, 218). Their products, however, were usually not pure, and modern work on this subject would indicate that the arsine oxides were not obtained (cf. Section IV-5-B). Merijanian and Zingaro (219) have recently shown that trialkylarsine oxides can be prepared by the oxidation of trialkylarsines with hydrogen peroxide or mercuric oxide.

In contrast to the reaction with molecular oxygen, there seems to be very little doubt that trialkylarsines react readily with sulfur to give trialkylarsine sulfides. Earlier work in this field (189, 218) has been confirmed by some recent work, in which trimethyl-, triethyl-, and tricyclohexylarsine sulfides were prepared from the corresponding arsine and sulfur (220). In another recent paper Horner and Fuchs (221) have shown that optically active tertiary arsines of the type RR′R″As react with sulfur in benzene to give optically active arsine sulfides. Trimethylarsine selenide has been reported (222) as an unstable compound, formed by the reaction between trimethylarsine and selenium. Recently, Zingaro and Merijanian (223) have prepared and characterized nine trialkylarsine selenides. These were obtained by refluxing the corresponding arsines with selenium powder for several hours. Triphenylarsine selenide, however, could not be prepared in this manner.

Tertiary alkylarsines are powerful reducing agents. Thus, trimethylarsine reduces both phosphorus and antimony pentachlorides to the corresponding trichlorides with the concomitant formation of trimethylarsenic dichloride (224).

Tertiary arsines can be readily converted to imides (Section V-4), dihalides (Section V-7), compounds of the type R_3AsXY (Section V-8), and arsonium compounds (Section V-9-A). They also react with sulfur monochloride to yield adducts of unknown structure (Section V-1-E). The use of tertiary arsines as ligands for transition metals is discussed in Section IV-5-C. The reaction of tertiary arsines with diborane, aluminum chloride, and several other Lewis acids not containing transition metals is discussed in Section VI-7. The formation of an unstable adduct $(Ph_3AsO)_2 \cdot H_2O_2$ by the reaction of triphenylarsine with hydrogen peroxide is mentioned in Section V-1-A. The direct conversion of triphenylarsine to an ylid is described in Section V-9-B.

B. Reactions in which a carbon–arsenic bond is ruptured

Ayscough and Emeléus (225) have studied the pyrolysis of trimethylarsine, tris(trifluoromethyl)arsine, and tris(pentafluoroethyl)arsine. The pyrolyses are homogeneous and first order. The principal product from trimethylarsine is methane; smaller amounts of other hydrocarbons together with hydrogen are formed. Tris(trifluoromethyl)arsine gives principally hexafluoroethane. It is suggested that methyl or trifluoromethyl radicals are first formed by rupture of the C—As bond and that the final products are those formed by reaction of these radicals. The activation energies for the decomposition of trimethyl- and tris(trifluoromethyl)arsines were 54.6 and 57.4 kcal/mole, respectively. If it is assumed that the activation energies of subsequent reactions are small, these values represent the approximate dissociation energy of the carbon–arsenic bond (cf. Section IV-6-C).

Unlike triarylarsines, trialkylarsines are not oxidized in the air to arsine oxides (220). When a solution of triethylarsine in ethyl acetate is allowed to stand in the air, diethylarsinic acid crystallizes from the solution (107). Butyldiethylarsine under similar conditions gives butylethylarsinic acid. Tri-α-furylarsine, when oxidized in alkaline solution, gives di-α-furylarsinic acid (226). Dimethylphenylarsine, after standing for several years, developed a crystalline deposit, which was identified as methylphenylarsinic acid (227). The reaction was written as:

$$C_6H_5As(CH_3)_2 + O_2 + H_2O \rightarrow C_6H_5(CH_3)AsO_2H + CH_3OH$$

By contrast methyldiphenylarsine was easily oxidized to methyldiphenylarsine oxide. The mechanism of these interesting degradations has not been worked out.

Introduction of the trifluoromethyl group greatly weakens the strength of the carbon–arsenic bond (16, 190, 228, 229). Tris(trifluoromethyl)arsine is rapidly hydrolyzed by alkali to fluoroform (228). Methylbis(trifluoromethyl)arsine requires longer heating, and the hydrolysis does not go to completion with 20% aqueous sodium hydroxide at 100° (16), while dimethyl(trifluoromethyl)arsine is only 9% hydrolyzed in 20% sodium hydroxide in 3 days (190). Diphenyl(trifluoromethyl)- and phenylbis(trifluoromethyl)-arsines are both hydrolyzed by hot alcoholic alkali at 85° (229). In contrast to the trifluoromethylarsines, tris(perfluorovinyl)arsine is stable to alkaline hydrolysis (230). The reaction between tris(trifluoromethyl)arsine and N-chlorobis(trifluoromethyl)amine leads to the elimination of trifluoromethyl chloride and the formation of the amides, $(CF_3)_2AsN(CF_3)_2$ and $CF_3As[N(CF_3)_2]_2$ (231).

The ethynyl group behaves in a similar manner to the trifluoromethyl group and renders the carbon–arsenic bond more susceptible to alkaline hydrolysis. Thus, Hartmann and co-workers (86, 87) have shown that such tertiary arsines as $As(C{\equiv}CC_6H_5)_3$, $(C_6H_5)_2AsC{\equiv}CC_6H_5$, and $(\alpha\text{-}C_{10}H_7)_2AsC{\equiv}CH$ are hydrolyzed to phenylacetylene (or acetylene in the last example) in alkaline solution.

Although triarylarsines are more stable than trialkylarsines, the carbon–arsenic bond in triarylarsines can be cleaved under certain reaction conditions. Triphenylarsine in tetrahydrofuran is cleaved by metallic lithium to give lithium diphenylarsenide (232). Similarly, potassium diphenylarsenide can be prepared by heating potassium with triphenylarsine in dioxane for several hours (13). Hewertson and Watson (233) have shown that triphenylarsine is cleaved rapidly with sodium in liquid ammonia. Addition of 1,2-dichloroethane to the solution results in the formation of 1,2-bis(diphenylarsino)ethane in good yield. Other examples of the reaction of organic halides with alkali arsenides are discussed in Section IV-4-C.

Jackson and Sasse (234) have studied the hydrogenolysis of triphenylarsine, -stibine, and -bismuth with W-7 Raney nickel; when boiled with the catalyst in methanol solution, all of these compounds were cleaved to yield benzene in yields of approximately 90%. The authors suggested that the compounds are chemiabsorbed by the catalyst through the heteroatoms. Triphenylarsine, heated in xylene to 60 atmospheres in the presence of hydrogen, gives benzene and metallic arsenic (235). In a nitrogen rather than in a hydrogen atmosphere, biphenyl and metallic arsenic are formed.

Dessy and Pohl (236) have examined the electrochemical behavior of pentaphenylarsole and a large number of related compounds. When one electron per molecule of arsole was added, a blue radical anion with a half-life of one minute was formed. Addition of more than one electron per molecule resulted in decomposition of the molecule.

Goddard (237) has found that certain trixylylarsines react with thallic chloride with cleavage of a carbon–arsenic bond. Thus, tris(2,5-dimethylphenyl)- and tris(2,4-dimethylphenyl)arsines when boiled with thallic chloride in xylene give the corresponding xylyldichloroarsines and thallous chloride.

Tri-α-furylarsine, when treated with mercuric chloride, yields 2-chloromercurifuran, while iodination gives 2-iodofuran (115). Similarly, tris(5-chlorofuryl)arsine and tris(5-bromofuryl)arsine when treated with mercuric chloride give 5-chloro- and 5-bromo-2-chloromercurifuran, respectively (238). These reactions demonstrate the relative weakness of the carbon–arsenic bond in the furan arsenicals, since the carbon–arsenic bond is not cleaved in triphenylarsine under the same conditions. α-Furyldichloro- and di-α-furylchloroarsine are even less stable than the tertiary arsine, since the carbon–arsenic bond is cleaved in these compounds by either boiling water or alcohol, whereas tri-α-furylarsine can be distilled with steam.

Chlorination of α-furyldichloroarsine, di-α-furylchloroarsine, or tri-α-furylarsine leads to rupture of the carbon–arsenic bond in all cases and the formation of 2-chlorofuran tetrachloride (238). In the case of di-α-furylchloroarsine and tri-α-furylarsine, however, it is possible to demonstrate that the arsenic is oxidized prior to rupture of the carbon–arsenic bond. Thus, from di-α-furylchloroarsine an 8% yield of di-α-furylarsinic acid can be obtained by hydrolysis of the reaction product obtained after partial chlorination of the chloroarsine. Similarly, tri-α-furylarsine dichloride was obtained in 1.7% yield by partial chlorination of tri-α-furylarsine. Attempts to obtain α-furanarsonic acid by partial chlorination of α-furyldichloroarsine and subsequent hydrolysis of the reaction product were unsuccessful.

Compounds in which two arsenic atoms are attached to a furan nucleus appear to be even less stable (115). Thus, when 2,5-bis(chloromercuri)furan was treated with arsenic trichloride, mercuric chloride was formed, but no arsenical could be isolated from the reaction mixture. It is possible, however, that a compound containing the carbon–arsenic bond was formed in this reaction, since treatment of the reaction mixture with iodine gave 2,5-diiodofuran.

The conversion of tertiary arsines to halo- and dihaloarsines has been discussed in Section III-6-A-(3). The reaction of a tertiary arsine with an alkyl halide to form a different tertiary arsine has been discussd in Section IV-4-H. Sisler and Stratton (239) have reported that the reaction of triphenylarsine with chlorine or with mixtures of chloramine and ammonia leads to partial cleavage of the C—As bond. These reactions are discussed in connection with triarylarsenic dihalides (Section V-7) and compounds of the type R_3AsXY (Section V-8).

C. Tertiary arsines as ligands in coordination chemistry

Tertiary arsines, which contain a lone pair of electrons, will form strong σ bonds with a large number of transition metals. For this reason arsines have been used extensively as ligands in coordination chemistry, and a tremendous number of papers have appeared on this subject. The present discussion will be largely limited to the donor properties of the arsines themselves. For the chemistry and properties of the coordination compounds, reviews on the subject should be consulted (240–242).

The ability of tertiary arsines to coordinate with metallic salts was known long before the development of modern coordination chemistry. Thus, Cahours and Gal (243) in 1870 prepared triethylarsine complexes of platinum, palladium, and gold chlorides, while Dehn and Williams (203) in 1908 used trimethylarsine as a ligand for mercuric chloride. In recent years, the use of tertiary arsines has been widely extended, and a number of interesting arsines have been synthesized specifically for use as ligands.

The trialkylarsines have not been used as extensively as triaryl- or mixed alkylarylarsines. Jensen (244) investigated the complexes of platinum(II) with a large series of trialkyl phosphines, arsines, and stibines. Both *cis* and *trans* planar compounds of the type $(R_3As)_2PtX_2$ were obtained, where R was an alkyl group and X a halogen, nitrate, or other anion. Chatt and co-workers (245, 246) have found that the reduction of *cis*-dichlorobis-(triethylarsine)platinum(II) with hydrazine hydrate yields *trans*-chlorohydrido-bis(triethylarsine)platinum(II), $(Et_3As)_2PtHCl$. The infrared spectrum contains a strong band at 2174 cm^{-1}, which has been assigned to Pt–H stretching. The compound shows considerable resistance to thermal decomposition, oxidation, and hydrolysis. Chatt (247) has also prepared a series of bridged platinum compounds of the following type:

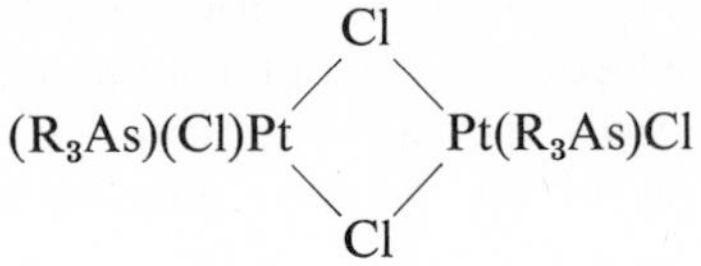

```
                 Cl
               /    \
(R3As)(Cl)Pt          Pt(R3As)Cl
               \    /
                 Cl
```

By means of X-ray diffraction studies, these compounds were shown to have the *trans* configuration (247, 248). Octahedral complexes of Pt(IV) of the type $[(C_2H_5)_3As]_2PtX_4$ have also been prepared (249). A number of bis-(trialkylarsine)palladium(II) dihalides (250) and bridged derivatives of trimethylarsine with palladium(II) halides (251) have been prepared.

Other coordination compounds prepared from trialkylarsines include complexes of cadmium (252), cobalt (253), copper (254, 255), gold (255, 256), iridium (257), manganese (258), mercury (252), molybdenum (259), nickel (259), rhodium (260), silver (254, 256), and zinc (253).

Triarylarsines and mixed alkylarylarsines of the types Ar_2AsR and $ArAsR_2$ have been used as ligands by a large number of investigators. Nyholm, Dwyer, and co-workers, in particular, have used these types of compounds to study the coordination chemistry of the transition elements. Among the many coordination compounds prepared by these workers or by other groups of investigators are complexes of chromium (261), cobalt (262), copper (263), gold (264), iridium (265–267), iron (268, 269), magnanese (270), mercury (271), molybdenum (259), nickel (259, 272), osmium (273), palladium (274, 275), platinum 274, 276), rhenium (277–279), rhodium (280–287), ruthenium (273, 288), silver (289, 290), and zinc (291). Nyholm (274) has stated that methyldiphenylarsine is a more satisfactory ligand for divalent platinum or palladium than either triphenylarsine or dimethylphenylarsine.

A number of tertiary arsines have been prepared specifically for use as ligands in coordination chemistry. The best known of these is *o*-phenylene-bis(dimethylarsine), first prepared by Chatt and Mann by the action of methylmagnesium bromide on *o*-phenylenebis(dichloroarsine) (292). A better method for preparing this di-tertiary arsine has been described in Section IV-4-C. This compound is a powerful bidentate chelating agent, and a large number of papers have been published on its use. It forms coordination compounds with a number of metals which are not complexed by trialkylarsines, triarylarsines, or the mixed alkylarylarsines. Metals which form complexes with *o*-phenylenebis(dimethylarsine) include cadmium (293), chromium (294), cobalt (295, 296), copper (297), gold (298), hafnium (299), iron (300, 301), manganese (270), mercury (293), molybdenum (302–304), nickel (295, 305), niobium (306, 307), osmium (308), palladium (309), platinum (310), rhenium (311), rhodium (312), ruthenium (313), silver (293), tantalum (307), technetium (314, 315), titanium (299, 316), vanadium (299), tungsten (304, 317), zinc (293, 318), and zirconium (299).

In addition to phenylenebis(dimethylarsine), a large number of other di-tertiary arsines have been used as chelating agents. These include phenylene-bis(diethylarsine) (319), ethylenebis(dibutylarsine) (320), and a variety of other compounds (321). Several bidendate chelating agents containing one tertiary arsenic atom and a second type of donor grouping have been prepared. For example, Jones and Mann (108) have described the preparation and use of the following phosphorus containing tertiary arsines:

$C_6H_4(AsMe_2)(PMe_2)$ $C_6H_4(AsEt_2)(PEt_2)$

In addition, diphenyl(*o*-diphenylarsinophenyl)phosphine (322, 323), the corresponding phosphine sulfide (322), *o*-dimethylaminophenyldimethyl-arsine (324), dimethyl-*o*-methylthiophenylarsine (325), and dimethyl-4-pentenylarsine (326) have been employed as bidentate chelating agents. The last compound forms the following coordination compound with platinum(II) bromide:

$Br_2Pt[CH_2{=}CH(CH_2)_3As(CH_3)_2]$

Another interesting bidentate chelating agent is α-picolyldimethylarsine, prepared by the reaction of α-picolyllithium with dimethyliodoarsine (101). Coordination compounds with copper, silver, palladium, platinum, and ruthenium were prepared from this arsine. Recently, Uhlig and Maaser (44) have studied 2-(2-diphenylarsinoethyl)pyridine as a ligand for compounds of cobalt, nickel, copper, and zinc. With cobalt and zinc, only the nitrogen acted as a donor; with copper both nitrogen and arsenic acted as donor atoms, while with nickel both types of complexes were formed. Bidentate chelating agents that contain arsenic were reviewed by Harris and Livingstone (327) in 1964.

Several tridentate chelating agents have been described. These include bis(3-dimethylarsinopropyl)methylarsine (328–330), bis(*o*-diphenylarsinophenyl)phenylphosphine (331), bis(*o*-diphenylarsinophenyl)phenylarsine (331–333), bis(*o*-dimethylarsinophenyl)methylarsine (334), and 1,1,1-tris-(dimethylarsinomethyl)ethane (4, 137, 335). The last compound has been found useful for preparing complexes containing copper-manganese bonds (335). Several other tri-tertiary arsines have been reported in the patent literature (336).

Only a few tetradentate arsines have been prepared (332). One of these is tris(3-dimethylarsinopropyl)arsine, $As(CH_2CH_2CH_2AsMe_2)_3$ (337). Octahedral complexes of iron, cobalt, and nickel have been formed with this arsine. A second tetradentate arsine is tris(*o*-diphenylarsinophenyl)arsine (338, 339). It forms a five-coordinate compound with platinum(II) salts of the type [(arsine)PtX]Y, where X is Cl, Br, I, or NCS, and Y is X, ClO_4, or $(C_6H_5)_4B$. Three of the arsenic atoms are approximately in the plane of the platinum atom, while the fourth arsenic atom and the X group are at the apices of a trigonal bipyramid. Palladium forms similar compounds with tris(*o*-diphenylarsinophenyl)arsine (340). With rhenium(II), both tris(*o*-diphenylarsinophenyl)arsine and the tridentate chelating agent, bis(*o*-diphenylarsinophenyl)phenylarsine, form complexes of the type ReX_2L, where X is Cl, Br, or I, and L is the ligand (341). The complexes formed by the reaction of these arsines with rhenium trihalides were found to contain oxygen. The tri-tertiary arsine yielded complexes of the type $[ReOX_3L]$, which were shown to contain Re—O bonds; the tetra-tertiary arsine gave a complex $[ReCl_2LO]Cl$, in which one arsenic atom had been oxidized to an arsine oxide. Complexes of the tetra-tertiary arsine with ruthenium (342) and rhodium (333) have also been reported. Other arsenic-containing tetradentate ligands that have been prepared include tris(*o*-diphenylarsinophenyl)phosphine (331) and tetrakis-(diphenylarsinomethyl)methane (144).

One sexadentate chelating agent, tetrakis(3-dimethylarsinopropyl)-*o*-phenylenediarsine, has been described (343). It was prepared by the reaction of *o*-phenylenebis(dichloroarsine) with the Grignard reagent obtained from (3-chloropropyl)dimethylarsine. This reagent forms complexes with iron, cobalt, nickel, palladium, and platinum, in which the six arsenic atoms probably occupy the corners of an octahedron.

In addition to the expected σ bonding between tertiary arsines and metal atoms, there is now a considerable body of evidence that d_π-d_π bonding can occur between *d* electrons of the metal and vacant *d* orbitals of the ligand. This suggestion was first advanced by Nyholm and his co-workers (344, 345). The importance of π-bonding in complexes of transition metals with phosphines, arsines, and stibines was reviewed by Booth (241) in 1964. This subject has also been considered by Ahrland, Chatt, and Davies (346) in a review of the relative coordinating affinities of group V atoms for various acceptor molecules and ions.

Although tertiary arsines normally behave as donors in coordination chemistry, a few cases are known in which arsines act as acceptors. Thus, tris(trifluoromethyl)arsine probably forms a weak 1:2 complex with pyridine (347). Solid 1:1 complexes of triphenylphosphine, -arsine, and -stibine with hexamethylbenzene have been isolated by Shaw and co-workers (348). Similar complexes with triphenylamine and triphenylbismuth could not be obtained. These results are consistent with the donation of electrons to vacant *d*-orbitals of the phosphorus, arsenic, and antimony atoms.

D. The resolution of tertiary arsines into enantiomers

Weston (349) has calculated the potential energy barrier to inversion for tertiary amines, phosphines, arsines, and stibines and concluded that arsines and stibines containing three different groups should be capable of resolution. The resolution of tertiary arsines was first achieved by Lesslie and Turner (350–354), who resolved a number of phenoxarsines of the following type:

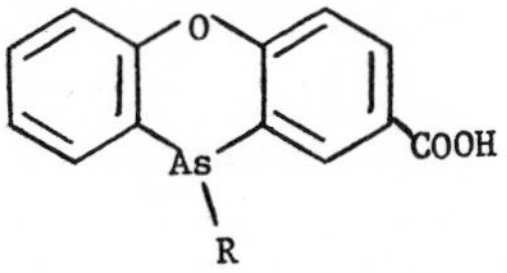

They proposed that the asymmetry in these cases should be ascribed to the fact that the two phenyl groups are folded in a "butterfly" conformation and that there is a considerable barrier to inversion by folding about the O—As axis. This concept has also been employed by Chatt and Mann (355) who prepared 5,10-di-*p*-tolyl-5,10-dihydroarsanthrene and concluded that it can theoretically exist in three isomeric forms, one *trans* and two *cis*. In one *cis* form, the *p*-tolyl groups are within the angle subtended by the two wings of the molecule; in the other *cis* form the *p*-tolyl groups are on the other side of the molecule and project toward one another. The latter form, however, was excluded for steric reasons. Chatt and Mann actually obtained two different forms of this compound with melting points of 178–179° and 179–181°, respectively. A mixed melting point of the two forms gave a value of 144–158°. Recently, Kennard, Mann, and co-workers (356) have shown by means of X-ray diffraction that the 5,10-dihydro-5,10-dimethylarsanthrene molecule is folded along the As—As axis to give a "butterfly" conformation. The two methyl groups were located within the angle (117°) subtended by the two *o*-phenylene groups. No other form of this molecule has been isolated.

Campbell and Poller (357) have resolved 9-*p*-carboxyphenyl-2-methoxy-9-arsafluorene (**30**) and 2-amino-9-phenyl-9-arsafluorene (**31**) into optically active forms. **30** was optically stable in pyridine at room temperature,

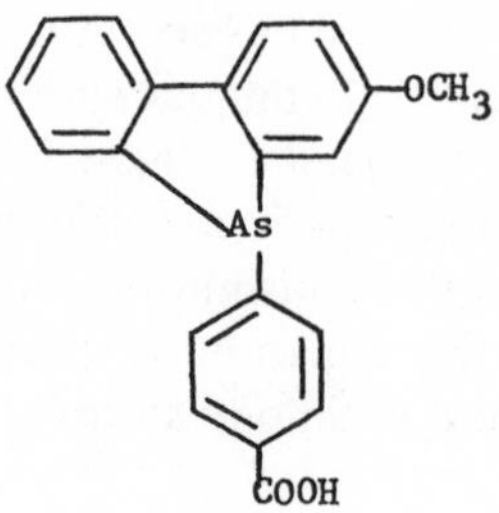

OCH_3 As COOH

30

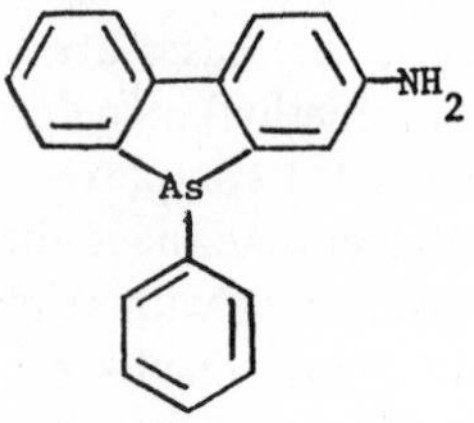

NH_2 As

31

$[\alpha]_D = \pm 160°$, but racemized when heated in ethanol–chloroform solution. **31** was even more stable, being unchanged after heating to 110° in ethanol for one hour. The authors were careful to point out that if the two *o*-phenylene rings were not planar then the optical activity could be attributed to this effect rather than to the asymmetry of the arsenic atom. However, they adduced considerable evidence, based on ultraviolet absorption spectra and other data, to show that the whole tricyclic system was essentially planar (see also Section IV-6-B).

The conception that the phenoxarsines and arsanthrenes owe their asymmetry to "butterfly" conformations has been challenged by Mislow and coworkers (358). These authors have calculated that the potential barrier to folding about the central O—As or As—As axis cannot exceed 6–7 kcal/mole, and accordingly have ascribed the optical activity of the phenoxarsines to the asymmetric arsenic atom. In support of this claim they prepared phenyl-*m*-carboxyphenyl-*p*-biphenylylarsine and resolved it by means of its amphetamine and α-phenylethylamine salts to give the two enantiomers, $[\alpha]_D = -6°$ and $[\alpha]_D = +4°$ in dioxane. These same authors also prepared 2-methyl-10-(*p*-carboxymethyl)-5,10-dihydroarsacridine, which was resolved

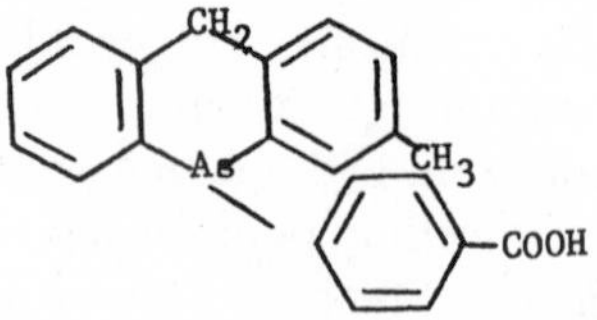

CH_2 CH_3 As COOH

into its enantiomers, $[\alpha]_D = -65°$ and $[\alpha]_D = +62°$ in dioxane. This compound was prepared and resolved in order to demonstrate that the resolution of phenoxarsines should not be attributed to some special function of the ring oxygen atom in these compounds.

Recent work by Horner and co-workers has also established the pyramidal stability of the trivalent arsenic atom. Horner and Fuchs (221) first resolved several quaternary arsonium salts containing four different organic groups. Reduction of the quaternary arsonium salts at a mercury cathode resulted in elimination of one of the organic groups and the formation of an optically active tertiary arsine. Thus, (+)benzylethylmethylphenylarsonium perchlorate was reduced to ethylmethylphenylarsine, $[\alpha]_D^{20} = +1.9°$; and (+)benzyl-*n*-butylmethylphenylarsonium perchlorate gave *n*-butylmethylphenylarsine, $[\alpha]_D^{20} = +12.4°$. It is interesting to note that reduction of (+)allylbenzylmethylphenylarsonium perchlorate gave a mixture of allylmethylphenylarsine and benzylmethylphenylarsine (359). By an involved series of reactions, Horner and Fuchs (359) were able to demonstrate that both the reduction of quaternary arsonium salts and the quaternization of the arsines occurred with retention of configuration. More recently, Horner and Hofer (360) have shown that cleavage of optically-active quaternary arsonium salts (to tertiary arsines) with potassium cyanide also occurred with retention of configuration.

The tertiary arsines were found to be optically stable when heated for as long as ten hours, but did racemize when exposed to ultraviolet radiation (361). When the arsines were heated with methanol containing hydrochloric acid, activity was rapidly lost (362). The rate of racemization was first order in arsine, and proportional to the square of the hydrogen chloride concentration. That protons must play a part in this racemization was shown by the fact that a methanol solution of sodium chloride did not effect racemization. However, the rate of racemization in methanolic hydrogen chloride was found to be first order both in hydrogen ion and in chloride ion concentrations. Racemization was also effected by a number of other mineral acids but was not brought about by a methanol solution of acetic acid at 20°. In order to account for these results the authors (362) proposed the following racemization mechanism:

$$a\text{-}As\langle^{b}_{c} \underset{-HX}{\overset{+HX}{\rightleftharpoons}} {}^{a}_{H}\rangle \overset{+}{As} \langle^{b}_{c} \quad X^-$$

$${}^{a}_{H}\rangle \overset{+}{As} \langle^{b}_{c} + X^- \rightleftharpoons {}^{a}_{H}\rangle \underset{}{\overset{X}{As}} \langle^{b}_{c} \underset{-X^-}{\overset{+X^-}{\rightleftharpoons}} [a, b, H, c; X, X]As$$

32

They concluded that either the second or the third step must be slow to account for the observed kinetics. The mechanism proposed by Horner and Hofer is open to some criticism. If the third step were slow, the rate of racemization would be proportional to the square of the chloride ion concentration. Moreover, it seems more likely to the present authors that a trigonal-bipyramidal intermediate (similar to structure **32**) is formed and undergoes rapid pseudorotation (363).

The optical activity of trivalent arsenical compounds was reviewed by Mann (364) in 1958.

E. Miscellaneous reactions

The metalation of triphenylarsine by *n*-butyllithium in ether has been found to yield (*m*-lithiophenyl)diphenylarsine (365). The identity of the product was established by reaction with dimethyl sulfate to yield *m*-tolyldiphenylarsine and by reaction with carbon dioxide to yield (*m*-carboxyphenyl)diphenylarsine, which was converted to the known arsine oxide.

Wittig and Benz (366) have found that treatment of triphenylarsine with $Na[(C_6H_5)_2Li]$ results in the elimination of sodium hydride and the formation of 5-phenyldibenzarsole. They suggested that the reaction proceeds via a benzyne-type intermediate:

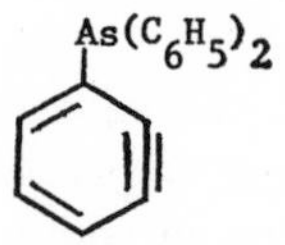

$As(C_6H_5)_2$

The surprising conversion of a di-tertiary arsine to a tri-tertiary arsine has been recently reported (367). When *o*-phenylenebis(dimethylarsine) is heated in diethylene glycol in the presence of nickel chloride hexahydrate, a 20% yield of bis(*o*-dimethylarsinophenyl)methylarsine is obtained. The trimethylarsine that was presumably formed in the disproportionation was not isolated.

6. PHYSICAL PROPERTIES OF TERTIARY ARSINES

A. Absorption spectra

The ultraviolet absorption of tertiary arsines has been studied by a number of workers (8, 16, 102, 103, 105, 138, 143, 148, 205, 207, 357, 358, 368–379). The most complete investigation to date has been reported by Cullen and coworkers (374, 376, 378), who examined the electronic spectra of a group of arsines of the general formula ArAsXY, where X, Y = aryl, CH_3, CF_3, or C_3F_7. Two distinct band systems were noted in the spectrum of every compound. One system consists of a very weak series of bands in the 245–270 mμ region. These bands appear only as inflections in the spectra of triphenylarsine and methyldiphenylarsine, but are distinct maxima in the spectra of compounds in which X or Y is CF_3 or C_3F_7. This band system does not shift appreciably as the polarity of the solvent is changed. Cullen and co-workers have attributed this system to the benzene $\pi \rightarrow \pi^*$ transition.

The second system noted in the spectra of the aromatic arsines appears as a strong, structureless band in the 220–250 mμ region. This band shifts progressively to lower wavelengths as phenyl groups are replaced by methyl or trifluoromethyl groups. Hypsochromic shifts are also observed when the solvent is changed from cyclohexane to more polar solvents. Cullen and co-workers have tentatively concluded that these bands are charge transfer transitions involving excitation of an electron from the nonbonding lone pair on the arsenic atom to an anti-bonding π-orbital of a phenyl group.

Campbell and Poller (357, 371) have found that the intense 248 mμ band of triphenylarsine is shifted to an appreciably longer wavelength in the spectrum of 5-*p*-tolyldibenzarsole. They have attributed this displacement to the presence of considerably increased conjugation between the heteroatom and the condensed ring system. This interpretation, they suggest, leads to the conclusion that the tricyclic system of the dibenzarsoles must be essentially planar. The present authors feel that the bathochromic shift noted by Campbell and Poller may be associated primarily with the presence of a biphenyl linkage in the dibenzarsoles and not necessarily with increased conjugation involving the arsenic atom. Since it *does* seem likely that the tricyclic system is planar (cf. Section IV-6-B), the possibility of resonance between the phenyl groups in the biphenyl linkage is enhanced; and, accordingly, a large bathochromic shift is not unexpected (380).

Weiner and Pasternack (379) have recently described the electronic spectra of a series of vinyl-substituted phosphines and arsines. With one exception, the spectra of these arsines were characterized by a maximum in the 220–235 mμ region (ε, 2000–5000) and the suggestion of a second, more intense absorption below 200 mμ. The position of the maximum appears to increase with the electron-donating inductive effect of the groups attached to the central atom. Thus in each series of compounds, $R_nM(C_2H_3)_{3-n}$, the wavelength of maximum absorption increases with R in the order $CH_3 < C_2H_5 \lesssim n\text{-}C_4H_9$. Furthermore, replacement of an alkyl group by a vinyl group results in a *decrease* of λ_{max}. The spectrum of the one perfluoroalkyl compound investigated (*n*-heptafluoropropyldivinylarsine) did not contain a peak in the 220–235 mμ region but only an inflection at about 215–220 mμ; this result is presumably associated with the powerful electron-withdrawing effect of a perfluoroalkyl group. Weiner and Pasternack believe that their data are best explained by the type of electron-transfer mechanism described by Cullen and co-workers for the arylarsines. In the case of the vinyl compounds, an electron is removed from the nonbonded pair on the central atom and transferred to an empty π^* orbital on one of the vinyl groups. This mechanism is consistent with the blue shift usually observed when methanol (rather than isooctane) was used as the solvent. The virtual disappearance of all absorption above 220 mμ, upon oxidation of the trivalent compounds to the corresponding oxides, appears to eliminate the possibility that the 220–235 mμ band is actually an ethylene π–π^* transition. Weiner and Pasternack have also described the ultraviolet absorption of triethylarsine; in isooctane there is a peak at 208 mμ (ε, 11, 700). The type of electronic transition responsible for this peak is not discussed. It is noted, however, that oxidation of the arsine to triethylarsine oxide causes the peak to disappear.

Significant absorption in the accessible ultraviolet has been observed (207) for the diarsabicycloöctatriene derivative **29**. The spectrum in acetonitrile exhibits a peak at 297 mμ (ε, 960) with a shoulder at 238 mμ (ε, 90). Analogous bicycloöctatrienes containing no heteroatoms show much less absorption in this spectral region.

The electronic absorption spectra of complexes of cobalt (381), nickel (325), ruthenium (342, 382), palladium (325, 340, 382), osmium (382), and platinum (339, 382) with a number of tertiary arsines have been reported.

There have been several investigations (383–389) of the Raman spectrum of trimethylarsine. The most recent work on this subject was done by Bouquet and Bigorgne (389) who reported that the C—As stretching frequencies occur at 567 and 582 cm^{-1}, while the C—As—C bending frequencies are at 236 and 221 cm^{-1}. These workers also studied the Raman spectrum of the nickel carbonyl complex $Ni(CO)_3(AsMe_3)$, and found that the C—As stretching frequencies increase to 584 and 602 cm^{-1} and the bending frequencies are shifted to 255 and 216 cm^{-1}. The stretching and bending force constants for trimethylarsine were calculated from the Raman data and compared with the larger values obtained for trimethylphosphine. In a later paper Bouquet and co-workers (390) reported a systematic study of the Raman and infrared spectra of the complexes $Ni(CO)_{4-n}L_n$, where L is PF_3, $(MeO)_3P$, Me_3P, Me_3As, or Me_3Sb. The Ni—C stretches were at 460 and 498 cm^{-1} for $Ni(CO)_2(AsMe_3)_2$ and at 418 and 457 cm^{-1} for $Ni(CO)_3(AsMe_3)$.

Miller and Lemmon (391) have made a detailed analysis of the infrared (35 to 4000 cm^{-1}) and Raman spectra of triethynylphosphine, -arsine, and -stibine. A strong band at 517 cm^{-1} in the vapor-phase infrared spectrum of the arsine was assigned to a C—As stretching mode, while a polarized Raman line at 526 cm^{-1} was assigned to the second C—As stretch. In all, thirteen of the fourteen spectroscopically-active fundamental vibrations of the arsenic compound were identified.

The infrared spectra of four trialkylarsines (triethyl-, tri-*n*-propyl-, tri-*n*-butyl-, and tri-*n*-pentylarsine) have been found to contain two strong bands between 546 and 654 cm^{-1} which were assigned to the C—As stretch (392). In contrast, this vibration gives rise to a *weak* band at 653 cm^{-1} in the spectrum of tris(perfluorovinyl)arsine (393). A peak at 595 cm^{-1} in the infrared spectra of both the platinum bromide and platinum iodide complexes of triethylarsine and a peak at 570 cm^{-1} in the spectrum of the platinum chloride complex of tri-*n*-propylarsine have also been attributed to this vibration (271). Jensen and Nielsen (271) have concluded that the C—As stretch in the infrared spectrum of triphenylarsine occurs at a much lower frequency, viz. 474 cm^{-1}. Zingaro and his co-workers (392), on the other hand, believe that the phenyl-As vibrations are to be found at *higher* frequencies than the aliphatic C—As frequencies. Deacon and Jones (394), who have examined the infrared spectra of several triarylarsines, have concluded that two very strong bands at 471 and 478 cm^{-1} in the spectrum of triphenylarsine should be assigned to an "X-sensitive" out-of-plane deformation. More recently, Ellermann and Dorn (144) have compared the infrared spectra of triphenylarsine and tetrakis-(diphenylarsinomethyl)methane. A large number of bands in the spectrum of the latter compound were classified as stretching, deformation, or combination vibrations. An analysis of the spectrum of *p*-phenylenebis(diphenylarsine) has also been published (143).

Nesmeyanov and co-workers (100) have studied the infrared spectra of *cis*- and *trans*-tripropenylarsine and of several related compounds. The *trans* isomer has a strong peak at 960 cm^{-1} which was assigned to an out-of-plane C—H vibration; the corresponding band for the *cis* isomer occurs at 925 cm^{-1}. The C=C stretching frequency gives rise to a strong band at 1620 cm^{-1} for the *trans* isomer and at 1610 cm^{-1} for the *cis* isomer. These results are similar to those reported earlier for the analogous antimony compounds (cf. Section VIII-2-C). The C=C stretching frequency in tris(perfluorovinyl)-arsine is in the 1720–1730 cm^{-1} region (395).

Cullen and his co-workers (45, 50, 54, 229, 376, 396–399) have described the infrared spectra of a large number of fluorine-containing tertiary arsines, but the frequencies associated with the C—As vibrations in these compounds have not been identified. The position of the C≡C stretching frequency in the spectra of various acetylene derivatives of arsenic has been determined in two laboratories (88, 123).

The infrared spectra of palladium, platinum, and gold complexes of trimethylarsine have been reported (400). In all cases studied, the metal–arsenic frequencies were found in the 242–276 cm^{-1} range.

In addition to the examples noted in this section, the literature (326, 401) contains several other reports concerned with the effect of complex formation on the infrared spectra of tertiary arsines. The effect of quaternization on infrared absorption has also been investigated (402, 403) (see also Section V-9-C).

No systematic investigation of the nuclear magnetic resonance spectra of a large number of arsenicals has yet been published. Allred and co-workers (404, 405) have measured the proton NMR spectra of the trimethyl derivatives of the nitrogen family elements as well as the spectra of the methyl derivatives of many other representative elements. They concluded that the chemical shift is a function of the electronegativity of the central atom.

Kostyanovskii and co-workers (406) have compared the proton NMR spectra of the trimethyl derivatives of nitrogen, phosphorus, arsenic, and antimony with the spectra of the corresponding dimethylphenyl compounds. In going from the trimethyl to the dimethylphenyl compounds, the following changes in the methyl chemical shifts were observed (in ppm): N, +0.62; P, −0.12; As, −0.25; Sb, −0.27. The sign change was attributed to a decrease in electron density at nitrogen because of delocalization of the non-bonded electrons, while the electron density at the other group V elements was thought to be increased because of conjugation of the π-electrons of the phenyl group with the unoccupied orbitals of the heteroatoms.

The proton NMR spectrum of triethylarsine has been studied by Massey and co-workers (407, 408). The difference between the chemical shifts of the CH_3 and CH_2 protons was found to be relatively small and was related to the rather low electronegativity of the arsenic atom. Aguiar and Archibald (139) have used proton NMR data for establishing the configurations of *cis*- and *trans*-β-styryldiphenylarsine. The literature also contains reports of the proton NMR spectra of tris(2-chlorovinyl)arsine (163), diphenyl(4-hydroxy-3,5-di-*tert*-butylphenyl)arsine (134), *cis*- and *trans*-(1,2-diphenylvinyl)di-phenyl arsine (140), 1,1,1-tris(dimethylarsinomethyl)ethane (4), several vinyl-substituted tertiary arsines (409), and a number of acetylene derivatives of arsenic (88, 410), while Cullen and co-workers (50, 54, 398, 399, 411) have obtained both proton and ^{19}F NMR data on some trifluoromethyl- and difluoromethylene-substituted tertiary arsines. The ^{19}F spectra of tris(per-fluorovinyl)arsine (412, 413) and hexakis(trifluoromethyl)-1,4-diarsabicyclo-[2,2,2]octa-2,5,7-triene (**29**) (207) have also been determined. The quadrupole resonances of ^{75}As in triphenylarsine and in a number of inorganic arsenic compounds have been investigated by Barnes and Bray (414).

Fritz and Schwarzhans (415) have investigated the effect of complex formation on the proton NMR spectrum of trimethylarsine and related compounds. They found that the proton signal was shifted to lower fields when the arsine was coordinated with either platinum(II) chloride or dimethyl-platinum. The observed deshielding is undoubtedly associated with the fact that there is a formal positive charge on the arsenic atoms in the coordination compounds. Similar deshielding effects were observed when *o*-phenylenebis-(diethylarsine) was complexed with titanium tetrachloride (319) and when *o*- dimethylarsinostyrene was complexed with platinum(II) bromide (416).

B. Molecular structure

The structure of trimethylarsine in the gas phase has been determined by electron diffraction (417). As expected, the molecule is pyramidal with the arsenic atom at the apex and a C—As—C angle of 96 ± 5°. The observed C—As distance of 1.98 ± 0.02 A agrees exactly with the theoretical bond length calculated from the sum of the normal single bond covalent radii. The above angles and distances are not appreciably different from those found by X-ray analysis of crystals of tri-*p*-tolylarsine (418) and tri-*p*-xylylarsine (419): C—As—C = 102 ± 2° and C—As = 1.96 ± 0.05 A. X-ray diffraction has also been used to determine the structure of the heterocyclic compound, 5-phenyldibenzarsole (420). The arrangement of bonds around the arsenic is pyramidal; and the C—As bond distances and C—As—C bond angles are normal except for an angle of 88° in the five-membered ring. The three rings of the dibenzarsole system are coplanar, while the 5-phenyl group appears to be orientated so as to allow overlap of the arsenic *d*-orbitals with the π-orbitals of the biphenyl and phenyl groups.

Mootz and co-workers (421) have reported that tris(phenylethynyl)arsine has three molecules in its unit cell and exhibits the expected pyramidal structure (C_3 symmetry) with the lone pair of electrons presumably occupying the fourth coordination position around the arsenic atom. The arrangement of the three phenyl groups in the molecule are described as propeller-like.

The C—As distance (determined by electron diffraction) in tris(trifluoromethyl)arsine has been found to be surprisingly long, namely 2.053 ± 0.019 A (422). Bowen (422) has suggested that the ease of hydrolysis of the trifluoromethyl compound may be associated with its unusually long C—As distance.

The structure of trimethylarsine has also been investigated by microwave spectroscopy (423). Analysis of the data obtained indicates that the C—As distance is 1.959 ± 0.010 A, if a C—As—C angle of 96 ± 3° is assumed. This result agrees remarkably well with the value found by electron diffraction. It was also possible to estimate from the microwave spectrum that the barrier to rotation of a methyl group in trimethylarsine is between 1.5 and 2.5 kcal/mole.

C. Other physical measurements

The dipole moments of tertiary arsines are relatively small. Thus, triphenylarsine has a moment of 1.07 (424) or 1.23 D (425), and the moments of triethyl-, tri-*n*-propyl-, and tri-*n*-butylarsines are 1.04, 1.00, and 0.92 D, respectively (426). Tris(2-chlorovinyl)arsine has been reported (427) to have a dipole moment of 0.39 D but there is no information about the isomeric composition of the sample used. Aguiar and co-workers (140) have reported that *cis*-(1,2-diphenylvinyl)diphenylarsine has a dipole moment of 1.37 D, while the *trans* isomer has a moment of 0.97 D. The largest dipole moments yet recorded for tertiary arsines are 1.51 D for *p*-phenylenebis-(diphenylarsine) (143) and 1.82 D for tris(perfluorovinyl)arsine (428). Kuz'min and Kamai (426) have estimated that the aromatic C—As bond moment is 0.69 D (assuming that the moment points from the ring to the arsenic atom). Smyth (429) has concluded that the small dipole moments characteristic of tertiary arsines indicate that the C—As linkage is an ordinary covalent bond with very little ionic character. Davies (430) has noted that the dipole moments of the triphenyl derivatives of phosphorus, arsenic, antimony, and bismuth decrease almost linearly with the period number of the central atom.

The dipole moments discussed in the above paragraph were based on dielectric constant measurements in dilute solution; the electronic and atomic polarizations were estimated in the usual way from the molar refractivities of the arsines. Microwave spectroscopy has been used for obtaining the dipole moment of trimethylarsine; the value found was 0.86 ± 0.02 D (431). Gibbs (431) has concluded that orbital hybridization, as determined by the intervalency angle of the central atom, accounts for the major variation found in the dipole moments of the hydrides and trialkyl derivatives of elements of group V of the periodic system.

Long and Sackman (60) have measured the vapor pressure, melting point, and liquid density of a carefully purified sample of trimethylarsine. The vapor pressure results, which were obtained over the range -25 to $+15°$, can be represented by an equation of conventional form, $\log p = A/T + B$ (where $A = -1563$ and $B = 7.7119$). The extrapolated normal boiling point was found to be 50.4°, and the heat of vaporization, 7.15 kcal/mole; these values yield a Trouton's constant of 22.1. Vapor pressure data are also available for triphenylarsine (432), trivinylarsine (67, 68), and dimethyl-(trifluoromethyl)arsine (190). The molecular weight of trimethylarsine in the vapor state has been found to agree well with the calculated value (433). Other tertiary arsines have been found to give normal molecular weights both in the vapor state (376, 396) and in solution (144, 355, 434).

Physical properties of a large number of tertiary arsines have been studied by Jones and co-workers (435). Liquid densities at 20°, indices of refraction at 20°, and boiling points at several pressures were determined. Equations were then derived, which relate the structure of an arsine to its molecular volume at 20° and its boiling point at 10 mm. An equation showing the variation of vapor pressure as a function of temperature was also reported. The molar refraction values of the arsines containing phenyl groups were found to exhibit optical exaltation. In other tertiary arsines, however, the refraction equivalents of carbon and hydrogen appear to be normal, and the atomic refractivity of arsenic in these compounds has been found (61, 436, 437) to lie in the range 10.97–12.09 (for the sodium D line). Recently, a value of 12.43 has been reported for the atomic refractivity of arsenic in acetylene derivatives of the type $R_2AsC{\equiv}CAsR_2$ (85). From the refractive index data of Dyke and Jones (61), Gillis (438) has calculated a bond refraction value of 4.52 for the C—As bond in trialkylarsines. Kamai and Chernokal'skii (439) have obtained an average value of 4.51 for the C—As bond refraction in trialkylarsines and 4.94 for the aromatic C—As bond in dialkylarylarsines and alkyldiarylarsines.

Only a few calorimetric investigations of tertiary arsines have been reported. Long and Sackman (440) have found that the heat of combustion (ΔU) of trimethylarsine to arsenic trioxide, carbon dioxide, and water is −664.6 kcal/mole at 25°. This corresponds to a heat of formation (ΔH_f°) of −3.5 kcal/mole; i.e., the reaction between the elements to form trimethylarsine is slightly exothermic. The heat of formation has been combined with the heat of volatilization of trimethylarsine, the heat of atomization of arsenic, and the heat of formation of the methyl radical to give the *mean* dissociation energy of the As—C bond as 51.5 kcal at 25°. This value may be in considerable error, however, because of uncertainties in the ΔH_f° values for As_2O_3 and As_2O_5 and in the heat of atomization for arsenic. It is of interest that the mean dissociation energy calculated by Long and Sackman is in reasonable agreement with the activation energy of 54.6 kcal found by Ayscough and Emeléus (225) for the rate of decomposition of trimethylarsine at 400–500°. Price and Trotman-Dickenson (441) have suggested that the pyrolysis of trimethylarsine is probably so complex that it is unwise to assume (as Ayscough and Emeléus have done) that the overall experimental activation energy is equal to the strength of the $(CH_3)_2As{-}CH_3$ bond. The heat of combustion of triethylarsine has been found by Lautsch and co-workers (442, 443) to be −1158.1 kcal/mole. This value leads to a C—As bond energy of 33.0 kcal.

Birr (444) has reported that the heat of combustion of triphenylarsine to arsenic pentoxide, carbon dioxide, and water is −2406.6 kcal/mole. Using this result Birr calculated the heat of formation (ΔH_f°) of triphenylarsine to be 81.2 kcal/mole. Skinner (445) has pointed out, however, that Birr made certain errors in deriving ΔH_f° and that the corrected value should be 96.2. This value is in poor agreement with the results of Mortimer and Sellers (446) who measured the heat of combustion of the arsine in a rotating-bomb calorimeter. From their data Mortimer and Sellers concluded that the heat of formation of triphenylarsine is 71.0 kcal/mole and that the mean dissociation energy of the C—As bond is 60.3 kcal/mole. Birr has suggested that the higher C—As bond energy of the phenyl compound (compared to the methyl compound) is to be expected, since the aromatic C—As linkage is strengthened by bonding involving 4*d* orbitals of the arsenic atom and the π-orbitals of the benzene ring. Mortimer (447) has noted that metal-phenyl bonds are in general stronger than metal–alkyl bonds.

The mass spectra of the triphenyl derivatives of the group V elements have recently been determined (448). The dominant peak in the spectrum of triphenylamine and of triphenylphosphine is that of the molecular ion in addition to an appreciable M-1 peak. The spectra of triphenylarsine and of triphenylstibine have the phenyl–metal cation as the dominant peak, while the Bi^+ ion is the dominant peak in the spectrum of triphenylbismuth. There is a regular decrease in the abundance of the molecular ion as one proceeds from triphenylamine to triphenylbismuth. This decrease is, presumably, a reflection of a corresponding decrease in the strength of the carbon–hetero-atom bond. The mass spectra of the phosphorus, arsenic, and antimony compounds indicate that a dibenzoheterocyclic cation is formed in a one-step process involving the simultaneous loss of two hydrogen atoms:

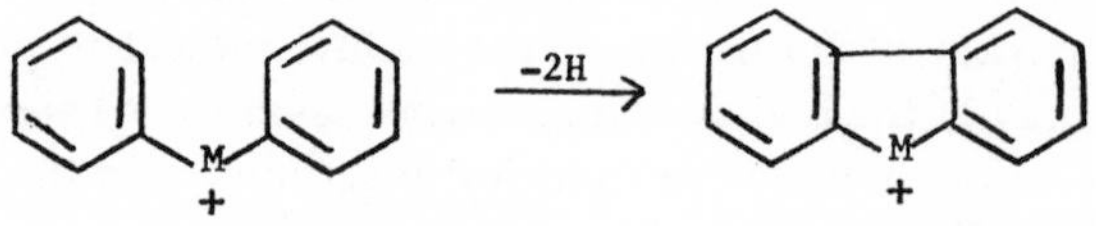

-2H

M +   M +

The $(C_6H_5)_2N^+$ ion apparently undergoes a similar transformation by a two-step process, while no "*ortho* coupling" at all is observed in the mass spectrum of triphenylbismuth.

Buu-Hoi and co-workers (449, 450) have investigated the mass spectrometry of 10-methyl-5,10-dihydrophenarsazine and of several closely related heterocyclic arsines. The most intense peak in the spectrum of 10-methyl-5,10-dihydrophenarsazine is at $m/e = 242$ and corresponds to the loss of a methyl group; there is also a doubly-charged ion at $m/e = 121$. The molecular ion peak at $m/e = 257$ is relatively abundant and there is in addition a doubly-charged molecular ion. The formation of carbazole by the loss of an arsenic atom is shown by a prominent peak at $m/e = 167$. There is also a small peak at $m/e = 241$ corresponding to the fully-conjugated species, phenarsazine:

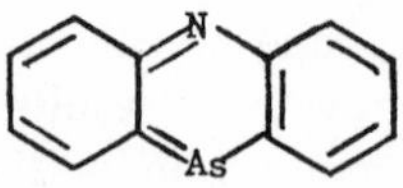

N

As

The other tertiary arsines studied by Buu-Hoi and co-workers show fragmentation patterns similar to that of 10-methyl-5,10-dihydrophenarsazine.

Dubov and co-workers (451) have described the mass spectra of the tris(perfluorovinyl) derivatives of phosphorus, arsenic, and antimony. In each case the principal ion was of the type $C_2F_3M^+$ (where M is P, As, or Sb). The relative abundances of molecular ions and of ions corresponding to the loss of one C_2F_3 group were very low. Cleavage of the C—M bond and the formation of fluorocarbon ions were observed with all three compounds. The mass spectra of trivinylphosphine, -arsine, and -stibine were also determined and showed moderately intense molecular ions.

Müller and co-workers (134) have determined the mass spectra of diphenyl-(4-hydroxy-3,5-di-*tert*-butylphenyl)arsine and several related compounds; in each case they observed the molecular ion and a number of the resulting fragments. The mass spectrum of tris(trifluoromethyl)arsine is discussed in Section III-8-D.

First ionization potentials of a number of trivalent arsenicals have been reported by Cullen and Frost (452). Their results, which were obtained by electron impact mass spectrometry, are given in Table 4-1. The magnitude

Table 4-1 First Ionization Potentials of Some Arsenic Compounds

Compound	Ionization Potential (eV)
AsH_3	10.6
CH_3AsH_2	9.7
$(CH_3)_2AsH$	9.0
$(CH_3)_3As$	8.3
$(CH_3)_2(CF_3)As$	9.2
$(CH_3)(CF_3)_2As$	10.5
$(CF_3)_2AsH$	10.9
$(CF_3)_3As$	11.0
$AsCl_3$	11.7
$(CF_3)_2AsCl$	11.0
CH_3AsCl_2	10.4
$(CH_3)_2AsCl$	9.9

of each ionization potential is, presumably, a measure of the energy required to remove an electron from the arsenic lone pair. It is seen that replacement of a hydrogen in arsine by a methyl group decreases the ionization potential by 0.7–0.9 eV. This result can be attributed to the lowering of the energy of the lone pair electrons by the electron-repelling effect of a methyl group. Conversely, the electron-withdrawing trifluoromethyl and chloro groups increase the ionization potential. However, in the series $(CF_3)_xAsCl_{3-x}$, the substitution of a chloro group by a trifluoromethyl group appears to cause a small but significant *lowering* of the ionization potential; this result suggests that the electron-withdrawing power of a trifluoromethyl group is less than that of a chloro group. Such a conclusion is, of course, contrary to the generally accepted opinion (453, 454).

By means of a photoionization method, Vilesov and Zaitsev (455) have found that the first ionization potential of triphenylarsine is 7.34 eV. They concluded that the positive ion was formed by removal of an electron from the lone pair on the arsenic atom.

The heat capacities of the triphenyl derivatives of the group V elements have been measured from 100 to 320°K (456). The results obtained suggest that there is a weakening of the binding force between the phenyl groups and the central atom as the atomic number of the latter increases.

The molar Kerr constant of triphenylarsine has been measured and used together with dipole moment and refractivity data to deduce the conformation of the molecule in benzene solution (425, 457). It was concluded that the phenyl rings are rotated $40 \pm 3°$ from their positions in an ideal model which would allow maximum overlap of the π orbitals of the aromatic groups with the arsenic lone pair. This result is in agreement with Trotter's X-ray data on tri-*p*-tolylarsine and tri-*p*-xylylarsine (418, 419).

From measurements of the density and surface tension of liquid triphenylarsine, Forward, Bowden, and Jones (432) were able to determine that the parachor of this compound is 622.2. Since this value was found to be essentially independent of temperature, it appears that liquid triphenylarsine is normal, i.e., unassociated. If Sugden's value of 190.0 for the phenyl group is used, the parachor equivalent for arsenic comes out to be 52.2. This is in reasonable agreement with the value of 54.0 recommended for arsenic by Quayle (458). The relationship between the parachors and structures of trialkylarsines has been investigated by Kuz'min and Kamai (437). Instead of using the simple system of atomic and structural constants devised by Sugden, the Russian workers analyzed their data by means of the rather complicated formulas reported by Gibling (459). Critical appraisals of Gibling's work have been previously published (460).

The diamagnetic susceptibility of triphenylarsine has been determined by Pascal (461), and the susceptibilities of several triarylarsines have been determined by Prasad and Mulay (462, 463). The latter authors noted a direct linear relationship between the molecular susceptibility and the number of electrons in the molecule.

Schleyer and West (464) have compared the effectiveness of a number of covalently-bonded atoms as proton acceptor groups in hydrogen bonding. One of the best acceptor molecules in their study was triethylarsine. Thus, this compound forms stronger hydrogen bonds with methyl alcohol than does tri-*n*-butylphosphine, di-*n*-butyl sulfide, di-*n*-butyl ether, or any of the *n*-butyl halides.

Lutskii and Obukhova (465) have shown that there are simple, precise correlations between the molar volumes and molar polarisabilities of the triphenyl derivatives of nitrogen, phosphorus, arsenic, and antimony and of the triethyl derivatives of all the group V elements.

The affinities of silver ion for a sulfonated triphenylarsine, a sulfonated triphenylphosphine, and several sulfonated aniline derivatives have been reported (466).

REFERENCES

1. G. Newbery and M. A. Phillips, *J. Chem. Soc.*, 2375 (1928).
2. F. F. Blicke and G. L. Webster, *J. Amer. Chem. Soc.*, **59,** 537 (1937).
3. A. B. Bruker, T. G. Spiridonova, and L. Z. Zaboroviskii, *Zh. Obshch. Khim.*, **28,** 350 (1958).
4. R. D. Feltham, A. Kasenally, and R. S. Nyholm, *J. Organometal. Chem.*, **7,** 285 (1967).
5. A. W. Palmer and W. M. Dehn, *Ber.*, **34,** 3594 (1901).
6. W. M. Dehn, *Amer. Chem. J.*, **33,** 101 (1905).
7. W. M. Dehn and B. B. Wilcox, *Amer. Chem. J.*, **35,** 1 (1906).
8. F. G. Mann and A. J. Wilkinson, *J. Chem. Soc.*, 3336 (1957).
9. N. Wigren, *J. Prakt. Chem.*, **126,** 223 (1930).
10. H. Albers, W. Künzel, and W. Schuler, *Chem. Ber.*, **85,** 239 (1952).
11. R. C. Cookson and F. G. Mann, *J. Chem. Soc.*, 67 (1949).
12. E. J. Cragoe, R. J. Andres, R. F. Coles, B. Elpern, J. F. Morgan, and C. S. Hamilton, *J. Amer. Chem. Soc.*, **69,** 925 (1947).
13. K. Issleib and A. Tzschach, *Angew. Chem.*, **73,** 26 (1961).
14. A. Tzschach and W. Lange, *Z. Anorg. Allg. Chem.*, **326,** 280 (1964).
15. F. G. A. Stone and A. B. Burg, *J. Amer. Chem. Soc.*, **76,** 386 (1954).
16. H. J. Emeléus, R. N. Haszeldine, and E. G. Walaschewski, *J. Chem. Soc.*, 1552 (1953).
17. E. Wiberg and K. Mödritzer, *Z. Naturforsch.*, **B, 11,** 751 (1956).
18. E. Wiberg and K. Mödritzer, *Z. Naturforsch.*, **B, 12,** 127 (1957).
19. W. R. Cullen, *Can. J. Chem.*, **39,** 1855 (1961).
20. R. G. Cavell and R. C. Dobbie, *J. Chem. Soc.*, **A,** 1308 (1967).
21. F. Fichter and E. Elkind, *Ber.*, **49,** 239 (1916).
22. K. Matsuyima and H. Nakata, *Mem. Coll. Sci., Kyoto Imp. Univ.*, **A, 12,** 63 (1929); through *C.A.*, **23,** 4939 (1929).
23. S. V. Vasil'ev and G. D. Vovchenko, *Vestn. Mosk. Univ.*, **5,** No. 3, *Ser. Fiz.-Mat. i Estestv. Nauk*, No. **2,** 73 (1950); through *C.A.*, **45,** 6594 (1951).
24. W. C. Johnson and A. Pechukas, *J. Amer. Chem. Soc.*, **59,** 2068 (1937).
25. F. G. Mann and B. B. Smith, *J. Chem. Soc.*, 4544 (1952).
26. A. Tzschach and G. Pacholke, *Chem. Ber.*, **97,** 419 (1964).
27. A. Tzschach and W. Lange, *Z. Anorg. Allg. Chem.*, **330,** 317 (1964).
28. F. F. Blicke and J. F. Oneto, *J. Amer. Chem. Soc.*, **57,** 749 (1935).
29. L. Schmerling (to Universal Oil Products Co.), U.S. Pat. 2,842,579 (July 8, 1958); *C.A.*, **55,** 497 (1961).
30. W. La Coste and A. Michaelis, *Justus Liebigs Ann. Chem.*, **201,** 184 (1880).
31. A. B. Bruker, E. I. Grinshtein, and L. Z. Soborovskii, *Zh. Obshch. Khim.*, **36,** 1133 (1966).
32. N. I. Wigren, *Justus Liebigs Ann. Chem.*, **437,** 285 (1924).
33. W. R. Cullen, *Can. J. Chem.*, **41,** 332 (1963).

34. E. Wiberg and K. Mödritzer, *Z. Naturforsch.*, **B, 12,** 135 (1957).
35. A. N. Nesmeyanov and R. Ch. Freidlina, *Ber.*, **67,** 735 (1934).
36. F. F. Blicke and L. D. Powers, *J. Amer. Chem. Soc.*, **55,** 1161 (1933).
37. W. Steinkopf and H. Dudek, *Ber.*, **62,** 2494 (1929).
38. G. I. Braz, A. Ya. Berlin, and Yu. V. Markova, *Zh. Obshch. Khim.*, **18,** 316 (1948); through *C.A.*, **42,** 6764 (1948).
39. F. F. Blicke and L. D. Powers, *J. Amer. Chem. Soc.*, **55,** 315 (1933).
40. F. F. Blicke and J. F. Oneto, *J. Amer. Chem. Soc.*, **56,** 685 (1934).
41. F. G. Mann and R. C. Cookson, *Nature*, **157,** 846 (1946).
42. R. C. Cookson and F. G. Mann, *J. Chem. Soc.*, 618 (1947).
43. M. J. Gallagher and F. G. Mann, *J. Chem. Soc.*, 5110 (1962).
44. E. Uhlig and M. Maaser, *Z. Anorg. Allg. Chem.*, **349,** 300 (1967).
45. W. R. Cullen, P. S. Dhaliwal, and G. E. Styan, *J. Organometal. Chem.*, **6,** 364 (1966).
46. R. Adams and C. S. Palmer, *J. Amer. Chem. Soc.*, **42,** 2375 (1920).
47. C. S. Palmer and R. Adams, *J. Amer. Chem. Soc.*, **44,** 1356 (1922).
48. W. R. Cullen and G. E. Styan, *J. Organometal. Chem.*, **4,** 151 (1965).
49. A. Tzschach and R. Schwarzer, *Justus Liebigs Ann. Chem.*, **709,** 248 (1967).
50. W. R. Cullen, D. S. Dawson, and G. E. Styan, *Can. J. Chem.*, **43,** 3392 (1965).
51. W. R. Cullen and G. E. Styan, *Can. J. Chem.*, **44,** 1225 (1966).
52. W. Steinkopf, I. Schubart, and S. Schmidt, *Ber.*, **61,** 678 (1928).
53. L. Anschütz and H. Wirth, *Naturwissenschaften*, **43,** 59 (1956); through *C.A.*, **51,** 14671 (1957).
54. W. R. Cullen and P. S. Dhaliwal, *Can. J. Chem.*, **45,** 719 (1967).
55. F. F. Blicke and L. D. Powers, *J. Amer. Chem. Soc.*, **54,** 3353 (1932).
55a. K. Issleib and R. Kümmel, *J. Organometal. Chem.*, **3,** 84 (1965).
56. W. C. Johnson and A. Pechukas, *J. Amer. Chem. Soc.*, **59,** 2068 (1937).
57. C. T. Mortimer and H. A. Skinner, *J. Chem. Soc.*, 3189 (1953).
58. A. B. Harvey and M. K. Wilson, *J. Chem. Phys.*, **44,** 3535 (1966).
59. M. T. Ryan and W. L. Lehn, Report No. ASD-TDR-63-832 of the AF Materials Laboratory, Wright-Patterson Air Force Base, Ohio, October, 1963; *C.A.*, **62,** 7276 (1965).
60. L. H. Long and J. F. Sackman, *Res. Correspondence*, **8,** S23, (1955).
61. W. J. C. Dyke and W. J. Jones, *J. Chem. Soc.*, 2426 (1930).
62. J. Seifter, *J. Amer. Chem. Soc.*, **61,** 530 (1939).
63. R. N. Meals, *J. Org. Chem.*, **9,** 211 (1944).
64. K. V. Vijayaraghavan, *J. Indian Chem. Soc.*, **22,** 141 (1945).
65. W. J. Jones, W. C. Davies, S. T. Bowden, C. Edwards, V. E. Davis, and L. H. Thomas, *J. Chem. Soc.*, 1446 (1947).
66. R. N. Sterlin, L. N. Pinkina, R. D. Yatsenko, and I. L. Knunyants, *Khim. Naukai Promy.*, **4,** 800 (1959); through *C.A.*, **54,** 14103 (1960).
67. L. Maier, D. Seyferth, F. G. A. Stone, and E. G. Rochow, *Z. Naturforsch.*, **B, 12,** 263 (1957).
68. L. Maier, D. Seyferth, F. G. A. Stone, and E. G. Rochow, *J. Amer. Chem. Soc.*, **79,** 5884 (1957).
69. W. Steinkopf, H. Dudek, and S. Schmidt, *Ber.*, **61,** 1911 (1928).
70. F. Challenger and A. T. Peters, *J. Chem. Soc.*, 2610 (1929).
71. J. Dodonow and H. Medox, *Ber.*, **68,** 1254 (1935).
72. D. Seyferth, *J. Amer. Chem. Soc.*, **80,** 1336 (1958).
73. J. R. Vaughan, Jr. and D. S. Tarbell, *J. Amer. Chem. Soc.*, **67,** 144 (1945).
74. M. Fild, O. Glemser, and G. Christoph, *Angew. Chem., Int. Ed. Engl.*, **3,** 801 (1964).
75. W. C. Davies and F. G. Mann, *J. Chem. Soc.*, 276 (1944).

76. H. E. Ramsden (to Metal and Thermit Corp.), Brit. Pat. 824, 944 (Dec. 9, 1959); through *C.A.*, **54,** 17238 (1960).
77. I. L. Knunyants and V. Ya. Pil'skaya, *Izv. Akad. Nauk SSSR, Otd. Khim. Nauk*, 472 (1955); through *C.A.*, **50,** 6298 (1956).
78. H. Gilman and S. Avakian, *J. Amer. Chem. Soc.*, **76,** 4031 (1954).
79. J. Chatt and F. G. Mann, *J. Chem. Soc.*, 610 (1939).
80. H. Heaney, F. G. Mann, and I. T. Millar, *J. Chem. Soc.*, 1 (1956).
81. H. Heaney, F. G. Mann, and I. T. Millar, *J. Chem. Soc.*, 3930 (1957).
82. I. Gorski, W. Schpanski, and L. Muljar, *Ber.*, **67,** 730 (1934).
83. J. J. Monagle, *J. Org. Chem.*, **27,** 3851 (1962).
84. H. Hartmann and G. Nowak, *Z. Anorg. Allg. Chem.*, **290,** 348 (1957).
85. K. I. Kuz'min and L. A. Pavlova, *Zh. Obshch. Khim.*, **36,** 1478 (1966).
86. H. Hartmann, H. Niemöller, W. Reiss, and B. Karbstein, *Naturwissenschaften*, **46,** 321 (1959).
87. H. Hartmann, W. Reiss, and B. Karbstein, *Naturwissenschaften*, **46,** 321 (1959).
88. J. Benaïm, *C. R. Acad. Sci., Paris*, **261,** 1996 (1965).
89. J. Benaïm, *C. R. Acad. Sci., Paris, Ser. C*, **262,** 937 (1966).
90. T. V. Talalaeva and K. A. Kocheshkov, *Zh. Obshch. Khim.*, **8,** 1831 (1938); through *C.A.*, **33,** 5819 (1939).
91. S. O. Lawesson, *Ark. Kemi*, **10,** 167 (1956).
92. F. F. Blicke, R. A. Patelski, and L. D. Powers, *J. Amer. Chem. Soc.*, **55,** 1158 (1933).
93. E. Gryszkiewicz-Trochimowski, *Rocz. Chem.*, **8,** 250 (1928); through *C.A.*, **22,** 4523 (1928).
94. F. Sachs and H. Kantorowicz, *Ber.*, **41,** 2767 (1908).
95. K. Matsumiya and M. Nakai, *Mem. Coll. Sci., Kyoto Imp. Univ., A*, **8,** 309 (1925); through *C.A.*, **19,** 3086 (1925).
96. K. Matsumiya and M. Nakai, *Mem. Coll. Sci., Kyoto Imp. Univ.*, **10,** 57 (1926); through *C.A.*, **21,** 904 (1927).
97. D. Seyferth and M. A. Weiner, *Chem. Ind.* (London), 402 (1959).
98. H. Gilman, W. A. Gregory, and S. M. Spatz, *J. Org. Chem.*, **16,** 1788 (1951).
99. E. Plazek and R. Tyka, *Zeszyty Nauk Politech. Wroclaw., Chem., No.* **4,** 79 (1957); through *C.A.*, **52,** 20156 (1958).
100. A. N. Nesmeyanov, A. E. Borisov, and A. I. Borisova, *Izv. Akad. Nauk SSSR, Otd. Khim. Nauk*, 1199 (1962); through *C.A.*, **58,** 9121 (1963).
101. H. A. Goodwin and F. Lions, *J. Amer. Chem. Soc.*, **81,** 311 (1959).
102. F. G. Mann, I. T. Millar, and B. B. Smith, *J. Chem. Soc.*, 1130 (1953).
103. H. Heaney, D. M. Heinekey, F. G. Mann, and I. T. Millar, *J. Chem. Soc.*, 3838 (1958).
104. F. C. Leavitt, T. A. Manuel, and F. Johnson, *J. Amer. Chem. Soc.*, **81,** 3163 (1959).
105. E. H. Braye, W. Hübel, and I. Caplier, *J. Amer. Chem. Soc.*, **83,** 4406 (1961).
106. F. C. Leavitt, T. A. Manuel, F. Johnson, L. U. Matternas, and D. S. Lehman, *J. Amer. Chem. Soc.*, **82,** 5099 (1960).
107. M. E. P. Friedrich and C. S. Marvel *J. Amer. Chem. Soc.*, **52,** 376 (1930).
108. E. R. H. Jones and F. G. Mann, *J. Chem. Soc.*, 4472 (1955).
109. H. Jenkner (to Kali-Chemie Akt.-Ges.), Ger. Pat. 1,064,513 (Sept. 3, 1959); *C.A.*, **55,** 11302 (1961).
110. Kali-Chemie Akt.-Ges., Brit. Pat. 820,146 (Sept. 16, 1959); *C.A.*, **54,** 6550 (1960).
111. Farbwerke Höchst Akt.-Ges., Brit. Pat. 839,370 (June 29, 1960); *C.A.*, **55,** 3435 (1961).
112. R. S. Dickson and B. O. West, *Aust. J. Chem.*, **15,** 710 (1962).
113. W. Stamm and A. Breindel, *Angew. Chem., Int. Ed. Engl.*, **3,** 66 (1964).
114. W. Steinkopf, *Justus Liebigs Ann. Chem.*, **413,** 310 (1917).
115. W. G. Lowe and C. S. Hamilton, *J. Amer. Chem. Soc.*, **57,** 1081 (1935).

116. J. F. Morgan, E. J. Cragoe, Jr., B. Elpern, and C. S. Hamilton, *J. Amer. Chem. Soc.*, **69,** 932 (1947).
117. W. W. Beck and C. S. Hamilton, *J. Amer. Chem. Soc.*, **60,** 620 (1938).
118. V. V. Kudinova, V. L. Foss, and I. F. Lutsenko, *Zh. Obshch. Khim.*, **36,** 1863 (1966).
119. A. W. Hofmann, *Justus Liebigs Ann. Chem.*, **103,** 357 (1857).
120. W. v. E. Doering and A. K. Hoffman, *J. Amer. Chem. Soc.*, **77,** 521 (1955).
121. B. D. Chernokal'skii, A. S. Gel'fond, and G. Kamai, *Zh. Obshch. Khim.*, **37,** 1396 (1967).
122. H. Schumann, H. Köpf, and M. Schmidt, *Z. Anorg. Allg. Chem.*, **331,** 200 (1964).
123. H. Hartmann, B. Karbstein, and W. Reiss, *Naturwissenschaften*, **52,** 59 (1965).
124. W. R. Cullen and P. E. Yates, *Can. J. Chem.*, **41,** 1625 (1963).
125. A. Michaelis, *Justus Liebigs Ann. Chem.*, **321,** 141 (1902).
126. W. J. Pope and E. E. Turner, *J. Chem. Soc.*, **117,** 1447 (1920).
127. R. L. Shriner and C. N. Wolf, *Org. Syn.*, *Coll. Vol.* **4,** 910 (1963).
128. D. E. Worrall, *J. Amer. Chem. Soc.*, **62,** 2514 (1940).
129. C. Finzi, *Gazz. Chim. Ital.*, **55,** 824 (1925).
130. A. Michaelis and L. Weitz, *Ber.*, **20,** 48 (1887).
131. H. N. Das-Gupta, *J. Indian Chem. Soc.*, **14,** 400 (1937).
132. D. R. Lyon and F. G. Mann, *J. Chem. Soc.*, 30 (1945).
133. D. R. Lyon, F. G. Mann, and G. H. Cookson, *J. Chem. Soc.*, 662 (1947).
134. E. Müller, B. Teissier, H. Eggensperger, A. Rieker, and K. Scheffler, *Justus Liebigs Ann. Chem.*, **705,** 54 (1967).
135. G. Kamai, *Zh. Obshch. Khim.*, **17,** 2178 (1947); through *C.A.*, **42,** 4521 (1948).
136. F. F. Blicke and E. L. Cataline, *J. Amer. Chem. Soc.*, **60,** 419 (1938).
137. J. R. Phillips and J. H. Vis, *Can. J. Chem.*, **45,** 675 (1967).
138. L. Sindellari and P. Centurioni, *Ann. Chim.* (Rome), **56,** 379 (1966).
139. A. M. Aguiar and T. G. Archibald, *J. Org. Chem.*, **32,** 2627 (1967).
140. A. M. Aguiar, T. G. Archibald, and L. A. Kapicak, *Tetrahedron Lett.*, 4447 (1967).
141. A. Tzschach and W. Lange, *Chem. Ber.*, **95,** 1360 (1962).
142. R. D. Feltham and W. Silverthorn, *Inorg. Syn.*, **10,** 159 (1967).
143. H. Zorn. H. Schindlbauer, and D. Hammer, *Monatsh. Chem.*, **98,** 731 (1967).
144. J. Ellermann and K. Dorn, *Chem. Ber.*, **100,** 1230 (1967).
145. A. Tzschach and W. Fischer, *Z. Chem.*, **7,** 196 (1967).
146. A. Job, R. Reich, and P. Vergnaud, *Bull. Soc. Chim. Fr.*, **35,** 1404 (1924).
147. M. H. Beeby, G. H. Cookson, and F. G. Mann, *J. Chem. Soc.*, 1917 (1950).
148. M. H. Beeby, F. G. Mann, and E. E. Turner, *J. Chem. Soc.*, 1923 (1950).
149. M. H. Beeby and F. G. Mann, *J. Chem. Soc.*, 886 (1951).
150. G. H. Cookson and F. G. Mann, *J. Chem. Soc.*, 2888 (1949).
151. D. A. Thornton, *J. S. Afr. Chem. Inst.*, **17,** 71 (1964).
152. A. Job and R. Reich, *C. R. Acad. Soc., Paris*, **177,** 56 (1923).
153. F. G. Mann and M. J. Pragnell, *J. Chem. Soc.*, 4120 (1965).
154. A. Tzschach and W. Deylig, *Chem. Ber.*, **98,** 977 (1965).
155. H. Bauer, *J. Amer. Chem. Soc.*, **67,** 591 (1945).
156. E. Gryszkiewicz-Trochimowski, M. Buczwinski, and J. Kwapiszewski, *Rocz. Chem.*, **8,** 423 (1928); through *C.A.*, **23,** 1614 (1929).
157. A. Cohen, H. King, and W. I. Strangeways, *J. Chem. Soc.*, 2505 (1932).
158. F. G. Mann and F. C. Baker, *J. Chem. Soc.*, 4142 (1952).
159. E. R. H. Jones and F. G. Mann, *J. Chem. Soc.*, 401 (1955).
160. E. R. H. Jones and F. G. Mann, *J. Chem. Soc.*, 405 (1955).
161. E. R. H. Jones and F. G. Mann, *J. Chem. Soc.*, 411 (1955).
162. D. M. Heinekey, I. T. Millar, and F. G. Mann, *J. Chem. Soc.*, 725 (1963).

163. M. H. Forbes, D. M. Heinekey, F. G. Mann, and I. T. Millar, *J. Chem. Soc.*, 2762 (1961).
164. D. A. Thornton, *J. S. Afr. Chem. Inst.*, **17,** 61 (1964).
165. A. N. Nesmejanow, *Ber.*, **62,** 1010 (1929).
166. A. N. Nesmejanow and E. J. Kahn, *Ber.*, **62,** 1018 (1929).
167. A. N. Nesmeyanov, L. G. Makarova, and T. P. Tolstaya, *Tetrahedron*, **1,** 145 (1957).
168. O. A. Reutov and Yu. G. Bundel, *Zh. Obshch. Khim.*, **25,** 2324 (1955).
169. O. A. Reutov and Yu. G. Bundel, *Vestn. Mosk. Univ.*, **10,** No. 8, *Ser. Fiz. Mat. i Estestv. Nauk*, No. **5,** 85 (1955); through *C.A.*, **50,** 11964 (1956).
170. W. A. Waters, *Nature*, **140,** 466 (1937).
171. W. A. Waters, *J. Chem. Soc.*, 2007 (1937).
172. W. A. Waters, *Nature*, **142,** 1077 (1938).
173. F. B. Makin and W. A. Waters, *J. Chem. Soc.*, 843 (1938).
174. W. A. Waters, *J. Chem. Soc.*, 864 (1939).
175. W. E. Hanby and W. A. Waters, *J. Chem. Soc.*, 1029 (1946).
176. W. A. Waters, *The Chemistry of Free Radicals*, 2nd ed., Oxford Univ. Press, London, 1948, p. 61.
177. L. G. Makarova and A. N. Nesmeyanov, *Izv. Akad. Nauk SSSR, Otd. Khim. Nauk*, 617 (1945); *C.A.*, **40,** 4686 (1946).
178. A. N. Nesmeyanov and L. G. Makarova, *Izv. Akad. Nauk SSSR, Otd. Khim. Nauk*, 213 (1947); *C.A.*, **42,** 5440 (1948).
179. A. B. Bruker, *Dokl. Akad. Nauk SSSR*, **58,** 803 (1947); *C.A.*, **46,** 8625 (1952).
180. A. B. Bruker, *Zh. Obshch. Khim.*, **18,** 1297 (1948); through *C.A.*, **43,** 4647 (1949).
181. M. Ya. Kraft and S. A. Rossina, *Dokl. Akad. Nauk SSSR*, **55,** 821 (1947); through *C.A.*, **42,** 531 (1948).
182. G. O. Doak and L. D. Freedman, *J. Amer. Chem. Soc.*, **73,** 5656 (1951).
183. O. A. Reutow, *Tetrahedron*, **1,** 67 (1957).
184. G. O. Doak and L. D. Freedman, unpublished results.
185. E. E. Turner and F. W. Bury, *J. Chem. Soc.*, **123,** 2489 (1923).
186. G. J. Burrows and E. E. Turner, *J. Chem. Soc.*, **119,** 426 (1921).
187. E. Roberts and E. E. Turner, *J. Chem. Soc.*, 1207 (1926).
188. E. R. H. Jones and F. G. Mann, *J. Chem. Soc.*, 1719 (1958).
189. H. Landolt, *Justus Liebigs Ann. Chem.*, **89,** 301 (1854).
190. R. N. Haszeldine and B. O. West, *J. Chem. Soc.*, 3631 (1956).
191. R. N. Haszeldine and B. O. West, *J. Chem. Soc.*, 3880 (1957).
192. W. R. Cullen, *Can. J. Chem.*, **41,** 2424 (1963).
193. W. R. Cullen and H. J. Emeléus, *J. Chem. Soc.*, 372 (1959).
194. F. Challenger, C. Higginbottom, and L. Ellis, *J. Chem. Soc.*, 95 (1933).
195. F. Challenger, *Chem. Ind.* (London), 657 (1935).
196. F. Challenger and A. A. Rawlings, *J. Chem. Soc.*, 868 (1937).
197. F. Challenger and P. T. Charlton, *J. Chem. Soc.*, 424 (1947).
198. S. Blackburn and F. Challenger, *J. Chem. Soc.*, 1872 (1938).
199. M. L. Bird and F. Challenger, *J. Chem. Soc.*, 163 (1939).
200. F. Challenger, D. B. Lisle, and P. B. Dransfield, *J. Chem. Soc.*, 1760 (1954).
201. G. Kamai and B. D. Chernokal'skiĭ, *Tr. Kazansk. Khim.-Tekhnol. Inst.*, 117 (1959); through *C.A.*, **54,** 24345 (1960).
202. V. Auger, *C. R. Acad. Sci., Paris*, **138,** 1705 (1904).
203. W. M. Dehn and E. Williams, *Amer. Chem. J.*, **40,** 103 (1908).
204. A. N. Nesmeyanov and E. G. Ippolitov, *Vestn. Mosk. Univ.*, **10,** No. 10, *Ser. Fiz. Mat. i Estestv. Nauk*, No. **7,** 87 (1955); through *C.A.*, **50,** 9906 (1956).
205. E. R. H. Jones and F. G. Mann, *J. Chem. Soc.*, 294 (1958).

206. F. F. Blicke and S. R. Safir, *J. Amer. Chem. Soc.*, **63**, 575 (1941).
207. C. G. Krespan, *J. Amer. Chem. Soc.*, **83**, 3432 (1961).
208. W. R. Cullen and L. G. Walker, *Can. J. Chem.*, **38**, 472 (1960).
209. Kh. R. Raver, A. B. Bruker, and L. Z. Soborovskii, U.S.S.R. Pat. 185,919 (Sept. 12, 1966); through *C.A.*, **66**, 10765 (1967).
210. W. C. Davies and H. W. Addis, *J. Chem. Soc.*, 1622 (1937).
211. R. J. Gillespie and E. A. Robinson, *Advances in Inorganic Chemistry and Radiochemistry*, Vol. **1**, Academic Press, New York, 1959, p. 385.
212. M. E. Peach and T. C. Waddington, *J. Chem. Soc.*, 1238 (1961).
213. M. Becke-Goehring and H. Thielemann, *Z. Anorg. Allg. Chem.*, **308**, 33 (1961).
214. B. J. Pullman and B. O. West, *J. Inorg. Nucl. Chem.*, **19**, 262 (1961).
215. N. N. Mel'nihov and M. S. Rokitskaya, *Zh. Obshch. Khim.*, **8**, 834 (1938); through *C.A.*, **33**, 1267 (1939).
216. J. F. Carson and F. F. Wong, *J. Org. Chem.*, **26**, 1467 (1961).
217. G. Tomaschewski, *J. Prakt. Chem.*, **33**, 168 (1966).
218. A. Cahours, *Justus Liebigs Ann. Chem.*, **112**, 228 (1859).
219. A. Merijanian and R. A. Zingaro, *Inorg. Chem.*, **5**, 187 (1966).
220. R. A. Zingaro and E. A. Meyers, *Inorg. Chem.*, **1**, 771 (1962).
221. L. Horner and H. Fuchs, *Tetrahedron Lett.*, 203 (1962).
222. R. R. Renshaw and G. E. Holm, *J. Amer. Chem. Soc.*, **42**, 1468 (1920).
223. R. A. Zingaro and A. Merijanian, *Inorg. Chem.*, **3**, 580 (1964).
224. R. R. Holmes and E. F. Bertaut, *J. Amer. Chem. Soc.*, **80**, 2983 (1958).
225. P. B. Ayscough and H. J. Emeléus, *J. Chem. Soc.*, 3381 (1954).
226. A. Étienne, *C. R. Acad. Sci., Paris*, **221**, 628 (1945).
227. G. J. Burrows, *J. Proc. Roy. Soc. N.S. Wales*, **68**, 72 (1935).
228. G. R. A. Brandt, H. J. Emeléus, and R. N. Haszeldine, *J. Chem. Soc.*, 2552 (1952).
229. W. R. Cullen, *Can. J. Chem.*, **38**, 445 (1960).
230. H. D. Kaesz, S. L. Stafford, and F. G. A. Stone, *J. Amer. Chem. Soc.*, **81**, 6336 (1959).
231. H. G. Ang and H. J. Emeléus, *Chem. Commun.*, 460 (1966).
232. D. Wittenberg and H. Gilman *J. Org. Chem.* **23**, 1063 (1958).
233. W. Hewertson and H. R. Watson, *J.Chem. Soc.*, 1490 (1962).
234. G. D. F. Jackson and W. H. F. Sasse, *J. Chem. Soc.*, 3746 (1962).
235. W. Ipatiew and G. Rasuwajew, *Ber.*, **63**, 1110 (1930).
236. R. E. Dessy and R. L. Pohl, *J. Amer. Chem. Soc.*, **90**, 1995 (1968).
237. A. E. Goddard, *J. Chem. Soc.*, **123**, 1161 (1923).
238. W. G. Lowe and C. S. Hamilton, *J. Amer. Chem. Soc.*, **57**, 2314 (1935).
239. H. H. Sisler and C. Stratton, *Inorg. Chem.*, **5**, 2003 (1966).
240. D. P. Mellor in F. P. Dwyer and D. P. Mellor, eds., *Chelating Agents and Metal Chelates*, Academic Press, New York, 1964, pp. 10–27.
241. G. Booth, *Advances in Inorganic Chemistry and Radiochemistry*, Vol. **6**, Academic Press, New York, 1964, pp. 1–69.
242. J. C. Bailar, Jr., and D. H. Busch, in J. C. Bailar, Jr., ed., *The Chemistry of the Coordination Compounds*, Reinhold, New York, 1956, pp. 78–84.
243. A. Cahours and H. Gal, *C. R. Acad. Sci., Paris*, **71**, 208 (1870).
244. K. A. Jensen, *Z. Anorg. Allg. Chem.*, **229**, 225 (1936).
245. J. Chatt, L. A. Duncanson, and B. L. Shaw, *Proc. Chem. Soc.*, 343 (1957).
246. J. Chatt and B. L. Shaw, *J. Chem. Soc.*, 5075 (1962).
247. J. Chatt, *J. Chem. Soc.*, 652 (1951).
248. A. F. Wells, *Proc. Roy. Soc., Ser. A*, **167**, 169 (1938).
249. R. S. Nyholm, *J. Chem. Soc.*, 843 (1950).
250. F. G. Mann and D. Purdie, *J. Chem. Soc.*, 1549 (1935).

251. F. G. Mann and A. F. Wells, *J. Chem. Soc.*, 702 (1938).
252. R. C. Evans, F. G. Mann, H. S. Peiser, and D. Purdie, *J. Chem. Soc.*, 1209 (1940).
253. W. E. Hatfield and J. T. Yoke, III, *Inorg. Chem.*, **1,** 475 (1962).
254. A. F. Wells, *Z. Kristallogr.*, **94,** 447 (1936).
255. F. G. Mann and A. F. Wells, *Nature*, **140,** 502 (1937).
256. F. G. Mann, A. F. Wells, and D. Purdie, *J. Chem. Soc.*, 1828 (1937).
257. J. Chatt, A. E. Field, and B. L. Shaw, *J. Chem. Soc.*, 3371 (1963).
258. G. E. Schroll (to Ethyl Corp.), U.S. Pat. 3,130,215 (Apr. 21, 1964); *C.A.*, **61,** 4396 (1964).
259. G. Bouquet and M. Bigorgne, *Bull. Soc. Chim. Fr.*, 433 (1962).
260. J. Chatt, N. P. Johnson, and B. L. Shaw, *J. Chem. Soc.*, 2508 (1964).
261. C. N. Matthews, T. A. Magee, and J. H. Wotiz, *J. Amer. Chem. Soc.*, **81,** 2273 (1959).
262. D. M. L. Goodgame, M. Goodgame, and F. A. Cotton, *Inorg. Chem.*, **1,** 239 (1962).
263. R. S. Nyholm, *J. Chem. Soc.*, 1257 (1952).
264. F. P. Dwyer and D. M. Stewart, *J. Proc. Roy. Soc. N.S. Wales*, **83,** 177 (1949).
265. F. P. Dwyer and R. S. Nyholm, *J. Proc. Roy. Soc. N.S. Wales*, **77,** 116 (1943).
266. F. P. Dwyer and R. S. Nyholm, *J. Proc. Roy. Soc. N.S. Wales*, **79,** 121 (1945).
267. J. Chat, N. P. Johnson, and B. L. Shaw, *J. Chem. Soc.*, **A,** 604 (1967).
268. R. S. Nyholm, *J. Proc. Roy. Soc. N.S. Wales*, **78,** 229 (1944).
269. L. Naldini, *Gazz. Chim. Ital.*, **90,** 1231 (1960).
270. R. S. Nyholm, S. S. Sandhu, and M. H. B. Stiddard, *J. Chem. Soc.*, 5916 (1963).
271. K. A. Jensen and P. H. Nielsen, *Acta Chem. Scand.*, **17,** 1875 (1963).
272. J. Chatt and F. A. Hart, *J. Chem. Soc.*, 1378 (1960).
273. L. Vaska and J. W. DiLuzio, *J. Amer. Chem. Soc.*, **83,** 1262 (1961).
274. R. S. Nyholm, *J. Chem. Soc.*, 848 (1950).
275. K. Vrieze, C. MacLean, P. Cossee, and C. W. Hilbers, *Rec. Trav. Chim. Pays-Bas*, **85,** 1077 (1966).
276. J. C. Bailar, Jr. and H. Itatani, *J. Amer. Chem. Soc.*, **89,** 1592 (1967).
277. J. Chatt and G. A. Rowe, *J. Chem. Soc.*, 4019 (1962).
278. N. P. Johnson, C. J. L. Lock, and G. Wilkinson, *J. Chem. Soc.*, 1054 (1964).
279. G. Rouschias and G. Wilkinson, *Chem. Commun.*, 442 (1967).
280. F. P. Dwyer and R. S. Nyholm, *J. Proc. Roy. Soc. N.S.Wales*, **75,** 127 (1941).
281. F. P. Dwyer and R. S. Nyholm, *J. Proc. Roy. Soc. N.S. Wales*, **75,** 140 (1941).
282. F. P. Dwyer and R. S. Nyholm, *J. Proc. Roy. Soc. N.S. Wales*, **76,** 129 (1942).
283. F. P. Dwyer and R. S. Nyholm, *J. Proc. Roy. Soc. N.S. Wales*, **76,** 133 (1942).
284. L. Vallarino, *J. Chem. Soc.*, 2287 (1957).
285. L. Vallarino, *J. Inorg. Nucl. Chem.*, **8,** 288 (1958).
286. H. C. Volger and K. Vrieze, *J. Organometal. Chem.*, **9,** 527 (1967).
287. P. R. Brookes and B. L. Shaw, *J. Chem. Soc.*, **A,** 1079 (1967).
288. F. P. Dwyer, J. E. Humpoletz, and R. S. Nyholm, *J. Proc. Roy. Soc. N.S. Wales*, **80,** 217 (1947).
289. G. J. Burrows and R. H. Parker, *J. Amer. Chem. Soc.*, **55,** 4133 (1933).
290. B. Chiswell and S. E. Livingstone, *J. Chem. Soc.*, 2931 (1959).
291. G. J. Burrows and A. Lench, *J. Proc. Roy. Soc. N.S. Wales*, **70,** 222 (1937).
292. J. Chatt and F. G. Mann, *J. Chem. Soc.*, 610 (1939).
293. J. Lewis, R. S. Nyholm, and D. J. Phillips, *J. Chem. Soc.*, 2177 (1962).
294. J. Lewis, R. S. Nyholm, C. S. Pande, S. S. Sandhu, and M. H. B. Stiddard, *J. Chem. Soc.*, 3009 (1964).
295. G. A. Rodley and P. W. Smith, *J. Chem. Soc.*, **A,** 1580 (1967).
296. F. W. B. Einstein and G. A. Rodley, *J. Inorg. Nucl. Chem.*, **29,** 347 (1967).
297. A. Kabesh and R. S. Nyholm, *J. Chem. Soc.*, 38 (1951).

298. C. M. Harris and R. S. Nyholm, *J. Chem. Soc.*, 63 (1957).
299. R. J. H. Clark, J. Lewis and R. S. Nyholm, *J. Chem. Soc.*, 2460 (1962).
300. R. S. Nyholm, *J. Chem. Soc.*, 851 (1950).
301. H. Nigam, R. S. Nyholm, and D. V. R. Rao, *J. Chem. Soc.*, 1397 (1959).
302. H. L. Nigam, R. S. Nyholm, and M. H. B. Stiddard, *J. Chem. Soc.*, 1806 (1960).
303. J. Lewis, R. S. Nyholm, and P. W. Smith, *J. Chem. Soc.*, 2592 (1962).
304. J. Lewis, R. S. Nyholm, C. S. Pande, and M. H. B. Stiddard, *J. Chem. Soc.*, 3600 (1963).
305. P. Kreisman, R. Marsh, J. R. Preer, and H. B. Gray, *J. Amer. Chem. Soc.*, **90,** 1067 (1968).
306. R. J. H. Clark, D. L. Kepert, J. Lewis, and R. S. Nyholm, *J. Chem. Soc.*, 2865 (1965).
307. R. J. H. Clark, D. L. Kepert, and R. S. Nyholm, *J. Chem. Soc.*, 2877 (1965).
308. R. S. Nyholm and G. J. Sutton, *J. Chem. Soc.*, 572 (1958).
309. R. Ettorre, A. Peloso, and G. Dolcetti, *Gazz. Chim. Ital.*, **97,** 968 (1967).
310. C. M. Harris, R. S. Nyholm, and D. J. Phillips, *J. Chem. Soc.*, 4379 (1960).
311. N. F. Curtis, J. E. Fergusson, and R. S. Nyholm, *Chem. Ind.* (London), 625 (1958).
312. R. S. Nyholm, *J. Chem. Soc.*, 857 (1950).
313. J. Chatt and R. G. Hayter, *J. Chem. Soc.*, 6017 (1963).
314. J. E. Fergusson and R. S. Nyholm, *Nature*, **183,** 1039 (1959).
315. J. E. Fergusson and R. S. Nyholm, *Chem. Ind.* (London), 347 (1960).
316. G. J. Sutton, *Aust. J. Chem.*, **12,** 122 (1959).
317. H. L. Nigam, R. S. Nyholm, and M. H. B. Stiddard, *J. Chem. Soc.*, 1803 (1960).
318. J. G. Noltes and J. W. G. van den Hurk, *J. Organometal. Chem.*, **1,** 377 (1964).
319. R. J. H. Clark, *J. Chem. Soc.*, 5699 (1965).
320. J. Chatt and F. G. Mann, *J. Chem. Soc.*, 1622 (1939).
321. M. Dub, *Organometallic Compounds*, Volume III, 2nd edition, Springer-Verlag, New York, 1968, pp. 98–104.
322. P. Nicpon and D. W. Meek, *Inorg. Chem.*, **6,** 145 (1967).
323. T. D. DuBois and D. W. Meek, *Inorg. Chem.*, **6,** 1395 (1967).
324. F. G. Mann and F. H. C. Stewart, *J. Chem. Soc.*, 1269 (1955).
325. S. E. Livingstone, *J. Chem. Soc.*, 4222 (1958).
326. H. W. Kouwenhoven, J. Lewis, and R. S. Nyholm, *Proc. Chem. Soc.*, 220 (1961).
327. C. M. Harris and S. E. Livingstone, in F. P. Dwyer and D. P. Mellor, eds., *Chelating Agents and Metal Chelates*, Academic Press, New York, 1964, pp. 129–141.
328. G. A. Barclay and R. S. Nyholm, *Chem. Ind.* (London), 378 (1953).
329. G. A. Barclay, R. S. Nyholm, and R. V. Parish, *J. Chem. Soc.*, 4433 (1961).
330. G. A. Barclay, I. K. Gregor, M. J. Lambert, and S. B. Wild, *Aust. J. Chem.*, **20,** 1571 (1967).
331. T. E. W. Howell, S. A. J. Pratt, and L. M. Venanzi, *J. Chem. Soc.*, 3167 (1961).
332. L. M. Venanzi, *Angew. Chem., Int. Ed. Engl.*, **3,** 453 (1964).
333. R. J. Mawby and L. M. Venanzi, *Experientia*, Suppl. No. 9, 240 (1964).
334. R. G. Cunninghame, R. S. Nyholm, and M. L. Tobe, *J. Chem. Soc., Suppl.*, 5800 (1964).
335. A. S. Kasenally, R. S. Nyholm, and M. H. B. Stiddard, *J. Amer. Chem. Soc.*, **86,** 1884 (1964).
336. R. S. Nyholm (to Ethyl Corp.), U.S. Pat. 3,037,037 (May 29, 1962); *C.A.*, **57,** 13806 (1962).
337. G. A. Barclay and A. K. Barnard, *J. Chem. Soc.*, 4269 (1961).
338. G. A. Mair, H. M. Powell, and L. M. Venanzi, *Proc. Chem. Soc.*, 170 (1961).
339. J. A. Brewster, C. A. Savage, and L. M. Venanzi, *J. Chem. Soc.*, 3699 (1961).
340. C. A. Savage and L. M. Venanzi, *J. Chem. Soc.*, 1548 (1962).

341. R. J. Mawby and L. M. Venanzi, *J. Chem. Soc.*, 4447 (1962).
342. J. G. Hartley and L. M. Venanzi, *J. Chem. Soc.*, 182 (1962).
343. G. A. Barclay, C. M. Harris, and J. V. Kingston, *Chem. Commun.*, 965 (1968).
344. F. H. Burstall and R. S. Nyholm, *J. Chem. Soc.*, 3570 (1952).
345. R. S. Nyholm and L. N. Short, *J. Chem. Soc.*, 2670 (1953).
346. S. Ahrland, J. Chatt, and N. R. Davies, *Quart. Rev.* (London), **12,** 265 (1958).
347. W. R. Cullen, *Can. J. Chem.*, **41,** 317 (1963).
348. R. A. Shaw, B. C. Smith, and C. P. Thakur, *Chem. Commun.*, 228 (1966).
349. R. E. Weston, Jr., *J. Amer. Chem. Soc.*, **76,** 2645 (1954).
350. M. S. Lesslie and E. E. Turner, *J. Chem. Soc.*, 1170 (1934).
351. M. S. Lesslie and E. E. Turner, *J. Chem. Soc.*, 1268 (1935).
352. M. S. Lesslie and E. E. Turner, *J. Chem. Soc.*, 730 (1936).
353. M. S. Lesslie, *J. Chem. Soc.*, 1001 (1938).
354. M. S. Lesslie, *J. Chem. Soc.*, 1183 (1949).
355. J. Chatt and F. G. Mann, *J. Chem. Soc.*, 1184 (1940).
356. O. Kennard, F. G. Mann, D. G. Watson, J. K. Fawcett, and K. A. Kerr, *Chem. Commun.*, 269 (1968).
357. I. G. M. Campbell and R. C. Poller, *J. Chem. Soc.*, 1195 (1956).
358. K. Mislow, A. Zimmerman, and J. T. Melillo, *J. Amer. Chem. Soc.*, **85,** 594 (1963).
359. L. Horner and H. Fuchs, *Tetrahedron Lett.*, 1573 (1963).
360. L. Horner and W. Hofer, *Tetrahedron Lett.*, 3321 (1966).
361. L. Horner and W. Hofer, *Tetrahedron Lett.*, 3323 (1966).
362. L. Horner and W. Hofer, *Tetrahedron Lett.*, 4091 (1965).
363. F. H. Westheimer, *Accounts Chem. Res.*, **1,** 70 (1968).
364. F. G. Mann, "The Stereochemistry of the Group V Elements," in W. Klyne and P. B. D. de la Mare, eds., *Progress in Stereochemistry*, Vol. **2,** Academic Press, New York, 1958, pp. 196–207.
365. H. Gilman and C. G. Stuckwisch, *J. Amer. Chem. Soc.*, **63,** 3532 (1941).
366. G. Wittig and E. Benz, *Chem. Ber.*, **91,** 873 (1958).
367. B. Bosnich, R. S. Nyholm, P. J. Pauling, and M. L. Tobe, *J. Amer. Chem. Soc.*, **90,** 4741 (1968).
368. J. E. Purvis and N. P. McCleland, *J. Chem. Soc.*, **101,** 1514 (1912).
369. H. Mohler, *Protar*, **7,** 78 (1941); through *C.A.*, **35,** 4868 (1941).
370. K. Bowden and E. A. Braude, *J. Chem. Soc.*, 1068 (1952).
371. I. G. M. Campbell and R. C. Poller, *Chem. Ind.* (London), 1126 (1953).
372. H. H. Jaffé, *J. Chem. Phys.*, **22,** 1430 (1954).
373. C. N. R. Rao, J. Ramachandran, M. S. C. Iah, S. Somasekhara, and T. V. Rajakumar, *Nature*, **183,** 1475 (1959).
374. W. R. Cullen and R. M. Hochstrasser, *J. Mol. Spectrosc.*, **5,** 118 (1960).
375. C. N. R. Rao, J. Ramachandran, and A. Balasubramanian, *Can. J. Chem.*, **39,** 171 (1961).
376. W. R. Cullen, *Can. J. Chem.*, **39,** 2486 (1961).
377. O. V. Kolninov and Z. V. Zvonkova, *Zh. Fiz. Khim.*, **36,** 2228 (1962).
378. W. R. Cullen, B. R. Green, R. M. Hochstrasser, *J. Inorg. Nucl., Chem.*, **27,** 641 (1965).
379. M. A. Weiner and G. Pasternack, *J. Org. Chem.*, **32,** 3707 (1967).
380. H. H. Jaffé and M. Orchin, *Theory and Applications of Ultraviolet Spectroscopy*, John Wiley and Sons, Inc., New York, 1962, pp. 397–407.
381. B. Chiswell and S. E. Livingstone, *J. Chem. Soc.*, 97 (1960).
382. J. G. Hartley, L. M. Venanzi, and D. C. Goodall, *J. Chem. Soc.*, 3930 (1963).
383. F. Fehér and W. Kolb, *Naturwissenschaften*, **27,** 615 (1939).
384. E. J. Rosenbaum, D. J. Rubin, and C. R. Sandberg, *J. Chem. Phys.*, **8,** 366 (1940).

385. G. Petit, *C. R. Acad. Sci., Paris*, **218,** 414 (1944).
386. H. Siebert, *Z. Anorg. Allg. Chem.*, **273,** 161 (1953).
387. G. P. van der Kelen and M. A. Herman, *Bull. Soc. Chim. Belges*, **65,** 350 (1956).
388. E. G. Claeys and G. P. van der Kelen, *Spectrochim. Acta*, **22,** 2095 (1966).
389. G. Bouquet and M. Bigorgne, *Spectrochim. Acta, Part A*, **23,** 1231 (1967).
390. G. Bouquet, A. Loutellier, and M. Bigorgne, *J. Mol. Structure*, **1,** 211 (1967–68).
391. F. A. Miller and D. H. Lemmon, *Spectrochim. Acta, Part A*, **23,** 1099 (1967).
392. R. A. Zingaro, R. E. McGlothin, and R. M. Hedges, *Trans. Faraday Soc.*, **59,** 798 (1963).
393. S. L. Stafford and F. G. A. Stone, *Spectrochim. Acta*, **17,** 412 (1961).
394. G. B. Deacon and R. A. Jones, *Aust. J. Chem.*, **16,** 499 (1963).
395. R. N. Sterlin and S. S. Dubov, *Zh. Vses. Khim. Obshchest.*, **7,** No. 1, 117 (1962); *C.A.*, **57,** 294 (1962).
396. W. R. Cullen, *Can. J. Chem.*, **38,** 439 (1960).
397. W. R. Cullen, *Can. J. Chem.*, **40,** 426 (1962).
398. W. R. Cullen and N. K. Hota, *Can. J. Chem.*, **42,** 1123 (1964).
399. W. R. Cullen, D. S. Dawson, P. S. Dhaliwal, and G. E. Styan, *Chem. Ind.* (London), 502 (1964).
400. G. E. Coates and C. Parkin, *J. Chem. Soc.*, 421 (1963).
401. G. A. Barclay, I. K. Gregor, and S. B. Wild, *Chem. Ind.* (London), 1710 (1964).
402. W. R. Cullen, G. B. Deacon, and J. H. S. Green, *Can. J. Chem.*, **43,** 3193 (1965).
403. W. R. Cullen, G. B. Deacon, and J. H. S. Green, *Can. J. Chem.*, **44,** 717 (1966).
404. A. L. Allred and A. L. Hensley, Jr., *J. Inorg. Nucl. Chem.*, **17,** 43 (1961).
405. C. R. McCoy and A. L. Allred, *J. Inorg. Nucl. Chem.*, **25,** 1219 (1963).
406. R. G. Kostyanovskii, V. V. Yakshin, I. I. Chervin, and S. L. Zimont, *Izv. Akad. Nauk SSSR, Ser. Khim.*, 2128 (1967); *C.A.*, **68,** 2451 (1968).
407. A. G. Massey, E. W. Randall, and D. Shaw, *Spectrochim. Acta*, **20,** 379 (1964).
408. A. G. Massey, E. W. Randall, and D. Shaw, *Spectrochim. Acta*, **21,** 263 (1965).
409. W. Brügel, T. Ankel, and F. Krückeberg, *Z. Electrochem.*, **64,** 1121 (1960).
410. W. Voskuil and J. F. Arens, *Rec. Trav. Chim. Pays-Bas*, **83,** 1301 (1964).
411. W. R. Cullen, D. S. Dawson, and G. E. Styan, *J. Organometal. Chem.*, **3,** 406 (1965).
412. T. D. Coyle, S. L. Stafford, and F. G. A. Stone, *Spectrochim. Acta*, **17,** 968 (1961).
413. S. S. Dubov, B. I. Tetel'baum, and R. N. Sterlin, *Zh. Vses. Khim. Obshchest.*, **7,** 691 (1962); through *C.A.*, **58,** 8538 (1963).
414. R. G. Barnes and P. J. Bray, *J. Chem. Phys.*, **23,** 407 (1955).
415. H. P. Fritz and K. E. Schwarzhans, *J. Organometal. Chem.*, **5,** 103 (1966).
416. M. A. Bennett, G. J. Erskine, and R. S. Nyholm, *J. Chem. Soc.*, **A,** 1260 (1967).
417. H. D. Springall and L. O. Brockway, *J. Amer. Chem. Soc.*, **60,** 996 (1938).
418. J. Trotter, *Can. J. Chem.*, **41,** 14 (1963).
419. J. Trotter, *Acta Crystallogr.*, **16,** 1187 (1963).
420. D. Sartain and M. R. Truter, *J. Chem. Soc.*, 4414 (1963).
421. D. Mootz, P. Holst, L. Berg, and K. Drews, *Z. Kristallogr.*, **117,** 233 (1962).
422. H. J. M. Bowen, *Trans. Faraday Soc.*, **50,** 463 (1954).
423. D. R. Lide, Jr., *Spectrochim. Acta*, **15,** 473 (1959).
424. E. Bergmann and W. Schütz, *Z. Phys. Chem.*, **B19,** 401 (1932).
425. M. J. Aroney, R. J. W. Le Fèvre, and J. D. Saxby, *J. Chem. Soc.*, 1739 (1963).
426. K. I. Kuz'min and G. Kamai, *Dokl. Akad. Nauk SSSR*, **73,** 709 (1950); *C.A.*, **44,** 10409 (1950).
427. H. Mohler, *Helv. Chim. Acta*, **21,** 789 (1938).
428. R. N. Sterlin, S. S. Dubov, W. K. Li, L. P. Vakhomchik, and I. L. Knunyants, *Zh. Vses. Khim. Obshchest.*, **6,** No. 1, 110 (1961).

429. C. P. Smyth, *J. Org. Chem.*, **6,** 421 (1941).
430. W. C. Davies, *J. Chem. Soc.*, 462 (1935).
431. J. H. Gibbs, *J. Phys. Chem.*, **59,** 644 (1955).
432. M. V. Forward, S. T. Bowden, and W. J. Jones, *J. Chem. Soc.*, S121 (1949).
433. E. J. Rosenbaum and C. R. Sandberg, *J. Amer. Chem. Soc.*, **62,** 1622 (1940).
434. F. Challenger and A. E. Goddard, *J. Chem. Soc.*, **117,** 762 (1920).
435. W. J. Jones, W. J. C. Dyke, G. Davies, D. C. Griffiths, and J. H. E. Webb, *J. Chem. Soc.*, 2284 (1932).
436. E. Gryszkiewicz-Trochimowski and S. F. Sikorski, *Bull. Soc. Chim. Fr.*, **41,** 1570 (1927).
437. K. I. Kuz'min and G. Kamai, *Sb. Statei Obshch. Khim.*, **1,** 223 (1953); through *C.A.*, **49,** 841 (1955).
438. R. G. Gillis, *Rev. Pure Appl. Chem.*, **10,** 21 (1960).
439. G. Kamai and B. D. Chernokal'skii, *Dokl. Akad. Nauk SSSR*, **149,** 850 (1963).
440. L. H. Long and J. F. Sackman, *Trans. Faraday Soc.*, **52,** 1201 (1956).
441. S. J. W. Price and A. F. Trotman-Dickenson, *Trans. Faraday Soc.*, **54,** 1630 (1958).
442. W. F. Lautsch, *Chem. Tech.* (Berlin), **10,** 419 (1958); *C.A.*, **53,** 43 (1959).
443. W. F. Lautsch, P. Erzberger, and A. Tröber, *Wiss. Z. Tech. Hochsch. Chem. Leuna-Merseburg*, **1,** 31 (1958–59); *C.A.*, **54,** 13845 (1960).
444. K. H. Birr, *Z. Anorg. Allg. Chem.*, **311,** 92 (1961).
445. H. A. Skinner, *Advances in Organometallic Chemistry*, Vol. **2,** 1964, Academic Press, New York, p. 84.
446. C. T. Mortimer and P. W. Sellers, *J. Chem. Soc.*, 1965 (1964).
447. C. T. Mortimer, *J. Chem. Educ.*, **35,** 381 (1958).
448. D. E. Bublitz and A. W. Baker, *J. Organometal. Chem.*, **9,** 383 (1967).
449. N. P. Buu-Hoi, M. Mangane, and P. Jacquignon, *J. Heterocycl. Chem.*, **3,** 149 (1966).
450. N. P. Buu-Hoi, M. Mangane, and P. Jacquignon, *J. Heterocycl. Chem.*, **3,** 374 (1966).
451. S. S. Dubov, F. N. Chelobov, and R. N. Sterlin, *Zh. Vses. Khim. Obshchest.*, **7,** 585 (1962); through *C.A.*, **58,** 2970 (1963).
452. W. R. Cullen and D. C. Frost, *Can. J. Chem.*, **40,** 390 (1962).
453. J. J. Lagowski, *Quart. Rev.* (London), **13,** 233 (1959).
454. E. A. Robinson, *Can. J. Chem.*, **39,** 247 (1961).
455. F. I. Vilesov and V. M. Zaitsev, *Dokl. Akad. Nauk SSSR*, **154,** 886 (1964).
456. R. H. Smith and D. H. Andrews, *J. Amer. Chem. Soc.*, **53,** 3661 (1931).
457. M. J. Aroney, R. J. W. Le Fèvre, and J. D. Saxby, *Can. J. Chem.*, **42,** 1493 (1964).
458. O. R. Quayle, *Chem. Rev.*, **53,** 439 (1953).
459. T. W. Gibling, *J. Chem. Soc.*, 236 (1945).
460. G. O. Doak and L. D. Freedman, *Chem. Rev.*, **61,** 31 (1961).
461. P. Pascal, *C. R. Acad. Sci., Paris*, **218,** 57 (1944).
462. M. Prasad and L. N. Mulay, *J. Chem. Phys.*, **19,** 1051 (1951).
463. M. Prasad and L. N. Mulay, *J. Chem. Phys.*, **20,** 201 (1952).
464. P. von R. Schleyer and R. West, *J. Amer. Chem. Soc.*, **81,** 3164 (1959).
465. A. E. Lutskii and E. M. Obukhova, *Zh. Fiz. Khim.*, **35,** 1960 (1961).
466. S. Ahrland, J. Chatt, N. R. Davies, and A. A. Williams, *J. Chem. Soc.*, 276 (1958).

V Arsine oxides, arsenic halides, and arsonium salts

1. PREPARATION OF ARSINE OXIDES

A. By oxidation of tertiary arsines

Oxidation of aromatic tertiary arsines yields either the arsine oxides, Ar_3AsO, or the dihydroxides, $Ar_3As(OH)_2$. When the latter are obtained, they may be readily dehydrated to the oxides by heating in the presence of a drying agent. Potassium permanganate and hydrogen peroxide probably are used the most for the oxidation of aromatic tertiary arsines. With potassium permanganate the arsine may either be dissolved in acetone and oxidized by the addition of the solid oxidizing agent (1, 2) or the arsine may be refluxed with aqueous potassium permanganate solution (3). Like potassium permanganate, hydrogen peroxide gives excellent results (4–7). There is some evidence that the reaction of triphenylarsine with hydrogen peroxide forms an adduct $(Ph_3AsO)_2{\cdot}H_2O_2$, which is readily decomposed to triphenylarsine oxide (8).

Iodine has also been used successfully for oxidizing aromatic arsines. Tris(4-aminophenyl)arsine oxide may be obtained in this manner (9). (β-Carboxyethyl)diphenylarsine oxide has been prepared from the corresponding arsine by the use of mercuric oxide in alcohol solution (10). Other oxidizing agents which have been used include perbenzoic acid (11), phenyl benzenethiolsulfinate (12), and selenium dioxide (13). When triphenylarsine is oxidized with selenium dioxide, a mixture of the arsine oxide and arsine selenide is obtained (13).

The oxidation of trialkylarsines and dialkylarylarsines does not always yield arsine oxides. For example, Friedrich and Marvel (14) demonstrated that in the oxidation of triethyl- or *n*-butyldiethylarsines, one ethyl group is cleaved from the arsenic atom to give diethylarsinic acid and *n*-butylethylarsinic acid, respectively. These oxidations were performed either by allowing the arsines to stand in the air or by passing a stream of oxygen through ether solutions of the arsines. In a similar manner, tribenzylarsine is rapidly oxidized in the air (particularly in the presence of traces of impurities) to give a mixture of dibenzylarsinic acid, benzaldehyde, and arsenic oxide (15).

The alkaline oxidation of tri-α-furylarsine cleaves a furyl group to give di-α-furylarsinic acid (16). Dimethylphenylarsine, when allowed to stand in the air for several years, yields methylphenylarsinic acid (17). It was demonstrated that the reaction proceeds according to the following equation:

$$C_6H_5As(CH_3)_2 + O_2 + H_2O \rightarrow C_6H_5(CH_3)AsO_2H + CH_3OH$$

However, the oxidation of methyldiphenylarsine with hydrogen peroxide does give the expected diphenylmethylarsine oxide. Similarly, the reaction of a number of aryldiethylarsines with hydrogen peroxide in acetone yields the corresponding arsine oxides (18).

Recent work by Merijanian and Zingaro (19) indicates that the prolonged exposure of trialkylarsines to the atmosphere or to an excess of strong oxidizing agents normally leads to the formation of dialkylarsinic acids. The arsine oxides *can* be prepared, however, by refluxing the arsine in acetone with mercuric oxide in the absence of air or by treating an excess of the arsine in ether with hydrogen peroxide. The physical constants of the arsine oxides prepared by these methods often differ significantly from the values reported in the earlier literature (20–23). It seems likely, therefore, that much of the older work on the oxidation of trialkylarsines to arsine oxides is open to question.

Whether the arsine oxide or the dihydroxide is obtained on oxidation depends both on the nature of the groups attached to the arsenic atom and on the method of isolation. No dihydroxides have been reported by the oxidation of aliphatic tertiary arsines. By contrast tertiary aromatic arsines may give either the dihydroxide or the oxide. For example, tris(3-acetylaminophenyl)arsine, when oxidized with potassium permanganate, gives the dihydroxide but alkaline hydrolysis of this dihydroxide gives tris(3-aminophenyl)arsine oxide (1).

Rasuwajew and Malinowski (24) studied the iodine oxidation of a number of phenarsazines of the following type:

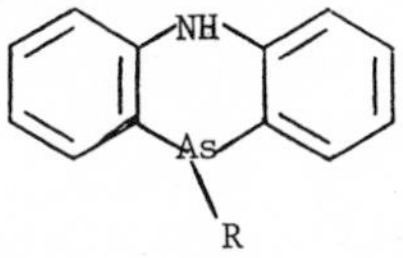

where R was methyl, ethyl, *n*-propyl, isoamyl, or phenyl. Regardless of the method of isolation, only the oxide is obtained when R is phenyl. When R is an aliphatic group, however, either the oxide or the hydroxide is obtained, depending upon the method used for isolation. When the HI, which is formed in the oxidation, is removed by reaction with silver oxide, the arsine oxide is invariably formed. Recrystallization of the dihydroxide, where R = ethyl, from aqueous alcohol gives a mixture of the oxide and dihydroxide.

B. By hydrolysis

Hydrolysis of the arsenic dihalides, R_3AsX_2, produces the arsine oxides or the corresponding dihydroxides. Worrall (25) prepared tri-*p*-biphenylylarsenic dihydroxide by boiling tri-*p*-biphenylylarsenic dibromide or dichloride in aqueous solution or by treatment of the dihalides with alcoholic ammonia. Tri-*o*-biphenylylarsine oxide has been prepared from the corresponding dibromide or dichloride by hydrolysis with alcoholic potassium hydroxide (26). Alkaline hydrolysis of *p*-biphenylyl(*p*-chlorophenyl)phenylarsenic dibromide gives the corresponding arsine oxide (27).

Hydrolysis of hydroxy salts of the type $R_3As(OH)X$ also gives the dihydroxides. Thus, Lyon and Mann (28) hydrolyzed 2-hydroxy-2-phenylisoarsindolinium nitrate (**1**) to the corresponding dihydroxide by treatment with

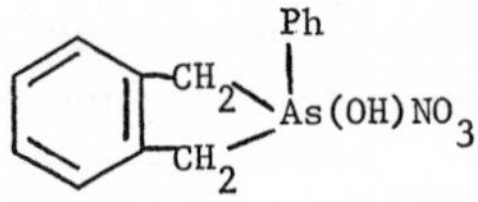

Ph

CH_2 As(OH)NO_3

CH_2

1

sodium hydroxide. The dihydroxide is readily dehydrated to the oxide, but exposure to moist air regenerates the dihydroxide. Vaughan and Tarbell (4) hydrolyzed tris(3-nitro-4-ethoxyphenyl)arsenic hydroxychloride with alkali to the corresponding oxide.

The mechanism of the hydrolysis of tertiary arsenic dihalides has been studied quantitatively by Nylén (29). By titration of aqueous solutions of trimethyl- and triethylarsenic dibromides with standard alkali, Nylén demonstrated that these compounds behave as dibasic acids. Then by determination of both the pH and the bromide ion concentration (with a silver–silver bromide electrode), he was able to demonstrate that the following reactions occur:

$$R_3AsBr_2 + H_2O \rightarrow R_3As(OH)Br + HBr$$

$$R_3As(OH)Br \rightleftharpoons [R_3AsOH]^+ + Br^-$$

Although the hydroxybromides are not completely ionized in aqueous solution, they may be titrated with alkali:

$$[R_3AsOH]^+Br^- + OH^- \rightarrow R_3AsO + H_2O + Br^-$$

In addition to hydrolysis, the tertiary arsenic dihalides can be converted to the oxides by treatment with suitable metallic oxides. Thus, tri-*n*-butylarsine oxide has been obtained from the dibromide by treatment with either mercuric or silver oxides in alcohol (21). Tris(trifluoromethyl)arsine oxide, however, can not be prepared either by the hydrolysis of tris(trifluoromethyl)-arsenic dichloride or by reaction of the dichloride with silver oxide (30). Hydrolysis leads to tris(trifluoromethyl)arsine and bis(trifluoromethyl)arsinic acid; reaction with silver oxide gives chlorobis(trifluoromethyl)arsine, carbon dioxide, fluoride ion, and bis(trifluoromethyl)arsinic acid.

The conversion of arsonium ylids and arsine imides to arsine oxides will be discussed in Sections V-9-B and V-4, respectively.

C. By the Bart or Meyer reaction

Although diphenylarsinic acids are readily prepared from alkali arylarsonites and diazonium salts, the corresponding reaction with an alkali diarylarsinite has apparently been investigated in only one case. This case is the preparation of tris(*p*-nitrophenyl)arsine oxide from sodium bis(*p*-nitrophenyl)arsinite and diazotized *p*-nitroaniline (31). Kraft and Rossina (32) have demonstrated, however, that an arsine oxide can be obtained as a by-product of the Bart reaction. From the crude sodium salt of *p*-hydroxybenzenearsonic acid (corresponding to 1200 g of the acid obtained from the Bart reaction), the authors (32) isolated 23 g of bis(*p*-hydroxyphenyl)arsinic acid and 5 g of tris(*p*-hydroxyphenyl)arsine oxide. Since the arsine oxide was obtained only in small yield, it may well have been overlooked in many other Bart reactions.

In several cases the arsine oxide has been obtained as the sole product of the Bart reaction. Thus, tris[*o*-(2-carboxyvinyl)phenyl]arsine oxide (33) has been obtained from *o*-aminocinnamic acid, and tris(2-oxo-1,2-benzopyran-6-yl)arsine oxide (24) from 6-aminocoumarin.

There are a few cases where the Meyer reaction has been used for the preparation of arsine oxides. Trimethylarsine oxide has been prepared from cacodyl oxide, methyl iodide, and sodium hydroxide in methyl alcohol solution (35). Braunholtz and Mann (36) prepared bis(carboxymethyl)-phenylarsine oxide from (carboxymethyl)phenylchloroarsine and sodium chloroacetate:

$$C_6H_5As(Cl)CH_2COOH + ClCH_2CO_2Na \rightarrow C_6H_5As(O)(CH_2COOH)_2 + NaCl$$

The product thus obtained was identical with the arsine oxide obtained by the oxidation of bis(carboxymethyl)phenylarsine with hydrogen peroxide. In a similar manner Cookson and Mann (37) prepared bis(γ-carboxypropyl)-phenylarsine oxide by refluxing (γ-carboxypropyl)phenylchloroarsine and 1-bromo-3-butyronitrile in sodium hydroxide solution.

D. By cyclodehydration

Just as a suitably oriented arsonic acid can be cyclized to a heterocyclic arsinic acid by cyclodehydration (cf. Section II-1-E-(3)), arsinic acids can be cyclized to heterocyclic arsine oxides. The reaction, however, has not been studied extensively, and there are only a few examples in the chemical literature. Jones and Mann (38) cyclized (β-phenylethyl)phenylarsinic acid to 1-phenylarsindoline oxide by heating the acid with concentrated sulfuric acid:

$$C_6H_5CH_2CH_2(C_6H_5)AsO_2H \xrightarrow{H_2SO_4}$$ [1-phenylarsindoline oxide: ring with CH_2–CH_2 and $O{=}As{-}C_6H_5$]

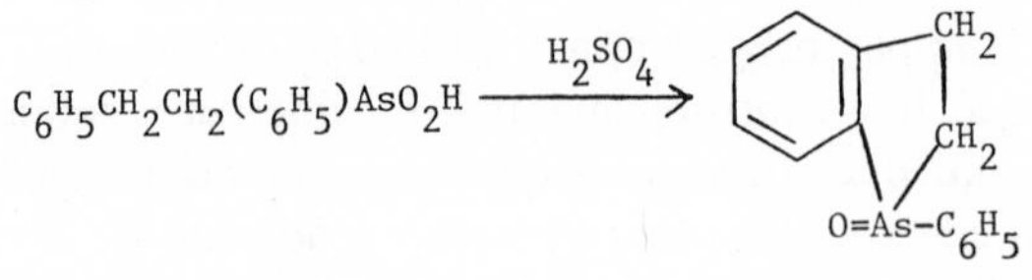

In a similar manner Campbell and Poller (39) have cyclized several (2-biphenylyl)phenylarsinic acids to heterocyclic arsine oxides:

[Structures: (2-biphenylyl)phenylarsinic acid with R on the 4'-position, As(=O)(OH) and R'-substituted phenyl → cyclized arsine oxide As=O with R'-phenyl and R]

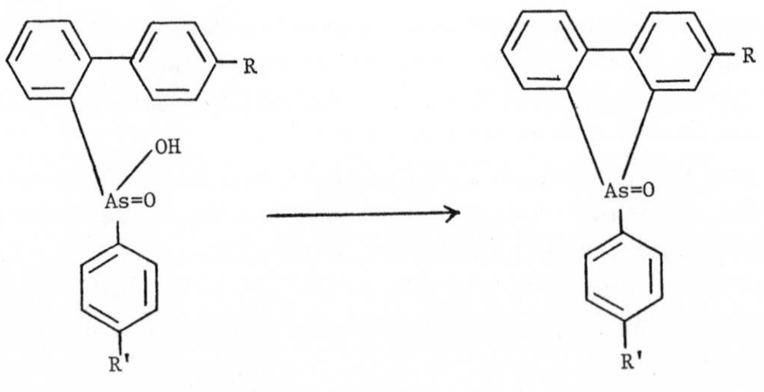

When R = NO_2 and R′ = H, the acid is easily cyclized by warming in concentrated sulfuric acid solution. When R = OCH_3 and R′ = COOH, however, sulfonation occurs, but cyclization can be effected by heating with polyphosphoric acid.

E. Miscellaneous methods

Cookson and Mann (40) allowed diazomethane to react with diphenylchloroarsine and obtained the interesting arsine oxide $(C_6H_5)_2AsCH_2As(C_6H_5)_2$.

$$(C_6H_5)_2AsCH_2\underset{\substack{\downarrow \\ O}}{As}(C_6H_5)_2$$

The compound was characterized by conversion to the hydroxypicrate, -styphnate, and -picrolonate and also by conversion to the sulfide with hydrogen sulfide. The authors suggested, for this peculiar reaction, that traces of moisture in the solvents converted diphenylchloroarsine to bis(diphenylarsenic) oxide which then reacted with diazomethane according to the following reaction sequence:

$$[(C_6H_5)_2As]_2O + CH_2N_2 \longrightarrow (C_6H_5)_2\underset{\substack{| \\ CH_2N_2^-}}{\overset{+}{As}}\text{—}OAs(C_6H_5)_2 \longrightarrow$$

$$(C_6H_5)_2As\overset{-}{C}H_2 + (C_6H_5)_2\overset{+}{As}O + N_2 \longrightarrow (C_6H_5)_2AsCH_2\underset{\substack{\downarrow \\ O}}{As}(C_6H_5)_2$$

Oxidation of this monoxide, either by hydrogen peroxide or simply by dissolving in ether containing peroxides, gives the dioxide,

$$(C_6H_5)_2\underset{\substack{\downarrow \\ O}}{As}CH_2\underset{\substack{\downarrow \\ O}}{As}(C_6H_5)_2$$

Triphenylarsine oxide has been obtained by the thermal decomposition of a triphenylarsine-sulfur trioxide adduct (41) and by the air oxidation of a triphenylarsine-aluminum chloride adduct (42). Several triarylarsenic dihydroxides have been prepared by the alkaline hydrolysis of 1:1 adducts of triarylarsines and sulfur monochloride (43).

2. REACTIONS OF ARSINE OXIDES

The arsine oxides (and dihydroxides) can be readily reduced to tertiary arsines by use of the same reducing agents that have proved successful for reducing arsonic and arsinic acids to arsonous or arsinous acids. Sulfur dioxide in the presence of a trace of iodide ion is the customary reducing agent (6, 36, 40). The arsine oxide may be suspended in chloroform containing a trace of potassium iodide and a stream of sulfur dioxide passed through the solution. As the arsine is formed, it dissolves in the chloroform and may be recovered by distillation. Hypophosphorous acid has also been used for reducing arsine oxides (2). Tris(*p*-nitrophenyl)arsine oxide has been reduced to tris(*p*-aminophenyl)arsine by stannous chloride (1).

An interesting reduction has been carried out by Vaughan and Tarbell (4). Tris(3-nitro-4-ethoxyphenyl)arsine oxide, when reduced at room temperature and low pressure (50 psi) with Raney nickel and hydrogen, gives the expected tris(3-amino-4-ethoxyphenyl)arsine oxide, but at 1000 psi and at a temperature of 80–100° tris(3-amino-4-ethoxyphenyl)arsine is formed.

Nitration of aromatic arsine oxides presumably gives only *meta* derivatives although this has not been proven. Michaelis (44) obtained a tris(nitrophenyl)arsine oxide by nitration of triphenylarsine with fuming nitric acid and sulfuric acid. Although Michaelis himself did not assign a structure to this compound, it is generally believed to be the *meta* derivative. Tris(*p*-bromophenyl)arsine oxide (2, 4) and methylbis(*p*-bromophenyl)arsine oxide (2) have been nitrated, and in both cases it has been assumed that the products contained the nitro group exclusively in *meta* position to the arsenic atom. Tribenzylarsine oxide, on nitration, gives principally the *para* derivative with traces of the *ortho* isomer (45). Proof of orientation in this case was demonstrated by oxidation to *o*- and *p*-nitrobenzoic acids.

The pyrolysis of arsine oxides has been studied by Kamai and Chernokal'skii (23, 46). Triethyl- and tri-*n*-propylarsine oxides, when heated to 230–250°, give triethyl- and tri-*n*-propylarsines together with ethyl and propyl alcohols. *n*-Butyldimethylarsine oxide, on heating, gives *n*-butyldimethylarsine and *n*-butyl *n*-butylmethylarsinite. Di-*n*-butylmethylarsine oxide gives a mixture of *n*-butyraldehyde, *n*-butyl alcohol, di-*n*-butylmethylarsine, tri-*n*-butylarsine, and methyl di-*n*-butylarsinite. The yields of the products, as given by the authors, do not show a material balance, and nothing can be said about the mechanism of these pyrolyses.

(β-Carboxyethyl)diphenylarsine oxide, heated just above its melting point, gives propionic acid and bis(diphenylarsenic) oxide (10). (β-Carboxypropyl)-diphenylarsine oxide, under similar conditions, gives isobutyric acid and bis(diphenylarsenic) oxide. No suggestions as to the mechanism of these interesting eliminations have been advanced.

The action of molten sodium hydroxide on both phosphine and arsine oxides has been studied by Horner, Hoffmann, and Wippel (47). They found that phosphinic or arsinic acids are formed in good yields:

$$R_3AsO + NaOH \rightarrow R_2AsO_2Na + RH$$

Their results suggest that the reaction involves the nucleophilic attack of hydroxide ion on the phosphorus or arsenic atom and subsequent elimination of a carbanion. Triphenylarsine oxide, when melted with sodium hydroxide, gives on acidification an 87% yield of diphenylarsinic acid.

Tertiary arsine oxides have received considerable attention as ligands in coordination chemistry. Nyholm (48) has demonstrated that when methyldiphenylarsine reacts with a hot aqueous solution of cupric chloride, the arsine is oxidized to the oxide, which forms the following coordination compound:

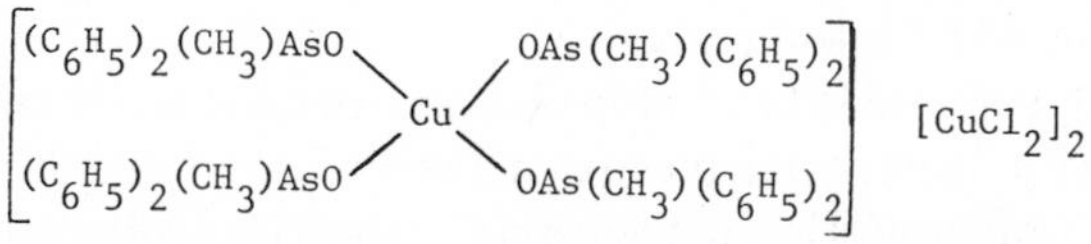

$$\left[\begin{matrix}(C_6H_5)_2(CH_3)AsO & & OAs(CH_3)(C_6H_5)_2 \\ & Cu & \\ (C_6H_5)_2(CH_3)AsO & & OAs(CH_3)(C_6H_5)_2\end{matrix}\right] [CuCl_2]_2$$

Goodwin and Lions (49) have shown that dimethyl-α-picolylarsine is oxidized in a similar manner to the arsine oxide by copper(II) with formation of the cation:

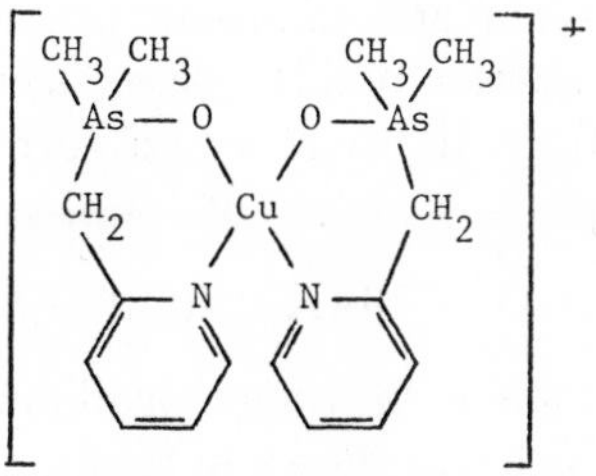

[Structure: Cu coordinated by two O and two pyridine N atoms of two $(CH_3)_2As(O)CH_2$-pyridyl ligands; overall charge +]

Oxidation of the copper(I) complex in the air gives the copper(II) complex.

More extensive studies of arsine oxides as ligands have been made by Goodgame, Cotton, and co-workers (50–54) and by Tyree and co-workers (55, 56). Triphenylarsine oxide forms the complexes $(Ph_3AsO)_2NiCl_2$ and $(Ph_3AsO)_2NiBr_2$ in which the nickel assumes a "pseudotetrahedral" structure (50). In a similar manner salts of manganese(II) form tetrahedral complexes of the type $(Ph_3AsO)_2MnX_2$, where X may be chlorine, bromine, or iodine (52). Copper(II) forms complexes of the type $(Ph_3AsO)_2CuX_2$, where X is chlorine or bromine (55). Cobalt forms complexes of the type $[Co(Ph_3AsO)_2]X_2$, where X is Cl, Br, I, NCS, or ClO_4 (53). Another compound prepared was $[Co(Ph_3AsO)_4]I_2$.

Molybdenum complexes of triphenylarsine and phosphine oxides have been investigated by Sheldon (57). From triphenylarsine oxide and chloromolybdic(II) acid two types of salt-like compounds are obtained. The first type is believed to be an arsonium salt with the structure

$$[(C_6H_5)_3AsOH]_2[(Mo_6Cl_8)Cl_6]$$

It is suggested that in the second type the cation exists as a hydrogen-bonded complex of the type $[(C_6H_5)_3AsOH \cdots OAs(C_6H_5)_3]^+$. When these salts are heated strongly *in vacuo*, or when ethanol solutions of molybdenum(II) chloride and triphenylarsine oxide or triphenylphosphine oxide are mixed, nonelectrolytes of the type $[(Mo_6Cl_8)Cl_4(Ph_3MO)_2]$ (where M is As or P) are obtained. It was suggested that these are octahedral complexes of the Mo_6Cl_8 polynucleus. These complexes are nonhydroscopic solids, stable to 200° *in vacuo*. They dissociate in nitrobenzene solution but can be recovered by the addition of petroleum ether.

Phillips and Tyree (55) have prepared an extended series of coordination compounds with triphenylarsine oxide (as the donor molecule) and such elements as chromium(III), manganese(II), iron(III), cobalt(II), nickel(II), zinc, mercury(II), tin(IV), and antimony(III) and antimony(V) (as acceptors). Salts used for this investigation included chlorides, bromides, nitrates, and perchlorates. Complexes of the halides, in general, were nonelectrolytes, whereas the nitrates and perchlorates appeared to be ionic.

The reactions of niobium and tantalum pentachlorides with triphenylphosphine and triphenylarsine oxides have been found (8) to involve abstraction of oxygen from the oxides and formation of adducts of the metal oxytrichlorides:

$$NbCl_5 + 3(C_6H_5)_3AsO \rightarrow NbOCl_3 \cdot 2(C_6H_5)_3AsO + (C_6H_5)_3AsCl_2$$

Trimethylarsine oxide reacts with trimethylgallium to form the 1:1 adduct, $(CH_3)_3AsOGa(CH_3)_3$, which was shown to have a normal cryoscopic molecular weight in benzene solution (58). The adduct is surprisingly stable and shows no tendency to dissociate even at 150°. The use of triphenylarsine oxide as an analytical reagent for a number of metals has been investigated by Pietsch and Nagl (59–61).

Zingaro and Meyers (62) have examined the reaction of triphenylarsine oxide and iodine by means of ultraviolet spectroscopy. Their results suggest the formation of a well-defined 1:1 complex in chloroform solution:

$$(C_6H_5)_3AsO + I_2 \rightleftharpoons (C_6H_5)_3AsO \cdot I_2$$

For the equilibrium defined by the above equation they found an equilibrium constant of 41.2 at 25°. More recently, Kolar, Zingaro, and Irgolic (63) have studied the 1:1 complexes formed by the reaction of iodine with tri-*n*-octyl- or tricyclohexylarsine oxide in carbon tetrachloride. These trialkylarsine oxides were found to possess a donor strength toward iodine considerably greater than that of triphenylarsine oxide, triphenylarsine, or trialkylphosphine oxides.

Hydrogen-bonded 1:1 adducts of triarylarsine oxides with primary amines or sulfonamides have been recently reported (64). Intermolecular hydrogen-bonding between the amino and arsine oxide groups of tris(*m*-aminophenyl)-arsine oxide has been suggested in order to explain the infrared spectrum of this compound (19).

Triphenylarsine oxide reacts with potassium in tetrahydrofuran to give first an intense red-brown solution followed by the precipitation of a similarly colored solid (65). Although the structure of this compound is not known, the reaction between triphenylphosphine oxide and alkali metals has been more thoroughly investigated and found to yield compounds of the type $(C_6H_5)_3POM$, $(C_6H_5)_3POM_2$, and $(C_6H_5)_2POM_3$ (66).

A number of unsuccessful attempts have been made to resolve arsine oxides of the type $R_1R_2R_3As \rightarrow O$ into optical isomers (67, 68). This is in contrast to amine and phosphine oxides which can be resolved. Mann (69) has suggested that the failure to resolve the arsine oxides may be due to the fact that, in the course of resolution, the recrystallized salt of the oxide with the optically active acid is always treated with alkali. This may give a covalent dihydroxide, $R_1R_2R_3As(OH)_2$, which possesses a plane of symmetry, and the resulting arsine oxide would then be a racemic mixture.

The reaction of arsine oxides with urea to form isocyanates of the type $R_3As(NCO)_2$ has been described in the patent literature (70). Monagle (71) has found that tertiary arsine oxides are effective catalysts for converting organic isocyanates to carbodiimides.

3. PHYSICAL PROPERTIES AND STRUCTURE OF ARSINE OXIDES AND DIHYDROXIDES

The arsine oxides and dihydroxides, in general, are high melting solids, insoluble in water, and usually only slightly soluble in organic solvents. Molecular weight determinations with a vapor pressure osmometer show that trialkylarsine oxides are not associated; the solvent used in this study was not specified (19). The dipole moment of triphenylarsine oxide has been found to be 5.50 D (72).

The question of conjugation of the arsenic atom with the phenyl rings in triphenylarsine oxide was first considered by Jaffé (73). He found that the ultraviolet absorption spectra of both triphenylarsine oxide and benzenearsonic acid, as well as related compounds of phosphorus, antimony, and sulfur, are characteristic of unperturbed or only weakly perturbed benzene rings. In compounds with several benzene rings, each ring appears to make an almost independent contribution to the molar extinction coefficient of the corresponding peaks. On the basis of this evidence Jaffé concluded that there is little or no conjugation between the arsenic (phosphorus, antimony, and sulfur) atom and the benzene ring in compounds such as triphenylarsine oxide, in which the arsenic atom does not contain any unshared valence electrons. Later workers have tended to agree with this conclusion (74).

The infrared spectrum of triphenylarsine oxide has been studied by a number of investigators (52, 54, 55, 57, 75, 76); a strong band at about 880 cm^{-1} has been assigned to the As=O stretching vibration. Similarly, the

As=O vibration of diphenyl(4-hydroxy-3,5-di-*tert*-butylphenyl)arsine oxide occurs at 893 cm^{-1} (7). The spectrum of methyldiphenylarsine oxide contains a strong band at 875 cm^{-1} and a shoulder at 866 cm^{-1}; these bands may arise from modes involving mixtures of As=O stretching and methyl rocking (54). The As=O stretching frequency in trialkylarsine oxides appears as a well-defined band in the 850–910 cm^{-1} region (19, 77). The effect of the electronegativity of the groups attached to the arsenic atom on the As=O frequency has been investigated by Shagidullin and co-workers (77) and has been discussed in Section II-3-A.

Rodley, Goodgame, and Cotton (54) have pointed out that one might expect that coordination of arsine oxides with metal salts would lead to an increase in the As=O frequency because of coupling of the arsenic–oxygen and metal–oxygen oscillators. It is found, however, that such coordination leads to comparatively little change in the position of the As=O band. In many cases, this frequency is actually *lowered* by a significant amount. Phillips and Tyree (55) have suggested that the As=O bond possesses some double-bond character arising from p_π-d_π overlap of filled p orbitals of the oxygen with vacant d orbitals of the arsenic. Coordination of the arsine oxide with an acceptor lowers the electron density in the π-orbital, i.e., reduces the amount of double-bond character and therefore tends to lower the As=O stretching frequency. In other words, the expected increase in the As=O frequency caused by coupling of the arsenic–oxygen and metal–oxygen oscillators is often more than offset by a lowering of the force constant of the As=O bond. The electronic spectra of coordination compounds of arsine oxides have also been reported (54, 78).

Müller and co-workers (7) have investigated the proton NMR and mass spectra of diphenyl(4-hydroxy-3,5-di-*tert*-butylphenyl)arsine oxide and the ESR spectrum of the radicals formed by treatment of the arsine oxide with alkaline potassium ferricyanide.

The structure of the tertiary arsenic dihydroxides has been the subject of some controversy and has not been decided with certainty. Three different structures have been advanced: $[R_3AsOH]OH$, $R_3As(OH)_2$, and $R_3AsO{\cdot}H_2O$; and there is some evidence in favor of each form.

Jensen has studied the dipole moments of triphenylphosphorus dihydroxide, triphenylarsenic dihydroxide, and triphenylantimony dihydroxide and obtained values of 4.56 D, 5.81 D, and 0, respectively (72, 79). On the basis of these results he concluded that the antimony compound exists in the pentacovalent form, with the two hydroxy groups at the apices of a trigonal bipyramid, but that the phosphorus and arsenic compounds exist as hydrated oxides. Further evidence for the hydrated oxide structure (in one compound) has been advanced by Mann and Wilkinson (80). These authors examined the infrared spectrum of compound **2** and found a broad absorption band in the 2.88–

2.96 μ region (3380–3470 cm^{-1}), which they ascribed to the presence of a water

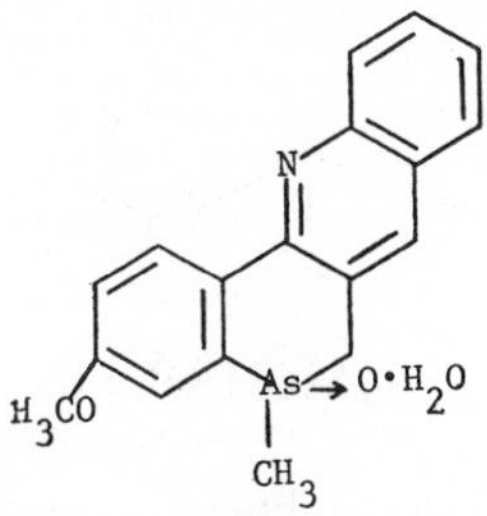

N
As→O·H_2O
H_3CO
CH_3
2

molecule; and thus the compound must be a hydrated oxide. They stated further that if the compound possessed the dihydroxide form it would give broad absorption above 3.3 μ, a property characteristic of the —As(OH) linkage.

The infrared spectrum of triphenylarsenic dihydroxide contains a strong band at 888 cm^{-1}, which has been assigned to the As=O stretch, and strong broad bands at ca. 3400 and 1665 cm^{-1}, which have been attributed to lattice water (81). These results suggest that this compound is a monohydrate, i.e., $Ph_3AsO \cdot H_2O$. It has also been found that acetonitrile solutions of this compound are virtually nonconducting.

Braunholtz and Mann (36) found that the oxidation of (2-carboxyethyl)-diphenylarsine gives a compound to which they assigned the following structure:

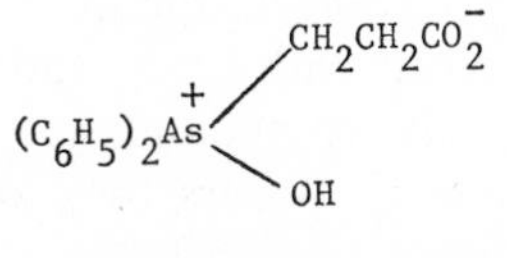

$(C_6H_5)_2\overset{+}{As}$ ($CH_2CH_2CO_2^-$) (OH)

This structure is based on the absence of any band corresponding to the As=O stretching frequency between 800 and 900 cm^{-1}, and on the absence of bands assignable to the carboxyl OH (which are present in the parent arsine). The oxidation of *o*-diphenylarsinobenzoic acid gives a similar type of zwitterion structure (3):

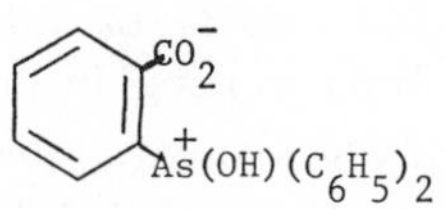

CO_2^-
$\overset{+}{As}(OH)(C_6H_5)_2$

Suggestive evidence for the existence of a pentacovalent dihydroxy structure has been found by Braunholtz and Mann (36). When bis(2-carboxyethyl)-phenylarsine or bis(2-carboxyethyl)(*p*-chlorophenyl)arsine was oxidized, both the analytical values and the infrared spectra of the products were in accord with a dilactone structure:

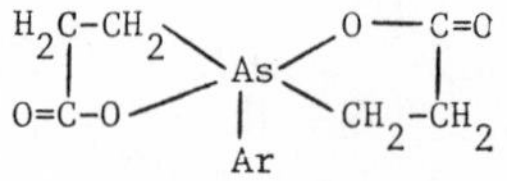

```
H2C-CH2         O—C=O
 |      \     /    |
 |       As<       |
O=C-O  /  |   \ CH2-CH2
          Ar
```

The third type of structure, $[R_3AsOH]OH$, has been proposed by Razuvaev, Malinovskii, and Godina (82). Their reasoning is based on the fact that tertiary arsenic dihydroxides will combine with many acids to give salts of the type $R_3As(OH)X$, and also on the ease with which the dihydroxides lose water to form the oxides. They further suggest that the dihydroxides are involved in an equilibrium of the type $[R_3AsOH]OH \rightleftharpoons R_3AsO{\cdot}H_2O$.

The evidence for these three types of structures was discussed at considerable length by Mann (83) in a Tilden lecture. Although he dismissed the hydrated arsine oxide structure as unlikely, later evidence from his laboratory (cf. ref. 80) is in accord with such a structure. The ionic form, $[R_3AsOH]OH$, and the pentacovalent form, $R_3As(OH)_2$, were considered, and evidence was advanced for both forms. Mann concluded that no single structure can account for all of the properties of the dihydroxides and that they may exist as resonance hybrids in which each hydroxy group possesses partial ionic character. It is evident that further investigation is necessary before the exact structure of the tertiary arsenic dihydroxides is known with certainty.

Closely related to the dihydroxides is the compound $(C_6H_5)_3As(OH)OC_6H_5$. It was prepared by allowing benzenediazonium chloride to react with metallic arsenic in acetone kept neutral with chalk and then treating the reaction product with water (84). The compound was nonconducting in alcohol solution and was considered by Waters (84) to be a covalent compound containing pentavalent arsenic.

4. TRIALKYL-, TRIARYLARSINE SULFIDES, AND RELATED COMPOUNDS

Sulfides of the type R_3AsS are usually prepared by the action of hydrogen sulfide on the oxides, dihydroxides, or dihalides. For example, tri-*p*-biphenylylarsine sulfide has been prepared by passing hydrogen sulfide into an alcoholic solution of tri-*p*-biphenylylarsenic dihydroxide (25). An alcoholic solution of triphenylarsine oxide similarly gives triphenylarsine sulfide (43). Kamai prepared ethylpropyl(*p*-carboxyphenyl)arsine sulfide by the action of hydrogen sulfide on the corresponding hydroxychloride (85). The same author (86), in an investigation of the stereochemisty of tertiary arsine oxides, was unable to crystallize either ethyl-α-naphthyl-*p*-tolylarsine oxide or ethyl-β-naphthyl-*p*-tolylarsine oxide. Both compounds, however, gave crystalline sulfides by the action of hydrogen sulfide.

Chatt and Mann (87) prepared 5,10-di-*p*-tolyl-5,10-dihydroarsanthrene, which can be separated into two isomeric forms (cf. Section IV-5-D). Both isomers give the same tetrabromide, which on treatment with sodium sulfide gives the monosulfide **3**. Only one form of this sulfide could be obtained. (β-

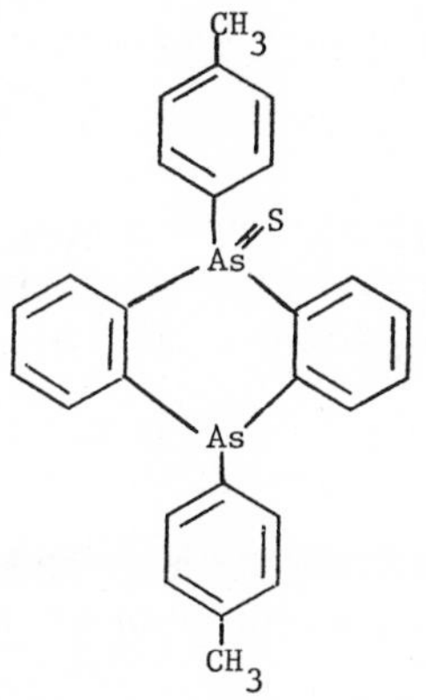

CH3 S As As CH3

3

Carboxyethyl)diphenylarsenic dibromide, on treatment with hydrogen sulfide, gives the sulfide $(C_6H_5)_2(HOOCCH_2CH_2)AsS$ (10).

Several other methods have been used to prepare these arsine sulfides. For example, there are reports of preparing them by the reaction of sulfur with a tertiary arsine (20, 62, 88, 89). The products, however, are sometimes obtained in impure form by this procedure. The reaction of Grignard reagents with arsenic trisulfide may give a mixture of R_3AsS, $(R_2As)_2S$, and R_3As (90). Thus, phenylmagnesium bromide and arsenic trisulfide give $(C_6H_5)_3AsS$ and $(C_6H_5)_3As$, while *p*-tolylmagnesium bromide gives all three products. Dehn (91) has prepared trimethylarsine sulfide by the disproportionation of methylarsenic disulfide:

$$3CH_3AsS_2 \xrightarrow{\Delta} (CH_3)_3AsS + As_2S_5$$

Trimethylarsine sulfide has also been obtained as a by-product of the reaction of dimethylarsinic acid and carbon disulfide (92). Dimethyl(trifluoromethyl)-arsine sulfide was probably formed in low yield in the reaction of cacodyl disulfide (see III-8-C) and trifluoroiodomethane, but the substance was not fully characterized (93).

The resolution of a tertiary arsine sulfide into optically active forms has been achieved by Mills and Raper (94). These authors prepared (*p*-carboxyphenyl)ethylmethylarsine sulfide, by the action of hydrogen sulfide on the corresponding oxide, and resolved it through the brucine and morphine salts. The two isomers gave the following molecular rotations in alcohol solution: $[M]^{20}_{5780} = -52°$; $[M]^{20}_{5780} = +51°$. Chatt and Mann (95) isolated two forms of ethylenebis(phenyl-*n*-butylarsine sulfide) (**4**), which they believed

$$\begin{array}{l} \quad\quad\quad S \\ \quad\quad\quad \uparrow \\ CH_2\text{—}As(C_4H_9)(C_6H_5) \\ | \\ CH_2\text{—}As(C_4H_9)(C_6H_5) \\ \quad\quad\quad \downarrow \\ \quad\quad\quad S \end{array} \quad \mathbf{4}$$

were meso and racemic forms. The lower melting form can be readily converted into the higher melting form on warming, either alone or in an organic solvent. Mann (83) later suggested that the interconversion may invove the following equilibrium:

$$\text{diarsine disulfide} \rightleftharpoons \text{diarsine monosulfide} + \text{sulfur}$$

The formation of optically active tertiary arsine sulfides by the reaction of optically active tertiary arsines with sulfur has been discussed in Section IV-5-A.

The infrared absorption spectra of a number of arsine sulfides have been investigated by Zingaro and co-workers (89). An intense band in the 470–490 cm^{-1} range was assigned to As=S stretching. Jensen and Nielsen (75) independently concluded that the stretching mode of the As=S bond in triphenylarsine sulfide occurs at 495 cm^{-1}. Similarly, the As=S vibration of diphenyl(4-hydroxy-3,5-di-*tert*-butylphenyl)arsine sulfide occurs at 485 cm^{-1} (7).

Zingaro and Meyers (62) have examined the ultraviolet absorption spectra of mixtures of iodine and arsine sulfides in chloroform. Their results suggest that the triiodide ion is present in these solutions. When carried out at higher concentrations ($\sim 0.1M$), the addition of iodine to arsine sulfides was found to be noticeably exothermic and leads to the almost instantaneous formation of crystalline solids. Analysis of these solids shows that they are iodoarsonium triodides, i.e., $[R_3AsI]^+I_3^-$. The fate of the sulfur was not determined.

Nicpon and Meek (96) have recently prepared triphenylarsine sulfide complexes of palladium bromide and silver perchlorate, and have studied both their conductivity and infrared spectra. The As=S vibration is apparently shifted to a lower frequency compared to the value (497 cm^{-1}) for the uncomplexed sulfide, but its position could not be unambiguously determined.

Müller and co-workers (7, 97) have reported that oxidation of diphenyl-(4-hydroxy-3,5-di-*tert*-butylphenyl)arsine sulfide in benzene yields a bright green solution, which contains radicals and which becomes light brown on decomposition of the radicals. The ESR spectrum of the green solution consists of four triplets indicating that the odd electron is coupled with the aromatic protons (spin, $\frac{1}{2}$) and the ^{75}As nucleus (spin, $\frac{3}{2}$). Müller and coworkers (7) also reported proton NMR and mass spectral data for the arsine sulfide; its molecular weight in benzene solution was found to be normal for the monomer (97).

Triphenylarsine selenide has been prepared by the action of selenium dioxide on triphenylarsine, but the product is contaminated with the oxide (13). Trimethylarsine selenide, prepared from trimethylarsine and powdered selenium, is unstable in the air (98). Recently, Zingaro and Merijanian (99) have prepared and characterized nine trialkylarsine selenides. The compounds were prepared by refluxing the corresponding arsines with selenium powder for several hours. Triphenylarsine selenide, however, could not be prepared in this manner.

Jensen and Nielsen (75) isolated a mixture of the oxide and selenide from the reaction of triphenylarsenic dichloride with hydrogen selenide at 0°. A reasonably pure sample of triphenylarsine selenide was obtained by recrystallization of the mixture from aqueous ethanol. By contrast, Zingaro and Merijanian (99) were unable to obtain the selenide by the reaction of hydrogen selenide with either triphenylarsenic dichloride or dibromide; triphenylarsine and selenium were the only products isolated.

The fundamental As—Se vibration in the trialkyl selenides is usually found to occur as a doublet near 348 cm^{-1} (99). The separation between the bands ranges from 8 to 26 cm^{-1}. There seems to be no obvious explanation for the observed splitting. The possibility of intermolecular association and the existence of an equilibrium between the monomer and associated molecules was ruled out by showing that the molecular weights in benzene are normal and that the infrared spectra are virtually temperature independent. Jensen and Nielsen (75) were unable to find an absorption band corresponding to the As=Se vibration in triphenylarsine selenide; this failure was tentatively attributed to the presence of a strong phenyl absorption at 460 cm^{-1}.

Closely related to the arsine oxides and sulfides are the arsine imides, $R_3As{=}NY$. Mann and Chaplin (100) allowed anhydrous chloramine-T to react with triphenylarsine and with the *ortho* and *para* isomers of tritolylarsine and obtained compounds of the structure *p*-$CH_3C_6H_4SO_2N{=}AsAr_3$. The imide of tri-*m*-tolylarsine, however, could not be obtained in crystalline form.

Wittig and Hellwinkel (101) have also synthesized several arsine imides by the reaction of chloramine-T with triarylarsines, while Tarbell and Vaughan (102) obtained an arsine imide from tri-*p*-tolylarsine and the potassium salt of *p*-(chlorosulfamyl)acetanilide. When hydrated chloramine-T is used, pentacovalent arsenic compounds of the type *p*-$CH_3C_6H_4SO_2NHAs(OH)Ar_3$ are isolated (28, 100, 103, 104). The same type of pentacovalent compound has been prepared by the reaction of arylsulfonamides with triarylarsenic dihydroxides or arsine oxides (102, 104).

Triphenylarsine imide itself has been prepared by the dehydrohalogenation of triphenylarsenic aminochloride with an alkali amide in liquid ammonia (105):

$$Ph_3As(NH_2)Cl + NH_2^- \rightarrow Ph_3As{=}NH + NH_3 + Cl^-$$

The requisite aminochloride can be obtained by the addition of chloramine to triphenylarsine (106) and is also formed by the addition of hydrogen chloride to the imide (105). Treatment of the imide with *p*-toluenesulfonyl chloride results in the formation of the same compound that can be obtained by the reaction of triphenylarsine with anhydrous chloroamine-T (100, 101).

In addition to the arsine imides described above, there are a few other compounds that may contain the As=N linkage. Thus, the sodium salt of methyl N-chlorocarbamate has been found to react with triphenylarsine to form a compound which has been formulated as $CH_3OCON{=}AsPh_3$ (107). N-Bromotriphenylphosphine imide and triphenylarsine interact to give a 97% yield of the interesting compound $[Ph_3P{\equiv}N{\equiv}AsPh_3]Br$ (108). Several compounds containing the As≡N≡As grouping have been described by Sisler and Stratton (106). The infrared spectra of these substances contain a strong band near 950 cm^{-1}, which was attributed to As≡N≡As vibrations. An intense band at 943 cm^{-1} has been found in the spectrum of the tetramer, $(C_6H_5AsN)_4$, and has been assigned to the As=N vibration (109). This compound may possess a cyclic structure similar to that of the phosphonitrilic chlorides (110).

5. ALKYL- AND ARYLARSENIC TETRAHALIDES

Compounds of the type $RAsX_4$ are comparatively rare, and very little modern work has been done with them. Methylarsenic tetrachloride is an unstable crystalline compound formed by the action of chlorine on methyldichloroarsine (111). It decomposes at −10°. The tetraiodide has been prepared by the action of concentrated hydriodic acid on disodium methanearsonate (112). Phenylarsenic tetrachloride and the isomeric tolyl compounds have been prepared by the action of chlorine on the dichloroarsines (113). No arylarsenic tetrabromides have been isolated, although Prat (114) states that the action of excess hydrobromic acid on benzenearsonic acid gives phenylarsenic tetrabromide, which breaks down spontaneously to phenyldibromoarsine, bromine, bromobenzene, and arsenic tribromide. *o*-Tolylarsenic tetrachloride has been prepared by the action of lead tetrachloride on *o*-tolyldichloroarsine (115). The *para* isomer, however, could not be obtained by this procedure. A number of tetrafluorides have been prepared by the action of sulfur tetrafluoride on arsonic acids at about 70° (116–118):

$$RAsO_3H_2 + 3SF_4 \rightarrow RAsF_4 + 3SOF_2 + 2HF$$

It has been claimed (118) that the oxyfluorides, $RAsOF_2$, are formed when the reaction is run at lower temperatures with less sulfur tetrafluoride.

Comparatively little is known concerning the properties of the tetrahalides. They are all readily hydrolyzed by water to the arsonic acids. When heated in the air phenylarsenic tetrachloride gives the dichloroarsine and chlorine, but when heated to 150° in a sealed tube it gives chlorobenzene and arsenic trichloride (113). It reacts with acetic acid to give chloroacetic acid, phenyldichloroarsine, and hydrogen chloride.

The structure of the tetrahalides has not been elucidated. Smith (117) determined the infrared spectrum as well as the nuclear magnetic resonance spectrum of phenylarsenic tetrafluoride, but was unable to draw any conclusions as to its structure. More recently, Muetterties and co-workers (119) included this compound in an extensive ^{19}F NMR investigation of five-coordinate stereochemistry. Since the five bonds around the arsenic atom are probably directed towards the corners of a trigonal bipyramid, one would expect two ^{19}F signals—one for the axial fluorine (or fluorines) and one for the equatorial fluorines. It was found, however, that the ^{19}F spectrum of phenylarsenic tetrafluoride consists of a single, relatively sharp peak; i.e., the four fluorine atoms appear to be equivalent. The reason for this spectroscopic equivalence is not obvious, but it has been suggested that a fast intramolecular fluorine exchange may be occurring.

6. DIALKYL- AND DIARYLARSENIC TRIHALIDES

Relatively few of these compounds have been prepared, and much of the work was performed over a century ago. They are usually prepared by the reaction of dry halogens on dialkyl- or diarylhaloarsines. Dimethylarsenic trichloride can be obtained either by the action of chlorine on dimethylchloroarsine in carbon disulfide solution or by the action of phosphorus pentachloride on cacodylic acid (111). Diphenylchloroarsine reacts with bromine to give $(C_6H_5)_2AsClBr_2$ or, with excess bromine, to give the perbromide, $(C_6H_5)_2AsClBr_4$ (120). Similarly, diphenylbromoarsine and bromine give a tribromide and a perbromide. The tribromide has also been prepared by the bromination of diphenylarsine in ether (121). Diphenylchloroarsine and chlorine give the trichloride $(C_6H_5)_2AsCl_3$. The latter compound can also be prepared by the action of thionyl chloride on diphenylarsinic acid (122) or on its hydrochloride, $(C_6H_5)_2AsO_2H{\cdot}HCl$ (120).

In a similar manner, thionyl chloride reacts with phenarsazinic acid to give the corresponding trichloride. Diphenylcyanoarsine can be readily chlorinated to give $(C_6H_5)_2As(CN)Cl_2$ (123). Bis(trifluoromethyl)arsenic trichloride has been obtained by allowing tris(trifluoromethyl)arsine to react with chlorine in a sealed tube for one month (124). Dibenzylarsenic trichloride can be prepared by heating tribenzylarsine with an excess of benzyl chloride at 200° (125); it is also one of the products formed when benzyl chloride and arsenic trichloride are condensed with sodium (126).

The dialkyl- and diarylarsenic trihalides are probably more stable than the tetrahalides, but they can be readily decomposed at relatively low temperatures. Dicyclohexylarsenic trichloride, when warmed to 80–90°, loses chlorocyclohexane to yield cyclohexyldichloroarsine (127). Diphenylarsenic trichloride, when warmed in a stream of carbon dioxide, yields diphenylchloroarsine, but when heated in a sealed tube to 200° gives chlorobenzene and phenyldichloroarsine (113). Di-*o*-biphenylylarsenic trichloride loses hydrogen chloride at 265° to form the following spirocyclic arsonium salt (101):

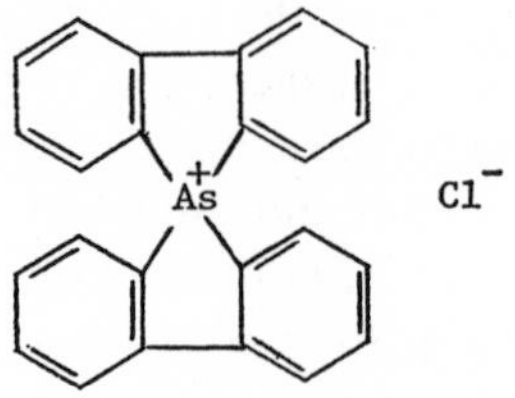

The trihalides are moisture-sensitive compounds and are readily hydrolyzed to arsinic acids (120). Treatment of diphenylarsenic trichloride with either lithium aluminum hydride or lithium borohydride has been found to yield diphenylarsine (128).

Very little is known concerning the structure of the trihalides. On the basis of their reactions, Kappelmeier regarded these compounds as ionic, with the structure $[R_2AsX_2]X$ (120). He regarded the two perbromides that he prepared as having a structure containing hexacovalent arsenic, namely, $[(C_6H_5)_2AsBr_4]Cl$ and $[(C_6H_5)_2AsBr_4]Br$. Kappelmeier advanced no physical evidence for any of the formulations he proposed; a decision on the structure of these compounds must await more detailed study.

Muetterties and co-workers (119) have examined the ^{19}F NMR spectra of diphenylarsenic trifluoride and dimethylarsenic trifluoride. The spectrum of the diphenyl compound was found to consist of a doublet and a triplet of relative intensity 2 and 1. It was concluded that this compound has a slightly distorted trigonal bipyramidal structure in which two fluorine atoms occupy (equivalent) axial positions and the other fluorine atom occupies an equatorial position.

Strangely enough, the structure of dimethylarsenic trifluoride appears to differ markedly from that of analogous alkyl and aryl derivatives of group V elements. Thus, unlike the diphenyl compound, the dimethyl compound shows evidence of association and is not very soluble in aromatic hydrocarbons. Solutions in acetone or acetonitrile showed a single ^{19}F peak at 25° and doublet-triplet resonances at low temperatures. It was suggested that the low temperature spectrum is representative of an octahedral species (in which the dimethyl compound is presumably acting as a Lewis acid) rather than a five-coordinate structure.

7. TRIALKYL-, TRIARYLARSENIC DIHALIDES, AND RELATED COMPOUNDS

This class of compound has been studied to a much greater extent than the tri- and tetrahalides described in Sections V-6 and V-5, respectively. The usual procedure for the preparation of the dihalides is the treatment of a trialkyl- or triarylarsine in a nonpolar solvent with a stoichiometric quantity of the halogen in the same solvent. Thus, Dyke, Davies, and Jones (21) have prepared a number of trialkylarsenic dichlorides and dibromides from the trialkylarsines in carbon tetrachloride solution; for the iodides, petroleum ether was used as the solvent. Similarly, triarylarsenic dichlorides, dibromides, and diiodides have been obtained by the reaction of triarylarsines with the appropriate halogen in a organic solvent (44, 129). Triphenylarsenic bromide iodide can be prepared by the interaction of triphenylarsine with iodine bromide in acetonitrile (129).

Tris(trimethylsilylmethyl)arsenic dibromide, $[(CH_3)_3SiCH_2]_3AsBr_2$, was prepared from the corresponding arsine in petroleum ether by the addition of bromine (130). The diiodide was prepared in methylene dichloride solution. Tris(trifluoromethyl)arsine and chlorine, when allowed to react in a sealed tube at room temperature for 84 hours, give a 34% yield of tris(trifluoromethyl)arsenic dichloride (124). In the vapor phase, however, the dichloride is not obtained, but rather bis(trifluoromethyl)arsenic trichloride. The dibromide, the diiodide, and the difluoride can not be prepared by the reaction of the appropriate halogens with tris(trifluoromethyl)arsine. Instead, one or more trifluoromethyl groups are cleaved from the arsenic, and a mixture of products, which include the arsenic trihalide, is obtained. In a similar manner Cullen (131) was unable to obtain a dibromide by the action of bromine on bis(trifluoromethyl)phenylarsine. The tendency to split the carbon–arsenic bond on treatment with a halogen has been observed with arsenicals other than those containing the trifluoromethyl group. Thus, tri-2-furylarsine (as well as di-2-furylchloroarsine and 2-furyldichloroarsine) gives 2-chlorofuran tetrachloride when treated with chlorine (132); and tri-2-furylarsine on treatment with iodine yields 2-iodofuran (133).

A number of mixed aliphatic–aromatic arsenic dihalides have been prepared from the arsines and the appropriate halogen. Dimethyl-α-naphthylarsine and bromine in carbon tetrachloride give dimethyl-α-naphthylarsenic dibromide (134); methyl-α-naphthylphenylarsine and chlorine give the dichloride (135). Similarly, methyldiphenylarsenic dibromide can be obtained by the bromination of the corresponding arsine in carbon tetrachloride (136). If two tertiary arsine groups are present in the molecule, both groups may be halogenated. Thus, Chatt and Mann (87) prepared 5,10-di-*p*-tolyl-5,10-dihydroarsanthrene tetrabromide by bromination of the di-arsine.

Instead of chlorine, lead tetrachloride has been used as the chlorinating agent. Thus, methyldiphenyl- and ethylmethylphenylarsines, when treated with lead tetrachloride at $-5°$, give the corresponding arsenic dichlorides and lead dichloride (115). From dimethylphenylarsine and lead tetrachloride, only the hydroxychloride could be isolated. This result was attributed to the ease of hydrolysis of the expected dichloride.

The reaction of acids with arsine oxides or dihydroxides generally leads to hydroxy compounds of the type $R_3As(OH)X$, where X is the anion of the acid employed. In a few cases, however, hydrohalic acids react with arsine oxides to give the arsenic dihalides. Thus, tribenzylarsine oxide and excess hydriodic acid give tribenzylarsenic diiodide (125). Similarly, Cookson and Mann (37) have prepared bis(γ-carboxypropyl)phenylarsenic dichloride by the action of hydrochloric acid on the corresponding oxide.

Reutov and Bundel (137) have isolated triarylarsenic dichlorides from the reaction of arenediazonium tetrachloroferrates on arsenic trichloride or phenyldichloroarsine in the presence of powdered iron. In addition to Ar_3AsX_2, compounds of the type $ArAsX_2$, Ar_2AsX, and Ar_3As are formed. If zinc rather than iron is used, no triarylarsenic dihalides are formed. The authors proposed that the triarylarsenic dichlorides are formed by the following reaction:

$$Ar_2AsCl + (ArN_2)(FeCl_4) \rightarrow Ar_3AsCl_2 + N_2 + FeCl_3$$

Comparatively few difluorides are known. Triphenylarsenic difluoride has been obtained by the reaction between triphenylarsine and sulfur tetrafluoride (117) and by the reaction between triphenylarsenic dichloride and silver fluoride (138, 139). The metathetical reaction between a dichloride and silver fluoride has also been used for the synthesis of the following difluorides: trimethyl (138, 139), triethyl (139), tribenzyl (139), tris(trifluoromethyl) (124), and tri-2-thienyl (139).

Under certain conditions, trialkyl- and triarylarsines react with bromine, iodine, and interhalogens to form colored substances of the type R_3AsX_4. Thus, the addition of an excess of bromine to an ethereal solution of trimethylarsine yields a red precipitate of trimethylarsenic tetrabromide (140).

Dimethylphenylarsine (141) and methyldiphenylarsine (136) have been converted to tetrabromides in a similar manner. Michaelis (44) prepared both triphenylarsenic tetraiodide and dibromide diiodide; the latter compound was obtained by the use of equimolar amounts of iodine and bromine. The dibromide diiodide has also been prepared by the action of iodine bromide on the tertiary arsine and by the interaction of iodine and triphenylarsenic dibromide (129). Similarly, triphenylarsenic tribromide iodide was prepared from triphenylarsenic dibromide and iodine bromide, while triphenylarsenic bromide triiodide was prepared from triphenylarsenic bromide iodide and iodine. The conversion of trialkyl- and triarylarsine sulfides to tetraiodides has been discussed in Section V-4.

Feigl and co-workers (142, 143) have shown that triphenylarsine (and the corresponding phosphine, stibine, and bismuth compound) reacts with iodine vapor to form a brown or yellow product. They suggested that the colored substance is a polyiodide of the type $Ph_3AsI_2{\cdot}xI_2$.

The arsenic dihalides are generally low melting crystalline solids, soluble in alcohol but only slightly soluble or insoluble in nonpolar solvents. With a few exceptions, they are readily hydrolyzed by water to the hydroxy halide (Section V-8), while alkali usually hydrolyzes them completely to the oxides or dihydroxides (29). Attempts to convert tris(trifluoromethyl)arsenic dichloride to the corresponding dihydroxide or oxide, however, have been unsuccessful (30). Treatment of the dichloride with water yields mainly tris(trifluoromethyl)arsine and a small amount of bis(trifluoromethyl)arsinic acid. Treatment with ethanol gives a quantitative yield of the tertiary arsine. Acetic acid converts the dichloride to bis(trifluoromethyl)chloroarsine and chlorotrifluoromethane, which are the usual thermal decomposition products of the dichloride. When shaken with silver oxide and water, the dichloride yields bis(trifluoromethyl)chloroarsine and bis(trifluoromethyl)arsinic acid; carbon dioxide and fluoride ion are also formed. Treatment of the dichloride with aqueous alkali liberates fluoroform rapidly and quantitatively (124). Diphenyl(trifluoromethyl)arsenic dibromide on hydrolysis with aqueous sodium hydroxide also yields fluoroform (131). It was postulated that a hydroxybromide is first formed, which then eliminates fluoroform.

Tertiary arsenic dihalides containing at least one alkyl group yield on pyrolysis an alkyl halide and a haloarsine:

$$R_3AsX_2 \rightarrow R_2AsX + RX$$

This reaction has been studied by a number of authors and has been used for the synthesis of haloarsines. Thus, Spada (144) prepared methyl-α-naphthylbromoarsine by brominating dimethyl-α-naphthylarsine, which was then converted to the dibromide. Tricyclohexylarsine on chlorination and subsequent pyrolysis gives dicyclohexylchloroarsine (127). Trivinylarsine on iodination gives trivinylarsenic diiodide, which on pyrolysis gives divinyliodoarsine (145). Cullen (131) has studied the relative ease of elimination of several different groups from the dibromides. Bromination of methylphenyl(trifluoromethyl)arsine gives an unstable dibromide, which decomposes to give principally methyl bromide with smaller amounts of bromotrifluoromethane. From these results the author (131) concluded that the ease of elimination is in the order $C_6H_5 < CF_3 < CH_3$.

An interesting example of alkyl halide elimination has been studied by Gorski, Schpanski, and Muljar (146). 1-Methylarsenane dichloride (**5**) on pyrolysis gives 1-chloroarsenane by elimination of methyl chloride. 1-(β-Chlorovinyl)arsenane dichloride on pyrolysis, however, gives (ε-chloroamyl)-(β-chlorovinyl)chloroarsine:

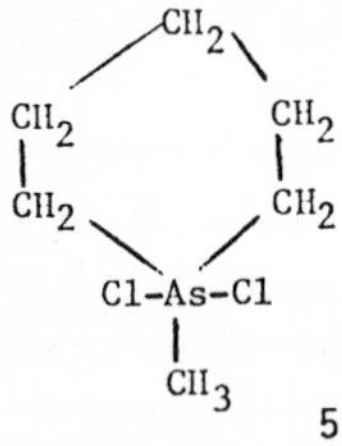

CH2, CH2, CH2, CH2, CH2, Cl–As–Cl, CH3

5

CH2—CH2, CH2, CH2—CH2, Cl, AsCH=CHCl, Cl $\longrightarrow$ Cl(CH2)5As(Cl)CH=CHCl

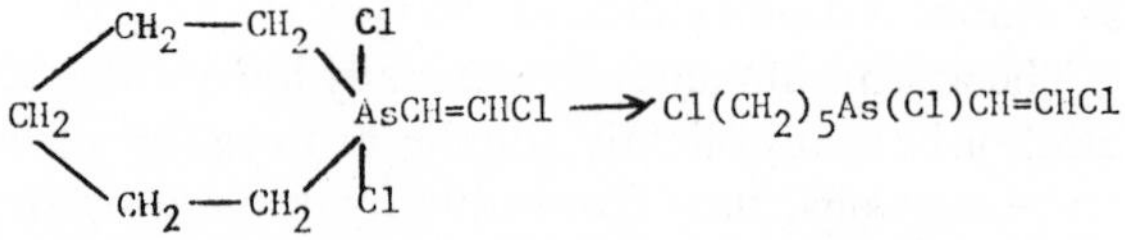

Das-Gupta (147) has reported that the pyrolysis of dimethylstyrylarsenic dichloride yields the unstable methylstyrylchloroarsine, which spontaneously loses hydrogen chloride to give 1-methylarsindole:

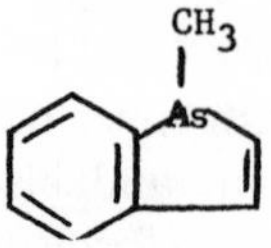

CH3, As

Tertiary arsenic dihalides can be readily reduced to tertiary arsines. For example, lithium aluminum hydride and lithium borohydride have been used for the reduction of triphenylarsenic dichloride (128); tris(trifluoromethyl)-arsenic dichloride has been reduced with mercury (124). The electrolytic reduction of triphenylarsenic dibromide appears to be a one-step, two-electron process which involves loss of bromine as bromide ion and formation of triphenylarsine (148):

$$Ph_3AsBr_2 + 2e^- \rightarrow Ph_3As + 2Br^-$$

Triarylarsenic dihalides react with appropriate methylene compounds to give ylids according to the equation (149):

$$(C_6H_5)_3AsCl_2 + H_2C(X)(Y) \xrightarrow{Et_3N} (C_6H_5)_3As{=}C(X)(Y)$$

The X and Y groups employed in this reaction are as follows:

$$X{=}Y{=}CN; \quad X{=}Y{=}SO_2C_6H_5;$$

$$X{=}CO_2CH_3, \quad Y{=}CN; \quad \text{and } X{=}NO_2, \quad Y{=}C_6H_5$$

Other methods for preparing arsenic ylids are discussed in Section V-9-B. Attempts to prepare pentamethylarsenic by the interaction of trimethylarsenic dibromide with methyllithium have been unsuccessful (150). The synthesis of pentaarylarsenic compounds from triarylarsenic dihalides will be discussed in Section V-10. The reactions of trialkyl- and triarylarsenic dihalides with hydrogen sulfide and hydrogen selenide have been described in Section V-4; metathetical reactions of the dihalides with various metal salts will be discussed later in the present section. The reaction of trimethylarsenic dichloride with the sodium salt of trimethylsilanol has been found to take the following course (151):

$$(CH_3)_3AsCl_2 + 2NaOSi(CH_3)_3 \rightarrow (CH_3)_3SiOSi(CH_3)_3 + (CH_3)_3AsO + 2NaCl$$

The preparation of a spirocyclic arsonium salt by the elimination of hydrogen chloride from a triarylarsenic dichloride is discussed in Section V-9-A.

The structure of the tertiary arsenic dihalides has not been established with certainty. Three possible structures can be suggested, namely, $[R_3As]^{2+}2X^-$, $[R_3AsX]^+X^-$, and R_3AsX_2, in which the arsenic atom is surrounded by six, eight, or ten valence electrons, respectively. Goubeau and Baumgärtner (152) have recently produced convincing evidence, based on infrared and Raman spectra, that the dihalotrimethylphosphoranes possess the $[R_3PX]^+X^-$ structure. Wells (153), on the other hand, has shown by means of X-ray data that the trimethylantimony dihalides exist as trigonal bipyramids with the two halogens at the apical positions. The antimony-halogen distances, however, are longer than the sum of the covalent radii and suggest that the Sb—X bond is partially ionic. Jensen (79) determined the dipole moments of triphenylantimony, -bismuth, and -arsenic dibromides in benzene solution. Although all three compounds gave dipole moments of zero, the arsenic compound was too hygroscopic to give reliable results. Jensen and Nielsen (75) reported the infrared spectra of triphenylarsenic dichloride and dibromide, but their samples were probably partially hydrolyzed (81, 138).

The structure $[R_3As]^{2+}2X^-$ can probably be dismissed since Nylén (29) has demonstrated that even in aqueous solution trimethyl- and triethylarsenic dibromides are not completely ionized. For the two remaining structural possibilities Mann (83) has concluded that the properties of the arsenic dibromides show that they are true covalent compounds. By contrast, Rochow, Hurd, and Lewis (154) have suggested that in the tertiary arsenic dihalides, one of the halogens is ionic.

In recent years there have been several investigations of the physical properties and structure of the arsenic dihalides. Thus, Muetterties and co-workers (119) have reported that triphenylarsenic difluoride is monomeric in benzene solution and that the ^{19}F NMR spectrum consists of a single, relatively sharp peak. Their results are consistent with (but do not establish) a trigonal bipyramidal structure with axial fluorine atoms. The conductivity (139) and infrared (138, 139) data of O'Brien and co-workers also indicate that this compound is covalent. Trimethyl-, triethyl-, tribenzyl-, and tri-2-thienylarsenic difluorides likewise are nonelectrolytes in acetonitrile solution (139). The infrared spectra of the trimethyl, triethyl, and tribenzyl compounds, and the ^{19}F NMR spectra of the trimethyl, triethyl, and tri-2-thienyl compounds are in agreement with the conclusion that the difluorides contain pentacovalent arsenic.

The conductivity studies of Beveridge and Harris (129, 155) on triphenylarsenic dichloride and of O'Brien (139) on trimethyl-, triethyl-, tribenzyl-, triphenyl-, and tri-2-thienylarsenic dichlorides show that these compounds are weak electrolytes in acetonitrile. The infrared spectra of the trimethyl (138, 139), triethyl (139), and tribenzyl (139) compounds are also consistent with a trigonal bipyramidal structure. The conductivities of the trimethyl- (139), triethyl- (139), and triphenylarsenic (129, 139) dibromides in acetonitrile are larger than those of the corresponding dichlorides but are rather low compared to the values observed for strong 1:1 electrolytes. The infrared spectrum of triethylarsenic dibromide also indicates that this compound is covalent (139). Unlike the spectra of the difluorides, dichlorides, and of triethylarsenic dibromide, the spectrum of trimethylarsenic dibromide contains *two* bands (at 582 and 642 cm^{-1}) that have been assigned to C—As stretching (138, 139). This result has been interpreted to mean that trimethylarsenic dibromide has an ionic structure in the solid state.

The infrared spectra of trimethyl- (138, 139) and triethylarsenic (139) diiodides also indicate that these compounds are ionic in the solid state. These diiodides (139) as well as triphenylarsenic diiodide (129, 139) form highly conducting solutions in acetonitrile. Beveridge and Harris (129) have concluded that the high conductance of triphenylarsenic diiodide does not arise from direct ionization of the compound in solution but rather is due to the following disproportionation:

$$2Ph_3AsI_2 \rightleftharpoons Ph_3As + Ph_3AsI^+ + I_3^-$$

The trialkylarsenic diiodides appear to undergo a similar reaction in acetonitrile (139). Triphenylarsenic bromide iodide also disproportionates in acetonitrile (129):

$$2Ph_3AsIBr \rightleftharpoons Ph_3As + Ph_3AsBr^+ + I_2Br^-$$

The infrared spectra of triphenylarsenic dichloride, dibromide, and a number of related compounds have been recently investigated by Mackay and co-workers (156). Their analysis of the spectra was based on the classic work of Whiffen (157) on the halobenzenes. Most of the bands in the spectra of the arsenic compounds correspond to the twenty-four phenyl modes, which involve only motions of the carbon and hydrogen atoms. The spectrum of the dichloride also contains a strong absorption at 364 cm^{-1} and a medium-strong doublet centered at 270 cm^{-1}. In the spectrum of the dibromide there is a band at 361 cm^{-1}, but there is no low-frequency absorption corresponding to the 270 cm^{-1} band of the dichloride. Assuming that the dihalide molecules are trigonal-bipyramidal, the authors (156) concluded that the bands at 364 and 361 cm^{-1} contain the equatorial stretching mode, while the doublet at 270 cm^{-1} contains the axial stretch. Thus, the 270 cm^{-1} band is associated with As—Cl stretching. The As—Br stretching frequency is presumably below the instrument limit (250 cm^{-1}). O'Brien and co-workers (138) reported a band at 360 cm^{-1} in the spectrum of triphenylarsenic dichloride and a band at 355 cm^{-1} in the spectrum of the dibromide, but failed to observe any other bands in the 400–250 cm^{-1} region.

Beveridge and Harris (129) have also investigated the structure of the triphenylarsenic tetrahalides. Unlike the triphenylarsenic dihalides, the tetrahalides appear to be ionic in the solid state. In acetonitrile solution these compounds are strong electrolytes and dissociate according to the following scheme:

$$Ph_3AsX_4 \rightarrow Ph_3AsX^+ + X_3^-$$

This conclusion is based on the results of conductometric titrations of triphenylarsine with iodine and iodine bromide and on ultraviolet examination of the resulting solutions. In the case of Ph_3AsBrI_3, $Ph_3AsBr_2I_2$, and Ph_3AsBr_3I, the positive ion always contains an As—Br bond. The preference for the Ph_3AsBr^+ ion is attributed to the fact that the As—Br bond is stronger than the As—I bond. A preliminary X-ray investigation of the tetrahalides indicates that the lattices are built of the same ions as those which have been found in the acetonitrile solutions.

Augdahl and co-workers (158) have studied the reaction between triphenylarsine and iodine in several organic solvents by means of spectroscopic methods. They showed that in carbon tetrachloride and dichloromethane a 1:1 adduct is formed, which they regarded as a charge-transfer complex $Ph_3As{\cdot}I_2$. Formation constants at several temperatures and other thermodynamic quantities were derived from the spectral data. In acetonitrile solutions of triphenylarsine and iodine, no charge-transfer complex was detected; instead, the ultraviolet absorption spectrum indicated the presence of triiodide ions. The authors suggested that the initially formed charge-transfer complex quickly undergoes an ionization, which is followed by the formation of triiodide ion from the reaction between iodide ion and iodine. If we assume that the positive ion is Ph_3AsI^+, the net result of the reaction of the complex with iodine can be represented as follows:

$$Ph_3As{\cdot}I_2 + I_2 \rightarrow Ph_3AsI^+ + I_3^-$$

This mechanism implies that the concentration of triiodide ion can not exceed the initial concentration of triphenylarsine. It was found, however, that the use of excess iodine results in the formation of more than the stoichiometric amount of triiodide ion. This result indicates that some other reaction is also involved.

The charge-transfer complex of triphenylarsine and iodine has been studied by Bhaskar and co-workers (159) in chloroform and by Bhat and Rao (160) in dichloromethane, chloroform, and carbon tetrachloride. It was found that the intensity of the charge-transfer band, which occurs at 320 mμ in chloroform, decreases at a rate that obeys first order kinetics. This result has been explained by assuming that the "outer complex," $Ph_3As{\cdot}I_2$, is slowly transformed to an "inner complex," $Ph_3AsI^+I^-$. The ultraviolet absorption spectrum of this complex is characterized by weak benzenoid absorption at about 265 mμ. Bhat and Rao reported that the decrease in the intensity of the 320 mμ charge-transfer band is accompanied by the appearance of triiodide ion absorption at 297 and 363 mμ. The formation of the triiodide ion was ascribed to a fast reaction of the "inner complex" with iodine:

$$Ph_3AsI^+I^- + I_2 \rightarrow Ph_3AsI^+I_3^-$$

The source of the excess iodine required for this reaction is not clear, since all the experiments described involved the use of a large excess of triphenylarsine. The formation of triiodide ion under these conditions was apparently not observed by Bhaskar and co-workers.

In addition to the halides, a few pentavalent derivatives of tertiary arsines have been prepared containing other negative groups. Both triethyl- (20) and triphenylarsenic dinitrates (43, 44) have been obtained from the corresponding oxide or hydroxide by treatment with concentrated nitric acid. The triphenyl compound has also been prepared by the reaction of triphenylarsenic dibromide with either silver nitrate or dinitrogen tetroxide (161). It is extremely hygroscopic and rapidly liquefies on exposure to traces of moisture. When a solution of triphenylarsenic dinitrate in moist acetone is treated with petroleum ether, triphenylarsenic hydroxynitrate is precipitated as a white crystalline solid. Dimethyl(*p*-phenoxyphenyl)arsenic diiodide has been reported to react with silver nitrate with the quantitative precipitation of silver iodide. The arsenic dinitrate, however, was not isolated (162).

The infrared spectrum of triphenylarsenic dinitrate has been found to include strong bands at 1550, 1285, 925, and 783 cm^{-1}, which have been assigned respectively to the ν_4, ν_1, ν_2, and ν_6 modes of covalently bonded, unidentate nitrate (161). The molar conductance in acetonitrile shows that the dinitrate is a weak electrolyte in this solvent. Accordingly, it can be concluded that this compound contains pentacovalent arsenic.

Tri-*n*-propylarsenic sulfate has been prepared from tri-*n*-propylarsenic dibromide and silver sulfate (21). Triphenylarsenic dibenzoate has also been reported (163). It is formed by the reaction of triphenylarsine and dibenzoyl peroxide. There have been no studies of the structure of these interesting compounds.

A different approach to compounds of this type has been taken by Becke-Goehring and Thielemann (41). These authors studied the reactions of sulfur trioxide as a Lewis acid and found that it formed adducts of the type $R_3MO{\cdot}SO_3$ where M is phosphorus, arsenic, or antimony and R is either phenyl or cyclohexyl. The two arsine oxide adducts, $(C_6H_5)_3AsO{\cdot}SO_3$ and $(C_6H_{11})_3AsO{\cdot}SO_3$, are extremely susceptible to hydrolysis. Recrystallization of these adducts from moist solvents gives compounds to which the authors assigned the structures $(C_6H_5)_3As(OH)OSO_3H$ and $(C_6H_{11})_3As(OH)OSO_3H$. The assignment of structure is based on the analytical results and on a comparison of the infrared spectra of $(C_6H_5)_3As(OH)OSO_3H$ and potassium bisulfate. There is, however, very little difference between the spectra of the SO_3 adducts and the hydrolysis products, and the evidence for the structures assigned to the latter would appear to be somewhat tenuous.

Trialkyl- and triarylarsenic diperoxides of the type $R_3As(O{-}OR')_2$ have been prepared by Rieche and co-workers (164, 165). These compounds were obtained by reaction of a trialkyl- or triarylarsenic dihalide with either an alkyl hydroperoxide in the presence of a tertiary amine or with the sodium salt of an alkyl hydroperoxide. Other methods for their preparation include the reaction between an aminohalide, $R_3As(NH_2)X$, and an alkyl hydroperoxide, and the exchange reaction:

$$R_3As(OR'')_2 + 2R'OOH \rightarrow R_3As(OOR')_2 + 2R''OH$$

The alkoxides required for the exchange reaction were obtained by the reaction of trialkyl- and triarylarsenic dihalides with either sodium alkoxides or with alcohols in the presence of ammonia and were used without actually being isolated. Polymeric peroxides of the empirical formula R_3AsO_2 were prepared by the interaction of the alkoxides with hydrogen peroxide (165). The trialkylarsenic diperoxides are stable at room temperature, but explode when heated in a flame. They are readily hydrolyzed even by atmospheric moisture.

8. COMPOUNDS OF THE TYPE R_3AsXY

A number of compounds of the R_3AsXY type are known in which X and Y are different groups. The best known of these are probably the hydroxy compounds, $R_3As(OH)X$, where X may be a halogen, a picrate, a nitrate, or other negative ion. These compounds are usually prepared by the reaction of the appropriate acid with the tertiary arsine oxide or dihydroxide (4, 40, 44, 102, 136, 166). For example, methyldiphenylarsenic hydroxychloride has been prepared from the corresponding oxide by treatment with 5*N* hydrochloric acid (136). Methylenebis(diphenylarsine) dioxide, when treated with cold concentrated hydrochloric acid, gives methylenebis(diphenylarsenic hydroxychloride), $Ph_2As(OH)(Cl)CH_2As(OH)(Cl)Ph_2$, while treatment with hydrobromic acid gives $Ph_2As(O)CH_2As(OH)(Br)Ph_2$ (40). Similarly, Michaelis (44) prepared triphenylarsenic hydroxynitrate by the interaction of triphenylarsenic dihydroxide and dilute nitric acid.

In addition to their preparation from the dihydroxides or oxides, hydroxyhalides and nitrates can be obtained by the partial hydrolysis of the dihalides and dinitrates, respectively. For example, when a solution of dimethyl-*o*-tolylarsenic dibromide in an ethanol-water mixture is evaporated to incipient crystallization, the corresponding hydroxybromide is obtained (22). Many other hydroxyhalides have been made by a similar procedure (81, 138, 139). Tranter and co-workers (161) prepared triphenylarsenic hydroxynitrate by the treatment of a solution of the dinitrate in moist acetone with petroleum ether. O'Brien and co-workers (138, 139) have made the same hydroxy compound by the reaction of triphenylarsenic dichloride with two equivalents of silver nitrate in aqueous ethanol.

A number of hydroxy compounds have been made directly from the tertiary arsines. Thus, the action of lead tetrachloride on dimethylphenylarsine gives the corresponding hydroxychloride (115). Presumably, the dichloride is first formed and then yields the hydroxychloride on partial hydrolysis. Dodonow and Medox (15) prepared tribenzylarsenic hydroxychloride by passing a stream of oxygen through a solution of tribenzylarsine in ether, water, and hydrochloric acid. The hydroxychloride was reported to react with potassium bromide to give $(C_6H_5CH_2)_3As(OH)OAs(Br)(C_6H_5CH_2)_3$, which was hydrolyzed by heating with hydrobromic acid to tribenzylarsenic hydroxybromide. A number of hydroxynitrates have been prepared by the treatment of tertiary arsines with nitric acid (82, 115, 161). For example, Lyon and Mann (28) prepared 2-phenylisoarsindoline hydroxynitrate by oxidation of 2-phenylisoarsindoline with concentrated nitric acid and recrystallization of the product from dilute nitric acid.

The preparation of hydroxy compounds of the type $R_3As(OH)NHSO_2R'$ has been described in Section V-4. The conversion of one such compound to a hydrochloride by treatment with 5% hydrochloric acid at 100° has been reported by Tarbell and Vaughan (102). Hydrolysis of compounds of the type $R_3As(CN)X$ (the preparation of which is described later in this section) also leads to the formation of hydroxyhalides (127, 167).

Harris and Inglis (81) have recently prepared triphenylarsenic hydroxyperchlorate, $Ph_3As(OH)ClO_4$, by the reaction of the hydroxybromide with silver perchlorate in dry ethanol. An attempt to prepare the hydroxytetrafluoroborate by a similar procedure, however, gave a boron trifluoride adduct, $Ph_3AsO{\cdot}BF_3$. It seems likely that the tetrafluoroborate compound is first formed and then eliminates hydrogen fluoride. Attempts to prepare triphenylarsenic hydroxytribromide, $Ph_3As(OH)Br_3$, by the addition of bromine to the hydroxybromide or by the partial hydrolysis of triphenylarsenic tetrabromide yielded a compound that was formulated as $[Ph_3AsOH]_2^+Br_3^-Br^-$. The desired hydroxytribromide was finally obtained by the freeze-drying of a solution of the tetrabromide in moist acetonitrile. Treatment of the hydroxychloride with iodine monochloride was found to yield $Ph_3As(OH)ICl_2$. The hydroxytetrachloromercurate, $[Ph_3As(OH)]_2HgCl_4$, and the corresponding tetrabromomercurate, $[Ph_3As(OH)]_2HgBr_4$, were readily prepared by the addition of the appropriate mercuric halide to an ethanol solution of the hydroxychloride and the hydroxybromide, respectively. In a later paper, Harris and co-workers (168) reported that the reaction of triphenylarsenic dihydroxide with mercuric chloride in hydrochloric acid also yields $[Ph_3As(OH)]_2HgCl_4$. Under similar conditions, however, the interaction of the dihydroxide, mercuric bromide, and hydrobromic acid yields a new bromo compound, which was shown to have the following hydrogen-bonded structure:

$$[Ph_3AsO\cdots\overset{+}{H}\cdots OAsPh_3]_2Hg_2Br_6^{2-}$$

O'Brien and co-workers (138, 139) have described a compound $[Ph_3AsO]_2$-$HClO_4$, which may have an analogous hydrogen-bonded structure.

Many of the hydroxy compounds are readily crystallizable substances with definite melting points and have been quite extensively used for the purification and characterization of tertiary arsines and arsine oxides (6, 82, 169, 170). Michaelis (44) reported a melting point of 160–161° for triphenylarsenic hydroxynitrate. This value agrees closely with that later reported by Razuvaev and co-workers (82) but differs materially from the value of 104° reported both by Tranter and co-workers (161) and by O'Brien (139). It is interesting to note that Michaelis (44) reported a melting point of 99–100° for triphenylarsenic dinitrate, whereas Tranter and co-workers (161) reported a melting point of 161–162° for this compound.

The structure of compounds of the type $R_3As(OH)X$ was first studied by Steinkopf and Schwen (171). They found that distillation of dimethylphenylarsenic hydroxybromide in vacuum gives a variety of products which includes trimethylphenylarsonium bromide, methyl bromide, and methyl alcohol. On the basis of their studies the authors were unable to decide between two structures $[R_3AsOH]^+X^-$ and $[R_3AsX]^+OH^-$. The structure of the trialkylarsenic hydroxyhalides was intensively studied by Nylén (29). He demonstrated that in aqueous solution trimethyl- and triethylarsenic dibromides hydrolyze completely to the hydroxybromides. These latter compounds behave as strong electrolytes in water; e.g., a 0.0107*M* solution of trimethylarsenic hydroxybromide is 93.5% dissociated into M_3AsOH^+ and Br^- in aqueous solution. A similar conclusion had been reached earlier by Hantzsch and Hibbert (140) who determined the molar conductivity of trimethylarsenic dibromide in aqueous solution. From their results they concluded that $(CH_3)_3AsBr_2$ is hydrolyzed completely to the hydroxybromide and that the latter behaved as a strong electrolyte in aqueous solution.

Further evidence for the "onium" structure for these hydroxy compounds was advanced by Mann and Watson (172). They demonstrated that both diphenyl-α-pyridylarsine oxide and tri-α-pyridylarsine oxide form only monopicrates when treated with picric acid. Since tri-α-pyridylarsine forms a dipicrate, the authors assumed that the structure of the cation of the monopicrates must be $[R_3AsOH]^+$. The failure of the oxides to form dipicrates was then attributed to the deactivating influence of the positively charged arsenic on the pyridine nitrogen atom or atoms. Jensen (72) determined the dipole moment of triphenylarsenic hydroxychloride and obtained a value of 9.2 D from which he concluded that the compound exists as a resonance hybrid:

$$(C_6H_5)_3\overset{+}{As}\cdots\overset{-}{O}\cdots HCl \leftrightarrow (C_6H_5)_3\overset{+}{As}—OH\ Cl^-$$

In the past few years there have been several investigations of the structure of the hydroxy compounds. The infrared spectrum of trimethylarsenic hydroxychloride contains bands at 847 and 303 cm^{-1}, which have been assigned to the As—O and As—Cl stretching frequencies, respectively (138, 139). Thus, it has been concluded that in the solid state this compound contains five groups covalently bonded to arsenic. The infrared spectrum of trimethylarsenic hydroxynitrate also suggests a structure containing a pentacovalent arsenic atom. The infrared spectra of trimethylarsenic hydroxybromide and of triphenylarsenic hydroxychloride, hydroxybromide, and hydroxynitrate have also been determined (81, 138, 139, 161), but the structures of these compounds could not be unequivocally deduced from the infrared data. Conductivity measurements on all the trimethyl and triphenyl compounds show that they are weak electrolytes in acetonitrile solution (81, 139, 161). By contrast, triphenyl- and trimethylarsenic hydroxyperchlorates are strong electrolytes in acetonitrile (81, 139) and have infrared spectra (81, 138, 139) that indicate the presence of the perchlorate ion. The molar conductances in acetonitrile and the infrared spectra in the solid state of a series of compounds of the type $Ph_3As(OH)X$, where X is Br_3, ICl_2, $HgCl_4$, or $HgBr_4$, also show that these compounds contain the Ph_3AsOH^+ cation (81).

Very recently, Ferguson and Macaulay (173) have determined the structures of triphenylarsenic hydroxychloride and hydroxybromide by means of X-ray diffraction. Their results show that the arsenic atoms in these hydroxyhalides are tetrahedrally coordinated to three phenyl rings and to an oxygen atom, which is in turn strongly hydrogen bonded to the halogen by an unusually short hydrogen bond. The As—O distances are significantly longer than the As=O bond in triphenylarsine oxide monohydrate and comparable in length to an ordinary As—O single bond. It was concluded that the following formulation, which is essentially similar to that earlier proposed by Jensen (72), is a reasonable representation of the structure of triphenylarsenic hydroxychloride and hydroxybromide:

$$Ph_3\overset{+}{As}—\overset{\frac{1}{2}-}{O}\cdots H\cdots \overset{\frac{1}{2}-}{Hal}$$

Ferguson and Macaulay (173) have also studied the proton NMR spectra of triphenylarsenic hydroxychloride and hydroxybromide. In both cases, the aromatic proton signal in deuteriochloroform was found to occur at τ 2.14 and to be unaffected by the addition of small amounts of protonating solvent. In contrast, the aromatic proton signal of triphenylarsine oxide occurs at τ 2.40 and is shifted to τ 2.14 by the introduction of a small amount of a protonating solvent, e.g., trifluoroacetic acid. It was suggested that the downfield shift in the NMR spectrum is associated with the presence of a hydrogen atom about 1.22 A from the arsenic oxygen atom and "can be attributed to either or both of electric field effect or $d\pi$-aromatic overlap."

Alkyl esters of dialkylarsinous (174) and alkylarylarsinous (175) acids react readily with alkyl iodides to give compounds of the type $R_3As(OR')I$. For example, the interaction of *n*-propyl iodide and *n*-propyl diethylarsinite yields diethyl-*n*-propylarsenic *n*-propoxyiodide (174). Compounds of this type have been obtained as by-products in the reaction of dialkyliodoarsines with sodium alkoxides to form alkyl dialkylarsinites (176). Triphenylarsenic benzyloxybromide, $Ph_3As(OCH_2Ph)Br$, has been prepared by the reaction of one mole of triphenylarsenic dibromide with one mole of benzyl alcohol in the presence of triethylamine (177). Several dialkylphenyl alkoxyiodides have been converted to the nitrates, $R_2ArAs(OR')NO_3$, by evaporation of nitric acid solutions of the corresponding alkoxyiodides (178).

The addition of a cyanogen halide to a tertiary arsine results in the formation of an arsenic cyanohalide, $R_3As(CN)X$ (135, 144, 167). The cyanohalides are extremely susceptible to hydrolysis with cleavage of the cyano group and formation of the arsenic hydroxyhalide. Thus, dimethylphenylarsenic cyanobromide on standing in the air gives dimethylphenylarsenic hydroxybromide (167).

When tertiary arsenic cyanohalides containing at least one alkyl group are heated, an alkyl halide is eliminated with the formation of the trivalent cyanoarsine. Thus, cyclohexylmethylphenylarsenic cyanobromide, when heated to 70–80°, gives methyl bromide and cyclohexylphenylcyanoarsine

(127). Triphenylarsenic cyanobromide, however, was found to yield triphenylarsine, cyanogen bromide, cyanogen, and an unidentified arsenic compound (167). The structure of the cyanohalides has not been investigated.

Only one compound of the type $R_3As(NH_2)X$ has been reported. This is triphenylarsenic aminochloride, prepared by the addition of chloramine to triphenylarsine (105, 106) or by the reaction of a triphenylarsenic dihalide with anhydrous ammonia in benzene (164, 165). Reactions of this compound are described in Sections V-4 and V-7.

9. ARSONIUM COMPOUNDS

A. Preparation

Trialkyl- and triarylarsines combine with alkyl halides to form arsonium halides of the type $[R_4As]^+X^-$ and $[Ar_3RAs]^+X^-$. The reaction occurs readily in the presence of a solvent (14, 21, 26) or in some cases simply by mixing the arsine and alkyl halide (179). The mechanism of the reaction has been studied by Davies and Lewis (180, 181) who demonstrated that arsines were quaternized faster than the corresponding amines but slower than the phosphines. With phenyldialkylarsines, electron-repelling groups on the aromatic ring increase the reaction rate, whereas electron-attracting groups decrease the rate. Methyl halides react faster than ethyl halides, and iodides react faster than bromides. These results are consistent with an S_N2 attack of the arsine on the alkyl halide. This is essentially the mechanism suggested by Davies and Lewis. It is not surprising to find, therefore, that aryl and vinyl halides cannot be used in this reaction. Tetraphenylarsonium tetrafluoroborate, however, has been prepared by the reaction between three moles of triphenylarsine and one mole of $(C_6H_5)_2IBF_4$ at 213° (182). The yield of arsonium compound was 74%. Dimethyl(trifluoromethyl)arsine reacts with methyl iodide in a sealed tube at 85° for 14 hours to give trimethyl(trifluoromethyl)arsonium iodide (183). Tris(trifluoromethyl)- (124), methylbis(trifluoromethyl)- (184), and diphenyl(trifluoromethyl)arsines (150), however, can not be quaternized with methyl iodide. Triphenylarsine and trifluoroiodomethane, when heated in a sealed tube, give fluoroform, benzene, benzotrifluoride, and arsenic triiodide (131). Methyl-*n*-butylphenylarsine has been quaternized with benzyl *p*-toluenesulfonate to the oily methyl-*n*-butylphenylbenzylarsonium *p*-toluenesulfonate (185).

Several interesting spirocyclic arsonium salts have been prepared by Mann and co-workers. 2-Methylisoarsindoline, when treated with o,α,α'-dibromoxylene, gives the quaternary arsonium salt **6**, which on pyrolysis loses methyl bromide and undergoes cyclization to give the spiran **7** (28).

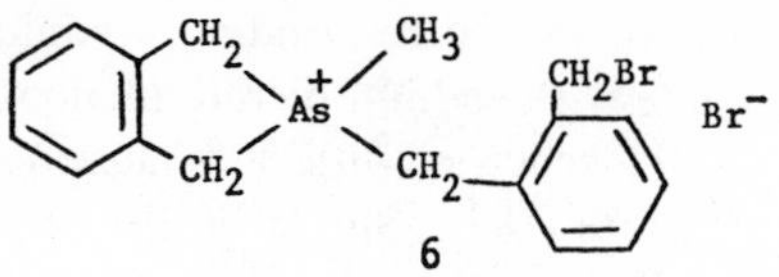

CH_2 CH_3 $\overset{+}{As}$ CH_2Br Br^- CH_2 CH_2

6

CH_2 CH_2 $\overset{+}{As}$ Br^- CH_2 CH_2

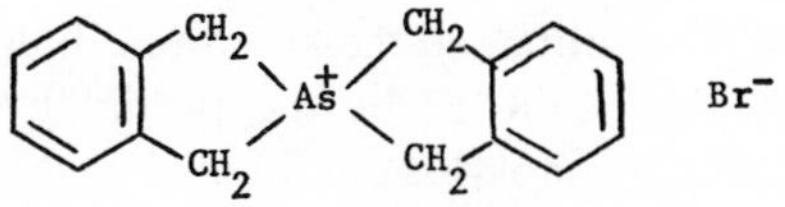

7

Similarly, the quaternary salt **8** on pyrolysis gives the spiran **9** (186).

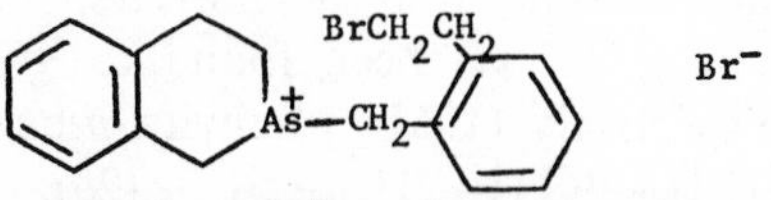

$BrCH_2CH_2$ Br^- $\overset{+}{As}$—CH_2

8

$\overset{+}{As}$ Br^-

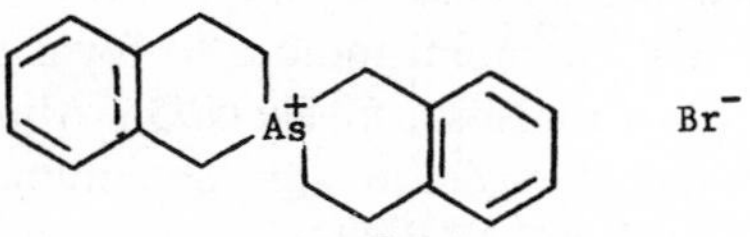

9

The isomeric spiran **10** was prepared by cyclization of **11** (187).

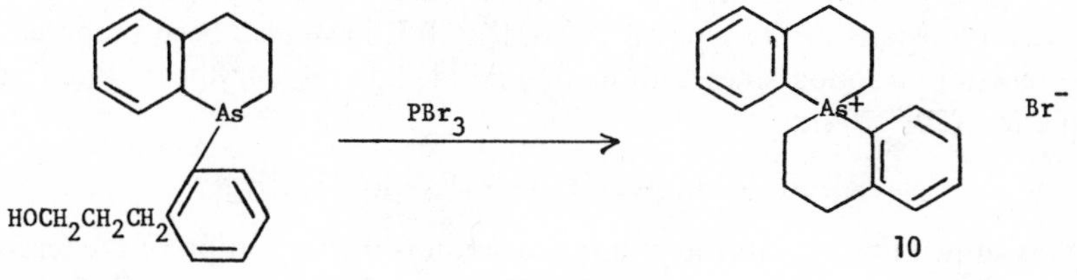

As PBr_3 As^+ Br^-

$HOCH_2CH_2CH_2$

11

10

Seyferth (130) has obtained arsonium salts containing the trimethylsilylmethyl group by reaction of the corresponding arsine with the appropriate halide. Thus, tris(trimethylsilylmethyl)arsine reacts with trimethylsilylmethyl, methyl, and ethyl iodides to produce the arsonium salts $[(Me_3SiCH_2)_4As]I$, $[(Me_3SiCH_2)_3AsMe]I$, and $[(Me_3SiCH_2)_3AsEt]I$, respectively.

Arsonium salts other than halides are generally prepared from the halides by reaction with the appropriate reagent. Thus, evaporation of an arsonium iodide with nitric acid leads to the corresponding arsonium nitrate (175); reaction of an arsonium halide with sodium picrate in alcohol solution leads to the arsonium picrate (188); reaction with 30% aqueous perchloric acid leads to the arsonium perchlorate (189). Similarly, benzyltriphenylarsonium nitrate has been prepared by heating the chloride with ammonium nitrate in aqueous solution and evaporating until crystallization occurs (190). Arsonium salts can also be prepared by the action of acids on arsonium hydroxides (82, 191) or by metathesis between arsonium halides and the appropriate silver salt (82, 192). Roesky (193) has obtained the arsonium dihalothiophosphates $[Ph_4As][SPOCl_2]$ and $[Ph_4As][SPOClF]$ by the partial hydrolysis of thiophosphoryl chloride or thiophosphoryl monofluoride dichloride, $PSCl_2F$, in the presence of tetraphenylarsonium chloride.

Arsonium halides react with halogens to give perhalides. Thus, Blicke and Monroe (194) have prepared tetraphenylarsonium trichloride, tribromide, and triiodide. The reaction of tetramethylarsonium iodide in methanol with chlorine or bromine has been found to give $[(CH_3)_4As]ICl_2$ and $[(CH_3)_4As]IBr_2$, respectively (195). Methyltri-*p*-biphenylylarsonium iodide and chlorine give $[CH_3(p\text{-}C_6H_5C_6H_4)_3As]ICl_2$ (25).

Arsonium halides react with a variety of transition metal compounds to yield arsonium salts of complex anions. Thus, diethyldimethylarsonium iodide reacts with mercuric and cadmium iodides to form the compounds $[Me_2Et_2As]HgI_3$ and $[Me_2Et_2As]CdI_3$, respectively (195). Aliphatic arsonium compounds react with platinic chloride to give compounds of the type $[R_4As]_2[PtCl_6]$ (21). A large number of similar compounds have been prepared, which contain such anions as $Co(SCN)_4^{2-}$ (196), $CoBr_4^{2-}$ (197), $Co(NO_3)_4^{2-}$ (198), $CuBr_4^{2-}$ (197), $FeCl_4^{2-}$ (199), $Fe(SCN)_6^{3-}$ (200), $MnCl_4^{2-}$ (197), $NiCl_4^{2-}$ (197), $ZnBr_4^{2-}$ (197), VCl_4^- (201), and BiI_4^- (202).

Arsonium salts of the type $[R_2As(CH_3)_2]HgI_3$ have also been prepared by the reaction of iodoarsines with methyl iodide and mercury according to the equation (203, 204):

$$R_2AsI + Hg + 2CH_3I \rightarrow [R_2As(CH_3)_2]HgI_3$$

It was shown that the first step in the reaction is the formation of the tertiary arsine mercuric iodide complex, $R_2AsCH_3 \cdot HgI_2$, which then reacts further with methyl iodide to form the arsonium salt.

Tetraethylarsonium iodide was found to react with excess mercuric iodide in alcohol solution in the cold to form the expected $[(C_2H_5)_4As]HgI_3$ (203). When this was heated with alcohol, or when the original reaction was carried out in hot alcohol, the compound $[(C_2H_5)_4As]_2HgI_4$ was obtained. Similar arsonium salts have been reported by Cass, Coates, and Hayter (205). They prepared these compounds by the reaction of tertiary arsine–mercuric iodide complexes of the type $(R_3As)_2HgI_2$ with two moles of methyl iodide.

Faraglia and co-workers (206) have described a number of arsonium salts of the following types:

$$[Ph_4As][R_2TlX_2],\ [Ph_4As]_2[R_2TlX_3],\ \text{and}\ [Ph_4As][RTlX_3]$$

They were obtained by the reaction of tetraphenylarsonium chloride, bromide, iodide, or thiocyanate with organothallium compounds of the generic formulas R_2TlX and $RTlX_2$, where R is CH_3 or C_6H_5 and X is Cl, Br, I, NO_3, or SCN. An attempt to prepare $[Ph_4As]_2[RTlCl_4]$, however, was unsuccessful.

Hellwinkel and Kilthau (207) have prepared the "onium-ate" compound

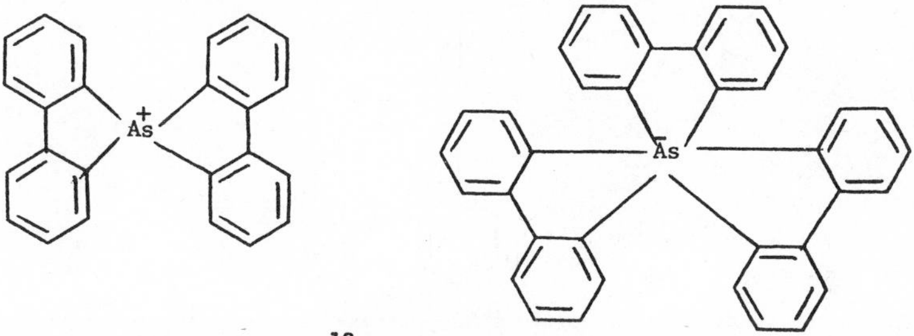

12

12 by treatment of the arsonium iodide **13** with two moles of 2,2′-dilithiobiphenyl. When one mole of the lithium reagent was used per mole of the iodide, the lithium salt **14** was obtained. This salt was converted to a brucine

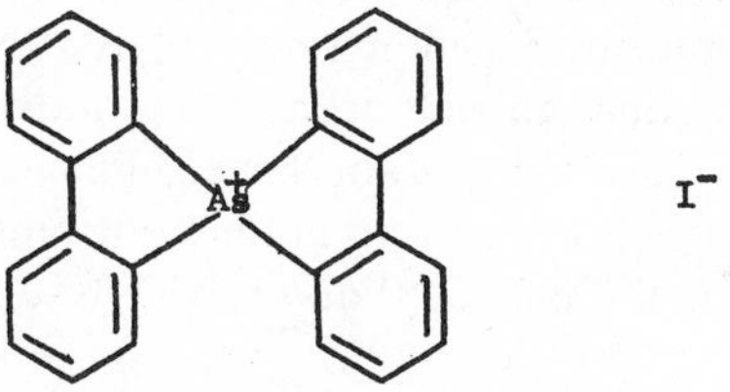

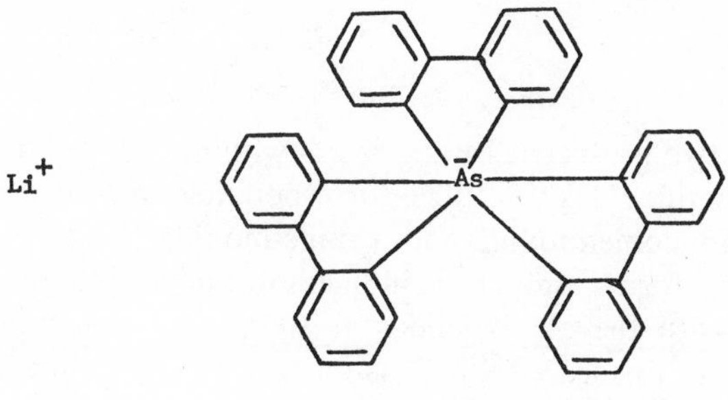

13

14

salt, which could be separated into diastereomers. When either diastereomer was treated with a mixture of acetone and hydrochloric acid, one arsenic–carbon bond was cleaved to give the pentacovalent compound **15**. This was

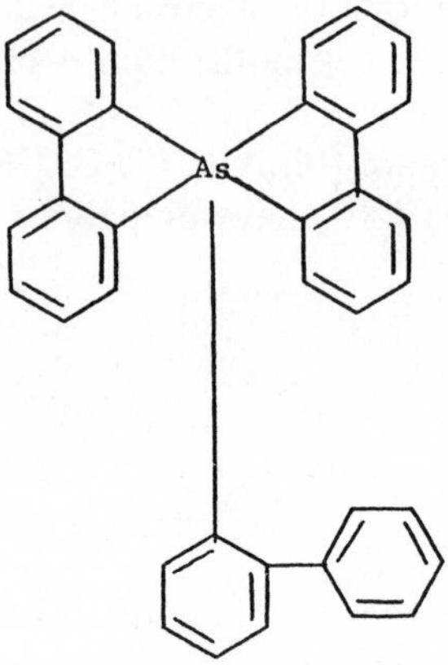

15

optically inactive, a fact attributed to pseudorotation.

When two atoms capable of quaternization are in the same molecule, varying results are obtained, depending not only on the nature of the basic groups but also on the quaternizing agent. Mann and his co-workers have studied this problem in great detail and have summarized their conclusions (172). If two arsenic atoms are present in the same molecule, quaternization of both atoms is difficult. Thus, 5,10-di-*p*-tolyl-5,10-dihydroarsanthrene **16**,

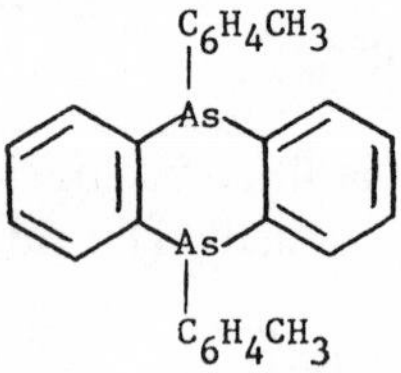

$C_6H_4CH_3$

As

As

$C_6H_4CH_3$

16

which exists in two geometric forms, reacts with methyl iodide to form two distinct monomethiodides (87). The corresponding 5,10-dimethyl compound forms both a monomethiodide and a dimethiodide (208). *o*-Phenylenebis-(dimethylarsine) forms only a monomethobromide and monomethiodide unless heated with the alkyl halides at 100° when dimethobromides and dimethiodides are formed (209). When a tertiary phosphine and arsine are present in the same molecule, the phosphine is preferentially quaternized. Thus *o*-diethylphosphinophenyldimethylarsine forms only a monomethiodide **17**,

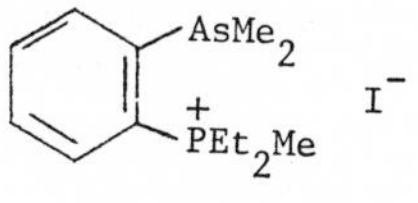

$$\left[o\text{-}C_6H_4(AsMe_2)(\overset{+}{P}Et_2Me) \right] I^-$$

17

but *o*-diethylphosphinophenyldiethylarsine forms a mono- and dimethiodide in the ratio of 5:1 (210). When a tertiary amine and arsine are present in the same molecule, the arsine is preferentially quaternized. *o*-Dimethylaminophenyldimethylarsine forms only a monomethiodide and a monomethopicrate, in both of which the arsenic atom has been quaternized (211). Similar results were obtained with a series of pyridyl arsines prepared by Mann and Watson (172). They explained the difficulty in quaternizing the second basic group in these molecules as follows. Once one group is quaternized, the strong electron attraction of the positive pole deactivates the second basic group so that it is incapable of nucleophilic attack on the alkyl group.

Different results are often obtained, however, if a dibromide such as ethylene dibromide is used for quaternization. *o*-Phenylenebis(dimethylarsine) reacts readily with ethylene dibromide to form the cyclic dibromide **18**

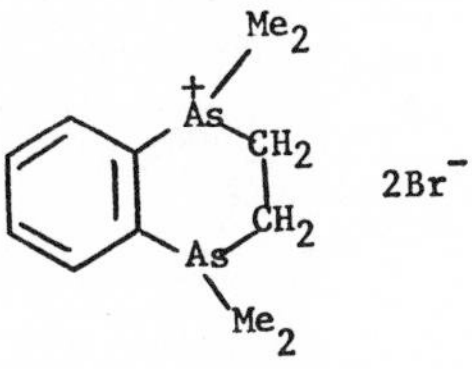

$$o\text{-}C_6H_4(\overset{+}{A}sMe_2CH_2CH_2\overset{+}{A}sMe_2)\ 2Br^-$$

18

(209). Similarly *o*-diethylphosphinophenyldimethylarsine combines with ethylene dibromide to form the cyclic dibromide **19** (210). Hexahydro-1,4-

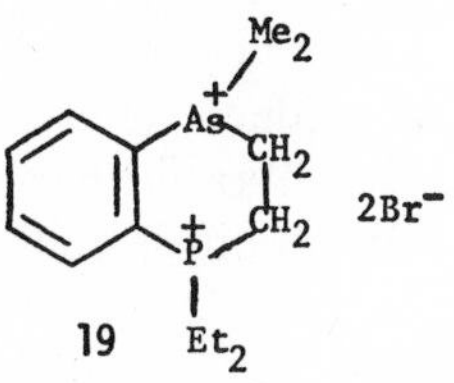

$$o\text{-}C_6H_4(\overset{+}{A}sMe_2CH_2CH_2\overset{+}{P}Et_2)\ 2Br^-$$

19

diphenylazarsine, however, reacts with ethylene dibromide to form the compound **20** in which only the arsenic atoms are quaternized. *o*-Phenyl-

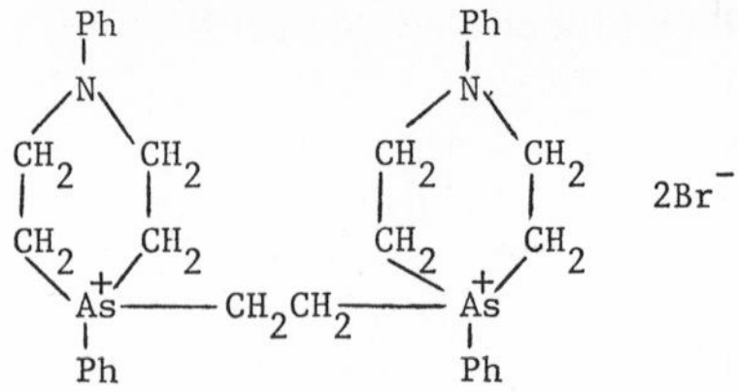

20

enebis(dimethylarsine) reacts with methylene bromide and with carbon tetrabromide to form the heterocyclic arsonium salts **21** and **22**, respectively (212).

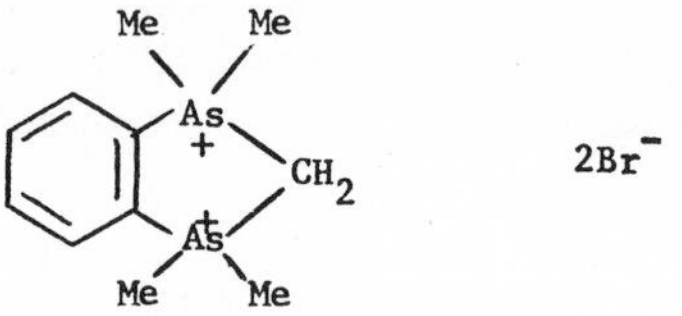

21

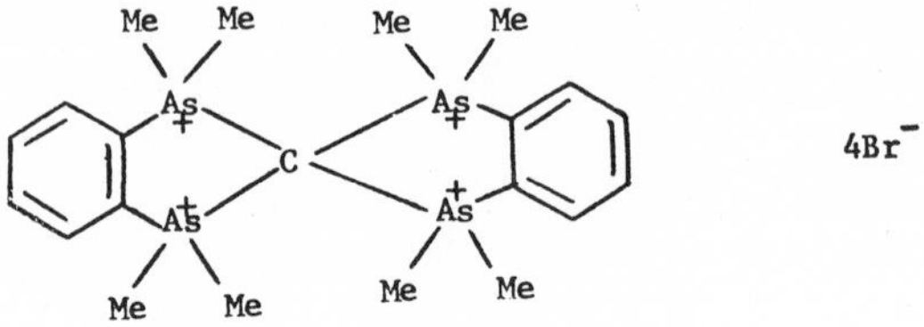

22

In addition to quaternization of tertiary arsines, several other methods of preparing arsonium salts have appeared in the older literature. Thus metallic arsenic and methyl iodide, when heated on the water bath, give a mixture of tetramethylarsonium iodide, methyldiiodoarsine, and arsenic triiodide (213). The same arsonium compound may be prepared from sodium arsenide and methyl iodide (214), from arsenomethane and methyl iodide (215), and from cacodyl and methyl iodide (216). Tetramethylarsonium chloride has been obtained by the thermal decomposition of addition products of trimethylarsine with arsenic trichloride or methyldichloroarsine (217). More recently Issleib and Tzschach (218) have obtained dicyclohexyldimethylarsonium iodide by the reaction of dicyclohexylarsine with methyl iodide. Arsenic ylids (the preparation of which is described in Section V-9-B) react with water (219) to form arsonium hydroxides, with perchloric acid (220) to form arsonium perchlorates, and with bromine (221) or trimethylbromosilane (219, 222) to form arsonium bromides. Boron trifluoride (219) and sulfur trioxide (221) add to ylids to give zwitterionic arsonium compounds. For example, boron trifluoride adds to methylenetriphenylarsenic, Ph_3AsCH_2, to yield $Ph_3\overset{+}{As}\overset{-}{C}H_2BF_3$ (219). Wittig and Laib (223) have reported that an ethereal solution of 9-fluorenylidenetrimethylarsenic and phenyllithium reacts with benzophenone to give (after treatment with water) phenyltrimethylarsonium hydroxide and 9-fluorenyldiphenylcarbinol. The reaction of trimethylarsenic dibromide with methyllithium at low temperatures has been found to give low yields of arsonium salts (150).

Two excellent methods for the preparation of tetraarylarsonium halides have been devised. The first of these is due to Blicke and co-workers and depends upon the reaction of a triarylarsine oxide with an aromatic Grignard reagent (192, 194, 224, 225). Thus, when triphenylarsine oxide is allowed to react with phenylmagnesium bromide and hydrobromic acid is added to the reaction mixture, a precipitate of tetraphenylarsonium bromide is obtained. Recrystallization of this from hydrochloric acid gives the peculiar hydrochloride $(C_6H_5)_4AsCl \cdot HCl$ from which tetraphenylarsonium chloride can be obtained by treatment with dilute alkali. The phenyl Grignard reagent also reacts with the anhydride of benzenearsonic acid to give tetraphenylarsonium bromide, but the yield is much smaller than when the arsine oxide is used (194). In many instances, however, the interaction of a triarylarsine oxide and an aryl Grignard reagent yields only gummy products (225). Several mixed aryl alkyl arsonium salts have been prepared both by the reaction of triarylarsine oxides with alkyl Grignard reagents and by the reaction of alkyldiarylarsine oxides with aryl Grignard reagents (194, 225). Triphenylcyclopentadienylarsonium salts have been obtained via the reaction of triphenylarsine oxide and cyclopentadienylmagnesium bromide (221).

The second method of preparation was discovered by Chatt and Mann and consists of heating a mixture of benzene, arsenic trichloride, and anhydrous aluminum chloride (226). Addition of potassium iodide to the reaction mixture precipitates tetraphenylarsonium iodide. Tetraphenylarsonium salts are also formed by heating aluminum chloride with phenyldichloroarsine, diphenylchloroarsine, or triphenylarsine. The best results, however, are obtained from a mixture of aluminum chloride, triphenylarsine, and bromobenzene (226, 227). The mechanism of this last reaction has been examined by Lyon and Mann (42) who demonstrated that aluminum trichloride reacts with triphenylarsine to form a crystalline addition product $[(C_6H_5)_3AsAlCl_3]$. This then reacts with bromobenzene to form the arsonium salt $[(C_6H_5)_4As]$-$[AlCl_3Br]$. Different mechanisms were suggested for the reactions of phenyldichloro- and diphenylchloroarsines with aluminum chloride. It is interesting to note that all attempts to prepare tetra-*o*-tolylarsonium salts by the above procedures failed. Instead, compounds formulated as **23** and **24** were obtained.

$$(o\text{-}CH_3C_6H_4)_3As\text{—}OH \quad \diagdown \quad O \quad \diagup \quad (o\text{-}CH_3C_6H_4)_3As\text{—}Br$$

23

$$(o\text{-}CH_3C_6H_4)_3As\text{—}Br \quad \diagdown \quad O \quad \diagup \quad (o\text{-}CH_3C_6H_4)_3As\text{—}Br$$

24

In addition to the syntheses of Blicke and co-workers and of Mann and co-workers, several other methods for the preparation of tetraarylarsonium salts have been devised. Thus, the spirocyclic arsonium chloride **25** has

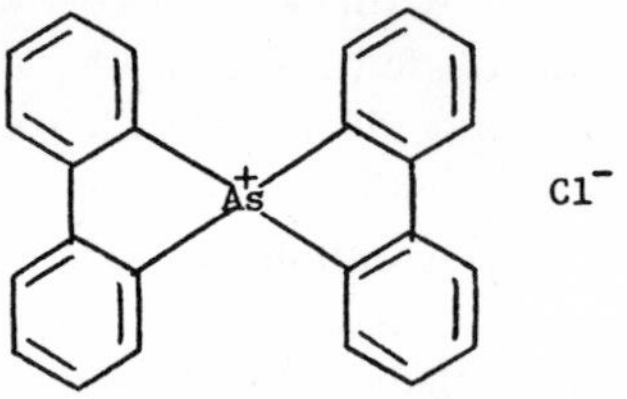

As Cl⁻

25

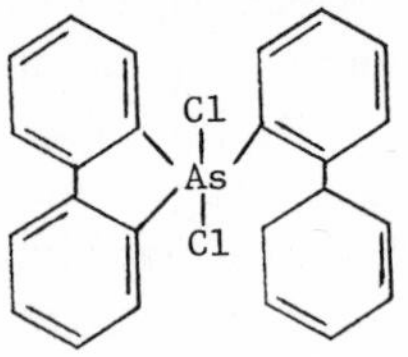

Cl As Cl

26

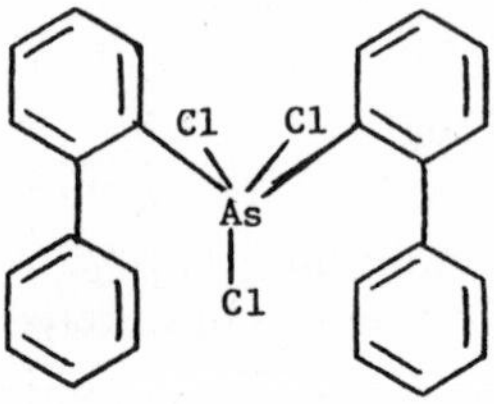

Cl Cl As Cl

27

been obtained by the cyclodehydrohalogenation of compound **26** or **27** at about 270° (101). Similarly, the iodide **28** has been prepared via

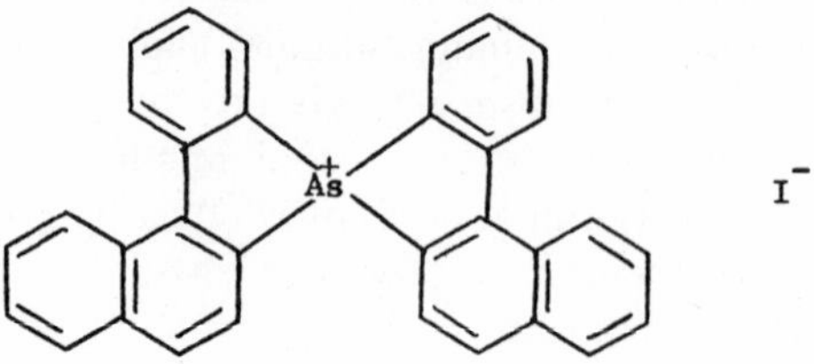

As I⁻

28

cyclodehydration of the arsinic acid **29** in polyphosphoric acid. Tetraaryl-

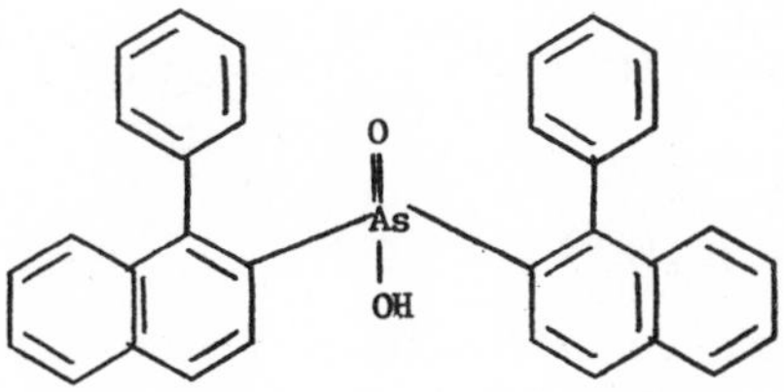

29

arsonium salts have also been obtained by the reaction of pentaarylarsenic compounds with a variety of electrophilic reagents (see Section V-10).

Arsonium hydroxides of the type R_4AsOH can be prepared by the action of silver oxide on an arsonium halide (188, 228). In one case aqueous sodium hydroxide was used instead of silver oxide (229). (2-Hydroxyethyl)trimethylarsonium hydroxide has been prepared by the reaction of trimethylarsine with ethylene oxide in aqueous solution (230). The arsonium hydroxides are strong bases, which precipitate heavy metal hydroxides from solutions of their salts and liberate ammonia from ammonium salts. They combine with acids to form arsonium salts.

B. Reactions

Many alkyl arsonium halides and hydroxides on pyrolysis lose an alkyl group and yield tertiary arsines. The cleavage undoubtedly involves the nucleophilic attack of a halide or hydroxide ion on the alkyl carbon atom bonded to the arsenic. This type of reaction has been summarized in Section IV-4-E and has been fully discussed by Mann in two reviews (231, 232).

The reaction of arsonium salts with organolithium compounds has been investigated by several workers. Triphenylmethylarsonium iodide reacts with phenyllithium to give an ylid as shown in the following equation (233):

$$(C_6H_5)_3CH_3AsI + C_6H_5Li \rightarrow (C_6H_5)_3As{=}CH_2 + LiI + C_6H_6$$

This ylid was not isolated but allowed to react with benzophenone to give, after hydrolysis with hydrochloric acid, a mixture of triphenylarsine, triphenylarsine oxide, diphenylacetaldehyde, and diphenylethylene. The latter two products were obtained in approximately equal amounts. The formation of diphenylacetaldehyde in addition to the expected olefin (the Wittig reaction) is surprising. Henry and Wittig (233) suggested that it may arise through a reaction involving an epoxide ring:

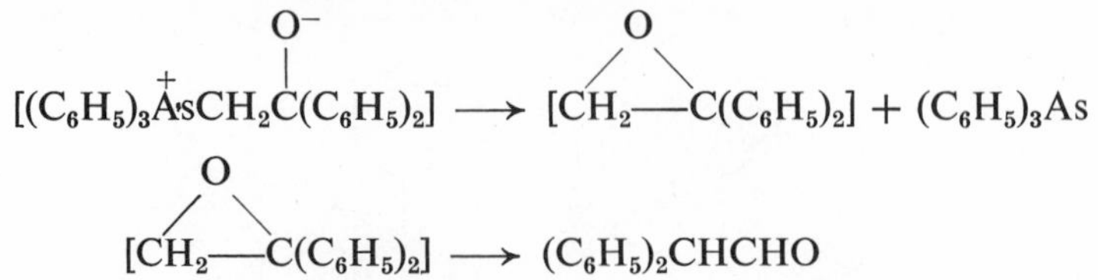

$$[(C_6H_5)_3\overset{+}{A}sCH_2\overset{O^-}{\overset{|}{C}}(C_6H_5)_2] \longrightarrow [\overset{O}{CH_2\text{—}C}(C_6H_5)_2] + (C_6H_5)_3As$$

$$[\overset{O}{CH_2\text{—}C}(C_6H_5)_2] \longrightarrow (C_6H_5)_2CHCHO$$

Johnson (234) has suggested that the epoxide rearranges to the aldehyde when the reaction mixture is subjected to acid hydrolysis. In later work Johnson and Martin (235) actually isolated epoxides from the reaction of arsonium ylids with carbonyl compounds. The ylid described above has been prepared independently by Seyferth and co-workers (219, 222) by the same procedure used by Henry and Wittig.

Wittig and Laib (223) have prepared the ylids **30**, **31**, and **32** by the reaction

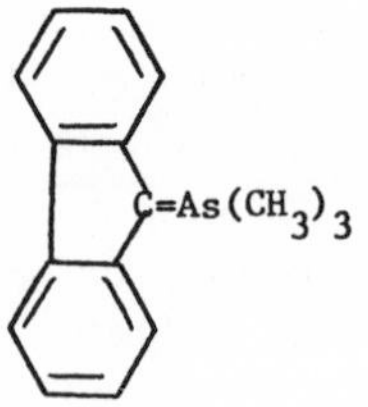

$C{=}As(CH_3)_3$

30

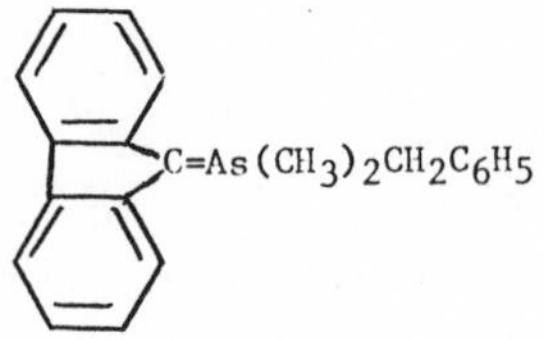

$C{=}As(CH_3)_2CH_2C_6H_5$

31

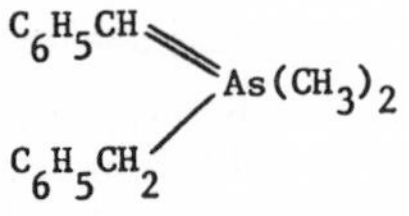

$C_6H_5CH{=}As(CH_3)_2$
$C_6H_5CH_2$

32

of phenyllithium with the appropriate arsonium bromide. None of these ylids was isolated and characterized. Compound **32** changed spontaneously into dimethyl(1,2-diphenylethyl)arsine. Compounds **30** and **31** were allowed to react with benzophenone, but, instead of the expected olefin, 9-fluorenyldiphenylcarbinol was isolated in both cases.

In contrast to these results Johnson (236) has obtained the expected olefinic products of the Wittig reaction from 9-fluorenylidenetriphenylarsenic. The ylid was prepared by the action of sodium hydroxide on triphenyl-9-fluorenylarsonium bromide and was characterized by analysis and ultraviolet absorption spectroscopy. Johnson compared the reaction of the arsonium ylid with that of the corresponding phosphonium ylid, which he had earlier described (237). Both ylids react with aldehydes to give the expected olefins and triphenylarsine oxide or triphenylphosphine oxide, and neither ylid reacted with ketones. There was a marked difference, however, between the arsonium ylid and the phosphonium ylid in their reactions with *p*-substituted benzaldehydes. Whereas the phosphonium compound gave 0, 40, and 100% conversions in its reaction with *p*-dimethylaminobenzaldehyde, anisaldehyde, and *p*-nitrobenzaldehyde, the arsenic analog gave 97, 89, and 92% conversions with these aldehydes. From these results he concluded that arsenic and phosphorus differ slightly in the extent of *d*-orbital resonance, with the phosphorus undergoing greater octet expansion than the arsenic. It is interesting to note that dipole moment studies with the phosphonium ylid indicate that the P—C bond possesses essentially 50% double bond character. Unfortunately, dipole moment studies were not conducted with the arsenic analog.

Kröhnke (238) has investigated the reactions of *o*-, *m*-, and *p*-nitrobenzyltriphenylphosphonium and arsonium salts with aqueous alkali. The *m*-nitro compounds, on treatment with alkali and extraction into chloroform, give only pale pink or yellow colors which the author ascribed to the ylids, $m\text{-}O_2NC_6H_4CH{=}MPh_3$, where M is phosphorus or arsenic. When the *o*- or *p*-nitrobenzyl compounds in dilute aqueous solution were treated with alkali, intensely colored red or violet precipitates were formed. The color could be extracted into chloroform. The marked increase in color intensity of these compounds compared with the *m*-isomers was attributed to resonance interaction between the nitro and the $Ph_3\overset{+}{M}CH$ groups.

Several arsonium ylids have been prepared by the action of base on arsonium salts derived from triphenylarsine and α-bromoketones. The preparation and reactions of ylids of this type, as well as the chemistry of arsonium ylids in general, have been reviewed by Johnson (234). More recently, some Chinese workers (239) have prepared carbomethoxymethylenetriphenylarsenic and showed that it reacts with both aldehydes and ketones to yield the expected methyl esters of α,β-unsaturated acids. Nesmeyanov and co-workers (240) have prepared similar carboalkoxy-substituted ylids as well as compounds of the types $Ph_3As{=}CHCONH_2$ and $Ph_3As{=}CHCOR$; all of these compounds react with aldehydes in the usual way. In a later paper (241), it was reported that the ylid $Ph_3As{=}CHCOPh$ can be stabilized by reaction with mercuric chloride:

$$Ph_3As{=}CHCOPh + HgCl_2 \longrightarrow Ph_3\overset{+}{As}{-}CH\begin{matrix} \diagup HgCl \\ \\ \diagdown COPh \end{matrix} \quad Cl^-$$

This adduct was found to react with *p*-nitrobenzaldehyde to give a high yield of *β*-(*p*-nitrostyryl)phenyl ketone.

Nesmeyanov and co-workers (242) have found that the ylid $Ph_3As{=}CHPh$ decomposes when heated in benzene–ether solution to yield triphenylarsine and a mixture of *cis*- and *trans*-stilbene in a ratio of 1 to 5.2. This type of reaction does not occur with the corresponding phosphorus ylid. It was suggested that the decomposition is catalyzed by traces of the arsonium salt remaining in the ylid when it is prepared from a benzyltriphenylarsonium salt and phenyllithium. A mechanism for the decomposition was also proposed.

Maccioni and Secci (243) have reported an astonishing reaction between ethylidenetriphenylarsenic, $Ph_3As{=}CHCH_3$, and *p*-tolualdehyde. Although a small amount of the expected *trans*-1-propenyl-4-methylbenzene was obtained, the main product was *p*-xylyl methyl ketone. There appears to be no precedent for this type of reaction.

Nesmeyanov and co-workers (221) have reported that they were unable to obtain an ylid by the treatment of a cyclopentadienyltriphenylarsonium salt with base. Recently, however, Lloyd and Singer (220) prepared the ylid **33** by heating a mixture of triphenylarsine and diazotetraphenylcyclopentadiene.

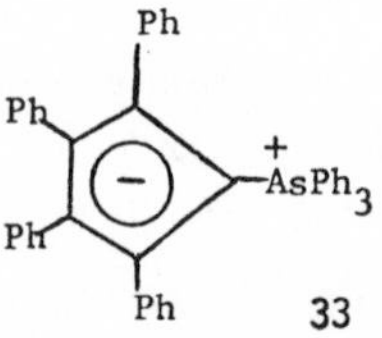

Ph Ph Ph Ph $\overset{+}{As}Ph_3$ (cyclopentadienide ring, −)

33

It is presumed that the thermal decomposition of the diazo compound yields a carbene, which reacts with the lone pair of electrons on the arsenic atom.

The preparation of ylids from arsonium salts appears to require stronger bases than are necessary for the conversion of phosphonium salts to ylids. This fact has been correlated with the greater acidity of the phosphonium compounds (244). Thus, Johnson and LaCount (245) have reported that the acidity of triphenyl-9-fluorenylphosphonium bromide ($pK_A = 7.5$) in aqueous dioxane is about twice that of the corresponding arsonium salt ($pK_A = 7.8$). More recently, Nesmeyanov and co-workers (244) have compared the acidity in aqueous ethanol of a series of arsonium and phosphonium compounds of the type (p-$XC_6H_4COCH_2MPh_3$)Br, where M is P or As and X is H, CH_3O, CH_3, Cl, Br, or NO_2. It was found that the phosphorus compounds were about 200–230 times more acidic than the analogous arsenic compounds. This result was attributed to the lower electronegativity of the arsenic atom and perhaps to the greater diffusion of its 4 *d*-orbitals, which prevents their effective overlapping with the 2 *p*-orbitals of carbon. In contrast, Doering and Hoffmann (246) concluded from a study of deuterium exchange in the tetramethyl onium salts of the group V elements that *d*-orbital resonance is roughly constant in going from phosphorus to antimony. Factors other than the electronegativity of the central atom and the extent of *d*-orbital participation must also play a part in determining the relative acidity of onium salts. Thus, the fact that phosphonium salts are stronger acids than the corresponding arsonium compounds has been confirmed by Issleib and Lindner (247), who compared the acid dissociation constants of a number of compounds of the type $[Et_3MCH_2R]X$, where M was P, As, or Sb and R was PhCO or CH_3CO_2. However, the one stibonium compound studied, $[Et_3SbCH_2CO_2CH_3][BPh_4]$, was a considerably stronger acid ($pK_A = 8.5$) in aqueous acetone than either the corresponding phosphonium ($pK_A = 10.6$) or arsonium compound ($pK_A = 10.9$). The relatively great acidity of the antimony compound was considered by Issleib and Lindner to be an anomaly and was attributed to preferential solvation of the stibonium ylid. Aksnes and Songstad (248) have determined the acid dissociation constants of several arsonium salts in a variety of ethanol-water mixtures and compared these results with the dissociation constants of a number of other onium compounds. The effect of the concentration of ethanol on the dissociation constant was found to vary with the type of onium compound studied and was attributed to the relative contributions of the $R_3As{=}CHR'$ and $R_3\overset{+}{A}s\overset{-}{C}HR'$ forms.

The nitration of trimethylphenyl- and trimethylbenzylarsonium picrates has been studied by Ingold, Shaw, and Wilson (249). There is a decrease in the amount of *meta* isomer with both compounds as compared with the corresponding ammonium and phosphonium salts. There is still less *meta* isomer with trimethylphenylstibonium picrate. This effect was attributed to the increased shielding effect of the additional electron shells on the positive field of the central atom as the atomic weight increases from nitrogen to antimony. More recently, the nitration of trimethylbenzylarsonium and trimethylbenzylphosphonium picrates was reinvestigated by Riley and Rothstein (250). Their results are in general agreement with those of the earlier workers.

The reaction between 3-hydroxypropyltriphenylarsonium iodide and sodium hydride has been found by Hands and Mercer (251) to give the interesting heterocycle 2,2,2-triphenyl-1,2-oxaars(V)olan:

$$[HOCH_2CH_2CH_2AsPh_3]I + NaH \longrightarrow \begin{array}{c} (C_6H_5)_3As\text{—}CH_2 \\ |\qquad\quad | \\ O\qquad CH_2 \\ \diagdown\;\diagup \\ CH_2 \end{array}$$

The structure of the product was established by proton NMR studies. When heated to 160–180°, the heterocyclic compound decomposes to triphenylarsine and allyl alcohol. In contrast to the results of Hands and Mercer, Schmidbaur and co-workers (151) were unable to prepare compounds of the type $R_4AsOSiMe_3$ (where R is methyl or phenyl) by the reaction between arsonium chlorides and the sodium salt of trimethylsilanol.

Triphenylphosphine can be alkylated by heating with triphenylalkylarsonium salts in dry dimethylformamide (252):

$$Ph_3P + [Ph_3AsR]X \rightarrow [Ph_3PR]X + Ph_3As$$

The benzyl group migrates more readily than the other alkyl groups tested; the reaction between the phosphine and triphenylbenzylarsonium iodide begins at 60° and is complete in ten hours at 80°.

The conversion of tetraarylarsonium salts to pentaarylarsenic compounds will be discussed in Section V-10.

The resolution of arsonium salts containing four different organic groups has been the subject of considerable investigation. Kamai (253) obtained some evidence of resolution with benzylethylpropyl-*p*-tolylarsonium iodide, but the activity was slight and the compound rapidly became inactive in solution. More recently Kamai and Gatilov (254) prepared a number of phenyltrialkylarsonium salts. The (+)-π-bromocamphorsulfonates were noncrystallizable with the exception of the amylbenzylethylphenyl salt. This gave two diastereoisomers with rotations of $[M]_D = +347.7°$ and $-222.3°$, respectively. Only the first diastereoisomer could be converted into an active bromide, which gave $[\alpha]_D = +16.5°$ and $[M]_D = +70.0°$. This rapidly racemized on standing. Attempts to resolve benzyl and allyl diarsonium salts of **34** by means of their dibenzoyltartrates were unsuccessful (255).

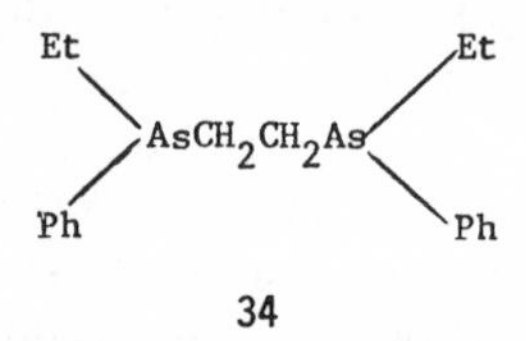

34

It was originally suggested by Burrows and Turner (67), and later by Mann (83), that the lack of success in resolving arsonium salts might be ascribed to a reversible dissociation of an alkyl group:

$$[R_3AsR']X \rightleftharpoons R_3As + R'X$$

For this reason Mann and Watson (256) prepared several arsonium salts containing four different aryl groups. With one exception, all of these compounds were glasses or oils which resisted all attempts at crystallization and resolution. This result was attributed to the physical properties of the compounds rather than to any lack of asymmetry in the molecules.

It now appears that the suggestion of a reversible dissociation of alkylarsonium salts as an explanation for the failure to resolve such compounds is incorrect. Burrows and Turner (67) had originally suggested, on rather shaky evidence, that 1,1-dimethyl-1,2,3,4-tetrahydroarsinolinium iodide undergoes the following reversible intramolecular reaction:

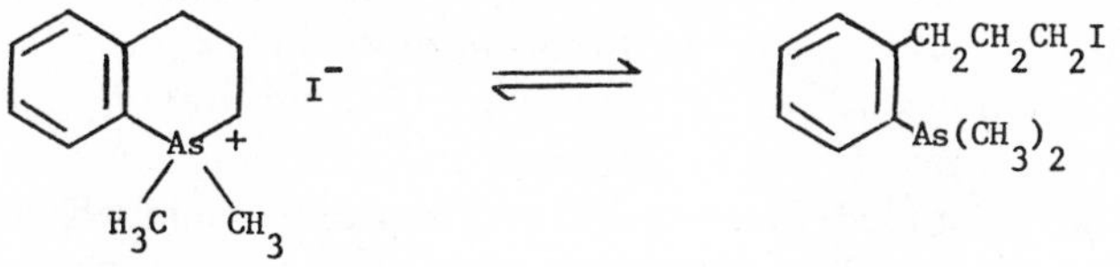

Thornton (257) has recently reported the synthesis of a number of quaternary arsinolinium salts. In no case was there any evidence of cleavage of the heterocyclic ring by the mechanism postulated by Burrows and Turner.

The first preparation of an optically stable quaternary arsonium salt was achieved by Holliman and Mann (258) who resolved several salts of the 2-phenyl-2-*p*-chlorophenacyl-1,2,3,4-tetrahydroisoarsinolinium ion **35**. The

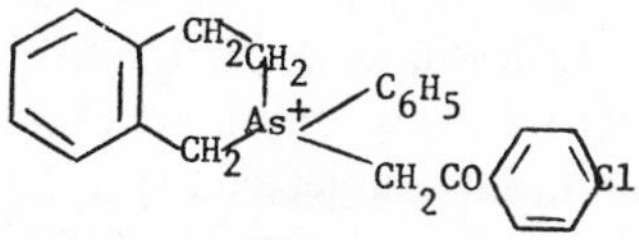

35

two picrates had molar rotations of $-450°$ and $+457°$, respectively, and were optically quite stable. The iodide, however, was slightly unstable. It lost 9% of its activity when refluxed for thirty minutes in aqueous acetone. Several of the spirocyclic arsonium salts prepared by Mann and his co-workers and described in Section V-9-A have also been resolved. More recently, enantiomers of several cyclic diarsonium dibromides have been obtained (259).

In 1962 Horner and Fuchs (185) resolved benzyl-*n*-butylmethylphenylarsonium and benzylethylmethylphenylarsonium ions by means of the corresponding diastereomeric dibenzoyltartrates, which were converted into optically active perchlorates. In a later paper Horner and co-workers (260) prepared optically active benzylmethyl-*n*-propylphenylarsonium tetraphenylborate and showed by the method of quasi-racemates (in which the corresponding phosphonium salt was used) that the (+)-arsonium salt has the S configuration. Reduction of the optically active arsonium salts at a mercury cathode results in cleavage of the benzyl group and the formation of optically active tertiary arsines (261). By an involved series of reactions, Horner and Fuchs (261) demonstrated conclusively that this reaction (as well as the quaternization of tertiary arsines) occurs with retention of configuration. Retention of configuration has also been observed in the cleavage of arsonium salts with potassium cyanide to yield tertiary arsines (262).

In 1964 Kamai and Gatilov (263) reported that trialkylphenylarsonium salts of the type [RR′R″ArAs]X can be resolved by column chromatography on optically active lactose or quartz. The optically active arsonium salts thus obtained were found to lose their activity in about an hour when they were dissolved in chloroform, ethyl alcohol, or ethyl acetate. It was suggested that the loss of activity was due to racemization of the arsonium cations by the dissociation mechanism of Burrows and Turner (67).

In sharp contrast to the conclusions of Kamai and Gatilov (254, 263), Horner and Hofer (264) found that enantiomeric arsonium salts do not undergo racemization readily. Thus, a 0.155*M* solution of benzyl-*n*-butylethylphenylarsonium chloride in chloroform loses only 31.5% of its optical activity on heating at 103° for 70 hours. The corresponding bromide under similar conditions loses 70% of its activity. Under these conditions, optically active arsonium iodides are somewhat less stable than the corresponding chlorides and bromides, while the perchlorates are not significantly racemized. The rate of racemization is solvent dependent and is slower in more polar solvents such as dimethylformamide and dimethyl sulfoxide. In the protic solvents, water, methanol, ethanol, acetic acid, or 1*N* hydrochloric acid, arsonium salts do not racemize. By means of thin-layer chromatography, Horner and Hofer showed that arsonium salts subjected to racemization are still chemically homogeneous; i.e., no redistribution of the attached groups occurs. The rate of racemization in aprotic solvents was found to be first order in arsonium salt and to be accelerated by the addition of bromide ion. Horner and Hofer proposed that the mechanism of the racemization involves attack of halide ions on the positively charged arsenic atom with the formation of an octahedral complex:

$$2[R_4As]^+X^- \rightarrow [R_4AsX_2]^-[R_4As]^+$$

It is difficult to understand how this mechanism is consistent with the observed first order kinetics.

Horner and Hofer have confirmed the finding of Kamai and Gatilov (263) that optically active arsonium salts can be produced by means of column chromatography on levorotatory quartz. The activity, however, was shown to be caused by the presence of fine particles of quartz admixed with the arsonium salts and not to the resolution of a racemic mixture. The loss of optical activity noted by Kamai and Gatilov appears to be due to the gradual sedimentation of the quartz particles.

Lesslie and Turner (265) have partly resolved 3′-bromo-2-biphenylyltrimethylarsonium iodide. In this case the optical activity is associated with restricted rotation around the bond connecting the two benzene rings.

C. Physical properties and structure of arsonium compounds

Mooney (266) has made an X-ray determination of the crystal structure of tetraphenylarsonium iodide. The arsonium ion is strictly tetrahedral with a

C—As distance of 1.95 A in close agreement with the sum of the corresponding covalent radii. More recently Collins, Sutor, and Mann (267) have made a thorough study of the crystal structure of tetramethylarsonium bromide. The As—C distance varies from 1.85 to 1.90 A with a standard deviation of 0.13 A, and the C—As—C angle varies from 107 to 112° with a standard deviation of 8.5°. The positions of arsenic and bromine are somewhat similar to those of sulfur and zinc in wurtzite (each tetrahedrally coordinated by the other), but there is significant deviation from regular tetrahedral coordination. The bromine-methyl distance varies between 3.84 and 4.17 A. The average distance, 3.94 A, is close to the sum of the van der Waals radii, 3.95A, but the deviations from the mean are larger than would be expected. In recent years a number of papers on the crystal and molecular structures of tetraphenylarsonium salts have been published (268–275). The findings concerning the C—As distances and the C—As—C angles in the tetraphenylarsonium ion are in general agreement with the earlier work of Mooney. The orientation of the planes of the phenyl rings with respect to one another, however, varies markedly from compound to compound. It has been suggested (270) that the versatility of form of the tetraphenylarsonium cation helps to account for its effectiveness as a precipitating agent (cf. Section V-9-D).

Bowden and Braude (276) were apparently the first to compare the ultraviolet absorption of an aromatic arsine and a corresponding arsonium compound. They noted that the spectrum of dimethylphenylarsine in alcohol has one broad, intense band at 241.5 mμ while the methiodide of this arsine exhibits four bands between 250 and 272 mμ, which resemble the spectrum of benzene in this region. Mann and his co-workers (277) reported similar results with triphenylarsine and its methiodide. The spectra of tetraphenylarsonium salts are also "benzene-like" (278–280). All of these observations are consistent with Jaffé's conclusion that there is little or no conjugation between a group V atom and a phenyl group attached to it unless the atom has a pair of unshared electrons (73). Recent work (281) indicates, however, that it may be necessary to modify this generalization when the benzene rings contain electron-repelling substituents.

The ultraviolet absorption spectrum and optical rotatory dispersion of a spiroarsonium salt have been investigated (187). It was noted that the molar rotation of the arsenic compound was about twice that of the corresponding phosphorus compound. The spectrum of the methiodide of 5-methyldibenzarsole has been found to exhibit a broad absorption centered at 317 mμ (log ε, 3.403), which was absent in the spectra of the other arsonium compounds studied (228). This band was attributed to the central 5-membered ring system. The ultraviolet absorption spectra of the methiodides of 5,6-dihydro-5-methylarsanthridin and 5,6-dihydro-5-phenylarsanthridin have been determined and compared with the spectra of non-heterocyclic analogs (282).

The Raman spectrum of tetramethylarsonium chloride has been determined by Siebert (283) and the force constants calculated. The Raman frequencies of the arsonium salt in aqueous solution are consistent with a tetrahedral configuration of the tetramethylarsonium ion. The infrared spectra of several tetramethyl-, tetraethyl-, and triethylmethylarsonium salts have been described (284, 285) and assignments have been made for the fundamental frequencies. The results for the tetramethylarsonium salts are in general agreement with the expected tetrahedral symmetry of the arsonium cation, although the CH_3—As symmetric deformation is observed as a weak or a very weak band (at about 1300 cm^{-1}) in the spectra of two of these salts. The asymmetric C—As stretching frequency is found in the 647–657 cm^{-1} range. Since the C—As stretching frequencies in trimethylarsine are at about 568 and 583 cm^{-1}, it appears that the positive charge on the arsonium ion leads to stronger carbon–metal bonding. The C—H stretching frequencies in the arsonium salts are also unusually high and suggest unusually strong C—H bonding. The infrared spectra of the tetraethylarsonium salts are consistent with tetrahedral symmetry of the C_4As skeleton, although the symmetric C—As stretching frequency appears as a very weak band (at 548 cm^{-1}) in the spectrum of tetraethylarsonium iodide; the asymmetric C—As stretch in these compounds occurs in the 611–613 cm^{-1} range. The spectra of the triethylmethylarsonium salts indicate that the skeletal symmetry is effectively C_{3v}. The CH_3—As frequency in these salts is virtually the same as that observed in the tetramethyl compounds; in addition, each triethylmethylarsonium salt has two infrared active C_2H_5—As stretching modes (at about 550 and 610 cm^{-1}).

The work described in the above paragraph has been recently extended by an investigation of the infrared spectra of some salts of mixed alkylarylarsonium cations (286). Replacement of one, two, or three methyl groups in $(CH_3)_4As^+$ by phenyl groups causes the value of the CH_3—As stretching frequency to drop only slightly; this fact suggests that there is little change in the strength of the CH_3—As bond in this series of arsonium compounds. As in the case of the tetraalkyl compounds, the CH_3—As and the C—H frequencies observed in the mixed alkylphenylarsonium compounds are higher than the corresponding frequencies in related arsines. Evidently onium ion formation again results in both stronger C—As and C—H bonding. Stretching of the phenyl-arsenic bond in both the arsines and the arsonium compounds gives rise to three "X-sensitive" frequencies (modes q, r, and t in Whiffen's notation) (157). As expected, it was found that these vibrations are shifted to higher wavenumbers on onium ion formation; i.e., the phenyl–arsenic bonding is stronger in the cations than in the arsines.

In addition to the work discussed above, the literature contains a number of other studies of the infrared spectra of arsonium compounds (6, 74, 156, 248, 287–291).

The proton NMR spectra of several tetraalkylarsonium salts have been determined (292–294). The spectrum of tetraethylarsonium bromide in chloroform is a normal A_3B_2 case in which the CH_3 peaks are upfield by 1.07 ppm relative to the CH_2 peaks. This comparatively large difference is presumably associated with the powerful electron-attracting effect of the positively-charged arsenic atom. In highly polar, ionizing solvents such as water or liquid sulfur dioxide, the peaks exhibited by tetraethylarsonium and tetramethylarsonium salts are significantly broadened. It has been suggested that this broadening is caused by spin coupling of the protons with the arsenic atom. From the relative degrees of broadening of the CH_3 and CH_2 peaks it appears that the long-range coupling, $J_{CH_3—As^+}$, is larger than the short-range coupling, $J_{CH_2—As^+}$.

The conductances of arsonium salts in water and in various polar solvents have been determined by a number of authors. Popov and Humphrey (278) made a thorough study of the conductance of tetraphenylarsonium chloride in water, acetonitrile, ethylene dichloride, and several dioxane-water mixtures. In water the limiting molar conductance, Λ_0, was 96.5; only a slight deviation from the Onsager equation was observed, a fact which indicates that the arsonium salt is essentially completely dissociated in this solvent. The conductance of tetraphenylarsonium chloride, picrate, and tetraphenylborate in acetonitrile again indicates essentially complete dissociation, but in solvents of lower dielectric constant there is considerable ion-pair formation.

Kraus and co-workers (295–297) have studied the conductances of a number of tetraalkyl- and alkylarylarsonium salts in ethylene dichloride and compared the conductances with those of the corresponding ammonium and phosphonium salts. They found, for example, that the dissociation constants for tetrabutylammonium, -phosphonium, and -arsonium picrates were 2.28×10^{-4}, 1.60×10^{-4}, and 1.42×10^{-4}, respectively. This indicates that there is a slight but significant decrease in the dissociation constant with increasing size of the central atom. The dissociation constants of three phosphonium salts were found to be very similar to those of the three corresponding arsonium salts (295). Thus the values ($\times 10^{-4}$) for the picrate, perchlorate, and nitrate of the ethyltriphenylphosphonium ion are 3.77, 3.60, and 2.18 compared with 3.43, 3.45, and 2.14 for the corresponding arsonium salts.

The equivalent conductances of tetraphenylarsonium chloride in sulfolane (298) and of tetraphenylarsonium iodide in 2,4-dimethylsulfolane (299) have also been reported.

The reduction potentials of a large number of arsonium salts have been determined polarographically in aqueous and alcoholic solution (300). It was found that those groups which are most stable as free radicals (and carbanions) are cleaved preferentially.

D. Uses in analytical chemistry

Arsonium salts have found considerable use in analytical chemistry. In particular, tetraphenylarsonium salts have been used as precipitants for a number of metallic ions. Blicke, Willard, and Taras (192) prepared a number of alkyltriphenylarsonium salts and compared them with tetraphenylarsonium chloride. For the determination of iodine and of perchlorate, perrhenate, and chlorocadmiate ions, the tetraphenylarsonium salt was preferable. It was also found that tetraphenylarsonium ion gives an insoluble permanganate, the only known insoluble compound of this anion.

Tetraphenylarsonium chloride in concentrations of 10^{-2} to $10^{-4}M$ has a half-wave potential of -1.42 V, and the potential is unaffected by changes in pH (301, 302). Bismuth (301), cadmium (301), mercuric (301, 303, 304), and stannic (304) ions are precipitated completely by tetraphenylarsonium chloride and can be determined amperometrically with this reagent.

Willard and Smith (305–307) have made considerable use of tetraphenylarsonium chloride as an analytical reagent. Either gravimetric methods or potentiometric titrations were used in these determinations. The arsonium salt reacts with triiodide ion to form an insoluble triiodide and can be standardized either gravimetrically or potentiometrically with standard iodine solution (305). Standard tetraphenylarsonium chloride solution has been used in the potentiometric determination of mercuric, stannic, cadmium, zinc (306), and perrhenate ions (307). This same reagent has also been used for separating rhenium from concentrated molybdenum solutions by extraction of tetraphenylarsonium perrhenate with chloroform (308, 309). Spectrophotometric methods have been described for the determination of cobalt as $[Ph_4As]_2Co(SCN)_4$ (310, 311). Bismuth can also be detected readily with tetraphenylarsonium chloride (312). Telluric acid, in the presence of hydrochloric acid, gives a precipitate of $[(C_6H_5)_4As]_2TeCl_6$ (313). This has been used for the separation of tellurium from selenium as well as for the gravimetric or volumetric determination of tellurium.

Murphy and Affsprung (314) have recently introduced tetraphenylarsonium chloride as a reagent for the separation and determination of gold. The compound $[(C_6H_5)_4As]AuCl_4$ is formed, which can be readily extracted from aqueous solution by chloroform, and the gold can be determined spectrophotometrically. Interference by ferric ion can be prevented by the addition of fluoride ion. Interference by osmium can be prevented by prior extraction of the aqueous solution with carbon tetrachloride. Other elements which have been determined by the use of tetraphenylarsonium chloride include iridium (315), osmium (316, 317), ruthenium (317), titanium (318), tantalum (319), germanium (320), niobium (321), technetium (322), and tungsten (323).

Methyltriphenylarsonium salts, particularly the thiocyanate and iodide, have also been used with considerable success for the detection and determination of various metallic ions. The thiocyanate has been used for the micro determination of copper (324), cobalt (196, 325), and iron (200). The iodide has been used for the detection and determination of antimony (326), cadmium (327, 328), osmium (329), palladium (330), rhodium (330), and ruthenium (330). The chloride has been used to determine nickel as the bis[dithiooxalatonickelate(II)] (331). Methyltriphenylarsonium salts have also been employed to aid in the detection of the endpoint in titrations with permanganate (332), dichromate (333), and iodine (334). Methyltriphenylarsonium permanganate has been used as an oxidant in non-aqueous aprotic solvents (335).

Cole and Pflaum (280) have determined the solubility and thermal decomposition characteristics of tetraphenylarsonium perchlorate, perrhenate, permanganate, tetraphenylborate, dichromate, trichlorocobaltate(II), and tetrachlorozincate. They concluded that tetraphenylarsonium chloride is valuable as a precipitant for perchlorate and perrhenate and that it shows promise for dichromate and tetraphenylborate.

10. QUINQUENARY ARSENIC COMPOUNDS

A number of attempts have been made to prepare pentaalkyl arsenic compounds by the reaction of tetraalkylarsonium salts with organometallic compounds. Cahours (336) originally claimed to have obtained pentamethylarsenic by the reaction of dimethylzinc with tetramethylarsonium iodide. Subsequent attempts to repeat this preparation, however, have met with failure (14, 150). Wittig and Torssell (150) found that methyllithium reacts with tetramethylarsonium iodide in ether solution in a nitrogen atmosphere to give methane with smaller amounts of ethane. Addition of methyl iodide to the residue gives diethyldimethylarsonium iodide. The authors postulated the following mechanism:

$$(CH_3)_4AsI + CH_3Li \longrightarrow (CH_3)_3As{=}CH_2 + CH_4 + LiI$$

$$(CH_3)_3As{=}CH_2 + CH_3Li \longrightarrow (CH_3)_2\underset{\displaystyle CH_2Li}{\underset{|}{As}}{=}CH_2 + CH_4$$

$$(CH_3)_2\underset{\displaystyle CH_2Li}{\underset{|}{As}}{=}CH_2 + 2CH_3I \longrightarrow (CH_3)_2(C_2H_5)_2AsI + LiI$$

Considerable success has been achieved in the preparation of pentaaryl compounds. Wittig and Clauss (337) were able to prepare pentaphenylarsenic in 65% yield by the reaction of phenyllithium and tetraphenylarsonium bromide. The same compound could also be prepared by the reaction of triphenylarsenic dichloride and phenyllithium, but there were accompanying side reactions leading to the formation of triphenylarsine and chlorobenzene so that the yield of the desired product was somewhat smaller. More recently, Wittig and Hellwinkel (101) have obtained pentaphenylarsenic by the reaction of phenyllithium with triphenylarsine oxide or with the imine $p\text{-}CH_3C_6H_4N{=}AsPh_3$. By allowing the imine to react with 2,2′-biphenylenedilithium, they were able to prepare the heterocyclic compound **36**. Spiro-

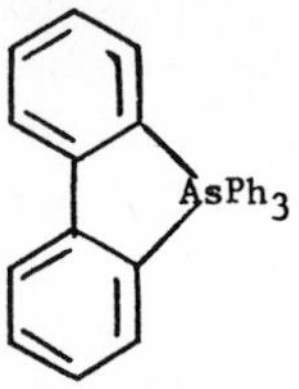

AsPh$_3$

36

cyclic compounds can be obtained by the reaction of the dilithium compound with imines of type **37** (R = alkyl or aryl) or by the reaction of spirocyclic

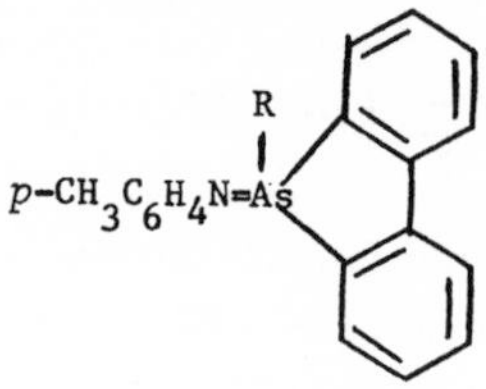

R

$p\text{-}CH_3C_6H_4N{=}As$

37

arsonium halides with either lithium (101, 338) or Grignard (338) reagents. The formation of the spirocyclic pentacovalent arsenical **15** by the acid cleavage of the brucine salt of **38** has been previously mentioned (see Section

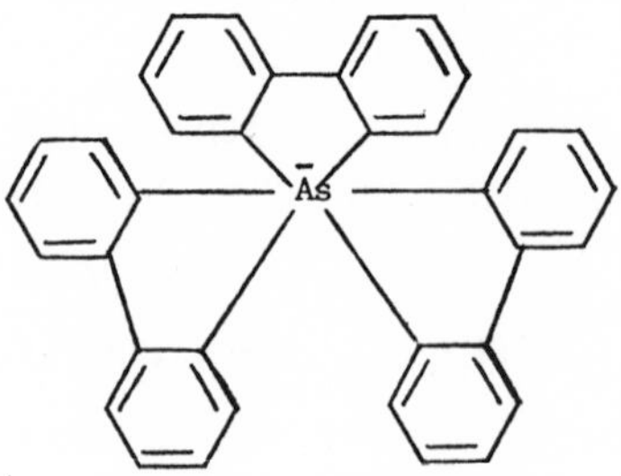

$\bar{A}s$

38

V-9-A). It is also possible to obtain spirocyclic pentacovalent arsenicals by means of the following exchange reaction (338):

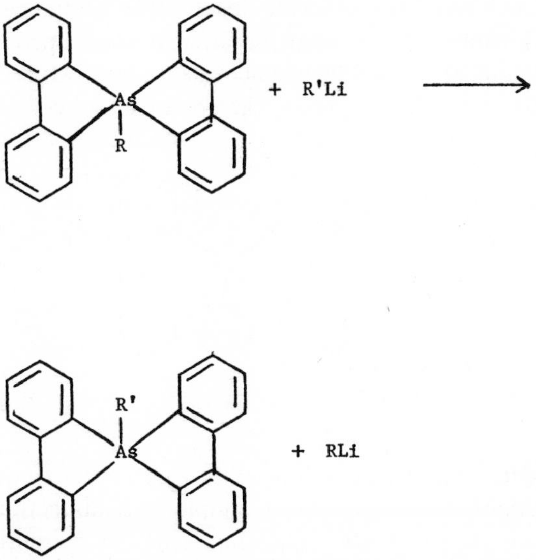

As—R + R'Li ⟶

As—R' + RLi

Hendrickson, Spenger, and Sims (339) originally reported that the reaction of dimethyl acetylenedicarboxylate and triphenylarsine gives a 50% yield of tetramethyl-1,1,1-triphenylarsole-2,3,4,5-tetracarboxylate:

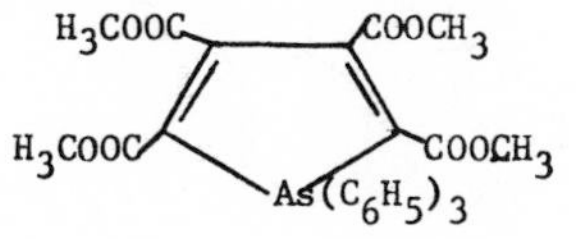

H_3COOC $COOCH_3$ H_3COOC $COOCH_3$ $As(C_6H_5)_3$

Later work by these authors (287), however, indicates that the material is an acyclic compound of the following structure:

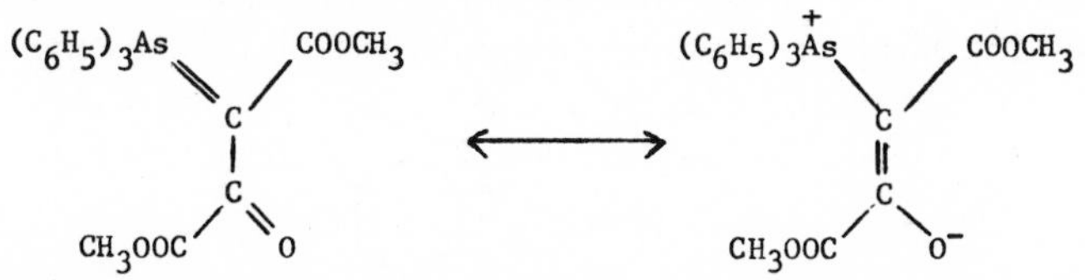

$(C_6H_5)_3As$ $COOCH_3$ C C CH_3OOC O ⟷ $(C_6H_5)_3\overset{+}{A}s$ $COOCH_3$ C C CH_3OOC O^-

Pentaphenylarsenic decomposes on heating to give principally triphenylarsine with smaller amounts of biphenyl and benzene (337). It reacts with excess chlorine, bromine, and iodine to give the corresponding tetraphenylarsonium trihalides and halobenzenes. It also is cleaved by acids. Thus, hydrobromic acid gives tetraphenylarsonium bromide and benzene; heating with acetic acid and subsequent addition of aqueous potassium iodide yield tetraphenylarsonium iodide. The interaction of pentaphenylarsenic and triphenylboron in ether results in the formation of tetraphenylarsonium tetraphenylborate. Mercuric chloride reacts with pentaphenylarsenic to give tetraphenylarsonium trichloromercurate and phenylmercuric chloride (101).

The reactions of spirocyclic pentacovalent arsenicals are similar to those of pentaphenylarsenic (338). Thus, the spiran **39** reacts with several electrophilic reagents with cleavage of one of the heterocyclic rings and the formation of an arsonium compound. For example, the interaction of methyl iodide and **39** yields the arsonium iodide **40**. Bromine, mercuric chloride, hydrochloric acid, and triphenylboron react in an analogous manner. The thermal decomposition of spirocyclic compounds of type **41** has also been investigated.

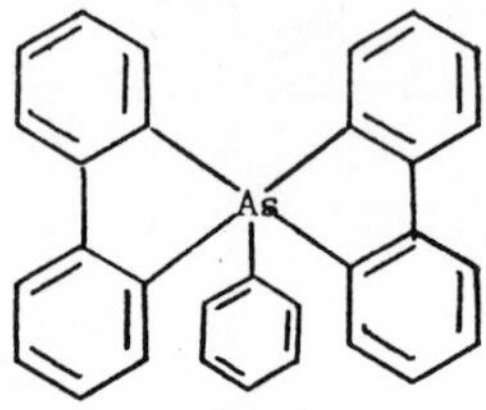

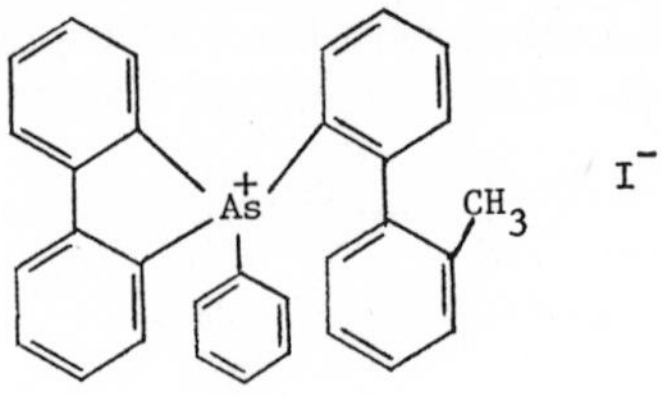

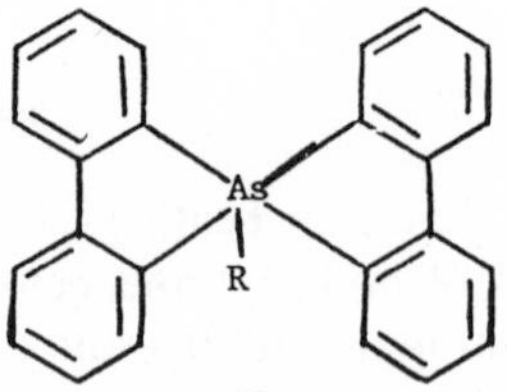

39

40

41

When R is ethyl or *n*-butyl, the tertiary arsine **42** is formed and the alkyl

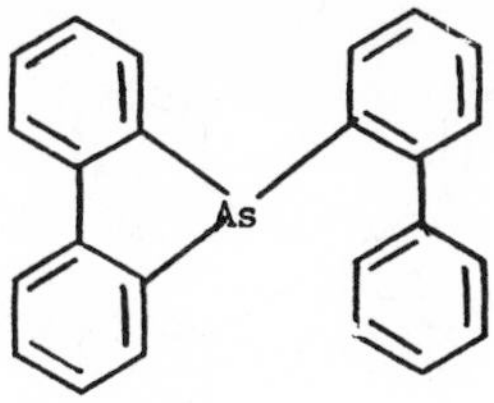

42

group is eliminated as an olefin. When R is methyl or phenyl (101), however, a ring expansion occurs and a compound containing a nine-membered ring results:

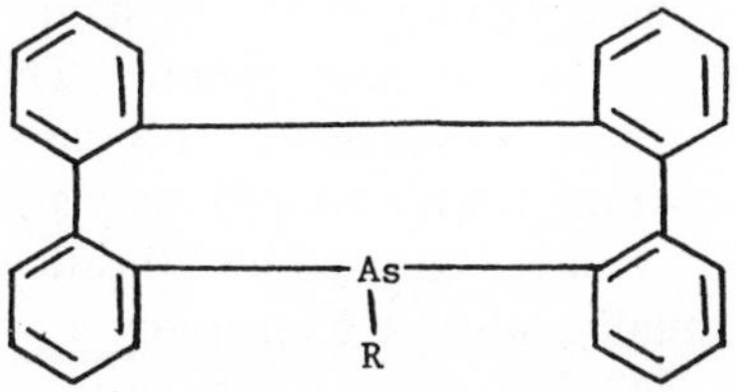

Several attempts have been made to obtain enantiomers of spirocyclic pentacovalent arsenicals. When Wittig and Hellwinkel (101) allowed compound **43** to react with silver hydrogen D-dibenzoyltartrate, one of the

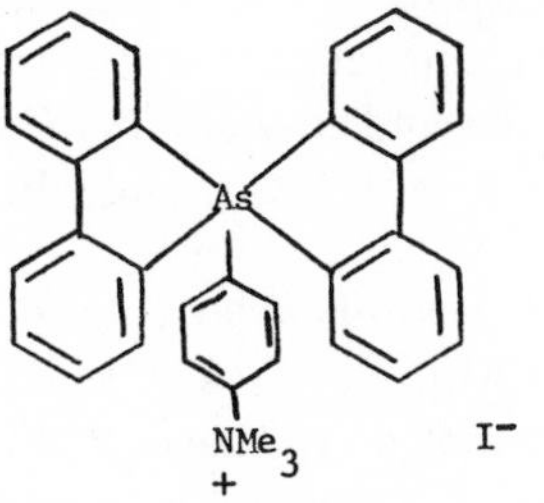

43

heterocyclic rings was cleaved and an arsonium salt was obtained. Further, they found that the reaction of enantiomers of the arsonium salt **44** with phenyllithium leads to the formation of the racemic pentaaryl compound. An unsuccessful attempt to prepare an optically-active pentaarylarsenic compound by the cleavage of one carbon–arsenic bond in a hexacoordinated arsenic derivative has been previously discussed (see Section V-9-A).

The structure of pentaphenylphosphorus, -arsenic, and -antimony was first investigated by Wheatley and Wittig (340) by means of X-ray diffraction

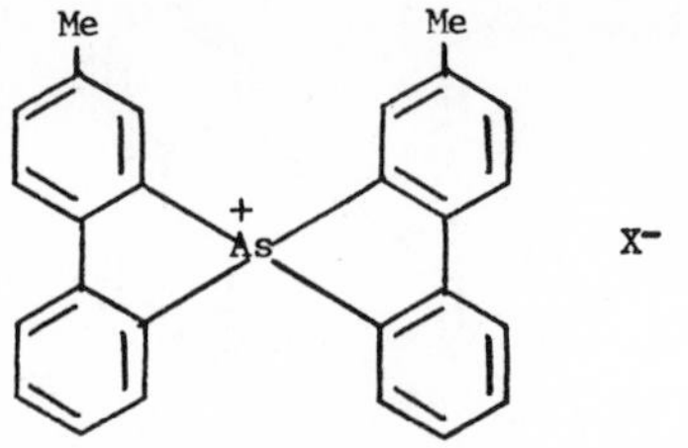

Me Me

$\overset{+}{As}$ X^-

44

and dipole moment measurements. The dipole moments in benzene solution were found to be small but were considered significant—1.25 D for the phosphorus compound, 1.32 D for the arsenic compound, and 1.59 D for the antimony compound. The crystallographic constants indicate that pentaphenylphosphorus and -arsenic are isomorphous. The structure of pentaphenylantimony was shown by two-dimensional X-ray diffraction techniques to be based on a square pyramid, with the antimony atom about 0.5 A above the base of the pyramid. It was suggested by Wheatley and Wittig that the phosphorus and arsenic analogs also have square-pyramidal structures. Later, more refined work (341, 342) has confirmed the proposed structure of pentaphenylantimony, but pentaphenylphosphorus has been shown by three-dimensional X-ray diffraction to have a trigonal bipyramidal configuration (343). Presumably, therefore, the isomorphous pentaphenylarsenic is also a trigonal bipyramid. The molecule of pentaphenylphosphorus in the solid state is not symmetrical because of the orientation of the benzene rings; according to Wheatley (343), this fact accounts for the finite dipole moment of this compound. Muetterties and co-workers (344) have suggested that the nonzero values of the dipole moments "could be due to a field induced moment."

The infrared spectrum of pentaphenylarsenic in the region between 4000 and 250 cm^{-1} has been recently investigated by Mackay and co-workers (156), who assigned the observed bands on the basis of Whiffen's work (157) on the halobenzenes. In the region of r, there is a strong absorption at 640 cm^{-1} and a weaker band at 663 cm^{-1}, which were assigned to axial and equatorial components, respectively. Vibration t is split into a doublet at 352 and 293 cm^{-1}, which involve equatorial and axial stretching, respectively. Complex absorption in the vicinity of 1060 cm^{-1} was ascribed to vibration q. Four strong bands between 447 and 473 cm^{-1} were assigned to y. The remaining two mass-sensitive vibrations (u and x) were outside the range of the spectrometer employed.

The proton NMR spectrum of penta-*p*-tolylarsenic (in carbon disulfide solution) exhibits a single methyl signal at $\delta = 2.27$ (345). The magnetic equivalence of the five methyl groups is not consistent with either a trigonal bipyramidal or a square pyramidal structure. It is assumed, therefore, that rapid pseudo-rotation averages the environment of the five groups attached to the arsenic atom.

REFERENCES

1. F. F. Blicke and E. L. Cataline, *J. Amer. Chem. Soc.*, **60,** 419 (1938).
2. F. F. Blicke and S. R. Safir, *J. Amer. Chem. Soc.*, **63,** 575 (1941).
3. E. R. H. Jones and F. G. Mann, *J. Chem. Soc.*, 294 (1958).
4. J. R. Vaughan, Jr., and D. S. Tarbell, *J. Amer. Chem. Soc.*, **67,** 144 (1945).
5. R. L. Shriner and C. N. Wolf, *Org. Syn.*, Coll. Vol. **4,** 910 (1963).
6. M. H. Beeby and F. G. Mann, *J. Chem. Soc.*, 886 (1951).
7. E. Müller, B. Teissier, H. Eggensperger, A. Rieker, and K. Scheffler, *Justus Liebigs Ann. Chem.*, **705,** 54 (1967).
8. D. B. Copley, F. Fairbrother, and A. Thompson, *J. Less-Common Metals*, **8,** 256 (1965).
9. H. Bauer, *J. Amer. Chem. Soc.*, **67,** 591 (1945).
10. R. C. Cookson and F. G. Mann, *J. Chem. Soc.*, 67 (1949).
11. H. Hartmann and G. Nowak, *Z. Anorg. Allg. Chem.*, **290,** 348 (1957).
12. J. F. Carson and F. F. Wong, *J. Org. Chem.*, **26,** 1467 (1961).
13. N. N. Mel'nikov and M. S. Rokitskaya, *Zh. Obshch. Khim.*, **8,** 834 (1938); through *C.A.*, **33,** 1267 (1939).
14. M. E. P. Friedrich and C. S. Marvel, *J. Amer. Chem. Soc.*, **52,** 376 (1930).
15. J. Dodonow and H. Medox, *Ber.*, **68,** 1254 (1935).
16. A. Étienne, *C. R. Acad. Sci., Paris*, **221,** 628 (1945).
17. G. J. Burrows, *J. Proc. Roy. Soc. N.S. Wales*, **68,** 72 (1935).
18. B. D. Chernokal'skii, A. S. Gel'fond, and G. Kamai, *Zh. Obshch. Khim.*, **37,** 1396 (1967).
19. A. Merijanian and R. A. Zingaro, *Inorg. Chem.*, **5,** 187 (1966).
20. H. Landolt, *Justus Liebigs Ann. Chem.*, **89,** 301 (1854).
21. W. J. C. Dyke, G. Davies, and W. J. Jones, *J. Chem. Soc.*, 185 (1931).
22. W. J. Jones, W. J. C. Dyke, G. Davies, D. C. Griffiths, and J. H. E. Webb, *J. Chem. Soc.*, 2284 (1932).
23. G. Kamai and B. D. Chernokal'skii, *Tr. Kazansk. Khim.-Tekhnol. Inst.*, 117 (1959); through *C.A.*, **54,** 24345 (1960).
24. G. A. Rasuwajew and W. S. Malinowski, *Ber.*, **64,** 120 (1931).
25. D. E. Worrall, *J. Amer. Chem. Soc.*, **52,** 664 (1930).
26. D. E. Worrall, *J. Amer. Chem. Soc.*, **62,** 2514 (1940).
27. F. G. Mann and J. Watson, *J. Chem. Soc.*, 505 (1947).
28. D. R. Lyon and F. G. Mann, *J. Chem. Soc.*, 30 (1945).
29. P. Nylén, *Z. Anorg. Allg. Chem.*, **246,** 227 (1941).
30. H. J. Emeléus, R. N. Haszeldine, and R. C. Paul, *J. Chem. Soc.*, 881 (1954).
31. H. Bart, Ger. Pat. 254,345 (Nov. 3, 1910); through *C.A.*, **7,** 1403 (1913).
32. M. Ya. Kraft and S. A. Rossina, *Dokl. Akad. Nauk SSSR*, **55,** 821 (1947); *C.A.*, **42,** 531 (1948).
33. H. N. Das-Gupta, *J. Indian Chem. Soc.*, **14,** 397 (1937).
34. M. Goswami and H. N. Das-Gupta, *J. Indian Chem. Soc.*, **8,** 417 (1931).
35. V. Auger, *C. R. Acad. Sci., Paris*, **137,** 925 (1903).
36. J. T. Braunholtz and F. G. Mann, *J. Chem. Soc.*, 3285 (1957).
37. R. C. Cookson and F. G. Mann, *J. Chem. Soc.*, 618 (1947).
38. E. R. H. Jones and F. G. Mann, *J. Chem. Soc.*, 1719 (1958).

39. I. G. M. Campbell and R. C. Poller, *J. Chem. Soc.*, 1195 (1956).
40. G. H. Cookson and F. G. Mann, *J. Chem. Soc.*, 2895 (1949).
41. M. Becke-Goehring and H. Thielemann, *Z. Anorg. Allg. Chem.*, **308,** 33 (1961).
42. D. R. Lyon and F. G. Mann, *J. Chem. Soc.*, 666 (1942).
43. F. Zuckerkandl and M. Sinai, *Ber.*, **54,** 2479 (1921).
44. A. Michaelis, *Justus Liebigs Ann. Chem.*, **321,** 1141 (1902).
45. F. Challenger and A. T. Peters, *J. Chem. Soc.*, 2610 (1929).
46. G. Kamai and B. D. Chernokal'skii, *Dokl. Akad. Nauk SSSR*, **128,** 299 (1959).
47. L. Horner, H. Hoffmann, and H. G. Wippel, *Chem. Ber.*, **91,** 64 (1958).
48. R. S. Nyholm, *J. Chem. Soc.*, 1767 (1951).
49. H. A. Goodwin and F. Lions, *J. Amer. Chem. Soc.*, **81,** 311 (1959).
50. D. M. L. Goodgame and F. A. Cotton, *J. Amer. Chem. Soc.*, **82,** 5774 (1960).
51. D. M. L. Goodgame and F. A. Cotton, *J. Chem. Soc.*, 2298 (1961).
52. D. M. L. Goodgame and F. A. Cotton, *J. Chem. Soc.*, 3735 (1961).
53. D. M. L. Goodgame, M. Goodgame, and F. A. Cotton, *Inorg. Chem.*, **1,** 239 (1962).
54. C. A. Rodley, D. M. L. Goodgame, and F. A. Cotton, *J. Chem. Soc.*, 1499 (1965).
55. D. J. Phillips and S. Y. Tyree, Jr., *J. Amer. Chem. Soc.*, **83,** 1806 (1961).
56. S. M. Horner, S. Y. Tyree, and D. L. Venezky, *Inorg. Chem.*, **1,** 844 (1962).
57. J. C. Sheldon, *J. Chem. Soc.*, 750 (1961).
58. F. Schindler, H. Schmidbaur, and G. Jonas, *Angew. Chem.*, **77,** 170 (1965).
59. R. Pietsch and G. Nagl, *Mikrochim. Ichnoanal. Acta*, 1085 (1965).
60. R. Pietsch and G. Nagl, *Z. Anal. Chem.*, **208,** 328 (1965).
61. R. Pietsch, *Mikrochim. Acta*, 708 (1967).
62. R. A. Zingaro and E. A. Meyers, *Inorg. Chem.*, **1,** 771 (1962).
63. F. L. Kolar, R. A. Zingaro, and K. Irgolic, *J. Inorg. Nucl. Chem.*, **28,** 2981 (1966).
64. B. D. Chernokal'skii, S. S. Nasybullina, R. R. Shagidullin, I. A. Lamanova, and G. Kamai, *Izv. Vyssh. Ucheb. Zaved., Khim. Khim. Tekhnol.*, **9,** 918 (1966); through *C.A.*, **67,** 1105 (1967).
65. F. Hein and H. Hecker, *Z. Naturforsch.*, **B, 11,** 677 (1956).
66. F. Hein, H. Plust, and H. Pohlemann, *Z. Anorg. Allg. Chem.*, **272,** 25 (1953).
67. G. J. Burrows and E. E. Turner, *J. Chem. Soc.*, **119,** 426 (1921).
68. J. A. Aeschlimann, *J. Chem. Soc.*, **127,** 811 (1925).
69. F. G. Mann, in W. Klyne and P. B. D. de la Mare, eds., *Progress in Stereochemistry*, Vol. 2, Academic Press, New York, 1958, pp. 215–216.
70. W. Stamm (to Stauffer Chem. Co.), Ger. Pat. 1,229,529 (Dec. 1, 1966); through *C.A.*, **66,** 2760 (1967).
71. J. J. Monagle, *J. Org. Chem.*, **27,** 3851 (1962).
72. K. A. Jensen, *Z. Anorg. Allg. Chem.*, **250,** 268 (1943).
73. H. H. Jaffé, *J. Chem. Phys.*, **22,** 1430 (1954).
74. C. N. R. Rao, J. Ramachandran, and A. Balasubramanian, *Can. J. Chem.*, **39,** 171 (1961).
75. K. A. Jensen and P. H. Nielsen, *Acta Chem. Scand.*, **17,** 1875 (1963).
76. J. Bernstein, M. Halmann, S. Pinchas, and D. Samuel, *J. Chem. Soc.*, 821 (1964).
77. R. R. Shagidullin, I. A. Lamanova, and A. K. Urazgil'deeva, *Dokl. Akad. Nauk SSSR*, **174,** 1359 (1967).
78. F. A. Cotton, D. M. L. Goodgame, and R. H. Soderberg, *Inorg. Chem.*, **2,** 1162 (1963).
79. K. A. Jensen, *Z. Anorg. Allg. Chem.*, **250,** 257 (1943).
80. F. G. Mann and A. J. Wilkinson, *J. Chem. Soc.*, 3346 (1957).
81. G. S. Harris and F. Inglis, *J. Chem. Soc.*, **A,** 497 (1967).
82. G. A. Razuvaev, V. S. Malinovskii, and D. A. Godina, *Zh. Obshch. Khim.*, **5,** 721 (1935); through *C.A.*, **30,** 1057 (1936).

83. F. G. Mann, *J. Chem. Soc.*, 65 (1945).
84. W. A. Waters, *J. Chem. Soc.*, 864 (1939).
85. G. Kamai, *Ber.*, **68,** 960 (1935).
86. G. Kamai, *Zh. Obshch. Khim.*, **12,** 104 (1942); through *C.A.*, **37,** 1997 (1943).
87. J. Chatt and F. G. Mann, *J. Chem. Soc.*, 1184 (1940).
88. A. Cahours, *Justus Liebigs Ann. Chem.*, **112,** 228 (1859).
89. R. A. Zingaro, R. E. McGlothin, and R. M. Hedges, *Trans. Faraday Soc.*, **59,** 798 (1963).
90. K. Matsumiya and M. Nakai, *Mem. Coll. Sci. Kyoto Imp. Univ.*, **10,** 57 (1926); through *C.A.*, **21,** 904 (1927).
91. W. M. Dehn, *Amer. Chem. J.*, **33,** 101 (1905).
92. W. T. Reichle, *Inorg. Chem.*, **1,** 650 (1962).
93. W. R. Cullen, *Can. J. Chem.* **41,** 2424 (1963).
94. W. H. Mills and R. Raper, *J. Chem. Soc.*, **127,** 2479 (1925).
95. J. Chatt and F. G. Mann, *J. Chem. Soc.*, 610 (1939).
96. P. Nicpon and D. W. Meek, *Chem. Commun.*, 398 (1966).
97. E. Müller, H. Eggensperger, B. Teissier, and K. Scheffler, *Z. Naturforsch.*, **B, 19,** 1079 (1964).
98. R. R. Renshaw and G. E. Holm, *J. Amer. Chem. Soc.*, **42,** 1468 (1920).
99. R. A. Zingaro and A. Merijanian, *Inorg. Chem.*, **3,** 580 (1964).
100. F. G. Mann and E. J. Chaplin, *J. Chem. Soc.*, 527 (1937).
101. G. Wittig and D. Hellwinkel, *Chem. Ber.*, **97,** 769 (1964).
102. D. S. Tarbell and J. R. Vaughan, Jr., *J. Amer. Chem. Soc.*, **67,** 41 (1945).
103. F. G. Mann and W. J. Pope, *J. Chem. Soc.*, **121,** 1754 (1922).
104. F. G. Mann, *J. Chem. Soc.*, 958 (1932).
105. R. Appel and D. Wagner, *Angew. Chem.*, **72,** 209 (1960).
106. H. H. Sisler and C. Stratton, *Inorg. Chem.*, **5,** 2003 (1966).
107. P. Chabrier, *C. R. Acad. Sci., Paris*, **214,** 362 (1914).
108. R. Appel and A. Hauss, *Z. Anorg. Allg. Chem.*, **311,** 290 (1961).
109. W. T. Reichle, *Tetrahedron Lett.*, 51 (1962).
110. N. L. Paddock and H. T. Searle, *Advances in Inorganic Chemistry and Radiochemistry*, Vol. 1, Academic Press, New York, 1959, p. 347.
111. A. Baeyer, *Justus Liebigs Ann. Chem.*, **107,** 257 (1858).
112. H. Klinger and A. Kreutz, *Justus Liebigs Ann. Chem.*, **249,** 147 (1888).
113. W. La Coste and A. Michaelis, *Justus Liebigs Ann. Chem.*, **201,** 184 (1880).
114. J. Prat, *Justus Liebigs Ann. Chem.*, **198,** 583 (1934).
115. G. J Burrows and A. Lench, *J. Proc. Roy. Soc. N.S. Wales*, **70,** 294 (1937).
116. W. C. Smith, C. W. Tullock, E. L. Muetterties, W. R. Hasek, F. S. Fawcett, V. A. Engelhardt, and D. D. Coffman, *J. Amer. Chem. Soc.*, **81,** 3165 (1959).
117. W. C. Smith, *J. Amer. Chem. Soc.*, **82,** 6176 (1960).
118. W. C. Smith (to E. I. duPont de Nemours and Company), U.S. Pat. 2,950,306 (Aug. 23, 1960); *C.A.*, **55,** 2569 (1961).
119. E. L. Muetterties, W. Mahler, K. J. Packer, and R. Schmutzler, *Inorg. Chem.*, **3,** 1298 (1964).
120. C. P. A. Kappelmeier, *Rec. Trav. Chim. Pays-Bas*, **49,** 57 (1930).
121. E. Wiberg and K. Mödritzer, *Z. Naturforsch.*, **B, 12,** 127 (1957).
122. C. S. Gibson, J. D. A. Johnson, and D. C. Vining, *Rec. Trav. Chim. Pays-Bas*, **49,** 1006 (1930).
123. A. McKenzie and J. K. Wood, *J. Chem. Soc.*, **117,** 406 (1920).
124. H. J. Emeléus, R. N. Haszeldine, and E. G. Walaschewski, *J. Chem. Soc.*, 1552 (1953).
125. A. Michaelis and U. Paetow, *Justus Liebigs Ann. Chem.*, **233,** 60 (1886).

126. A. Michaelis and U. Paetow, *Ber.*, **18,** 41 (1885).
127. W. Steinkopf, H. Dudek, and S. Schmidt, *Ber.*, **61,** 1911 (1928).
128. E. Wiberg and K. Mödritzer, *Z. Naturforsch.*, **B, 11,** 751 (1956).
129. A. D. Beveridge and G. S. Harris, *J. Chem. Soc.*, 6076 (1964).
130. D. Seyferth, *J. Amer. Chem. Soc.*, **80,** 1336 (1958).
131. W. R. Cullen, *Can. J. Chem.*, **38,** 445 (1960).
132. W. G. Lowe and C. S. Hamilton, *J. Amer. Chem. Soc.*, **57,** 2314 (1935).
133. W. G. Lowe and C. S. Hamilton, *J. Amer. Chem. Soc.*, **57,** 1081 (1935).
134. A. Spada, *Atti Soc. Nat. Mat. Modena*, **71,** 155 (1940).
135. J. Klippel , *Rocz. Chem.*, **10,** 777 (1930).
136. G. J. Burrows and A. Lench, *J. Proc. Roy. Soc. N.S. Wales*, **70,** 437 (1937).
137. O. A. Reutov and Yu. G. Bundel, *Zh. Obshch. Khim.*, **25,** 2324 (1955).
138. M. H. O'Brien, G. O. Doak, and G. G. Long, *Inorg. Chim. Acta*, **1,** 34 (1967).
139. M. H. O'Brien, Ph.D. Thesis, North Carolina State University, 1968.
140. A. Hantzsch and H. Hibbert, *Ber.*, **40,** 1508 (1907).
141. T. F. Winmill, *J. Chem. Soc.*, **101,** 718 (1912).
142. F. Feigl, D. Goldstein, and D. Haguenauer-Castro, *Z. Anal. Chem.*, **178,** 419 (1961).
143. F. Feigl and D. Goldstein, *Mikrochim. Acta*, 1 (1966).
144. A. Spada, *Atti Soc. Nat. Mat. Modena*, **72,** 34 (1941).
145. L. Maier, D. Seyferth, F. G. A. Stone, and E. G. Rochow, *Z. Naturforsch.*, **B, 12,** 263 (1957).
146. I. Gorski, W. Schpanski, and L. Muljar, *Ber.*, **67,** 730 (1934).
147. H. N. Das-Gupta, *J. Indian Chem. Soc.*, **14,** 400 (1937).
148. R. E. Dessy, T. Chivers, and W. Kitching, *J. Amer. Chem. Soc.*, **88,** 467 (1966).
149. L. Horner and H. Oediger, *Chem. Ber.*, **91,** 437 (1958).
150. G. Wittig and K. Torssell, *Acta Chem. Scand.*, **7,** 1293 (1953).
151. H. Schmidbaur, H. S. Arnold, and E. Beinhofer, *Chem. Ber.*, **97,** 449 (1964).
152. J. Goubeau and R. Baumgärtner, *Z. Electrochem.*, **64,** 598 (1960).
153. A. F. Wells, *Z. Kristallogr.*, **99,** 367 (1938).
154. E. G. Rochow, D. T. Hurd, and R. N. Lewis, *The Chemistry of Organometallic Compounds*, Wiley, New York, 1957, p. 215.
155. G. S. Harris, *Proc. Chem. Soc.*, 65 (1961).
156. K. M. Mackay, D. B. Sowerby, and W. C. Young, *Spectrochim. Acta*, **A, 24,** 611 (1968).
157. D. H. Whiffen, *J. Chem. Soc.*, 1350 (1956).
158. E. Augdahl, J. Grundnes, and P. Klaboe, *Inorg. Chem.*, **4,** 1475 (1965).
159. K. R. Bhaskar, S. N. Bhat, S. Singh, and C. N. R. Rao, *J. Inorg. Nucl. Chem.*, **28,** 1915 (1966).
160. S. N. Bhat and C. N. R. Rao, *J. Amer. Chem. Soc.*, **88,** 3216 (1966).
161. G. C. Tranter, C. C. Addison, and D. B. Sowerby, *J. Organometal. Chem.*, **12,** 369 (1968).
162. W. C. Davies and C. W. Othen, *J. Chem. Soc.*, 1236 (1936).
163. E. Jürgens, Dissertation, Univ. Mainz, 1957; quoted by L. Horner, H. Oediger, and H. Hoffmann, reference 177.
164. A. Rieche, J. Dahlmann, and D. List, *Angew. Chem.*, **73,** 494 (1961).
165. A. Rieche, J. Dahlmann, and D. List, *Justus Leibigs Ann. Chem.*, **678,** 167 (1964).
166. D. Hadži, *J. Chem. Soc.*, 5128 (1962).
167. W. Steinkopf and G. Schwen, *Ber.*, **54,** 2791 (1921).
168. G. S. Harris, F. Inglis, J. McKechnie, K. K. Cheung, and G. Ferguson, *Chem. Commun.*, 442 (1967).
169. F. G. Mann and A. J. Wilkinson, *J. Chem. Soc.*, 3336 (1957).

170. D. R. Lyon, F. G. Mann, and G. H. Cookson, *J. Chem. Soc.*, 662 (1947).
171. W. Steinkopf and G. Schwen, *Ber.*, **54,** 2802 (1921).
172. F. G. Mann and J. Watson, *J. Org. Chem.*, **13,** 502 (1948).
173. G. Ferguson and E. W. Macaulay, *Chem. Commun.*, 1288 (1968).
174. G. Kamai and B. D. Chernokal'skii, *Zh. Obshch. Khim.*, **30,** 1536 (1960).
175. I. M. Starshov and G. Kamai, *Zh. Obshch. Khim.*, **24,** 2044 (1954).
176. G. Kamai and B. D. Chernokal'skii, *Zh. Obshch. Khim.*, **29,** 1596 (1959).
177. L. Horner, H. Oediger, and H. Hoffmann, *Justus Liebigs Ann. Chem.*, **626,** 26 (1959).
178. G. Kamai and I. M. Starshov, *Tr. Kazansk. Khim.-Tekhnol. Inst.*, No. 21, 159 (1956); through *C.A.*, **51,** 12018 (1957).
179. L. Maier, D. Seyferth, F. G. A. Stone, and E. G. Rochow, *J. Amer. Chem. Soc.*, **79,** 5884 (1957).
180. W. C. Davies and W. P. G. Lewis, *J. Chem. Soc.*, 1599 (1934).
181. W. C. Davies, *J. Chem. Soc.*, 462 (1935).
182. A. N. Nesmeyanov and L. G. Makarova, *Uch. Zap. Mosk. Gos. Univ.*, No. 132, *Org. Khim.*, **7,** 109 (1950); through *C.A.*, **49,** 3903 (1955).
183. R. N. Haszeldine and B. O. West, *J. Chem. Soc.*, 3631 (1956).
184. R. N. Haszeldine and B. O. West, *J. Chem. Soc.*, 3880 (1957).
185. L. Horner and H. Fuchs, *Tetrahedron Lett.*, 203 (1962).
186. F. G. Holliman and F. G. Mann, *J. Chem. Soc.*, 45 (1945).
187. F. G. Holliman, F. G. Mann, and D. A. Thornton, *J. Chem. Soc.*, 9 (1960).
188. K. V. Vijayaraghavan, *J. Indian Chem. Soc.*, **22,** 141 (1945).
189. F. G. Mann and H. R. Watson, *J. Chem. Soc.*, 3945 (1957).
190. G. V. Medoks and E. M. Soshestvenskaya, *Zh. Obshch. Khim.*, **27,** 271 (1957); through *C.A.*, **51,** 12845 (1957).
191. F. F. Blicke and S. R. Safir, *J. Amer. Chem. Soc.*, **63,** 1493 (1941).
192. F. F. Blicke, H. H. Willard, and J. T. Taras, *J. Amer. Chem. Soc.*, **61,** 88 (1939).
193. H. W. Roesky, *Chem. Ber.*, **100,** 1447 (1967).
194. F. F. Blicke and E. Monroe, *J. Amer. Chem. Soc.*, **57,** 720 (1935).
195. L. Capatos, *C. R. Acad. Sci., Paris*, **194,** 1658 (1932).
196. F. P. Dwyer, N. A. Gibson, and R. S. Nyholm, *J. Proc. Roy. Soc. N.S. Wales*, **79,** 118 (1945).
197. N. S. Gill and R. S. Nyholm, *J. Chem. Soc.*, 3997 (1959).
198. D. K. Straub, R. S. Drago, and J. T. Donoghue, *Inorg. Chem.*, **1,** 848 (1962).
199. N. S. Gill, *J. Chem. Soc.*, 3512 (1961).
200. F. P. Dwyer and N. A. Gibson, *Analyst* (London), **76,** 548 (1951).
201. R. S. Nyholm, *Croat. Chem. Acta*, **33,** 157 (1961); through *C.A.*, **57,** 1838 (1962).
202. F. P. Dwyer, N. A. Gibson, and R. S. Nyholm, *J. Proc. Roy. Soc. N. S. Wales*, **78,** 118 (1944).
203. M. M. Baig and W. R. Cullen, *Can. J. Chem.*, **39,** 420 (1961).
204. M. M. Baig, W. R. Cullen, and D. S. Dawson, *Can. J. Chem.*, **40,** 46 (1962).
205. R. C. Cass, G. E. Coates, and R. G. Hayter, *J. Chem. Soc.*, 4007 (1955).
206. G. Faraglia, L. R. Fiorani, B. L. Pepe, and R. Barbieri, *J. Organometal. Chem.*, **10,** 363 (1967).
207. D. Hellwinkel and G. Kilthau, *Justus Liebigs Ann. Chem.*, **705,** 66 (1967).
208. E. R. H. Jones and F. G. Mann, *J. Chem. Soc.*, 411 (1955).
209. F. G. Mann and F. C. Baker, *J. Chem. Soc.*, 4142 (1952).
210. E. R. H. Jones and F. G. Mann, *J. Chem. Soc.*, 4472 (1955).
211. F. G. Mann and F. H. C. Stewart, *J. Chem. Soc.*, 1269 (1955).
212. R. N. Collinge, R. S. Nyholm, and M. L. Tobe, *Nature*, **201,** 1322 (1964).
213. V. Auger, *C. R. Acad. Sci., Paris*, **145,** 808 (1907).

214. A. Cahours and A. Riche, *C. R. Acad. Sci., Paris*, **39,** 541 (1854).
215. V. Auger, *C. R. Acad. Sci., Paris*, **138,** 1705 (1904).
216. A. Cahours, *Justus Liebigs Ann. Chem.*, **122,** 192 (1862).
217. A. Valeur and P. Gailliot, *Bull. Soc. Chim. Fr.*, **41,** 1318 (1927).
218. K. Issleib and A. Tzschach. *Angew. Chem.*, **73,** 26 (1961).
219. D. Seyferth and H. M. Cohen, *J. Inorg. Nucl. Chem.*, **20,** 73 (1961).
220. D. Lloyd and M. I. C. Singer, *Chem. Ind.* (London), 510 (1967).
221. N. A. Nesmeyanov, V. V. Pravdina, and O. A. Reutov, *Dokl. Akad. Nauk SSSR*, **155,** 1364 (1964).
222. S. O. Grim and D. Seyferth, *Chem. Ind.* (London) 849 (1959).
223. G. Wittig and H. Laib, *Justus Liebigs Ann. Chem.*, **580,** 57 (1953).
224. F. F. Blicke and C. Marzano, *J. Amer. Chem. Soc.*, **55,** 3056 (1933).
225. F. F. Blicke and E. L. Cataline, *J. Amer. Chem. Soc.*, **60,** 423 (1938).
226. J. Chatt and F. G. Mann, *J. Chem. Soc.*, 1192 (1940).
227. V. D. Nefedov, M. A. Toropova, I. V. Krivokhatskaya, and O. V. Kesarev, *Radiokhimiya*, **6,** 112 (1964).
228. M. H. Forbes, D. M. Heinekey, F. G. Mann, and I. T. Millar, *J. Chem. Soc.*, 2762 (1961).
229. G. Petit, *Ann. Chim.* (Paris), **16,** 5 (1941).
230. J. L. Brannon (to Bakelite Corp.), U.S. Pat. 2,475,005 (July 5, 1949); *C.A.*, **43,** 8208 (1949).
231. F. G. Mann, *The Heterocyclic Derivatives of Phosphorus, Arsenic, Antimony, Bismuth, and Silicon*, Interscience, New York, 1950, pp. 32–36, 48–49.
232. F. G. Mann, in *Progress in Organic Chemistry*, J. W. Cook, ed., Vol. 4, Butterworths, London, 1958, pp. 235–246.
233. M. C. Henry and G. Wittig, *J. Amer. Chem. Soc.*, **82,** 563 (1960).
234. A. W. Johnson, *Ylid Chemistry*, Academic Press, New York, 1966, pp. 284–299.
235. A. W. Johnson and J. O. Martin, *Chem. Ind.* (London), 1726 (1965).
236. A. W. Johnson, *J. Org. Chem.*, **25,** 183 (1960).
237. A. W. Johnson, *J. Org. Chem.*, **24,** 282 (1959).
238. F. Kröhnke, *Ber.*, **83,** 291 (1950).
239. Y. T. Huang, W. Y. Ting, and H. S. Cheng, *Hua Hsueh Hsueh Pao*, **31,** 38 (1965); through *C.A.*, **63,** 629 (1965).
240. N. A. Nesmeyanov, V. V. Pravdina, and O. A. Reutov, *Izv. Akad. Nauk SSSR, Ser. Khim.*, 1474 (1965).
241. N. A. Nesmeyanov, V. M. Novikov, and O. A. Reutov, *Zh. Org. Khim.*, **2,** 942 (1966).
242. N. A. Nesmeyanov, V. V. Pravdina, and O. A. Reutov, *Zh. Org. Khim.*, **3,** 598 (1967).
243. A. Maccioni and M. Secci, *Rend. Seminario Fac. Sci. Univ. Cagliari*, **34,** 328 (1964); through *C.A.*, **63,** 5674 (1965).
244. N. A. Nesmeyanov, V. V. Mikulshina, and O. A. Reutov, *J. Organometal. Chem.*, **13,** 263 (1968).
245. A. W. Johnson and R. B. LaCount, *Tetrahedron*, **9,** 130 (1960).
246. W. von E. Doering and A. K. Hoffmann, *J. Amer. Chem. Soc.*, **77,** 521 (1955).
247. K. Issleib and R. Lindner, *Justus Liebigs Ann. Chem.*, **707,** 120 (1967).
248. G. Aksnes and J. Songstad, *Acta Chem. Scand.*, **18,** 655 (1964).
249. C. K. Ingold, F. R. Shaw, and I. S. Wilson, *J. Chem. Soc.*, 1280 (1928).
250. F. L. Riley and E. Rothstein, *J. Chem. Soc.*, 3872 (1964).
251. A. R. Hands and A. J. H. Mercer, *J. Chem. Soc.*, **C,** 1099 (1967).
252. N. A. Nesmeyanov and O. A. Reutov, *Zh. Org. Khim.*, **2,** 1716 (1966).
253. G. Kamai, *Ber.*, **66,** 1779 (1933).
254. G. Kamai and Yu. F. Gatilov, *Dokl. Akad. Nauk SSSR*, **137,** 91 (1961).

255. G. Kamai and Yu. F. Gatilov, *Zh. Obshch. Khim.*, **33,** 1189 (1963).
256. F. G. Mann and J. Watson, *J. Chem. Soc.*, 505 (1947).
257. D. A. Thornton, *J. S. Afr. Chem. Inst.*, **17,** 61 (1964).
258. F. G. Holliman and F. G. Mann, *J. Chem. Soc.*, 550 (1943).
259. M. H. Forbes, F. G. Mann, I. T. Millar, and E. A. Moelwyn-Hughes, *J. Chem. Soc.*, 2833 (1963).
260. L. Horner, H. Winkler, and E. Meyer, *Tetrahedron Lett.*, 789 (1965).
261. L. Horner and H. Fuchs, *Tetrahedron Lett.*, 1573 (1963).
262. L. Horner and W. Hofer, *Tetrahedron Lett.*, 3321 (1966).
263. G. Kamai and Yu. F. Gatilov, *Zh. Obshch. Khim.*, **34,** 782 (1964).
264. L. Horner and W. Hofer, *Tetrahedron Lett.*, 3281 (1965).
265. M. S. Lesslie and E. E. Turner, *J. Chem. Soc.*, 1588 (1933).
266. R. C. L. Mooney, *J. Amer. Chem. Soc.*, **62,** 2955 (1940).
267. E. Collins, D. J. Sutor, and F. G. Mann, *J. Chem. Soc.*, 4051 (1963).
268. B. Zaslow and R. E. Rundle, *J. Phys. Chem.*, **61,** 490 (1957).
269. R. C. L. M. Slater, *Acta Crystallogr.*, **12,** 187 (1959).
270. F. A. Cotton and S. J. Lippard, *Inorg. Chem.*, **5,** 416 (1966).
271. G. J. Palenik, *Acta Crystallogr.*, **20,** 471 (1966).
272. T. E. Hopkins, A. Zalkin, D. H. Templeton, and M. G. Adamson, *Inorg. Chem.*, **5,** 1427 (1966).
273. B. R. Penfold and W. T. Robinson, *Inorg. Chem.*, **5,** 1758 (1966).
274. J. G. Scane, *Acta Crystallogr.*, **23,** 85 (1967).
275. B. D. Faithful and S. C. Wallwork, *Chem. Commun.*, 1211 (1967).
276. K. Bowden and E. A. Braude, *J Chem. Soc.*, 1068 (1952).
277. F. G. Mann, I. T. Millar, and B. B. Smith, *J. Chem. Soc.*, 1130 (1953).
278. A. I. Popov and R. E. Humphrey, *J. Amer. Chem. Soc.*, **81,** 2043 (1959).
279. H. Schindlbauer, *Spectrochim. Acta*, **20,** 1143 (1964).
280. J. J. Cole and R. T. Pflaum, *Proc. Iowa Acad. Sci.*, **21,** 145 (1964).
281. H. H. Hsieh, Dissertation, Univ. of Pittsburgh, 1964.
282. G. H. Cookson and F. G. Mann, *J. Chem. Soc.*, 2888 (1949).
283. H. Siebert, *Z. Anorg. Allg. Chem.*, **273,** 161 (1953).
284. W. R. Cullen, G. B. Deacon, and J. H. S. Green, *Can. J. Chem.*, **43,** 3193 (1965).
285. G. B. Deacon, J. H. S. Green, and W. Kynaston, *Aust. J. Chem.*, **19,** 1603 (1966).
286. W. R. Cullen, G. B. Deacon, and J. H. S. Green, *Can. J. Chem.*, **44,** 717 (1966).
287. J. B. Hendrickson, R. E. Spenger, and J. J. Sims, *Tetrahedron*, **19,** 707 (1963).
288. G. B. Deacon and R. A. Jones, *Aust. J. Chem.*, **16,** 499 (1963).
289. J. Benaïm, *C. R. Acad. Sci., Paris*, **261,** 1996 (1965).
290. M. F. A. Dove, *Chem. Commun.*, 23 (1965).
291. B. D. Faithful and D. G. Tuck, *Chem. Ind.* (London), 992 (1966).
292. A. G. Massey, E. W. Randall, and D. Shaw, *Spectrochim. Acta*, **20,** 379 (1964).
293. A. G. Massey, E. W. Randall, and D. Shaw, *Spectrochim. Acta*, **21,** 263 (1965).
294. E. W. Randall and D. Shaw, *Spectrochim. Acta*, **A, 23,** 1235 (1967).
295. L. F. Gleysteen and C. A. Kraus; *J. Amer. Chem. Soc.*, **69,** 451 (1947).
296. D. J. Mead, J. B. Ramsey, D. A. Rothrock, Jr., and C. A. Kraus, *J. Amer. Chem. Soc.*, **69,** 528 (1947).
297. E. R. Kline and C. A. Kraus, *J. Amer. Chem. Soc.*, **69,** 814 (1947).
298. R. L. Burwell , Jr., and C. H. Langford, *J. Amer. Chem. Soc.*, **81,** 3799 (1959).
299. J. Eliassaf, R. M. Fuoss, and J. E. Lind, Jr., *J. Phys. Chem.*, **67,** 1724 (1963).
300. L. Horner, F. Röttger, and H. Fuchs, *Chem. Ber.*, **96,** 3141 (1963).
301. M. Shinagawa, H. Matsuo, and F. Nakashima, *Kogyo Kagaku Zasshi*, **60,** 1409 (1957); through *C.A.*, **53,** 14781 (1959).

302. H. Matsuo, *J. Sci. Hiroshima Univ., Ser. A*, **22,** 281 (1958); through *C.A.*, **54,** 4206 (1960).
303. O. Menis, R. G. Ball, and D. L. Manning, *Anal. Chem.*, **29,** 245 (1957).
304. I. M. Kolthoff and R. A. Johnson, *J. Electrochem. Soc.*, **98,** 231 (1951).
305. H. H. Willard and G. M. Smith, *Ind. Eng. Chem., Anal. Ed.*, **11,** 186 (1939).
306. H. H. Willard and G. M. Smith, *Ind. Eng. Chem., Anal. Ed.*, **11,** 269 (1939).
307. H. H. Willard and G. M. Smith, *Ind. Eng. Chem., Anal. Ed.*, **11,** 305 (1939).
308. S. Tribalat, *Ann. Chim.* (Paris), **4,** 289 (1949).
309. S. Tribalat and C. Piolet, *Ann. Chim.* (Paris), **7,** 31 (1962).
310. H. E. Affsprung, N. A. Barnes, and H. A. Potratz, *Anal. Chem.*, **23,** 1680 (1951).
311. M. A. Khattak and R. J. Magee, *Microchem. J.*, **8,** 51 (1964).
312. H. A. Potratz and J. M. Rosen, *Anal. Chem.*, **21,** 1276 (1946).
313. H. Bode, *Z. Anal. Chem.*, **134,** 100 (1951).
314. J. W. Murphy and H. E. Affsprung, *Anal. Chem.*, **33,** 1658 (1961).
315. R. Neeb, *Z. Anal. Chem.*, **177,** 420 (1960).
316. R. Neeb, *Z. Anal. Chem.*, **154,** 23 (1957).
317. R. Neeb, *Z. Anal. Chem.*, **179,** 21 (1961).
318. M. Ziegler and O. Glemser, *Z. Anal. Chem.*, **157,** 17 (1957).
319. I. P. Alimarin and S. V. Makarova, *Zh. Anal. Khim.*, **17,** 1072 (1962).
320. J. P. Labbé, *Mikrochim. Acta*, 283 (1962).
321. H. E. Affsprung and J. L. Robinson, *Anal. Chim. Acta*, **37,** 81 (1967).
322. F. Jasim, R. J. Magee, and C. L. Wilson, *Mikrochim. Acta*, 721 (1960).
323. H. E. Affsprung and J. W. Murphy, *Anal. Chim. Acta*, **30,** 501 (1964).
324. K. W. Ellis and N. A. Gibson, *Anal. Chim. Acta*, **9,** 368 (1953).
325. K. W. Ellis and N. A. Gibson, *Anal. Chim. Acta*, **9,** 275 (1953).
326. B. Figgis and N. A. Gibson, *Anal. Chim. Acta*, **7,** 313 (1952).
327. F. P. Dwyer, N. A. Gibson, and R. S. Nyholm, *J. Proc. Roy. Soc. N.S. Wales*, **78,** 226 (1944).
328. F. P. Dwyer and N. A. Gibson, *Analyst* (London), **75,** 201 (1950).
329. R. Neeb, *Z. Anal. Chem.*, **152,** 158 (1956).
330. F. Jasim, R. J. Magee, and C. L. Wilson, *Rec. Trav. Chim. Pays-Bas*, **79,** 541 (1960).
331. A. J. Cameron and N. A. Gibson, *Anal. Chim. Acta*, **24,** 360 (1961).
332. N. A. Gibson and R. A. White, *Anal. Chim. Acta*, **12,** 115 (1955).
333. N. A. Gibson and R. A. White, *Anal. Chim. Acta*, **12,** 413 (1955).
334. N. A. Gibson and R. A. White, *Anal. Chim. Acta*, **13,** 546 (1955).
335. N. A. Gibson and J. W. Hosking, *Aust. J. Chem.*, **18,** 123 (1965).
336. A. Cahours, *Justus Liebigs Ann. Chem.*, **122,** 327 (1862).
337. G. Wittig and K. Clauss, *Justus Liebigs Ann. Chem.*, **577,** 26 (1952).
338. D. Hellwinkel and G. Kilthau, *Chem. Ber.*, **101,** 121 (1968).
339. J. B. Hendrickson, R. E. Spenger, and J. J. Sims, *Tetrahedron Lett.*, 477 (1961).
340. P. J. Wheatley and G. Wittig, *Proc. Chem. Soc.*, 251 (1962).
341. P. J. Wheatley, *J. Chem. Soc.*, 3718 (1964).
342. A. L. Beauchamp, M. J. Bennett, and F. A. Cotton, *J. Amer. Chem. Soc.*, **90,** 6675 (1968).
343. P. J. Wheatley, *J. Chem. Soc.*, 2206 (1964).
344. E. L. Muetterties, W. Mahler, and R. Schmutzler, *Inorg. Chem.*, **2,** 613 (1963).
345. D. Hellwinkel, *Angew. Chem.*, **78,** 749 (1966).

VI Arseno compounds $(RAs)_x$, diarsines $(R_2As)_2$, and related compounds

Arseno compounds were at one time widely used as therapeutic agents, particularly in the treatment of syphilis. In 1942, penicillin was found to be an effective agent against this disease, and the use of arsenical drugs soon came to a halt. Even before the introduction of penicillin, however, better arsenical drugs than arseno compounds had been introduced, and the latter had been largely abandoned. Most research work on the arseno compounds was performed in the period from 1910 to 1930. Modern work on these compounds has been confined largely to the elucidation of their structure. They were originally formulated as RAs=AsR. There is considerable doubt, however, that arsenic and other elements beyond the second period of the periodic system can form p_π-p_π bonds; and experimental evidence has been obtained that arseno compounds are polymeric. Their structure is considered in Section VI-3.

1. PREPARATION OF ARSENO COMPOUNDS

A. By reduction

The most important method for the preparation of arseno compounds is by reduction of arsonic acids, arsenoso compounds, arsonous acids, or dihaloarsines. The reagents commonly employed for this purpose are phosphorous or hypophosphorous acid, sodium dithionite ($Na_2S_2O_4$), stannous chloride, and sodium amalgam. Phosphorous acid and hypophosphorous acid are among the most satisfactory reducing agents for preparing arseno compounds, since the products are usually obtained in a relatively pure state. Thus,

Blicke and Smith (1) have obtained analytically pure samples of *p*-arsenoanisole, *p*-arsenotoluene, and arsenobenzene by reduction of the corresponding dichloroarsines in acetone solution with hypophosphorous acid. β-Arsenonaphthalene has been prepared by the reduction of β-naphthyldichloroarsine or β-arsenosonaphthalene with phosphorous acid (2); α-arsenofuran is obtained when α-furanearsonic acid is reduced with hypophosphorous acid (3). Arsenobenzene can be prepared by the reduction of benzenearsonic acid with hypophosphorous acid in aqueous solution at 40–50° (4) or by reduction with phosphorous acid at 180° (5). Aliphatic acids are also reduced to arseno compounds with phosphorous or hypophosphorous acid. Thus, arsenomethane admixed with some metallic arsenic is obtained by reduction of methanearsonic acid with hypophosphorous acid (6). Sodium hypophosphite has been used for the reduction of magnesium 1-propanearsonate to 1-arsenopropane (7). The product in this case was contaminated with a small amount of tetra-*n*-propyldiarsine.

In contrast to the reduction of dichloroarsines, the reduction of aryldiiodoarsines with phosphorous acid does not yield arseno compounds. By the action of solid phosphorous acid on aryldiiodoarsines in acetone solution, Blicke and Smith (8) prepared a series of 1,2-diiodo-1,2-diaryldiarsines, ArAs(I)As(I)Ar, in which the aryl group was phenyl, *p*-tolyl, *p*-anisyl, α-naphthyl, and *p*-biphenylyl. The phenyl compound was also prepared by the reaction of mercury with phenyldiiodoarsine. In this case, however, the diiododiarsine reacts further with mercury with the production of arsenobenzene:

$$C_6H_5As(I)As(I)C_6H_5 + Hg \rightarrow (C_6H_5As)_n + HgI_2$$

The preparation of arseno compounds by the treatment of dihaloarsines with mercury or sodium is further discussed in Section VI-1-D.

Ehrlich and Bertheim (9) investigated the use of sodium amalgam in methanol and stannous chloride in methanol or hydrochloric acid for the preparation of arseno compounds. Thus, they prepared *p*-arsenoaniline by reduction of the corresponding arsenoso compound with sodium amalgam in methanol or with stannous chloride in hydrochloric acid. They also obtained this arseno compound by the treatment of *p*-aminobenzenearsonic acid with a methanol solution of stannous chloride. More recently, other workers (10, 11) have used zinc and hydrochloric or sulfuric acid for the reduction of arsonic acids to arseno compounds.

For commercial preparation of drugs of the arseno type, sodium dithionite (also called sodium hydrosulfite) has been extensively used. The reduction of 3-nitro-4-hydroxybenzenearsonic acid in aqueous alkaline solution with sodium dithionite gives 3,3′-diamino-4,4′-dihydroxyarsenobenzene (12). This material, known as arsphenamine base or salvarsan base, was first prepared by

Ehrlich and Bertheim (13) in 1907 and introduced into medicine by Ehrlich for the treatment of syphilis. As the free base is extremely unstable, it had to be marketed as the dihydrochloride; this was then neutralized immediately prior to injection.

Arsphenamine, when prepared by reduction with sodium dithionite, is invariably contaminated with sulfur-containing impurities. The amount of sulfur in the product depends on the reaction conditions employed (14). The presence of sulfur-containing by-products is not, however, restricted to compounds of the arsphenamine type. Thus, Linsker and Bogert (15) found that the reduction of *p*-cyanobenzenearsonic acid with sodium dithionite gave an arseno compound contaminated with sulfur impurities, which could not be separated from *p*-arsenobenzonitrile. Since the presence of these sulfur-containing impurities was believed to increase the toxicity of arsphenamine, considerable experimental work was devoted to determining their structure and to find ways of preventing their formation. In one investigation of the reduction of 3-nitro-4-hydroxybenzenearsonic acid with sodium dithionite, King (16) isolated two sulfur-containing compounds which he believed to be 3-amino-4-hydroxy-5-sulfinobenzenearsonic acid and the corresponding arseno compound. There is also some evidence that 3-amino-4-hydroxythioarsenosobenzene may be formed in the reduction process (17).

Arsphenamine has been prepared by the use of many reducing agents other than sodium dithionite, although it possesses the advantage of reducing both the nitro group and the arsenic atom in one step. Vagenberg (18) reduced 3-amino-4-hydroxybenzenearsonic acid with sodium bisulfite; this same acid has also been reduced with hypophosphorous acid (19). A number of investigations have been made on the electrolytic reduction of both 3-nitro-4-hydroxy- and 3-amino-4-hydroxybenzenearsonic acids (20–23). The formation of the desired arseno compound is a function of the current density, the type of electrodes used, the pH of the solution, and the temperature of reduction.

A number of attempts have been made to modify the structure of arsphenamine in efforts to improve the stability and lower the toxicity of this compound. A vast number of compounds, both organic and inorganic, have been condensed with arsphenamine. Although structures were usually assigned to these products in the older chemical literature, there was little or no evidence for many of the proposed formulas. Two compounds which at one time achieved considerable therapeutic importance are neoarsphenamine (neosalvarsan) and sulfarsphenamine. The former was obtained by the condensation of sodium formaldehyde sulfoxylate with arsphenamine and was believed to be a mixture of compounds **1** and **2** (24). Sulfarsphenamine was obtained by the condensation of arsphenamine dihydrochloride with formaldehyde and sodium bisulfite and was believed to consist principally of **3** (25).

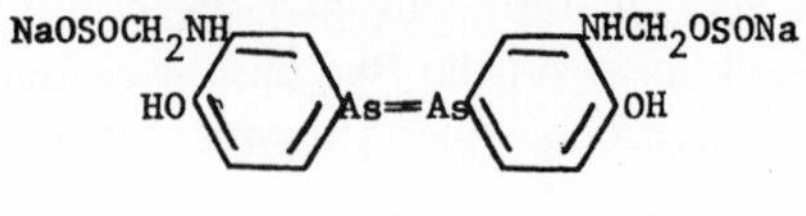

NaOSOCH$_2$NH NHCH$_2$OSONa
HO As=As OH

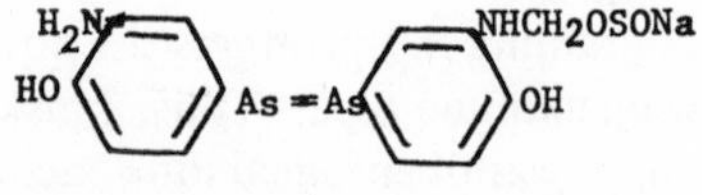

1

H$_2$N NHCH$_2$OSONa
HO As=As OH

2

NaO$_2$SOCH$_2$NH NHCH$_2$OSO$_2$Na
HO As=As OH

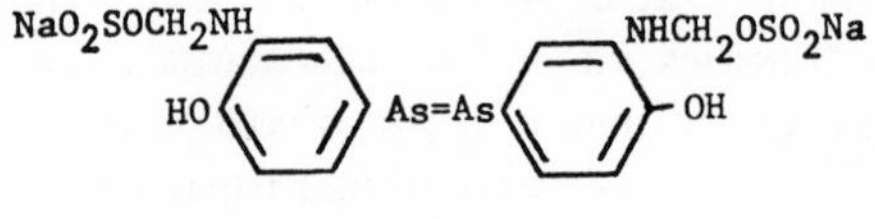

3

Dyke and King (26, 27), however, obtained evidence which convinced them

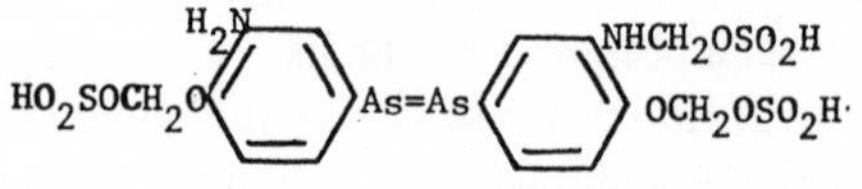

H$_2$N NHCH$_2$OSO$_2$H
HO$_2$SOCH$_2$O As=As OCH$_2$OSO$_2$H·

4

that sulfarsphenamine consists principally of the sodium salt of **4**. Obviously, materials such as neoarsphenamine, sulfarsphenamine, and a host of similar compounds once used in medicine are mixtures of variable composition.

The simultaneous reduction of a mixture of two different arsonic acids or arsenoso compounds leads to the so-called unsymmetrical arseno compounds. Although these have generally been formulated as RAs=AsR′, their exact structure is unknown. When a mixture of 4-aminobenzenearsonic acid and benzenearsonic acid is reduced with sodium dithionite in alkaline solution, an arseno compound is obtained, the analysis of which corresponds to the formula $H_2NC_6H_4As{=}AsC_6H_5$ (28). This material forms a hydrochloride which is insoluble in water, but condensation of the hydrochloride with formaldehyde and sodium bisulfite causes complete solution of the compound. From this fact it was argued that the arseno compound was a single product, since a mixture of arsenobenzene and *p*-arsenoaniline would only partially dissolve. Many other so-called unsymmetrical arseno compounds have been prepared. By reduction of a mixture of N-(*p*-arsonophenyl)glycine and N-(*p*-arsonophenyl)glycinamide with stannous chloride, Palmer and Kester (29) obtained a product to which the structure **5** was assigned. Similarly,

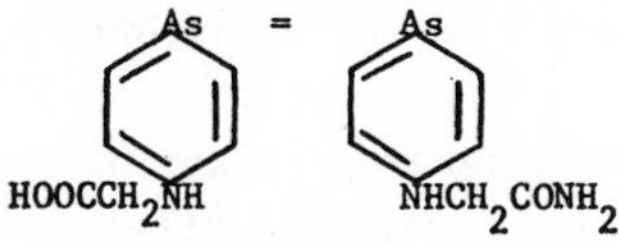

$HOOCCH_2NH$ ... $NHCH_2CONH_2$

5

reduction of a mixture of N-(2-hydroxyethyl)-*p*-arsanilic acid and *p*-arsonophenoxyacetic acid gives an unsymmetrical arseno compound. Both arseno compounds dissolve completely in alkali, from which fact the authors argued that mixtures of arseno compounds were not obtained. The reduction of some arsonic acid mixtures, however, does lead to a mixture of arseno compounds. Thus, when a mixture of benzenearsonic acid and 3-amino-4-hydroxybenzenearsonic acid is reduced, the product is invariably a mixture (29). Other workers have also obtained a mixture of arseno compounds when two arsonic acids are reduced simultaneously. Hart and Payne obtained a mixture of arseno compounds from the reduction of 3-amino-4-hydroxybenzenearsonic acid and 4-hydroxybenzenearsonic acid (30) or from the reduction of 3-amino-4-hydroxybenzenearsonic acid and N-(*p*-arsonophenyl)-glycine (31). In both of these cases, however, the authors claimed to have obtained the pure unsymmetrical arseno compounds by reduction of a mixture of the two corresponding arsenoso compounds. The arseno compound obtained by the reduction of a mixture of 3-amino-4-hydroxyarsenosobenzene and N-(*p*-arsenosophenyl)glycine with sodium dithionite is completely soluble in both sodium bicarbonate and dilute hydrochloric acid. Kraft and Katyshkina (32) reduced a mixture of 4-amino- and 4-hydroxybenzenearsonic acids with sodium dithionite. The resulting amorphous arseno compound has an arsenic:nitrogen ratio of 3:2. By a similar reduction of 4-hydroxy- and 3-amino-4-hydroxybenzenearsonic acids, arseno compounds are obtained in which the arsenic:nitrogen ratio varies between 3:1 and 1:1. The authors (32) formulated the unsymmetrical arseno compounds as linear polymers.

Reduction of a mixture of two arsonic acids sometimes leads to products containing more arsenic than that calculated for the arseno structure. Edee (33) reduced a mixture of 3-amino-4-hydroxybenzenearsonic acid and 2-hydroxyethanearsonic acid and obtained an amorphous product containing 69% arsenic to which structure **6** was assigned. The material was soluble in

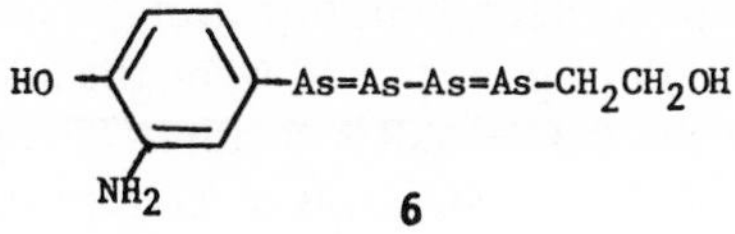

HO—$C_6H_3(NH_2)$—As=As–As=As–CH_2CH_2OH

6

both dilute aqueous alkali and in concentrated hydrochloric acid. Other so-called "tetraarseno" compounds have been obtained by other workers. There has been no modern work on these compounds, and their structure is unknown.

In addition to the unsymmetrical aromatic arseno compounds, mixed arseno compounds containing aryl and alkyl groups have been prepared. By the reduction of arsenosobenzene and arsenosoethane with phosphorous acid, Steinkopf, Schmidt, and Smie (34) obtained a product to which they assigned the formula $C_2H_5As{=}AsC_6H_5$. It melted at 177° and was identical with the material obtained by the condensation of phenylarsine and arsenosoethane. Reduction of a mixture of arsenomethane and 3-amino-4-hydroxyarsenosobenzene in methanol solution with sodium dithionite gives an unsymmetrical arseno compound, which was assigned structure 7 (35). This same product

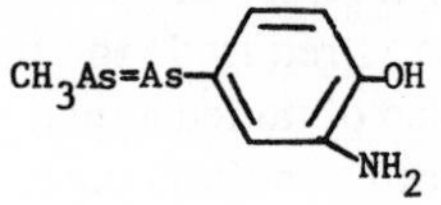

$CH_3As{=}As{-}C_6H_3(NH_2)(OH)$

7

was also obtained by reducing a mixture of sodium methanearsonate and 3-amino-4-hydroxybenzenearsonic acid or by reducing a mixture of the latter acid and arsenosomethane. The arseno compound is completely soluble in dilute hydrochloric acid or in aqueous sodium hydroxide solution.

A number of other alkyl-aryl arseno compounds were prepared by Edee (33) and Palmer and Edee (36) by reducing mixtures of aliphatic and aromatic arsonic acids with hypophosphorous acid or with stannous chloride. On the basis of analytical results and the solubility properties of the arseno compounds, the authors concluded that the pure unsymmetrical compounds are usually formed; under some conditions, however, mixtures of arseno compounds are obtained. The reduction of a mixture of arsanilic and arsonoacetic acids with hypophosphorous acid in sulfuric acid gives a precipitate of 4,4′-diammonioarsenobenzene sulfate after the solution stands for 10 days. On further standing arsenoacetic acid slowly separates from solution. The authors (36) concluded that arsanilic acid is reduced at a faster rate than arsonoacetic acid. When the same reduction is run in hydrochloric acid with the addition of pyridine, the product gives analytical values corresponding to the unsymmetrical arseno compound; it is completely soluble in cold aqueous alkali.

Since the structure of the mixed arseno compounds is not known, it is difficult to present a mechanism for their formation. A number of mechanisms, however, were suggested during the period when the arseno compounds were regarded as possessing the RAs=AsR structure. The theory was advanced that reduction of two arsonic acids gives a mixture of symmetrical arseno compounds which subsequently rearrange to the unsymmetrical arseno compound (29, 37). Evidence for such a mechanism was advanced by Palmer and Kester (29) who demonstrated that when 4,4′-bis(β-hydroxyethyl-amino)arsenobenzene is mixed with *p*-arsonophenoxyacetic acid and the mixture is treated with hypophosphorous acid, the resulting product possesses the properties of the unsymmetrical arseno compound. It is also possible to obtain the unsymmetrical arseno compounds by the interaction of two symmetrical arseno compounds in solution (38). Thus, a mixture of 3,3′-di-amino-4,4′-dihydroxyarsenobenzene and 3,3′-diamino-4,4′-dihydroxy-5,5′-diacetamidoarsenobenzene gives the unsymmetrical arseno compound when the two are dissolved in a solvent and then treated with methanolic hydro-chloric acid or with ether. By contrast, Hart and Payne (30) were unable to obtain evidence for rearrangement to the unsymmetrical arseno compound when a mixture of 3,3′-diamino-4,4′-dihydroxyarsenobenzene and 4-arseno-phenol is dissolved in alkali and the solution heated to 55° for some time.

The reduction of arsonic acids containing two arsono groups in the same molecule also leads to the production of arseno compounds. Lieb (39) reduced both *p*-benzene- and *m*-benzenediarsonic acids with phosphorous acid in a sealed tube at temperatures over 200° and obtained amorphous arseno compounds which decomposed without melting at high temperatures. Lieb and Wintersteiner (40, 41) also reduced *o*-benzenediarsonic acid and a number of substituted *p*-benzenediarsonic acids with hypophosphorous acid and obtained in each case amorphous arseno compounds, which usually decomposed at high temperatures without melting. Since the arseno compounds are insoluble in organic solvents, molecular weight data could not be obtained.

Karrer (42) reduced 2,2′-diarsono-4,4′-diaminostilbene with sodium dithionite and obtained an arseno compound which was formulated as **8**.

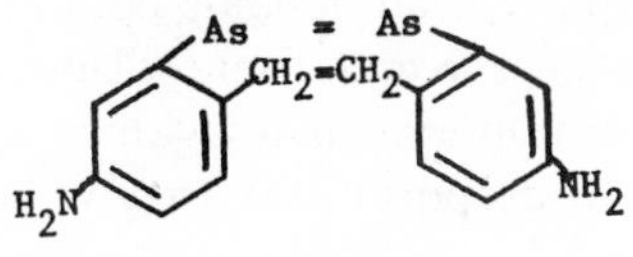

As = As
CH_2=CH_2
H_2N NH_2

8

Finzi and Bartelli (43) reduced 4,4′-diarsono-3,3′-biphenyldicarboxylic acid with hypophosphorous acid. The resulting arseno compound was assigned structure **9**. This compound is insoluble in organic solvents, but the disodium

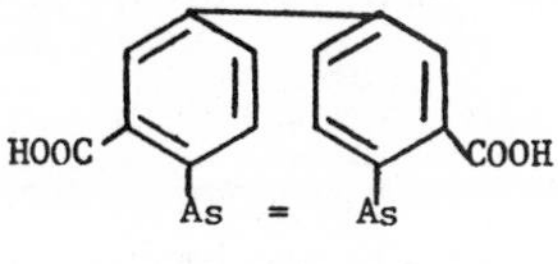

HOOC COOH
As = As

9

salt is readily soluble in water. A determination of the molecular weight by cryoscopic measurements in water gave very high values. Ebullioscopic measurements in water, however, gave a value one-third of the formula weight, from which fact it was concluded that the salt was monomolecular in hot water.

B. By oxidation

Aromatic arseno compounds are obtained as the first oxidation products of primary aromatic arsines (cf. Section IV-2). Since arsines are difficult to prepare and must be handled in an inert atmosphere, this reaction is not of preparative importance. Arseno compounds are themselves readily oxidized, so that the product is usually contaminated with the arsonic acid. Aliphatic arseno compounds cannot be obtained by the oxidation of aliphatic arsines.

Dehn (44) allowed an ether solution of phenylarsine to stand in the air and obtained arsenobenzene and benzenearsonic acid. Palmer and Adams (45) have investigated the oxidation of phenylarsine with aldehydes. They were able to obtain arsenobenzene in good yield by boiling an aldehyde and phenylarsine in glacial acetic acid in a carbon dioxide atmosphere. A small amount of benzenearsonic acid was always obtained from this reaction. Both aliphatic and aromatic aldehydes could be used; with chloral, however, phenyldichloroarsine and acetaldehyde were the sole products.

Phenylarsine has been oxidized to arsenobenzene by the use of organometallic compounds (46). For example, when phenylarsine and phenylmercuric chloride are heated in benzene solution in a nitrogen atmosphere, a quantitative yield of arsenobenzene is formed according to the equation:

$$n\mathrm{C_6H_5AsH_2} + n\mathrm{C_6H_5HgCl} \rightarrow n\mathrm{C_6H_6} + n\mathrm{Hg} + n\mathrm{HCl} + (\mathrm{C_6H_5As})_n$$

Diphenylmercury and diphenyllead diiodide both react with phenylarsine to give a mixture of arsenobenzene and arsenosobenzene. With the lead compound it was postulated that intermediates with an arsenic–lead bond may be formed and that these disproportionate to give arsenobenzene and lead dichloride:

$$n\mathrm{C_6H_5AsPbX_2} \rightarrow (\mathrm{C_6H_5As})_n + n\mathrm{PbX_2}$$

Anschutz and Wirth (47) have oxidized phenylarsine with thionyl chloride to obtain arsenobenzene, sulfur dioxide, and hydrogen chloride. If an excess of thionyl chloride is used, phenydichloroarsine and sulfur dioxide are obtained.

C. By condensation

Since at one time arsenobenzene was believed to have a structure similar to azobenzene, it is only natural that earlier workers in the field would attempt to prepare arseno compounds by reactions analogous to the condensation of nitroso compounds with amines. Such reactions do indeed succeed with organic arsenicals. Thus, the condensation of 4-aminophenylarsine and 4-arsenosoaniline gave 4-arsenoaniline (48). Unsymmetrical arseno compounds were also prepared in this manner. From 4-hydroxyarsenosobenzene and 4-aminophenylarsine the unsymmetrical arseno compound was obtained. Dichloroarsines will also condense with primary arsines. Thus, the condensation of N-(*p*-dichloroarsinophenyl)glycine with 3-amino-4-hydroxyphenylarsine gives a product to which the formula **10** was assigned. This

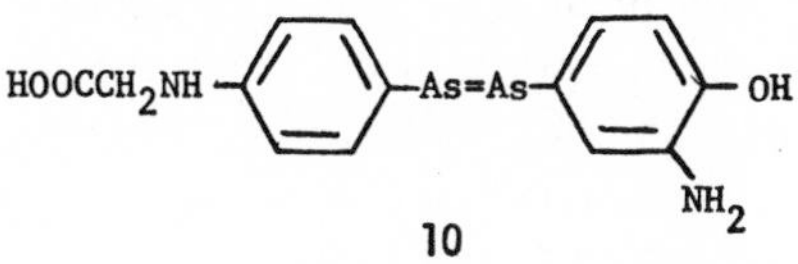

HOOCCH$_2$NH—C_6H_4—As=As—C_6H_3(OH)(NH_2)

10

same unsymmetrical arseno compound was also prepared by Hart and Payne (31) by condensing 3-amino-4-hydroxyphenylarsine with either N-(*p*-arsenosophenyl)glycine or N-(*p*-dichloroarsinophenyl)glycine. The condensation of phenylarsine and arsenosoethane in methanol solution and in an inert atmosphere also leads to the formation of an unsymmetrical arseno compound (34). The product is a white crystalline solid which may be recrystallized from benzene or toluene.

Blicke and Powers (49) have prepared arsenobenzene by various condensations related to the above reactions. The condensation of phenyldichloroarsine and diphenylarsine in ether solution gives arsenobenzene and diphenylchloroarsine. The authors demonstrated that tetraphenyldiarsine and arsenobenzene are the initial products of the reaction and that the diarsine reacts further with phenyldichloroarsine to produce diphenylchloroarsine and arsenobenzene. If phenyldiiodoarsine is condensed with diphenylarsine, 1,2-diiodo-1,2-diphenyldiarsine is produced rather than arsenobenzene. The condensation of diphenylchloro- or diphenyliodoarsine and phenylarsine gives arsenobenzene and tetraphenyldiarsine. The following reactions are given by Blicke and Powers to summarize these condensations:

$$2C_6H_5AsCl_2 + 4(C_6H_5)_2AsH \rightarrow C_6H_5As{=}AsC_6H_5 + 2(C_6H_5)_2As{-}As(C_6H_5)_2 + 4HCl$$

$$2(C_6H_5)_2As{-}As(C_6H_5)_2 + 2C_6H_5AsCl_2 \rightarrow 4(C_6H_5)_2AsCl + C_6H_5As{=}AsC_6H_5$$

$$(C_6H_5)_2As{-}As(C_6H_5)_2 + 2C_6H_5AsI_2 \rightarrow C_6H_5As(I)As(I)C_6H_5 + 2(C_6H_5)_2AsI$$

$$4(C_6H_5)_2AsCl + 2C_6H_5AsH_2 \rightarrow C_6H_5As{=}AsC_6H_5 + 2(C_6H_5)_2As{-}As(C_6H_5)_2 + 4HCl$$

Closely related to the above condensations is the reaction of arylarsines with bis(diarylarsenic) oxides (50). Thus, the reaction of *p*-aminophenylarsine with bis(diphenylarsenic) oxide has been formulated as follows:

$$[(C_6H_5)_2As]_2O + 2p\text{-}H_2NC_6H_4AsH_2 \rightarrow [(C_6H_5)_2As]_2 + p\text{-}H_2NC_6H_4As{=}AsC_6H_4NH_2\text{-}p$$

Steinkopf and Dudek (51) argued that since arsines and arsenoso compounds form arseno compounds, the condensation of arsines and nitroso compounds should lead to compounds of the type ArAs=NAr. Accordingly they condensed nitrosobenzene with phenylarsine in ether solution in an inert atmosphere. Arsenobenzene slowly crystallized from the solution. From the filtrate arsenosobenzene, aniline, and azobenzene were obtained. The authors suggested that the desired $C_6H_5As{=}NC_6H_5$ was obtained and then disproportionated to azobenzene and arsenobenzene. The formation of arsenosobenzene and aniline was attributed to the following reaction:

$$C_6H_5As{=}NC_6H_5 + H_2O \rightarrow C_6H_5AsO + C_6H_5NH_2$$

More recent evidence for arsenic–nitrogen double bonds has been summarized in Section V-4.

The mechanism of the condensation of arsines with arsenoso compounds or dihaloarsines is not known. Since arsenoso compounds are polymeric and hence not analogous in structure to nitroso compounds, it is obviously incorrect to assume that the formation of arseno compounds from arsines and arsenoso compounds is analogous to the condensation of amines and nitroso compounds. Palmer and Adams (45) have suggested that arsines and arsenoso compounds undergo simultaneous oxidation and reduction to a mixture of arseno compounds, which then rearrange:

$$2RAsH_2 + 2R'AsO \rightarrow RAs{=}AsR + R'As{=}AsR'$$

$$RAs{=}AsR + R'As{=}AsR' \rightarrow 2RAs{=}AsR'$$

As an alternative to the above mechanism they postulated a mechanism somewhat analogous to an aldol condensation followed by elimination of water:

$$RAsO + R'AsH_2 \rightarrow RAs(OH){-}As(R')H \rightarrow RAs{=}AsR' + H_2O$$

There is no experimental evidence for either mechanism.

D. Miscellaneous methods

Job, Reich, and Vergnaud (52), in a study of the reaction of $C_6H_5As(MgBr)_2$ with various organic halides, found that the magnesium compound reacts with phosgene or with ethylene dibromide with the formation of arsenobenzene. Somewhat similar results have been obtained by Beeby, Cookson, and Mann (53), who reported that the reaction of $C_6H_5As(MgBr)_2$ with *o*-(2-bromoethyl)benzyl bromide in benzene solution at 70° gives arsenobenzene and the isoarsinoline **11**. When the reaction is run in the cold, however, the

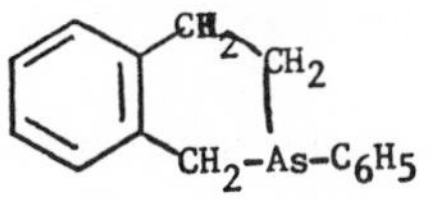

CH_2 CH_2
CH_2–As–C_6H_5

11

isoarsinoline is the principal product and only a trace of arsenobenzene is formed. The authors suggested a plausible mechanism for the formation of arsenobenzene:

$$2\ o\text{-}C_6H_4(CH_2CH_2Br)(CH_2Br) + C_6H_5As(MgBr)_2 \longrightarrow 2\ o\text{-}C_6H_4(CH_2CH_2Br)(CH_2MgBr) + C_6H_5AsBr_2$$

$$C_6H_5AsBr_2 + C_6H_5As(MgBr)_2 \longrightarrow C_6H_5As{=}AsC_6H_5 + 2MgBr_2$$

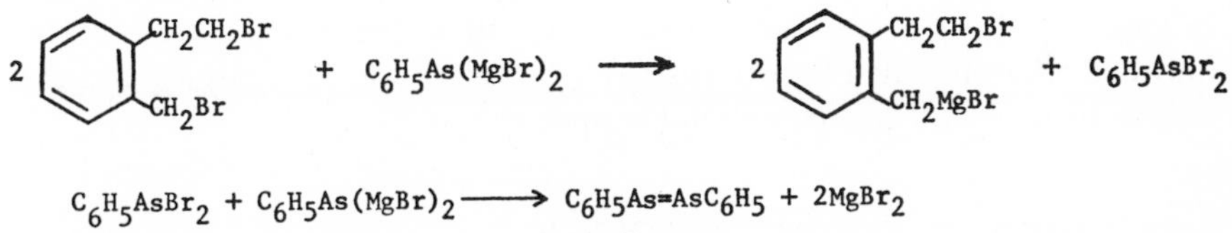

In support of this mechanism they demonstrated that phenyldichloroarsine is formed in the reaction between *o*,α,α'-dichloroxylene and $C_6H_5As(MgBr)_2$, and they cited uncompleted work to the effect that there is a slow reaction between phenyldibromoarsine and the magnesium reagent.

In studies on the preparation of heterocyclic arsenic compounds, Lyon and Mann (54) condensed *o*-xylene dibromide and phenyldichloroarsine in the presence of sodium to obtain 2-phenylisoarsindoline **12**. In addition to

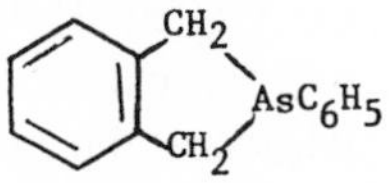

CH_2
AsC_6H_5
CH_2

12

this heterocycle, a considerable amount of arsenobenzene was isolated which was attributed to the condensation of phenyldichloroarsine in the presence of sodium. They then demonstrated that arsenobenzene or *p*-arsenotoluene could be readily formed in approximately 50% yield by refluxing phenyldichloroarsine or *p*-tolyldichloroarsine, respectively, with sodium in ether solution and in a nitrogen atmosphere. The arseno compounds thus prepared were quite pure, and it was suggested that this method might be of practical importance for the preparation of arseno compounds. More recently, several other arseno compounds have been prepared by the treatment of dihaloarsines with mercury (55–58).

Kraft and Batshchouk (59) claimed to have prepared arseno compounds by the reaction between arseno and arsenoso compounds according to the equation:

$$RAs{=}AsR + 2R'AsO \rightarrow 2RAsO + R'As{=}AsR'$$

Thus, a mixture of 3,3′-diamino-4,4′-dihydroxyarsenobenzene and 4-hydroxyarsenosobenzene gives 4-arsenophenol in 80% yield. Similarly, 3,3′-diamino-4,4′-dihydroxyarsenobenzene and arsenosomethane give arsenomethane and 3-amino-4-hydroxyarsenosobenzene.

The preparation of *p*-arsenothiophenol has been claimed by Krishna and Krishna (60). These authors treated diazotized arsanilic acid with potassium ethyl xanthate in an attempt to prepare *p*-mercaptobenzenearsonic acid. Acidification of the reaction mixture gave a solid which the authors believed to be *p*-arsenothiophenol rather than the arsonic acid. The only evidence presented for the suggested structure was an analysis, and the arseno structure for this compound is open to question.

2. REACTIONS OF ARSENO COMPOUNDS

Arseno compounds are highly reactive substances. It is generally considered that they are oxidized rapidly in the air to arsenoso compounds, but Blicke and Smith (1) demonstrated that pure arsenobenzene in solution was quite stable to oxygen. The addition of a trace of acid, iodine, or many other substances caused immediate oxidation. The toxicity (and the activity) of arsphenamine and its derivatives is believed to be largely dependent on oxidation products, principally 3-amino-4-hydroxyarsenosobenzene. In an interesting series of experiments, Eagle (61) demonstrated that when arsphenamine is dissolved in deaerated water in a nitrogen atmosphere, it is practically without effect on *Treponema pallidum*, the causative agent of syphilis. When solution is effected in the air, however, a highly treponemicidal solution is obtained. Kondo (62) demonstrated that when aqueous solutions of arsphenamine are allowed to stand, the amount of 3-amino-4-hydroxyarsenosobenzene formed is at first directly proportional to the absorption of oxygen. On longer standing, however, other oxidation products are formed as evidenced by the absorption of larger amounts of oxygen.

Kraft and Batshchouk (59) have reported that arsonic acids will oxidize arseno compounds to arsenoso compounds. From 3,3′-diamino-4,4′-dihydroxyarsenobenzene and 3-amino-4-hydroxybenzenearsonic acid in hydrochloric acid solution, they obtained 3-amino-4-hydroxyarsenosobenzene. The reaction is catalyzed by traces of hydriodic acid. The oxidation of arseno compounds with oxidizing agents such as hydrogen peroxide leads to the formation of arsonic acids. This fact has been used to advantage in the purification of these acids. Binz and Räth (63) purified 2-hydroxy-5-pyridinearsonic acid (obtained by the Bart reaction) by reduction to the arseno compound with sodium hypophosphite and subsequent reoxidation with 3% hydrogen peroxide. Sulfur dioxide oxidizes arsphenamine to 3-amino-4-hydroxyarsenosobenzene and the corresponding thioarsenosobenzene (17).

The reduction of arseno compounds with zinc dust and hydrochloric acid leads to the formation of arsines (64). When arsenobenzene is heated to 255° in a carbon dioxide atmosphere, it decomposes to give tetraphenyldiarsine and arsenic (7). An attempt to distill 1-arsenopropane at 13 mm similarly gave tetra-*n*-propyldiarsine, although the arseno compound could be distilled at lower pressure. Arsenomethane, when distilled in a stream of hydrogen, gives trimethylarsine and arsenic (6). It is probable that tetramethyldiarsine is first formed, which subsequently is further pyrolyzed to trimethylarsine and arsenic (7). Arseno compounds can be hydrolyzed in aqueous solution under pressure to give arsenic, arsenic trioxide, and the hydrocarbon (65). The hydrolysis appears to be acid catalyzed. Karrer (66) heated 2,2′-arsenobis(4-amino-5-hydroxybenzoic acid) with aqueous sodium acetate in a sealed tube and obtained 3-hydroxy-4-aminobenzoic acid.

The reaction of arseno compounds with sulfur has been known for many years. Michaelis and Schulte (5) demonstrated that heating arsenobenzene with 2 equivalents of sulfur leads to the formation of thioarsenosobenzene. With excess sulfur, however, diphenyl sulfide and arsenic trisulfide are formed. Cohen, King, and Strangeways (67) found that disulfides would react with arseno compounds to give compounds of the type $RAs(SR')_2$. Thus, the reaction of 4-arsenobenzoic acid and cystine gives a product identical with that obtained from the reaction of *p*-arsenosobenzoic acid and cysteine. The reaction of sulfides with arseno compounds has been reinvestigated by Kary (68, 69). This author allowed hydrogen sulfide to react with an arseno compound in acid solution, or reduced an arsonic acid to an arseno compound in acid solution in the presence of an alkali sulfide. By either reaction he obtained compounds to which he assigned the following structure:

$$\begin{array}{l} RAs{=}S \\ \;\;\| \\ RAs{=}S \end{array}$$

In support of this structure a molecular weight of 242 was reported for the compound where R is methyl. The details of the molecular weight determination, however, were not given. These sulfur-containing arsenicals, particularly the methyl compound, have shown considerable promise as insecticides (68–71). Kary's structure for these compounds in which each arsenic atom possesses two double bonds cannot be considered seriously. The substances may possibly be thioarsenoso compounds (cf. Sections III-4 and III-8-A). More data, however, will be necessary before a definite structure can be assigned to these compounds.

The reaction of alkali metals with arseno compounds has been investigated by Wittig and co-workers (72) and by Reesor and Wright (73). By the action of lithium on arsenobenzene in tetrahydrofuran, a dark red solution is obtained which slowly becomes lighter in color (72). Addition of methyl iodide to the lighter-colored solution gives 1,2-dimethyl-1,2-diphenyldiarsine. Hydrolysis of the same lighter-colored solution gives phenylarsine and arsenobenzene. Treatment with benzoic acid followed by benzoyl chloride gives arsenobenzene and benzaldehyde.

Wittig and co-workers (72) suggested that the dark-red color of the solution is due to the presence of a monolithium compound $C_6H_5As(Li)\dot{A}sC_6H_5$, which then adds more lithium to form the dilithium adduct. These authors, however, were aware of the polymeric nature of arsenobenzene, and did not regard the reaction as addition of lithium to an arsenic–arsenic double bond. The formation of phenylarsine and arsenobenzene on hydrolysis was attributed to the formation of $C_6H_5As(H)As(H)C_6H_5$, which then disproportionates to give the two products. Reaction of benzoic acid supposedly gives a monolithium compound $C_6H_5As(H)As(Li)C_6H_5$, which on treatment with benzoyl chloride gives arsenobenzene and benzaldehyde.

Reesor and Wright (73) added sodium rather than lithium to a solution of arsenobenzene in 2,5-dioxahexane(1,2-dimethoxyethane) and obtained a deep red solution which they attributed to the formation of a disodium adduct, $C_6H_5As(Na)As(Na)C_6H_5$. Treatment of this solution with methyl chloride yields 1,2-dimethyl-1,2-diphenyldiarsine. By the addition of excess sodium to the red solution a yellowish-green solution is obtained after several days. This was attributed to formation of $C_6H_5AsNa_2$, since treatment of this solution with methyl chloride gives dimethylphenylarsine. The authors (73) demonstrated that an equilibrium exists between the yellow $C_6H_5AsNa_2$ and the red compound, since addition of arsenobenzene to the yellow solution regenerates a red solution. The latter when treated with dimethyl sulfate gives 1,2-dimethyl-1,2-diphenyldiarsine. By the action of carbon dioxide on the yellow solution in an inert atmosphere and subsequent removal of the solvent *in vacuo*, a product is obtained which was believed to be $C_6H_5As(COONa)_2$.

Metal carbonyls have been found to react with arsenobenzene and with a number of analogous phosphorus compounds to give coordination compounds. Thus, Ang and West (74) reported that arsenobenzene reacts with molybdenum hexacarbonyl to form $(PhAs)_6Mo(CO)_4$, in which the arsenic compound acts as a bidentate ligand, and with tungsten hexacarbonyl to form $(PhAs)_6W(CO)_5$, in which the arsenic compound acts as a monodentate ligand. These results are in contrast to the earlier work of Fowles and Jenkins (75), who reported that arsenobenzene reacts with molybdenum hexacarbonyl to yield both a mononuclear complex $(PhAs)_4Mo(CO)_4$ and a binuclear complex $(PhAs)_4[Mo(CO)_4]_2$. Both complexes were found to be monomeric in benzene and to show CO stretching frequencies typical of *cis*-disubstituted hexacarbonyls. Fowles and Jenkins suggested that the binuclear complex has the following structure:

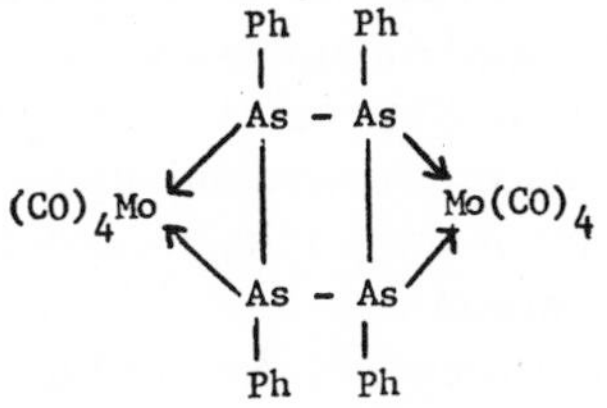

```
            Ph     Ph
            |      |
           ,As  -  As
          /  |     |  \
(CO)4Mo <    |     |   > Mo(CO)4
          \  |     |  /
           `As  -  As'
            |      |
            Ph     Ph
```

3. STRUCTURE AND PHYSICAL PROPERTIES OF ARSENO COMPOUNDS

Early workers in the field of organic arsenicals assumed that arseno compounds possess arsenic–arsenic double bonds, and these compounds are so formulated in many textbooks on organic chemistry even at the present time. It has long been believed, however, on empirical grounds that elements beyond the first row (Li to Ne) are reluctant to form multiple bonds. A qualitative explanation for this phenomenon was advanced by Pitzer (76). He concluded that the overlap of the p orbitals of one atom with the inner shell of a second atom would produce a repulsive force sufficient to prevent multiple bonding. Thus, in a double bond involving two phosphorus atoms, the 10 inner shell electrons of one atom and the $3p_x$ electrons of the second atom produce an important repulsive force. By contrast in nitrogen, with only 2 inner shell electrons, the repulsive force is considerably smaller. These qualitative considerations have been criticized by Mulliken (77), who calculated overlap integrals for single and multiple bonds for elements of the first two rows of the periodic system. The lack of p_π-p_π bonding in elements in the second row and above was attributed, not to weakness of the p_π-p_π bonds, but to the increased strength of p_σ bonds. The reason for this additional strength of p_σ bonds was attributed to hybridization of p and d orbitals.

The actual structure of the arseno compounds has been the subject of considerable attention. Earlier workers approached the problem principally by means of molecular weight determinations; the results, particularly with arsenobenzene, are confusing. Michaelis and Schäfer (78) originally obtained a single value of 399.8 for this compound by ebullioscopic measurements in benzene. Since the calculated value for the monomer $C_6H_5As{=}AsC_6H_5$ is

304, they concluded that arsenobenzene is not appreciably associated. Somewhat later Palmer and Scott (79) made a careful study of the molecular weight of a purified sample of arsenobenzene in three different solvents. Cryoscopically in naphthalene and ebullioscopically in benzene they obtained mean values of 642 and 402, respectively, but ebullioscopically in carbon disulfide they obtained a mean value of 334. From this data they naturally assumed that arsenobenzene is associated in the first two solvents, but is monomolecular in carbon disulfide. The association of arsenobenzene in boiling benzene has been qualitatively confirmed by several other workers; Blicke and Smith (1) gave values of 895 and 915, Blicke and Powers (80) a value of 867, while Lyon and Mann (54) gave values of 881 and 917. Strangely, these other workers did not repeat Palmer and Scott's determination in carbon disulfide. Blicke and Smith (1) found that *p*-arsenotoluene and *p*-arsenoanisole are also associated in boiling chloroform.

The structure of arsenobenzene, at least in the solid state, has been unequivocally determined by means of X-ray diffraction (81—83). These studies show that the compound is a 6-membered non-planar ring of arsenic atoms, arranged in the chair form. The As—As distances are 2.456 $\pm$ 0.005 A and the As—As—As angles average 91°, but there is considerable deviation from this mean value. The phenyl groups are arranged equatorially with the average As—As—C angle 100.1°, and the different angles do not differ significantly from this mean value. The compound possesses a center of symmetry, which is in accord with the fact that arsenobenzene does not possess an appreciable dipole moment (65).

Although these results establish the structure of crystalline arsenobenzene, the possibility of some dissociation in solution is not ruled out. The careful results of Palmer and Scott (79) would indicate this. The dissociated form need not necessarily involve p_π-p_π bonding but may exist as a diradical.

The structure of arsenomethane differs somewhat from that of arsenobenzene. It exists in two forms, a heavy yellow oil, which can be distilled *in vacuo*, and a red solid. Auger (6) originally obtained molecular weight values between 300 and 340 on the yellow oil and assumed that it exists as a tetramer, $(CH_3As)_4$. Later workers, however, showed definitely that this product is a pentamer. Thus, Steinkopf, Schmidt, and Smie (34) obtained molecular weight values between 425 and 469 cryoscopically in benzene and nitrobenzene and ebullioscopically in benzene; these values were confirmed by Waser and Schomaker (85) by means of vapor density measurements. Palmer and Scott (79) obtained a mean value of 474 for the yellow oil by ebullioscopic measurements in carbon disulfide. These latter authors also obtained a mean value of 476 for the molecular weight of a mixture of the yellow oil and red solid. An X-ray diffraction study of a single crystal of the yellow form of arsenomethane (mp 12°) has been carried out by Burns and Waser (86). They demonstrated that this isomer exists as a puckered 5-membered ring of arsenic atoms, with a methyl group attached to each arsenic atom. The average As—As value is almost exactly twice the As single bond radius assigned by Pauling (87), but the As—C bond distance is somewhat shorter that the sum of the covalent radii. Burns and Waser did not obtain X-ray data on the red form but suggested that it exists as a long chain of arsenic atoms.

The structure of $(CH_3As)_5$ and $(C_6H_5As)_6$ has been discussed in considerable detail by Donohue (88). He attributed the puckering of the 5-membered ring to two effects. One is relief of Pitzer (torsion) strain, which is at a maximum in planar 5-membered rings. The second effect is that puckering of the ring reduces the As—As—As bond angle to approximately 101°, which is much closer to the normal arsenic angle (about 100°) than would be the case in a planar ring. Donohue was unable to offer a satisfactory explanation for the existence of arsenomethane as a pentamer and arsenobenzene as a hexamer.

Although the simple arseno compounds, such as arsenobenzene and arsenomethane, undoubtedly are relatively low molecular weight compounds, this is not true of arseno compounds of the arsphenamine type. Early workers in the field were well aware of the fact that arsphenamine and its derivatives are polymeric. Klemensiewicz (89) determined the increase in viscosity of aqueous arsphenamine solutions as a function of time; he demonstrated that the viscosity increases quite rapidly to reach limiting values after 20 to 30 minutes. The rate of increase in viscosity can be accelerated by an increase in temperature, but at lower temperatures the viscosity reaches a somewhat higher value. Wright and co-workers (90), using membranes of various porosity, dialyzed arsphenamine and neoarsphenamine in aqueous solution under nitrogen gas and showed that these substances could be separated into various fractions with molecular weights varying from 5000 or less to values greater than 35,000. The low molecular weight fractions were more curative and less toxic than the high molecular weight fractions.

Kraft and his co-workers (23, 82, 91–96) have published a series of papers on the structure of arsphenamine and related arseno compounds. They pointed out (91) that the colloidal nature of arsphenamine solutions indicates a large molecular weight for this compound and suggested that it exists as a linear polymer with structure **12** (where R = 3-NH_2-4-HOC_6H_3). As

$$\mathrm{HOAs(R)(AsR)}_{n-2}\mathrm{As(R)OH}$$

12

evidence for such a structure they found that the oxidation of arsphenamine always consumes less iodine than required for the equation:

$$\mathrm{RAs{=}AsR} + 4\mathrm{I}_2 + 6\mathrm{H}_2\mathrm{O} \rightarrow 2\mathrm{RAsO}_3\mathrm{H}_2 + 8\mathrm{HI}$$

They concluded that such a result could only be obtained with a linear polymer in which the two end arsenic atoms possess a different oxidation number than the remaining arsenic atoms in the chain. They further pointed out that titration with iodine can be used to obtain the molecular weight of the arseno compounds. By titration of arseno compounds prepared by different methods, they obtained results which indicate that the average value of n in formula **12** varies between 6.9 and 36. The average molecular weight of arsphenamine prepared by reduction of 3-amino-4-hydroxybenzenearsonic acid with hypophosphorous acid is always considerably greater than the average molecular weight of the product obtained by electrolytic reduction with a lead cathode. When sodium dithionite is used as the reducing agent, the molecular weight varies over a wide range and depends largely on the amount of reducing agent employed. Kraft (96) also cited the color of arsphenamine as evidence of its existence as a high-molecular weight linear polymer. Thus, arsenobenzene and arsenomethane with low molecular weights are almost colorless, while arsphenamine is deep yellow in color. This yellow color Kraft attributed to "interaction of the electronic systems of the atoms that we call 'conjugation of the bonds' in the case of a chain of unsaturated carbon atoms." In a molecule possessing the *o*-aminophenol grouping, which so readily yields colored compounds by oxidation, this argument can hardly be considered as weighty evidence for the linear polymeric structure.

Kraft and Katyshkina (92) treated arsenomethane with iodine in benzene solution and found that the amount of iodine in the resulting products varies with the amount of benzene used in their preparation. They believed that their products are linear polymers of the following type:

$$IAs(CH_3)(AsCH_3)_{n-2}As(CH_3)I$$

From the arsenic:iodine ratio they calculated that the molecular weights for these products lie between 2693 and 4343.

Kraft and co-workers (82, 94, 95) consider that the change in viscosity of arsphenamine solutions on standing involves dehydration, which they express by the following equation:

$$m\text{HOAs(R)(AsR)}_{n-2}\text{As(R)OH} \rightleftharpoons$$

$$\text{HOAs(R)(AsR)}_{n-2}\text{As(R)[OAs(R)(AsR)}_{n-2}\text{As(R)]}_{m-2}\text{OAs(R)(AsR)}_{n-2}\text{As(R)OH}$$

$$+ \frac{m-2}{2} H_2O$$

They measured the viscosity of arsphenamine solutions prepared by several different procedures and calculated that arsphenamine can attain a molecular weight as large as 10^6 under certain conditions (95). No correlation was found between the viscosity of the arsphenamine solutions and the molecular weight obtained by iodine titration (23). Kraft offers no evidence for the polymerization by means of dehydration, and this portion of his work remains unproved.

The ultraviolet absorption of arseno compounds has not been systematically investigated. Eisenbrand (97) observed that arsenobenzene in chloroform has peaks at about 268 mμ (log $\varepsilon = 3.3$) and 340 mμ (log $\varepsilon = 2.2$), and noted that the spectrum of azobenzene in alcohol has maxima at longer wavelengths, viz., 318 mμ (log $\varepsilon = 4.3$) and 440 mμ (log $\varepsilon = 2.7$). The spectrum of arsphenamine in dilute aqueous sodium hydroxide was found to have a maximum at about 300 mμ (log $\varepsilon = 4.2$) and a shoulder at about 400 mμ (log $\varepsilon = 2.5$). Eisenbrand was unable to demonstrate any significant differences between the ultraviolet absorption of "toxic" and "nontoxic" samples of this drug. The light absorption of several other arsphenamine derivatives has also been reported (98).

More recently Badger, Drewer, and Lewis (99) have determined the ultraviolet absorption of arsenobenzene, *m*-arsenotoluene, and *p*-arsenotoluene in a mixture of cyclohexane and chloroform. The three spectra were found to be quite similar, each having a shoulder around 240–255 mμ (log ε, 4.80–5.05) and a second shoulder at 315 mμ (log ε, 4.10–4.15). Since the methyl groups appeared to have no significant effect on the 315 mμ band, this band was tentatively assigned to a transition of a non-bonding electron to an As—As d_π orbital.

The ultraviolet absorption spectrum of the arseno compound $(CF_3As)_4$ contains broad and approximately equally high peaks at 197 mμ and 224 mμ (ε, 3900) (55). Since these wavelengths are lower than the corresponding ones in $(CF_3P)_4$, it has been suggested that the delocalization of lone-pair electrons is more restricted in the arsenic compound. The vapor-molecular weight of the arseno compound was found to agree closely with the molecular formula.

The infrared spectra of several arseno compounds have been reported (55, 100, 101). Absorption bands at 310 and 295 cm^{-1} in the spectrum of $(CF_3As)_4$ have been assigned to As—CF_3 stretching vibrations (55).

The ^{19}F NMR spectrum of the arseno compound $(C_6F_5As)_4$ exhibits three peaks of relative intensity 2:1:2, which are centered at 125.9, 152.1, and 161.8 ppm, respectively (upfield relative to CCl_3F) (56, 57). The *ortho*-fluorine peak at 125.9 ppm shows evidence of As^{75} quadrupolar broadening. The molecular weight of the arseno compound was determined in chloroform solution with a vapor-pressure osmometer as well as by mass spectroscopy. The ^{19}F NMR spectra of the cyclic tetramer $(CF_3As)_4$ and pentamer $(CF_3As)_5$ and the proton NMR spectrum of arsenomethane $(CH_3As)_5$ have been recently investigated (58). (The molecular weight of $(CF_3As)_5$ was established by mass spectroscopy (102).) The NMR data suggest that the 5-membered rings have an effective plane of symmetry in the liquid phase.

The magnetic susceptibility of a variety of organoarsenic compounds has been studied by Prasad and Mulay (103, 104). The molar susceptibility of arsenic in arseno compounds was found to have an average value of -19.36×10^{-6} cgs units. This value does not appear to differ significantly from the average of -20.72×10^{-6} cgs units found for arsines. Prasad and Mulay also reported that there is a good linear relationship between the molar susceptibility of an arseno compound and the total number of electrons in the molecule. It is of interest that these workers believed that arseno compounds contain arsenic–arsenic double bonds.

4. PREPARATION OF DIARSINES

A. By reduction

Under certain conditions, the reduction of compounds with two organic groups attached to arsenic gives rise to tetraalkyl- or tetraaryldiarsines, $R_2As—AsR_2$. Hypophosphorous or phosphorous acid has been used for this reduction. Dimethylarsinic acid, when reduced with an excess of hypophosphorous acid in hydrochloric acid solution, gives tetramethyldiarsine (cacodyl) (105). Blicke and co-workers (106) have prepared tetrakis(3-aminophenyl)diarsine by reduction of bis(3-aminophenyl)arsinic acid with hypophosphorous acid and a trace of hydriodic acid. The reduction of bis(3-nitrophenyl)arsinous acid with hypophosphorous acid gives tetrakis(3-nitrophenyl)diarsine (106); reduction of bis(diphenylarsenic) oxide with phosphorous acid gave tetraphenyldiarsine (107). Diarsines of the type $RR'As—AsRR'$ are readily produced by reducing the appropriate unsymmetrical disubstituted compound. Thus, the reduction of (4-aminophenyl)(4-acetamidophenyl)arsinic acid with hypophosphorous acid gives 1,2-bis(4-aminophenyl)-1,2-bis(4-acetamidophenyl)diarsine (108). 1,2-Bis-(4-aminophenyl)-1,2-diphenyldiarsine is obtained by reducing the corresponding bis(diarylarsenic) oxide with hypophosphorous acid (109).

A number of reducing agents other than phosphorous and hypophosphorous acids have also been used for the preparation of diarsines. Thus, electrolytic reduction has been used for the conversion of dimethylarsinic acid (110) or bis(dimethylarsenic) oxide (111) to tetramethyldiarsine. According to Blicke and co-workers (112, 113), the method of choice for the preparation of aromatic diarsines is the reduction of diaryliodoarsines with mercury. Silver or zinc (but not magnesium) can be used in place of mercury, but the yields are inferior. Tetrakis(trifluoromethyl)diarsine has been prepared from bis(trifluoromethyl)iodoarsine and mercury in a sealed tube (114, 115). This diarsine can be obtained in 99% yield by shaking bis(trifluoromethyl)-chloroarsine with mercury for 24 hours at 20° (116). It is formed quantitatively by the reaction of bis[bis(trifluoromethyl)arsenic] sulfide with mercury for 21 days at 20° (117). Similarly, tetrakis(pentafluorophenyl)-diarsine has been prepared by the reaction of mercury with either bis-(pentafluorophenyl)chloroarsine or bis[bis(pentafluorophenyl)arsenic] sulfide (56, 57). It can also be obtained by the reduction of the chloroarsine with phosphine or trimethylsilane. Tetramethyldiarsine has been prepared by the reaction of dimethylchloroarsine with zinc dust at 100° (115).

B. By the Cadet reaction

The reaction between potassium acetate and arsenic trioxide gives cacodyl as one of the principal products. This reaction was discussed in some detail in Section III-1-F. Two modern patents give improved methods for the preparation of cacodyl by modifications of the Cadet method. In one patent a mixture of arsenic trioxide and acetic acid, both in the vapor state, is passed over a heated catalyst in a stream of nitrogen or carbon dioxide (111). The catalyst is an alkali metal salt or alkali metal hydroxide. The resulting solution of cacodyl oxide in acetic acid is then fed into an electrolytic cell where the cacodyl oxide is reduced to cacodyl in the cathode compartment. The second patent (118) is similar in many respects to the first patent. A mixture of arsenic trioxide and acetic acid, both in the vapor state, is passed over a heated catalyst. The catalyst is an alkali metal hydroxide or carbonate maintained at 300 to 400°. The reaction products are cacodyl and cacodyl oxide, but the ratio of the two products varies in different runs. By using propionic acid in place of acetic acid, tetraethyldiarsine and bis(diethyl-arsenic) oxide are obtained.

C. Miscellaneous methods

Blicke and his co-workers at the University of Michigan prepared aryl diarsines by a number of different methods during the course of their investigations of these compounds. Phenyldichloroarsine and excess diphenylarsine react to form tetraphenyldiarsine and arsenobenzene (49):

$$C_6H_5AsCl_2 + (C_6H_5)_2AsH \rightarrow (C_6H_5As)_6 + (C_6H_5)_2As{-}As(C_6H_5)_2 + HCl$$

Unless the diphenylarsine is in excess, the diarsine reacts further with phenyldichloroarsine to produce diphenylchloroarsine and arsenobenzene. Diphenylchloroarsine and diphenylarsine react in ether solution to yield tetraphenyldiarsine:

$$(C_6H_5)_2AsCl + (C_6H_5)_2AsH \rightarrow (C_6H_5)_2As{-}As(C_6H_5)_2 + HCl$$

Diphenyliodoarsine and diphenylarsine react similarly. Diphenylchloroarsine and phenylarsine react to give a mixture of arsenobenzene and tetraphenyldiarsine. The addition of one equivalent of iodine to two equivalents of diphenylarsine also gives tetraphenyldiarsine. With excess iodine, however, the diarsine reacts further to form diphenyliodoarsine (119). Triphenylarsenic hydroxychloride reacts with an excess of diphenylarsine in alcohol solution to form triphenylarsine and tetraphenyldiarsine:

$$4(C_6H_5)_2AsH + 2(C_6H_5)_3As(OH)Cl \rightarrow 2(C_6H_5)_3As + 2(C_6H_5)_2As{-}As(C_6H_5)_2 + 2HCl + 2H_2O$$

With excess hydroxychloride, however, the reaction proceeds further:

$$2(C_6H_5)_2As{-}As(C_6H_5)_2 + 2(C_6H_5)_3As(OH)Cl \rightarrow (C_6H_5)_2As{-}O{-}As(C_6H_5)_2 + 2(C_6H_5)_2AsCl + 2(C_6H_5)_3As + H_2O$$

Bis(diarylarsenic) oxides react with diarylarsines to give tetraaryldiarsines (109, 119). If the oxide and the arsine contain different substituents, a mixture of two different diarsines is produced rather than the mixed diarsine (50, 106):

$$[(3\text{-}NH_2C_6H_4)_2As]_2O + 2(C_6H_5)_2AsH \rightarrow [(3\text{-}NH_2C_6H_4)_2As]_2 + [(C_6H_5)_2As]_2 + H_2O$$

$$[(C_6H_5)_2As]_2O + 2(3\text{-}NH_2C_6H_4)_2AsH \rightarrow [(3\text{-}NH_2C_6H_4)_2As]_2 + [(C_6H_5)_2As]_2 + H_2O$$

Diphenylarsine and bis(diphenylarsenic) sulfide react to form tetraphenyldiarsine and hydrogen sulfide.

Blicke and Oneto (120) have prepared tetraphenyldiarsine by the reaction of $(C_6H_5)_2AsMgBr$ and a variety of different substances. Thus, this magnesium compound and mercuric chloride react as follows:

$$2(C_6H_5)_2AsMgBr + HgCl_2 \rightarrow [(C_6H_5)_2As]_2 + 2MgClBr + Hg$$

The diarsine was also obtained by reaction of the magnesium compound with iodine, with diphenylbromomethane, or with dichlorodiphenylmethane.

Several tetraalkyldiarsines have been prepared by the condensation of a chloroarsine with a secondary arsine. Thus, tetrakis(trifluoromethyl)diarsine is obtained by the reaction of bis(trifluoromethyl)chloroarsine and bis(trifluoromethyl)arsine at 100° (116). The interaction of phenyl(trifluoromethyl)chloroarsine and bis(trifluoromethyl)arsine at 100° yields a mixture of tetrakis(trifluoromethyl)diarsine, 1,2-diphenyl-1,2-bis(trifluoromethyl)diarsine, and a number of by-products (121). Cacodyl is formed in the condensation of dimethylarsine with dimethylchloroarsine or bis(trifluoromethyl)chloroarsine (116).

The reaction of the methyl radical (generated by heating tetramethyllead) with a cold mirror of metallic arsenic has been shown by Paneth (122) to give principally cacodyl together with traces of trimethylarsine. Tetraethyldiarsine and triethylarsine are similarly obtained from the ethyl radical. Issleib and Tzschach (123) have obtained tetracyclohexyldiarsine by the reaction of lithium dicyclohexylarsenide with iodine:

$$2(C_6H_{11})_2AsLi + I_2 \rightarrow [(C_6H_{11})_2As]_2 + 2LiI$$

The reaction of sodium diphenylarsenide and diphenylchloroarsine has been used for the preparation of tetraphenyldiarsine (124). This type of reaction has been investigated polarographically by Dessy and co-workers (125). Tetraphenyldiarsine has also been obtained by the reaction of potassium diphenylarsenide and 1,2-dibromoethane (126). The reaction between equimolar quantities of bis[bis(trifluoromethyl)arsenic] oxide and bis(trifluoromethyl)chlorophosphine has been found to yield tetrakis(trifluoromethyl)diarsine as well as bis(trifluoromethyl)chloroarsine and nonvolatile products (127). The preparation of diarsines by the reaction of alkyl halides with metal adducts of arsenobenzene has been described in Section VI-2.

5. REACTIONS OF DIARSINES

The diarsines are extremely reactive compounds. Cacodyl and tetraethyldiarsine inflame spontaneously in the air. The aromatic compounds are not spontaneously inflammable but are rapidly oxidized in the air and must be handled in an inert atmosphere (112). Borgstrom and Dewar (128) found that tetraphenyldiarsine reacts essentially quantitatively with moist air to give diphenylarsinic acid and bis(diphenylarsenic) oxide:

$$2(Ph_2As)_2 + 2O_2 + H_2O \rightarrow 2Ph_2AsO_2H + (Ph_2As)_2O$$

It was suggested that the initial step in the reaction is formation of a peroxide $Ph_2As—O—O—AsPh_2$, which is subsequently hydrolyzed to the acid and oxide. Tetrakis(trifluoromethyl)diarsine is stable to water, but when it is treated with aqueous sodium hydroxide at room temperature, it is hydrolyzed to fluoroform, fluoride ion, and carbonate ion (114). By contrast, tetraphenyldiarsine is not hydrolyzed in alkaline solution (129). It has been suggested by Emeléus and his co-workers (114, 130) that the hydrolysis of tetrakis(trifluoromethyl)diarsine involves hydrolytic cleavage of the As—As bond as the initial step of the reaction:

$$(CF_3)_2As—As(CF_3)_2 + H_2O \rightarrow (CF_3)_2AsH + (CF_3)_2AsOH$$

The arsinous acid is then converted to fluoroform, while the arsine is hydrolyzed to fluoroform, fluoride ion, and carbonate ion. Cullen and Emeléus (131) have shown that ammonolysis of tetrakis(trifluoromethyl)-diarsine follows a similar course, yielding 89% of the reaction product as fluoroform, presumably from the decomposition of $(CF_3)_2AsH$ and $(CF_3)_2AsNH_2$. Reaction of the diarsine with sodium methoxide in methanol produces 90% fluoroform (132).

Tetraphenyldiarsine reacts with ethanol to form diphenylarsine (133), and with methyl iodide to form dimethyldiphenylarsonium iodide (128). When heated with sulfur in carbon disulfide solution, tetraphenyldiarsine gives bis(diphenylarsenic) sulfide; with thionyl chloride or sulfur chloride (presumably S_2Cl_2), diphenylchloroarsine is formed (129). Both arsenic and phosphorus trichlorides react with the diarsine to yield diphenylchloroarsine. With arsenic trichloride, metallic arsenic is one of the products of the reaction. Phenylmagnesium bromide reacts with tetraphenyldiarsine to give triphenylarsine. In benzene solution tetraphenyldiarsine reacts with sodium-potassium alloy to form a yellow-brown solution, presumably containing $(C_6H_5)_2AsK$ or $(C_6H_5)_2AsNa$, since the addition of bromobenzene results in the formation of triphenylarsine. Cyanogen bromide reacts with the diarsine to form diphenylbromoarsine. The pyrolysis of tetraphenyldiarsine at 300° in an inert atmosphere results in the formation of triphenylarsine and metallic arsenic (133).

Tetramethyldiarsine reacts with trifluoroiodomethane to give dimethyl-(trifluoromethyl)arsine and dimethyliodoarsine (134). Tetrakis(trifluoromethyl)diarsine is also cleaved by trifluoroiodomethane to give tris-(trifluoromethyl)arsine and bis(trifluoromethyl)iodoarsine, but the rate is somewhat slower. Cleavage of this diarsine with methyl iodide to give methylbis(trifluoromethyl)arsine and bis(trifluoromethyl)iodoarsine is extremely slow.

Cullen and Hota (121) have found that cacodyl adds to certain fluorine-containing olefins and acetylenes. Thus, the diarsine reacts readily with hexafluoropropene or hexafluoro-2-butyne to form 1:1 adducts:

$$Me_2AsAsMe_2 + CF_3CF{=}CF_2 \rightarrow Me_2AsCF(CF_3)CF_2AsMe_2$$

$$Me_2AsAsMe_2 + CF_3C{\equiv}CCF_3 \rightarrow Me_2AsC(CF_3){=}C(CF_3)AsMe_2$$

Diarsines have been used for the preparation of coordination compounds of transition elements (124, 135–141). The reduction of diarsines to secondary arsines has been discussed in Section IV-1-A. The reaction of cacodyl with tetramethyldiphosphine disulfide will be mentioned in Section VI-7.

6. STRUCTURE AND PHYSICAL PROPERTIES OF DIARSINES

The possibility that tetraaryldiarsines might dissociate into free radicals has been considered by several investigators. Schlenk (142) in 1912 determined the molecular weight of tetraphenyldiarsine by ebullioscopic measurement in benzene and obtained a value of 462 for a calculated value of 458. By contrast Borgstrom and Dewar (128) obtained molecular weight values which indicated that tetraphenyldiarsine is associated and that the degree of association increases with time. Thus after 10–15 minutes, they obtained a value of 520 for the molecular weight by cryoscopic measurement in naphthalene, but after 303 minutes this value had increased to 766. These results, however, were not confirmed by Blicke and co-workers (113). In a series of determinations with several different diarsines in various solvents, molecular weights were obtained which indicated that no association of these compounds had occurred. In benzene, ethyl acetate, and naphthalene the molecular weights agree with the theoretical value for the monomer within experimental error. In biphenyl, however, four different diarsines give values which indicate slight dissociation, although the degree of dissociation never exceeds 10%. In spite of this rather nebulous evidence, the authors concluded that the extreme rapidity with which tetraaryldiarsines absorb an amount of oxygen corresponding to the formation of a peroxide (as well as their great reactivity with other substances) strongly suggests dissociation into Ar_2As radicals.

Tetrakis(trifluoromethyl)diarsine has been shown to have a normal molecular weight, and is presumably not dissociated to a significant extent (114). Allen and Sugden (143) found that 10,10′(5H,5′H)-diphenarsazine is diamagnetic and concluded that in the solid state no detectable quantities of free radicals are present.

The proton NMR spectrum of tetramethyldiarsine in deuterated chloroform at 37° has been found to consist of a single, rather broad line with a τ value of 8.89 (144). Tetrakis(pentafluorophenyl)diarsine has been separated into two isomers which have different ^{19}F NMR spectra and which are believed to be rotamers, possibly the *trans*- and *gauche*-forms (56, 57). Lambert and co-workers (145, 146) have observed two methyl peaks in the proton NMR spectrum of 1,2-dimethyl-1,2-diphenyldiarsine and have concluded that this compound exists in *meso* and *dl* modifications. The collapse of the two methyl peaks to a clean singlet above 200° was attributed to rapid interconversion of the two modifications.

Dessy and co-workers (147) have studied the electrolytic reduction of tetraphenyldiarsine and other compounds containing metal–metal bonds, and have concluded that the reduction potential parallels the strength of the metal–metal bond. It has also been reported that polarography can be used to determine the concentration of tetraphenyldiarsine or tetracyclohexyldiarsine in methanol (148).

Measurement of the heat of reaction of iodine with tetramethyldiarsine has led to an estimate of 38 kcal/mole for the dissociation energy of the As—As bond.

7. COMPOUNDS IN WHICH ARSENIC IS BONDED TO METALS OR METALLOIDS

During the period when arsphenamine was widely used in medicine and when arseno compounds were believed to contain arsenic–arsenic double bonds, a number of attempts were made to prepare compounds of the type RAs═M—R, where M is an element such as phosphorus, antimony, or bismuth. For example, Ehrlich and Karrer (150) in 1913 condensed 3-amino-4-hydroxyphenylarsine with phenyldichlorostibine and obtained an amorphous product to which structure **13** was assigned. Condensation of

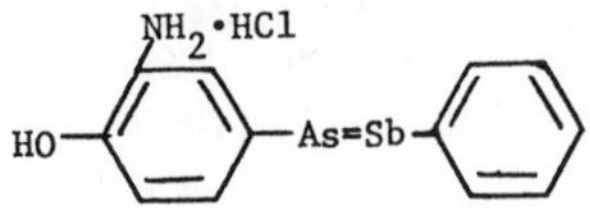

$NH_2 \cdot HCl$

HO–C_6H_3–As═Sb–C_6H_5

13

antimony trichloride or bismuth trichloride with 3-amino-4-hydroxyphenylarsine gives amorphous products, which were believed to contain As═Sb and As═Bi bonds. Many similar compounds have been described in the patent literature (151). In the light of modern work on the structure of the arseno compounds, the possibility of the existence of compounds in which arsenic is linked to metals such as antimony and bismuth by double bonds appears to be remote, and no modern work on such compounds has been attempted. Somewhat later Steinkopf and Dudek (51) condensed phenylarsine and phenyldichlorophosphine in ether solution. On removal of the ether they obtained a microcrystalline material, the analysis of which corresponded to the formula $(C_6H_5AsPC_6H_5)_n$. The material melted at 181°, but no molecular weight data or other physical constants were obtained which might help in determining the structure of the compound. An attempt to recrystallize the substance from benzene gave arsenobenzene and presumably phosphobenzene. The compound of Steinkopf and Dudek may possibly exist as a ring structure similar to arsenobenzene.

The first unequivocal preparation of compounds containing As—P bonds was achieved by Kamai and co-workers (152, 153). By the reaction of triethyl phosphite with ethylbutyliodoarsine, they obtained $(C_2H_5)(C_4H_9)AsP(O)(OC_2H_5)_2$ which with ethyl iodide gives the phosphonium salt, $[(C_2H_5)(C_4H_9)AsP(OC_2H_5)_3]I$. Other compounds were prepared from sodium diethyl phosphite, $(C_2H_5O)_2P(O)Na$, and haloarsines. Thus, from butylethyliodoarsine, $(C_2H_5)(C_4H_9)AsP(O)(OC_2H_5)_2$ is obtained. From diphenylchloroarsine, $(C_6H_5)_2AsP(O)(OC_2H_5)_2$ is obtained. Esters of this type have also been described in the patent literature (154). Kamai and co-workers were unable to obtain the phosphonic acids by acid hydrolysis of the esters; cleavage of the arsenic–phosphorus bond occurs when the compounds are heated with 15% hydrochloric acid.

Phosphonium salts of the type $[R_2AsPR_3']I$ have been prepared by the reaction of dialkylhaloarsines and tertiary phosphines (155, 156). From dimethylchloroarsine and triethylphosphine, $[(CH_3)_2AsP(C_2H_5)_3]Cl$ is obtained as a hygroscopic solid, readily crystallized from acetone. Although triphenylphosphine does not react with dimethylchloroarsine, dimethylphenylphosphine gives $[(CH_3)_2AsP(C_6H_5)(CH_3)_2]Cl$. The corresponding phosphonium iodides have also been prepared.

Schumann and co-workers (157) have recently described a number of compounds which contain several As—P bonds. These substances were prepared by the reaction of diphenylchlorophosphine or diphenylchloroarsine with trimethylstannyl-substituted phosphines or arsines. Thus, the chlorophosphine or chloroarsine reacts with tris(trimethylstannyl)phosphine or -arsine in the following manner:

$$(Me_3Sn)_3M + Ph_2M'Cl \rightarrow (Ph_2M')_3M + 3Me_3SnCl$$

where M and M′ = As or P. Several compounds were also obtained by the reaction of bis(trimethylstannyl)phenylphosphine or -arsine:

$$(Me_2Sn)_2MPh + Ph_2M'Cl \rightarrow (Ph_2M')_2MPh + 2Me_3SnCl$$

Harris and Hayter (144) have reported that cacodyl reacts with tetramethyldiphosphine disulfide in a sealed tube at 220°:

$$Me_2AsAsMe_2 + Me_2P(S)P(S)Me_2 \rightarrow 2Me_2AsP(S)Me_2$$

The resulting substance was not purified or analyzed, but the proton NMR spectrum appears to be consistent with the assumed structure.

A number of compounds have been described in which arsenic is bonded to silicon, germanium, tin, or lead. The first compound of this type was an arsonium iodide $[Me_3AsSiH_3]I$, prepared by the reaction of trimethylarsine and iodosilane at low temperatures (158). This substance is a solid, mp

8.1–9.6°, which in the vapor phase above 5° is dissociated into trimethylarsine and iodosilane. More recently, Russ and MacDiarmid (159) prepared dimethyl(trimethylsilyl)arsine by the reaction between lithium dimethylarsenide and trimethylchlorosilane:

$$LiAsDe_2 + Me_3SiCl \rightarrow Me_3SiAsMe_2 + LiCl$$

This compound reacts readily with water to form dimethylarsine and hexamethyldisiloxane:

$$2Me_3SiAsMe_2 + H_2O \rightarrow Me_3SiOSiMe_3 + 2Me_2AsH$$

With alcohols it forms dimethylarsine and alkoxytrimethylsilanes. At −96° it adds hydrogen bromide to give an arsonium compound, which decomposes above −50°:

$$[Me_3SiAsMe_2H]Br \rightarrow Me_3SiBr + Me_2AsH$$

The silylarsine is cleaved by methyl iodide to yield trimethyliodosilane, trimethylarsine, and tetramethylarsonium iodide. Boron trifluoride reacts to give a compound containing As—B bonds:

$$Me_3SiAsMe_2 + BF_3 \rightarrow Me_3SiF + Me_2AsB(F)AsMe_2$$

The silylarsine undergoes an insertion reaction with carbon disulfide:

$$Me_3SiAsMe_2 + CS_2 \longrightarrow Me_3SiC(=S)AsMe_2$$

Phosphorus pentafluoride reacts with the silylarsine to form a cacodyl complex:

$$2Me_3SiAsMe_2 + 2PF_5 \rightarrow PF_3 + 2Me_3SiF + Me_2AsAsMe_2{\cdot}PF_5$$

Maier (160) has reported the preparation of several silylarsines by the reaction of triphenylsilyllithium with methyldibromoarsine, dimethylbromoarsine, or diethylbromoarsine, but the chemical properties of these compounds were not described.

Compounds containing As—Ge, As—Sn, or As—Pb bonds have been prepared by the reaction of sodium diphenylarsenide with organogermanium (161), organotin (162, 163), or organolead (161) halides. Thus, this type of reaction was used by Campbell and co-workers (162) to prepare three compounds of the general structure Ph_2AsSnR_3, where R is ethyl, *n*-propyl, or *n*-butyl. In a similar manner, Schumann and co-workers (161, 163) obtained diphenyl(triphenylgermyl)arsine, diphenyl(triphenylplumbyl)arsine, and a variety of compounds containing As—Sn bonds. Schumann and co-workers (163) also prepared stannylarsines by the interaction of triphenylstannyllithium and haloarsines. Jones and Lappert (164) have described the synthesis of diphenyl(trimethylstannyl)arsine by means of the following reaction:

$$Me_2NSnMe_3 + Ph_2AsH \rightarrow Ph_2AsSnMe_3 + Me_2NH$$

The proton NMR spectrum of dimethyl(trimethylstannyl)arsine has been described (165), but the method used for preparing this compound has not yet been reported.

The compounds in which arsenic is bonded to germanium, tin, or lead are readily oxidized (161–163). For example, diphenyl(trialkylstannyl)arsines react with atmospheric oxygen to form esters (162):

$$Ph_2AsSnR_3 + O_2 \rightarrow Ph_2AsO_2SnR_3$$

Esters of this type can also be prepared by the reaction of diphenylarsinic acid with trialkyltin oxides or hydroxides (162) or with triphenylchlorogermane (161), triphenyltin chloride (163), or triphenyllead chloride (161).

Compounds containing the As—B linkage were first prepared by Stone and Burg (166). They allowed methyl-, dimethyl-, and trimethylarsines to react with diborane at $-78.5°$. With methylarsine the very unstable compound $CH_3AsH_2 \cdot BH_3$ is formed, which dissociates even at $-78.5°$ and is thus far less stable than the corresponding phosphorus compound. Dimethylarsine forms the compound $(CH_3)_2AsH \cdot BH_3$, more stable than the product from methylarsine, but again much less stable than the corresponding phosphorus compound. The compound $(CH_3)_3As \cdot BH_3$ is a white solid, mp 73.5–74.5°.

In addition to the above monomeric compounds, polymeric compounds containing As—B bonds were prepared by Stone and Burg. Thus, methylarsine and diborane in a sealed tube for 13 days at room temperature form a polymeric material with the approximate composition CH_3AsBH_3. From dimethylarsine and diborane at 50°, three products were obtained: $[(CH_3)_2AsBH_2]_3$, $[(CH_3)_2AsBH_2]_4$, and a material of higher molecular weight. The three products were separated by fractional sublimation. More recently, Lane and Burg (167) have prepared a new arsinoboron compound, $[(CF_3)_2AsBH_2]_3$, by the reaction of bis(trifluoromethyl)arsine with diborane. The infrared spectrum in the vapor phase exhibited a relatively weak peak at 332 cm^{-1} which was assigned to CF_3—As stretching. Although the structure of this substance was not unequivocally established, it was suggested that it consists of a hexatomic ring with alternating arsenic and boron atoms.

Coates and Livingstone (168) have studied compounds of the type Ar_2BMR_2, where M is nitrogen, phosphorus, or arsenic. Dipole moment and infrared studies of the aminodiarylboranes indicate that these compounds possess the structure $Ar_2B{=}NR_2$ (where R is either aliphatic or aromatic) in which the nitrogen acts as a strong electron donor to the boron. By contrast, in both the phosphorus and the arsenic compounds, the phosphino or arsino group is the negative end of the dipole. The authors suggested that in compounds of the type Ar_2BPAr_2 and Ar_2BAsAr_2 there is considerable contribution from resonance forms of the following type:

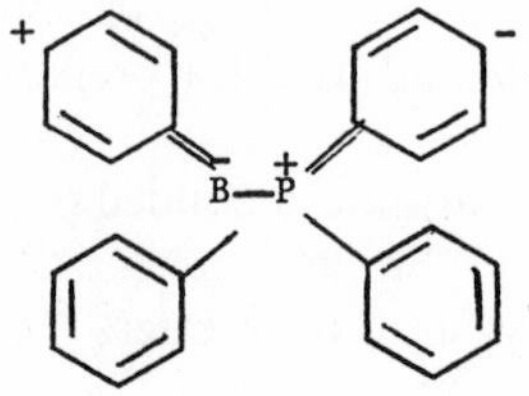

The compound $(C_6H_5)_3As{\cdot}BH_3$ has been prepared by the reaction between diborane and triphenylarsine in ether solution in a sealed tube (169). It is much less stable than the corresponding triphenylphosphine-BH_3 adduct, and when dissolved in organic solvents (such as ether, petroleum ether, ethanol, acetone, or dioxane) it rapidly decomposes with the formation of diborane.

Coates and Graham (170) have investigated the reaction of diphenylarsine with trimethylgallium, trimethylaluminum, and trimethylindium. In each case methane was evolved with the formation of dimers of the type $(Me_2M{-}AsPh_2)_2$. Similar compounds were formed when diphenylamine and diphenylphosphine were used. In contrast to these results, the reactions of dimethylarsine with trimethylaluminum, -gallium, and -indium yield compounds of the type $(Me_2M{-}AsMe_2)_x$, which are polymeric glasses in the condensed phase at room temperature but are cyclic trimers in benzene (171). The proton NMR spectra of these compounds have only two very sharp peaks of equal intensity, which correspond to the methyl groups on the arsenic atom and on the Group III metal atom. These observations are consistent with the rapid inversion between the two chair forms of a 6-membered ring; i.e., the axial and equatorial methyl groups become magnetically equivalent. At $-90°$ the ring inversion is considerably slower, and a broadened resonance is observed. Methyl- and phenylarsine were found to react with the trimethyl derivatives of aluminum, gallium, and indium to yield only nonvolatile polymers of the type $(MeM{-}AsR)_x$.

Triphenylarsine reacts with aluminum chloride to give a crystalline addition complex, $(C_6H_5)_3As{\cdot}AlCl_3$ (172). This complex, when dissolved in a nonpolar solvent, reacts with air to give triphenylarsine oxide and reacts with bromobenzene to give tetraphenylarsonium bromide. Nyholm and Ulm (173) have described the preparation and properties of a series of di-tertiary arsine complexes of gallium and indium halides.

Issleib and Tzschach (123) have isolated $(C_6H_5)_2AsK{\cdot}(dioxane)_2$, $(C_6H_5)_2$-$AsNa{\cdot}$dioxane, and $(C_6H_5)_2AsLi{\cdot}$dioxane by reaction of triphenylarsine and the corresponding metals in dioxane solution. Other examples of the cleavage of triarylarsines with alkali metals are given in Section IV-5-B.

Sodium (123) and lithium (123, 174) diaryl- and alkylarylarsenides have also been prepared by the treatment of secondary arsines with metallic sodium or phenyllithium. Similarly, dilithium arylarsenides have been prepared by interaction of primary arsines and phenyllithium (175). Metallic arsenides of the type C_6H_5AsHNa and C_6H_5AsHK have been prepared from phenylarsine and the corresponding metals in liquid ammonia (176). The structure of the alkali metal derivatives of primary and secondary arsines, as well as that of the Grignard-like compounds $C_6H_5As(MgBr)_2$ (120) and $(C_6H_5)_2AsMgBr$ (52), has not been investigated. Their usefulness for the preparation of primary and secondary arsines and of tertiary arsines is discussed in Sections IV-1-A and IV-4-C, respectively.

REFERENCES

1. F. F. Blicke and F. D. Smith, *J. Amer. Chem. Soc.*, **52,** 2937 (1930).
2. A. Michaelis, *Justus Liebigs Ann. Chem.*, **320,** 271 (1902).
3. A. Étienne, *C. R. Acad. Sci., Paris*, **221,** 628 (1945).
4. A. Binz, H. Bauer, and A. Hallstein, *Ber.*, **53,** 416 (1920).
5. A. Michaelis and C. Schulte, *Ber.*, **15,** 1952 (1882).
6. V. Auger, *C. R. Acad. Sci., Paris*, **138,** 1705 (1904).
7. W. Steinkopf and H. Dudek, *Ber.*, **61,** 1906 (1928).
8. F. F. Blicke and F. D. Smith, *J. Amer. Chem. Soc.*, **52,** 2946 (1930).
9. P. Ehrlich and A. Bertheim, *Ber.*, **44,** 1260 (1916).
10. M. Ya. Kraft, E. B. Agracheva, and E. N. Sytina, *Dokl. Akad. Nauk SSSR*, **99,** 259 (1954).
11. K. Uematsu and I. Nakaya (to Sankyo Co.), Jap. Pat. 1279 (Mar. 26, 1953); through *C.A.*, **48,** 12802 (1954).
12. P. A. Kober, *J. Amer. Chem. Soc.*, **41,** 442 (1919).
13. P. Ehrlich and A. Bertheim, *Ber.*, **40,** 3292 (1907).
14. W. G. Christiansen, *J. Amer. Chem. Soc.*, **44,** 847 (1922).
15. F. Linsker and M. T. Bogert, *J. Amer. Chem. Soc.*, **65,** 932 (1943).
16. H. King, *J. Chem. Soc.*, **119,** 1107 (1921).
17. M. Ya. Kraft and I. A. Batshchouk, *Dokl. Akad. Nauk SSSR*, **55,** 723 (1947); *C.A.*, **42,** 3742 (1948).
18. D. Vagenberg, *Zh. Obshch. Khim.*, **7,** 808 (1937); through *C.A.*, **31,** 5777 (1931).
19. R. G. Fargher and F. L. Pyman, *J. Chem. Soc.*, **117,** 370 (1920).
20. K. Nakada, *Collect. Lectures Inst. Chem. Res.* (Japan), **1,** 94 (1929); through *C.A.*, **23,** 4411 (1929).
21. K. Matsuyima and H. Nakata, *Mem. Coll. Sci. Kyoto Imp. Univ.*, **A, 12,** 63 (1929); through *C.A.*, **23,** 4939 (1929).
22. S. V. Vasil'ev and G. D. Vovchenko, *Vestn. Mosk. Univ.*, **5,** No. 3, *Ser. Fiz. Mat. i Estestv. Nauk*, No. **2,** 73 (1950); through *C.A.*, **45,** 6594 (1951).
23. M. Ya. Kraft, O. I. Korzina, and A. S. Morozova, *Sb. Statei Obshch. Khim.*, **2,** 1356 (1953); through *C.A.*, **49,** 5347 (1955).
24. G. W. Raiziss and M. Falkov, *J. Biol. Chem.*, **46,** 209 (1921).

25. C. Voegtlin and J. M. Johnson, *J. Amer. Chem. Soc.*, **44**, 2573 (1922).
26. W. J. C. Dyke and H. King, *J. Chem. Soc.*, 1003 (1933).
27. W. J. C. Dyke and H. King, *J. Chem. Soc.*, 1745 (1935).
28. G. Newbery and M. A. Phillips, *J. Chem. Soc.*, 116 (1928).
29. C. S. Palmer and E. B. Kester, *J. Amer. Chem. Soc.*, **50**, 3109 (1928).
30. M. C. Hart and W. B. Payne, *J. Amer. Pharm. Ass.*, **12**, 688 (1923).
31. M. C. Hart and W. B. Payne, *J. Amer. Pharm. Ass.*, **12**, 759 (1923).
32. M. Ya. Kraft and V. V. Katyshkina, *Dokl. Akad. Nauk SSSR*, **66**, 393 (1949); through *C.A.*, **44**, 128 (1950).
33. R. H. Edee, *J. Amer. Chem. Soc.*, **50**, 1394 (1928).
34. W. Steinkopf, S. Schmidt, and P. Smie, *Ber.*, **59**, 1463 (1926).
35. Farbwerke vorm. Meister Lucius & Brüning, Ger. Pat. 253,226 (Oct. 10, 1911); through *C.A.*, **7**, 683 (1913).
36. C. S. Palmer and R. H. Edee, *J. Amer. Chem. Soc.*, **49**, 998 (1927).
37. P. Karrer, *Ber.*, **49**, 1648 (1916).
38. G. Newbery and May and Baker, Ltd., Brit. Pat., 269,647 (Jan. 19, 1926); through *C.A.*, **22**, 1367 (1928).
39. H. Lieb, *Ber.*, **54**, 1511 (1921).
40. H. Lieb and O. Wintersteiner, *Ber.*, **56**, 425 (1923).
41. H. Lieb and O. Wintersteiner, *Ber.*, **56**, 1283 (1923).
42. P. Karrer, *Ber.*, **48**, 305 (1915).
43. C. Finzi and E. Bartelli, *Gazz. Chim. Ital.*, **62**, 545 (1932).
44. W. M. Dehn, *Amer. Chem. J.*, **33**, 101 (1905).
45. C. S. Palmer and R. Adams, *J. Amer. Chem. Soc.*, **44**, 1356 (1922).
46. A. N. Nesmejanow and R. Ch. Freidlina, *Ber.*, **67**, 735 (1934).
47. L. Anschutz and H. Wirth, *Naturwissenschaften*, **43**, 59 (1956).
48. Farbwerke vorm. Meister Lucius & Brüning, Ger. Pat. 254,187 (Feb. 4, 1911); through *Chem. Zentr.*, **84**, I, 134 (1913).
49. F. F. Blicke and L. D. Powers, *J. Amer. Chem. Soc.*, **54**, 3353 (1932).
50. F. F. Blicke and J. F. Oneto, *J. Amer. Chem. Soc.*, **56**, 685 (1934).
51. W. Steinkopf and H. Dudek, *Ber.*, **62**, 2494 (1929).
52. A. Job, R. Reich, and P. Vergnaud, *Bull. Soc. Chim. Fr.*, **35**, 1404 (1924).
53. M. H. Beeby, G. H. Cookson, and F. G. Mann, *J. Chem. Soc.*, 1917 (1950).
54. D. R. Lyon and F. G. Mann, *J. Chem. Soc.*, 30 (1945).
55. A. H. Cowley, A. B. Burg, and W. R. Cullen, *J. Amer. Chem. Soc.*, **88**, 3178 (1966).
56. M. Green and D. Kirkpatrick, *Chem. Commun.*, 57 (1967).
57. M. Green and D. Kirkpatrick, *J. Chem. Soc.*, **A**, 483 (1968).
58. E. J. Wells, R. C. Ferguson, J. G. Hallett, and L. K. Peterson, *Can. J. Chem.*, **46**, 2733 (1968).
59. M. Ya. Kraft and I. A. Batshchouk, *Dokl. Akad. Nauk SSSR*, **55**, 419 (1947); *C.A.*, **42**, 576 (1948).
60. S. Krishna and R. Krishna, *J. Indian Chem. Soc.*, **6**, 665 (1929).
61. H. Eagle, *J. Pharmacol. Exp. Ther.*, **66**, 423 (1939).
62. K. Kondo, *Nippon Yakurigaku Zasshi*, **27**, 235 (1939); through *C.A.*, **34**, 819 (1940).
63. A. Binz and C. Räth, *Justus Liebigs Ann. Chem.*, **455**, 127 (1927).
64. G. Newbery and M. A. Phillips, *J. Chem. Soc.*, 2375 (1928).
65. S. Orlić, *Arhiv. Hem. i Tehnol.*, **12**, 153 (1938); through *C.A.*, **34**, 6407 (1940).
66. P. Karrer, *Ber.*, **48**, 1058 (1915).
67. A. Cohen, H. King, and W. I. Strangeways, *J. Chem. Soc.*, 3043 (1931).
68. R. M. Kary (to American Smelting and Refining Co.), U.S. Pat. 2,646,440 (July 21, 1953); *C.A.*, **48**, 7049 (1954).

69. R. M. Kary, in *Metal-Organic Compounds*, Advances in Chemistry Series, No. 23, American Chemical Society, Washington, D.C., 1959, pp. 319–327.
70. T. Riedeburg, *Agr. Chem.*, **7,** No. 4, 52 (1952).
71. J. D. Early and J. H. Cochran, *J. Econ. Entomol.*, **49,** 239 (1956).
72. G. Wittig, M. A. Jesaitis, and M. Glos, *Justus Liebigs Ann. Chem.*, **577,** 1 (1952).
73. J. W. B. Reesor and G. F. Wright, *J. Org. Chem.*, **22,** 382 (1957).
74. H. G. Ang and B. O. West, *Aust. J. Chem.*, **20,** 1133 (1967).
75. G. W. A. Fowles and D. K. Jenkins, *Chem. Commun.*, 61 (1965).
76. K. S. Pitzer, *J. Amer. Chem. Soc.*, **70,** 2140 (1948).
77. R. S. Mulliken, *J. Amer. Chem. Soc.*, **72,** 4493 (1950).
78. A. Michaelis and A. Schäfer, *Ber.*, **46,** 1742 (1913).
79. C. S. Palmer and A. B. Scott, *J. Amer. Chem. Soc.*, **50,** 536 (1928).
80. F. F. Blicke and L. D. Powers, *J. Amer. Chem. Soc.*, **55,** 315 (1933).
81. S. E. Rasmussen and J. Danielsen, *Acta Chem. Scand.*, **14,** 1862 (1960).
82. M. Ya. Kraft, G. M. Borodina, I. N. Strel'tsova, and Yu. T. Struchkov, *Dokl. Akad. Nauk SSSR*, **131,** 1074 (1960).
83. K. Hedberg, E. W. Hughes, and J. Waser, *Acta Crystallogr.*, **14,** 369 (1961).
84. R. J. W. Le Fèvre and C. A. Parker, *J. Chem. Soc.*, 677 (1939).
85. J. Waser and V. Schomaker, *J. Amer. Chem. Soc.*, **67,** 2014 (1945).
86. J. H. Burns and J. Waser, *J. Amer. Chem. Soc.*, **79,** 859 (1957).
87. L. Pauling, *The Nature of the Chemical Bond*, 3rd ed., Cornell University Press, Ithaca, 1960, p. 225.
88. J. Donohue, *Acta Crystallogr.*, **15,** 708 (1962).
89. M. Z. Klemensiewicz, *Bull. Soc. Chim. Fr.*, **27,** 820 (1920).
90. H. N. Wright, A. Biedermann, E. Hanssen, and C. I. Cooper, *J. Pharmacol. Exp. Ther.*, **73,** 12 (1941).
91. M. Ya. Kraft and I. A. Bashchuk, *Dokl. Akad. Nauk SSSR*, **65,** 509 (1949); through *C.A.*, **45,** 2890 (1951).
92. M. Ya. Kraft and V. V. Katyshkina, *Dokl. Akad. Nauk SSSR*, **66,** 207 (1949); through *C.A.*, **44,** 127 (1950).
93. M. Ya. Kraft, O. P. Al'bitskaya, and A. S. Morozova, *Sb. Statei Obshch. Khim.*, **2,** 1360 (1953); through *C.A.*, **49,** 5347 (1955).
94. M. Ya. Kraft and E. B. Agracheva, *Dokl. Akad. Nauk SSSR*, **100,** 279 (1955); through *C.A.*, **50,** 1644 (1956).
95. M. Ya. Kraft and E. N. Sytina, *Dokl. Akad. Nauk SSSR*, **116,** 89 (1957).
96. M. Ya. Kraft, *Dokl. Akad. Nauk SSSR*, **131,** 1342 (1960).
97. J. Eisenbrand, *Arch. Pharm.*, **269,** 683 (1931).
98. K. Brand and E. Rosenkranz, *Pharm. Zentralh. Deut.*, **79,** 489 (1938).
99. G. M. Badger, R. J. Drewer, and G. E. Lewis, *Aust. J. Chem.*, **16,** 285 (1963).
100. G. P. Sollott and W. R. Peterson, Jr., *J. Org. Chem.*, **30,** 389 (1965).
101. T. J. Bardos, N. Datta-Gupta, and P. Hebborn, *J. Med. Chem.*, **9,** 221 (1966).
102. R. C. Dobbie and R. G. Cavell, *Inorg. Chem.*, **6,** 1450 (1967).
103. M. Prasad and L. N. Mulay, *J. Chem. Phys.*, **19,** 1051 (1951).
104. M. Prasad and L. N. Mulay, *J. Chem. Phys.*, **20,** 201 (1952).
105. V. Auger, *C.R. Acad. Sci., Paris*, **142,** 1151 (1906).
106. F. F. Blicke, U. O. Oakdale, and J. F. Oneto, *J. Amer. Chem. Soc.*, **56,** 141 (1934).
107. W. Schlenk and G. Racky, *Justus Liebigs Ann. Chem.*, **394,** 216 (1912).
108. H. Bauer, *J. Amer. Chem. Soc.*, **67,** 591 (1945).
109. F. F. Blicke and G. L. Webster, *J. Amer. Chem. Soc.*, **59,** 537 (1937).
110. F. Fichter and E. Elkind, *Ber.*, **49,** 239 (1916).

111. B. Witten (to the United States of America), U.S. Pat. 2,531,487 (Nov. 28, 1950); *C.A.*, **45**, 2799 (1951).
112. F. F. Blicke and F. D. Smith, *J. Amer. Chem. Soc.*, **51**, 2272 (1929).
113. F. F. Blicke, O. J. Weinkauff, and G. W. Hargreaves, *J. Amer. Chem. Soc.*, **52**, 780 (1930).
114. G. R. A. Brandt, H. J. Emeléus, and R. N. Haszeldine, *J. Chem. Soc.*, 2552 (1952).
115. W. R. Cullen, *Can. J. Chem.*, **38**, 439 (1960).
116. W. R. Cullen, *Can. J. Chem.*, **41**, 322 (1963).
117. W. R. Cullen, *Can. J. Chem.*, **41**, 2424 (1963).
118. R. C. Fuson and W. Shive (to the United States of America), U.S. Pat. 2,756,245 (July 14, 1956); *C.A.*, **51**, 2020 (1957).
119. F. F. Blicke and L. D. Powers, *J. Amer. Chem. Soc.*, **55**, 1161 (1933).
120. F. F. Blicke and J. F. Oneto, *J. Amer. Chem. Soc.*, **57**, 749 (1935).
121. W. R. Cullen and N. K. Hota, *Can. J. Chem.*, **42**, 1123 (1964).
122. F. A. Paneth and H. Loleit, *J. Chem. Soc.*, 366 (1935).
123. K. Issleib and A. Tzschach, *Angew. Chem.*, **73**, 26 (1961).
124. J. Chatt and D. A. Thornton, *J. Chem. Soc.*, 1005 (1964).
125. R. E. Dessy, T. Chivers, and W. Kitching, *J. Amer. Chem. Soc.*, **88**, 467 (1966).
126. A. Tzschach and W. Lange, *Chem. Ber.*, **95**, 1360 (1962).
127. J. Singh and A. B. Burg, *J. Amer. Chem. Soc.*, **88**, 718 (1966).
128. P. Borgstrom and M. M. Dewar, *J. Amer. Chem. Soc.*, **44**, 2915 (1922).
129. F. F. Blicke, R. A. Patelski, and L. D. Powers, *J. Amer. Chem. Soc.*, **55**, 1158 (1933).
130. H. J. Emeléus, R. N. Haszeldine, and E. G. Walaschewski, *J. Chem. Soc.*, 1552 (1953).
131. W. R. Cullen and H. J. Emeléus, *J. Chem. Soc.*, 372 (1959).
132. W. R. Cullen, *Can. J. Chem.*, **40**, 575 (1962).
133. C. W. Porter and P. Borgstrom, *J. Amer. Chem. Soc.*, **41**, 2048 (1919).
134. W. R. Cullen, *Can. J. Chem.*, **39**, 2486 (1961).
135. R. G. Hayter, *J. Amer. Chem. Soc.*, **85**, 3120 (1963).
136. R. G. Hayter, *Inorg. Chem.*, **2**, 1031 (1963).
137. R. G. Hayter, *J. Amer. Chem. Soc.*, **86**, 823 (1964).
138. R. G. Hayter and L. F. Williams, *Inorg. Chem.*, **3**, 717 (1964).
139. R. G. Hayter and L. F. Williams, *Inorg. Chem.*, **3**, 613 (1964).
140. R. G. Hayter, *Inorg. Chem.*, **3**, 711 (1964).
141. W. R. Cullen and R. G. Hayter, *J. Amer. Chem. Soc.*, **86**, 1030 (1964).
142. W. Schlenk, *Justus Liebigs Ann. Chem.*, **394**, 178 (1912).
143. F. L. Allen and S. Sugden, *J. Chem. Soc.*, 440 (1936).
144. R. K. Harris and R. G. Hayter, *Can. J. Chem.*, **42**, 2282 (1964).
145. J. B. Lambert and G. F. Jackson, III, *J. Amer. Chem. Soc.*, **90**, 1350 (1968).
146. J. B. Lambert, G. F. Jackson, III, and D. C. Mueller, *J. Amer. Chem. Soc.*, **90**, 6401 (1968).
147. R. E. Dessy, P. M. Weissman, and R. L. Pohl, *J. Amer. Chem. Soc.*, **88**, 5117 (1966).
148. H. Matschiner and A. Tzschach, *Z. Chem.*, **5**, 144 (1965).
149. C. T. Mortimer and H. A. Skinner, *J. Chem. Soc.*, 4331 (1952).
150. P. Ehrlich and P. Karrer, *Ber.*, **46**, 3564 (1913).
151. G. W. Raiziss and J. L. Gavron, *Organic Arsenical Compounds*, The Chemical Catalog Company, New York, 1923, pp. 177–180.
152. G. Kamai and O. N. Belorossova, *Izv. Akad. Nauk SSSR, Otd. Khim. Nauk*, 191 (1947); through *C.A.*, **42**, 4133 (1948).
153. G. Kamai and E. M. Sh. Bastanov, *Dokl. Akad. Nauk SSSR*, **89**, 693 (1953); through *C.A.*, **48**, 6374 (1954).

154. S. J. Strycker (to Dow Chemical Co.), U.S. Pat. 3,108,129 (Oct. 22, 1963); through *C.A.*, **60,** 1776 (1964).
155. G. E. Coates and J. G. Livingstone, *Chem. Ind.* (London), 1366 (1958).
156. J. M. F. Braddock and G. E. Coates, *J. Chem. Soc.*, 3208 (1961).
157. H. Schumann, A. Roth, and O. Stelzer, *Angew. Chem.*, **80,** 240 (1968).
158. B. J. Aylett, H. J. Emeléus, and A. G. Maddock, *J. Inorg. Nucl. Chem.*, **1,** 187 (1955).
159. C. R. Russ and A. G. MacDiarmid, *Angew. Chem., Int. Ed. Engl.*, **5,** 418 (1966).
160. L. Maier, *Helv. Chim. Acta*, **46,** 2667 (1963).
161. H. Schumann and M. Schmidt, *Inorg. Nucl. Chem. Lett.*, **1,** 1 (1965).
162. I. G. M. Campbell, G. W. A. Fowles, and L. A. Nixon, *J. Chem. Soc.*, 3026 (1964).
163. H. Schumann, T. Östermann, and M. Schmidt, *Chem. Ber.*, **99,** 2057 (1966).
164. K. Jones and M. F. Lappert, *Proc. Chem. Soc.*, 22 (1964).
165. E. W. Abel and D. B. Brady, *J. Organometal. Chem.*, **11,** 145 (1968).
166. F. G. A. Stone and A. B. Burg, *J. Amer. Chem. Soc.*, **76,** 386 (1954).
167. A. P. Lane and A. B. Burg, *J. Amer. Chem. Soc.*, **89,** 1040 (1967).
168. G. E. Coates and J. G. Livingstone, *J. Chem. Soc.*, 1000 (1961).
169. M. Becke-Goehring and H. Thielemann, *Z. Anorg. Allg. Chem.*, **308,** 33 (1961).
170. G. E. Coates and J. Graham, *J. Chem. Soc.*, 233 (1963).
171. O. T. Beachley and G. E. Coates, *J. Chem. Soc.*, 3241 (1965).
172. D. R. Lyon and F. G. Mann, *J. Chem. Soc.*, 666 (1942).
173. R. S. Nyholm and K. Ulm, *J. Chem. Soc.*, 4199 (1965).
174. A. Tzschach and W. Lange, *Z. Anorg. Allg. Chem.*, **330,** 317 (1964).
175. A. Tzschach and G. Pacholke, *Chem. Ber.*, **97,** 419 (1964).
176. W. C. Johnson and A. Pechukas, *J. Amer. Chem. Soc.*, **59,** 2068 (1935).

VII Pentavalent organoantimony compounds

1. STIBONIC AND STIBINIC ACIDS

A. Aliphatic

There have been no confirmed reports of the preparation of any aliphatic stibonic acid in the chemical literature. Some chemical reactions of both 1-propane- and 1-butanestibonic acids are mentioned in a patent (1), but the source of the acids is not given. Schmidt (2) briefly mentioned some of the properties of ethanestibonic acid in a paper appearing in 1920. He stated that the compound would be described in a later paper, but this work apparently has never been published. All attempts to prepare aliphatic stibonic acids under the conditions of the Michaelis-Arbuzov reaction have failed. Thus, no compound containing a carbon–antimony bond is formed when triethoxystibine is heated with either methyl or ethyl iodide (3). Mann (4) has stated that the Meyer reaction apparently fails with antimony, but his statement is not documented. A number of attempts to prepare methanestibonic acid under the conditions of the Meyer reaction (refluxing an alkyl iodide with antimony trioxide in alkaline solution) have met with failure in the laboratory of the present authors (5). Since hydrated antimony oxide is a very weak acid, it dissolves only in concentrated alkali solutions in which alkyl iodides are quickly hydrolyzed. An attempt was also made to prepare aliphatic stibonic acids by the hydrolysis of alkylantimony tetrachlorides, which in turn were prepared by chlorination of alkyldichlorostibines. Although materials containing appreciable amounts of alkyl groups (as judged by infrared spectra and elementary analyses) were obtained, no analytically pure aliphatic stibonic acid was ever isolated.

Several dialkylstibinic acids (now listed as dialkylhydroxystibine oxides by *Chemical Abstracts*) are known. Dimethylstibinic acid was first prepared by

Morgan and Davies (6) by the hydrolysis of dimethylantimony trichloride. The same acid was later obtained by Paneth and Loleit (7) in the course of experiments designed to demonstrate the existence of alkyl free radicals. From a mirror of metallic antimony that was allowed to react with methyl radicals formed by the pyrolysis of tetramethyllead, they obtained trimethylstibine and tetramethyldistibine. The latter compound, on oxidation in benzene solution, gave dimethylstibinic acid. Diethylstibinic acid was obtained in a similar manner. Bis(1-chloroethyl)stibinic acid has been obtained by the oxidation and hydrolysis of bis(1-chloroethyl)chlorostibine with aqueous hydrogen peroxide (8, 9).

B. Aromatic

(*1*) *Preparation*

In contrast to the paucity of aliphatic stibonic and stibinic acids, a large number of aromatic stibonic and stibinic acids have been prepared. Aromatic stibonic acids can be prepared under the conditions of the Bart reaction (cf. Section II-1-A); this procedure for preparing stibonic acids was first described in a patent in 1911 (10) and hence antedates the similar preparation of aromatic arsonic acids. The description given in the patent was amplified in 1920, when Schmidt described several different procedures for preparing aromatic stibonic acids and diarylstibinic acids (2). Probably the best technique utilized a diazonium tetrachloroantimonate(III). This type of compound (often referred to as a May double salt) precipitates almost quantitatively when a solution of antimony trichloride in hydrochloric acid is added to a diazonium chloride solution. When this precipitated salt is then added to aqueous alkali, nitrogen is evolved. Subsequent acidification of the solution leads to precipitation of the crude stibonic acid. This procedure was used by most of the earlier workers in this field, and a large number of aromatic stibonic acids were reported (11–14). There are many difficulties in the procedure as outlined. Large amounts of stiff foam are produced, which necessitates the use of large reaction vessels. The evolution of nitrogen often takes several days, and heating usually markedly lowers the yield. The use of glycerol or some other polyhydric alcohol is said to be helpful (15, 16), but this modification has not been widely accepted.

The difficulties encountered in alkaline solution have led to the introduction of a variety of methods in which the reaction is carried out either in aqueous acids or in an organic solvent. One such method has been described by Campbell (17) in which the precipitated diazonium tetrachloroantimonate(III) is first washed with alcohol and then suspended in this same solvent; loss of nitrogen is effected by the addition of a cuprous salt. When the resulting alcoholic solution is poured into water, the aromatic stibonic acid precipitates. The reaction can be carried out without isolating the diazonium tetrachloroantimonate(III) by diazotizing the amine in alcohol in the presence of antimony trichloride and subsequently adding a cuprous salt (18–20). In still further modifications, the amine is diazotized either in aqueous sulfuric (21) or in hydrochloric acid (22). Antimony trichloride is then added, and the evolution of nitrogen is effected by addition of a cuprous salt (21). Another method involves the use of dry diazonium tetrafluoroborates, antimony trichloride, and an organic solvent; and again a cuprous salt is used to effect the decomposition (23). In this last reaction the aromatic stibonic acids may be contaminated with diarylstibinic acids, which are difficult to separate from the stibonic acids. In spite of this difficulty *p*-sulfonamidobenzenestibonic acid (apparently in a pure state) has been prepared by this method (24).

Tomono and co-workers (25) have compared the yields of three different stibonic acids prepared by several of these modifications. The best yields were obtained by the decomposition of diazonium tetrachloroantimonates in acetone. Ordinary Bart reaction conditions gave lower yields, while the use of diazonium tetrafluoroborates in alkaline solution was even less satisfactory. The poorest results were obtained when the diazonium tetrafluoroborates were decomposed in acetone or methanol. With the three acids studied the order of increasing yield was *p*-nitrobenzene-, benzene-, *p*-toluenestibonic acid. This order, however, has not always been found by other workers using different reaction conditions.

The crude stibonic acids prepared by any of the above methods employing diazonium salts are usually contaminated with inorganic antimony compounds, although it has been claimed that pure stibonic acids can be obtained if only 60 to 70% of the stoichiometric amount of antimony trioxide is used (26). Because aromatic stibonic acids are generally not very soluble either in water or in organic solvents, it is usually difficult to effect purification by conventional crystallization techniques. *p*-Toluenestibonic (16) and *p*-biphenylstibonic (27) acids, however, have been recrystallized from alcohol. Doak and Steinman (28) recrystallized benzene-, *o*-bromobenzene-, and the three isomeric toluenestibonic acids from glacial acetic acid; the recrystallized acids were heated to 100° over sodium hydroxide to remove acetic acid from the crystals. A few workers (29–31) have claimed the preparation of analytically pure stibonic acids simply by reprecipitation from alkaline solution.

Because of the difficulties in purifying stibonic acids by more conventional techniques, it has become almost universal practice to effect purification through the use of quaternary ammonium salts of the type $[R_4N][ArSbCl_5]$. These compounds are readily formed by dissolving the crude stibonic acid in concentrated hydrochloric acid and adding either ammonium chloride or an amine hydrochloride, whereupon the antimony-containing salt precipitates from solution. The product is washed with hydrochloric acid and then hydrolyzed by water or alkali to give the analytically pure stibonic acid. When pyridine hydrochloride is used, the resulting pentachloroantimonate can be readily recrystallized from a mixture of methanol (or ethanol) and hydrochloric acid, and the product can be obtained in a high state of purity. The melting points of these pyridinium salts are sharp and reproducible and are considerably lower than those of the ammonium salts. Thus, the pyridinium salts are useful as derivatives for the characterization of stibonic acids, since the latter do not usually give satisfactory melting points. Quinolinium salts (13, 20, 32) have also been used, but these generally melt higher than pyridinium salts.

When the stibonic acid possesses an amino group, the arylpentachloroantimonate obtained has a zwitterion structure. Thus, Pfeiffer and Böttcher (33) obtained the compound $p\text{-}(CH_3)_3\overset{+}{N}C_6H_4SbCl_5^-$ by diazotizing $[p\text{-}(CH_3)_3NC_6H_4NH_2]Cl$ in the presence of antimony trichloride. The corresponding *meta* derivative has also been prepared (32). Addition of perchloric acid to the stibonic acids leads to the formation of the perchlorates $[p\text{-}(CH_3)_3NC_6H_4SbO_3H_2]ClO_4$ and $[m\text{-}(CH_3)_3NC_6H_4SbO_3H_2]ClO_4$.

The reaction between an inorganic trivalent antimony compound and a diazonium salt has been used with a wide variety of amines, and a large number of aromatic stibonic acids have been reported in the literature. *p*-Aminobenzenestibonic acid (stibanilic acid), which in the form of derivatives has found considerable medicinal use, has been prepared from *p*-aminoacetanilide (11, 29). From *p*-phenylenediamine, 1,4-benzenedistibonic acid has been obtained (30), while benzidine yields 4,4′-biphenyldistibonic acid (12). Both α-naphthalene- (13) and β-naphthalenestibonic acids (13, 34) have been prepared. Quinoline-5-, quinoline-6-, and quinoline-8-stibonic acids have been described (35); β-aminopyridine (22), 2-aminothiazole (15), 3-aminoacridone (36), and 2-aminopyrimidine (37) have also been used in preparing stibonic acids. *m*-Trifluoromethylbenzenestibonic acid was obtained in 19% yield by O'Donnell (14), who also prepared a number of interesting heterocyclic stibonic acids.

The reaction is capable of giving yields as high as 70–80%, although often somewhat lower yields are obtained, particularly where *ortho*-substituted amines are used. Nakai (38) has claimed that N-acetylbenzidine fails to yield a stibonic acid. Mistry and Guha (39) claim to have obtained *stiboso* compounds rather than stibonic acids when some diamines were used. Thus, from

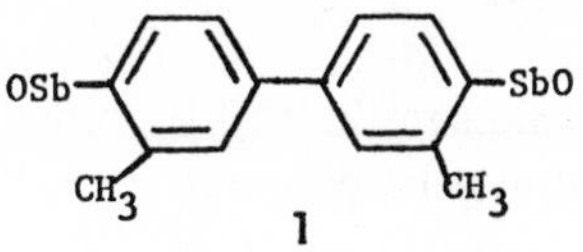

OSb— —SbO and OSb— —SbO
CH_3 CH_3 OCH_3 OCH_3
1 2

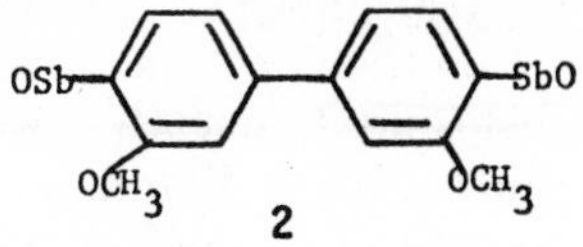

o-tolidine and *o*-dianisidine they claim to have obtained **1** and **2**, respectively. They also prepared the interesting compounds $P(-C_6H_4-SbO_3H_2)_3$, $As(-C_6H_4-SbO_3H_2)_2$, and $Sb(-C_6H_4-SbO_3H_2)_3$ from the tris(*p*-aminophenyl)phosphine, -arsine, and -stibine.

The mechanism of the reaction between diazonium salts and inorganic trivalent antimony compounds, either in aqueous alkaline solution or in an organic solvent, probably does not differ significantly from the similar reaction of trivalent arsenic compounds already described (cf. II-1-A). With arsenic compounds, however, the diazonium salts $[ArN_2][AsCl_4]$ are seldom isolated (but, cf. Bruker (40)). Tomono and co-workers in a series of papers (25, 41–45) have reported on the preparation of aromatic stibonic acids by the diazo reaction under various reaction conditions. Their results may be summarized as follows: when a powdered diazonium tetrachloroantimonate(III) is added to aqueous alkali at 0°, nitrogen is evolved and the stibonic acid can be readily isolated. In the absence of alkali only phenols and substituted biphenyls are obtained. In a wide variety of organic solvents no stibonic acid is obtained in the absence of a catalyst, and the diazonium salt is converted to a hydrocarbon or chlorinated hydrocarbon. Thus, benzenediazonium tetrachloroantimonate(III), when warmed in acetone, gives 19% benzene, 50% chlorobenzene, and 10% biphenyl. Under the same reaction conditions, except that cuprous chloride was added, a 60% yield of benzenestibonic acid, together with 10% benzene and a trace of biphenyl, is obtained. Qualitatively similar results are obtained in methanol, acetic acid, or acetic anhydride. In ether (with a catalyst) or in benzene (without a catalyst) chlorobenzene, biphenyl, and a trace of benzene are the only products.

Reutov and Ptitsyna (46) have investigated the reaction between arylazoformates and antimony trichloride in ethyl acetate solution. Derivatives of both arylstibonic and diarylstibinic acids may be obtained. For example, *p*-$CH_3C_6H_4N_2CO_2K$ gives an 11.5% yield of *p*-$CH_3C_6H_4SbOCl_2$ and a 44% yield of *p*-$CH_3C_6H_4SbO$, while $C_6H_5N_2CO_2K$ gives only derivatives of $(C_6H_5)_2SbO_2H$. Nesmeyanov and co-workers (47) have developed a method for preparing arylstibonic acids from diazonium hexachloroantimonate(V) salts. The reaction has been formulated as follows:

$$[ArN_2][SbCl_6] + 2CuCl \rightarrow ArSbCl_4 + N_2 + 2CuCl_2$$

Since the yields are frequently lower than those obtained by other methods and since the necessary diazonium salts cannot be prepared directly, in only a few cases does this method offer any advantages over the customary procedures.

Bruker and co-workers have developed methods for preparing both aromatic stibonic and diarylstibinic acids from phenylhydrazines, antimony trichloride, and cuprous chloride. Phenylhydrazine gives a 20–22% yield of benzenestibonic acid and a 15–18% yield of diphenylstibinic acid (48); *p*-tolylhydrazine gives principally *p*-toluenestibonic acid, while *β*-naphthylhydrazine gives only *β*-naphthalenestibonic acid (49). A mixture of *p*-fluorobenzenestibonic and bis(*p*-fluorophenyl)stibinic acids is obtained from *p*-fluorophenylhydrazine (50). The method involves the air oxidation of the arylhydrazine hydrochloride to a diazonium chloride and thus differs from the conventional procedure only in the use of an arylhydrazine as the source of the diazonium cation.

Before the discovery of the diazonium method for preparing stibonic acids, a few such compounds had been prepared by the hydrolysis of arylantimony tetrahalides, which in turn had been prepared by the halogenation of aryldihalostibines. As an example of this procedure, which is now seldom used, might be mentioned the preparation of *p*-biphenylstibonic (51) and *o*-biphenylstibonic acids (52) by the hydrolysis of the corresponding biphenylylantimony tetrachlorides. In neither case, however, were the tetrachlorides isolated and characterized. More recently, Yakubovich and Motsarev (53) have prepared benzenestibonic acid in 55% yield by treatment of diphenyldichlorosilane with antimony pentachloride at $-20°$ and subsequent hydrolysis of the reaction mixture.

The preparation of diarylstibinic acids is also usually accomplished by means of the diazo reaction. Frequently small amounts of diarylstibinic acids are found as by-products in the preparation of arylstibonic acids, and their presence often makes it difficult to obtain the stibonic acids in an analytically pure state (23). Voigt (54) found that the decomposition of an arenediazonium tetrachloroantimonate(III) in acetone or in aqueous acetone invariably gives some of the diaryl and triaryl derivative in addition to the desired stibonic acid.

The preparation of diarylstibinic acids, where the two aryl groups may be the same or different, can be accomplished by the reaction between a diazonium salt and a trivalent monoarylantimony compound. In his original publication on arylstibonic and diarylstibinic acids, Schmidt (2) described the preparation of both diphenylstibinic and (*p*-aminophenyl)phenylstibinic acid from benzenediazonium chloride and C_6H_5SbO or *p*-$H_2NC_6H_4SbO$, respectively. Diphenylstibinic acid has also been prepared from phenyldichlorostibine and benzenediazonium tetrachlorozincate in alcohol solution (19). The total number of diarylstibinic acids that have been prepared directly in this manner is somewhat limited; nor have studies been made to determine the best reaction conditions. A number of diarylstibinic acids have been obtained by the hydrolysis of diarylantimony trichlorides, which in turn are prepared by a number of methods including various modifications of the

Nesmeyanov reaction. Both symmetrical and unsymmetrical diarylstibinic acids have been obtained in this manner. Since the reaction

$$Ar_2SbO_2H + 3HCl \rightleftharpoons Ar_2SbCl_3 + 2H_2O$$

is freely reversible and since the arylantimony trichlorides are readily purified by recrystallization whereas the diarylstibinic acids are generally amorphous compounds with high molecular weights, the acids are usually characterized by conversion to the diarylantimony trichlorides. The latter class of compounds is described in Section VII-4-A.

Several methods not involving the use of diazonium salts have been developed for preparing diarylstibinic acids. The procedure of Bruker in which arylhydrazines are used has already been mentioned (48, 50). Worrall (52) has prepared di-*p*-biphenylylstibinic acid by hydrolysis of the corresponding trichloride; the latter was prepared by chlorination of di-*p*-biphenylylchlorostibine. Another method of potential value involves the simultaneous oxidation and dearylation of a triarylstibine. This method was originally discovered by Schmidt (55), who prepared diphenylstibinic and bis(*p*-aminophenyl)stibinic acids from triphenylstibine and tris(*p*-aminophenyl)stibine, respectively. For this purpose the stibine is treated in dilute alkali with 3% hydrogen peroxide. Woods (56) also obtained diphenylstibinic acid by this procedure, while Goddard and Yarsley (57) prepared di-*p*-tolylstibinic acid in small yield from tri-*p*-tolylstibine. In contrast to these results Yusunov and Manulkin (58) have stated that several unsymmetrical tertiary stibines of the type $(C_6H_5)_2ArSb$ give the corresponding triarylantimony dihydroxides on oxidation with 4% hydrogen peroxide in aqueous alkaline solution. Thus, (*p*-ethoxyphenyl)diphenylstibine and (*p*-bromophenyl)diphenylstibine give the corresponding triarylantimony dihydroxides, and (*p*-allylphenyl)diphenylstibine gives (*p*-carboxyphenyl)diphenylantimony dihydroxide. On the other hand, these same authors state that diphenyl(*o*-tolyl)stibine, (*p*-allylphenyl)diphenylstibine, and (*p*-cyclohexylphenyl)diphenylstibine decompose on standing in the air for one month to give in each case diphenylstibinic acid.

The preparation of heterocyclic stibinic acids by cyclodehydration of suitably substituted aromatic stibonic acids has been achieved by Morgan and Davies (27, 59). Thus *o*-biphenylstibonic acid, when warmed with concentrated sulfuric acid, gives 5-hydroxy-5H-dibenzostibole 5-oxide (**3**).

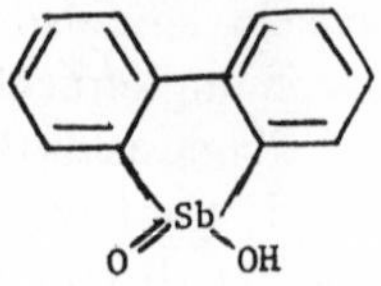

Sb
O OH

3

o-Benzylbenzenestibonic acid does not cyclize when warmed with sulfuric acid but cyclizes when warmed with acetic anhydride and a trace of sulfuric acid to yield 5-hydroxy-5,10-dihydrodibenz[*b*,*e*]antimonin 5-oxide (**4**) (59).

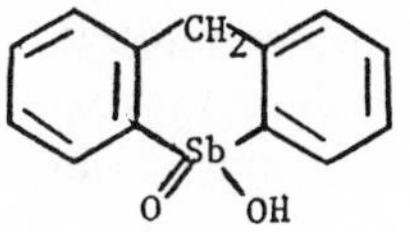

4

(*2*) *Structure*

Aromatic stibonic acids and both diaryl- and dialkylstibinic acids are entirely different in their physical properties from the corresponding arsenic and phosphorous acids. When precipitated from alkaline solution by the addition of excess mineral acid, the stibonic and stibinic acids are obtained as fine flocculent precipitates which only slowly settle. These precipitates are difficult to separate by filtration. When washed with water they readily peptize and pass through the filter. They are insoluble in water and insoluble or only slightly soluble in most organic solvents, and relatively few have been obtained in crystalline form. They generally do not possess definite melting points but decompose on heating. All of these properties suggest that they are polymeric in nature with possibly quite large molecular weights. In view of these properties and in the light of further experimental data described below, Schmidt (2, 60) proposed that these acids exist in the solid state as trimers with one of two structures, **5** or **6**:

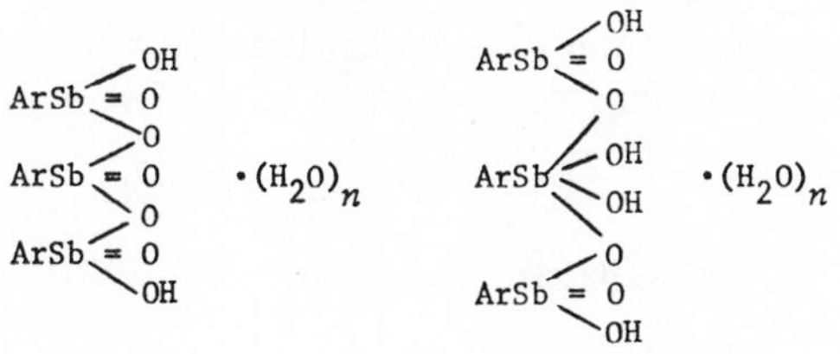

5 **6**

He further postulated that on treatment with alkali the trimeric form is slowly converted to the sodium salt of the monomer, $ArSb(O)(OH)_2$. The amount of associated water (*n* in formulas **5** and **6**) varies with the method of preparation. Thus, benzenestibonic acid is obtained in two different forms which were written as $[3(C_6H_5SbO_2)H_2O]\cdot2H_2O$ and $[3(C_6H_5SbO_2)H_2O]\cdot3H_2O$. These two formulations are supported both by analytical data and by a study of the weight losses obtained on drying *in vacuo* over sulfuric acid and on heating at 110°, 130°, and 150°. Schmidt (2) also found that when electron-attracting groups (*p*-Cl, *p*-NO_2, and *m*-NO_2) are attached to the benzene ring, the weight losses are in accord only with formula **6**. Schmidt's suggestion that arylstibonic acids depolymerize slowly on treatment with alkali was necessitated by the observation that arylstibonic acids dissolve in $\frac{1}{3}$ of an equivalent of aqueous sodium hydroxide to form a solution which is at first neutral to phenolphthalein, but which becomes acid on standing and which eventually requires one equivalent of alkali before a solution which is permanently neutral to phenolphthalein is obtained.

Schmidt's concept of the structure of aromatic stibonic acids was supported by evidence obtained by Fargher and Gray (61) and by Gray and Lamb (62), who advanced as further evidence for these structures their finding that sodium salts could be isolated in which the ratio of Na:Sb was considerably less than 1:1 and approached in some cases 1:3. These sodium salts were reported to give neutral solutions in water. On the other hand Macallum (63), on the basis of molecular weight data in phenol, rejected Schmidt's concept and suggested that the aromatic stibonic acids are monomeric.

The problem was subsequently investigated by Doak (64) and by several Japanese workers (65, 66) who confirmed Schmidt's findings with regard to the behavior of arylstibonic acids as pseudo acids and also the behavior of these compounds on drying. Doak found, however, that the apparent molecular weights of the three isomeric toluenestibonic acids, as determined cryoscopically in benzene, exceed the values expected for a trimer, and further that the molecular weight values vary linearly with concentration. By contrast, molecular weight values for a number of aromatic stibonic acids in formic and acetic acid are either in agreement with those expected for a monomer or give values somewhat lower than those expected. This latter result was interpreted as due to dissociation of the following type:

$$ArSbO_3H_2 \rightleftharpoons ArSbO_2 + H_2O$$

$$ArSbO_3H_2 \cdot H_2O \rightleftharpoons ArSbO_2 + 2H_2O$$

Doak also isolated several sodium salts in crystalline form by the addition of alcohol to a solution of the stibonic acid in aqueous alkali and showed that the analytical values obtained for both antimony and sodium are in agreement with the formula $[ArSb(OH)_5]Na$. On the basis of these results he concluded that arylstibonic acids are associated in the solid state by hydrogen bonding; when dissolved in alkaline solution they dissociate to give sodium salts of the monomeric acid $[ArSb(OH)_5]H$.

The Japanese workers (65, 66) suggested that aromatic stibonic acids precipitate from alkaline solution on acidification as $[ArSb(OH)_5]H$, and that on drying water molecules are lost both by an intra- and intermolecular process as follows:

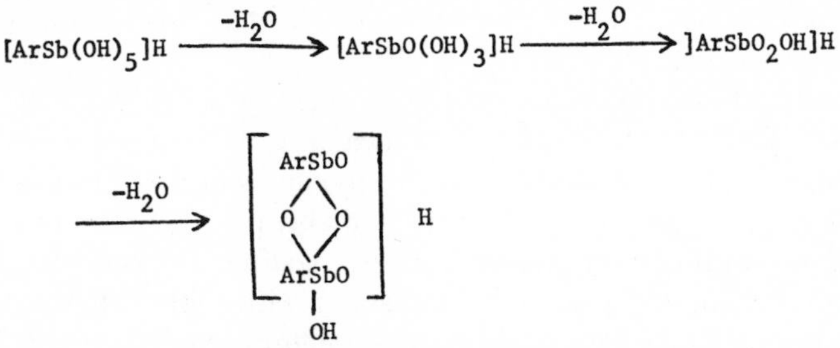

$$[ArSb(OH)_5]H \xrightarrow{-H_2O} [ArSbO(OH)_3]H \xrightarrow{-H_2O}]ArSbO_2OH]H$$

$$\xrightarrow{-H_2O} \left[\begin{array}{c} ArSbO \\ O \quad O \\ ArSbO \\ | \\ OH \end{array}\right] H$$

There is little or no evidence for the proposed structures.

The above mentioned work on the structure of stibonic and stibinic acids has recently been brought into doubt by studies on various compounds which had been believed to contain the Sb=O group. On the basis of infrared spectra (67–69) and X-ray diffraction data (70), a number of workers have questioned the existence of the Sb=O grouping in such compounds as $(C_6H_5)_3SbO$ and SbOCl. Prompted by these findings Madamba (71) examined the infrared spectra of a variety of arylstibonic and both diaryl- and dialkylstibinic acids, in Nujol mulls and in KBr pellets, and failed to find any bands in the region from 1000 to 300 cm^{-1} that could be assigned unequivocally to the Sb=O grouping. In the light of these findings and the results obtained with antimonic acid in aqueous solution at various pH values (discussed in the following section), it can only be concluded that aromatic stibonic acids exist as polymeric species of unknown structure.

(*3*) *Formation of esters, salts, anhydrides, and oxychlorides*

No esters of organic stibonic or stibinic acids are known with certainty. Nakai and co-workers (72) have reported a compound formed by the condensation of benzenestibonic acid with tartaric acid; the compound was formulated as an ester with the following structure:

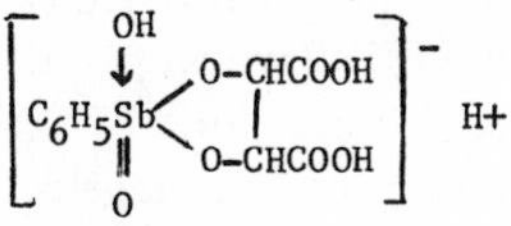

$$\left[\begin{array}{c} OH \\ \downarrow \\ C_6H_5Sb \\ \| \\ O \end{array} \begin{array}{l} O-CHCOOH \\ \quad | \\ O-CHCOOH \end{array}\right]^- H^+$$

They also obtained a sodium salt of this compound by the use of sodium acid tartrate rather than tartaric acid. It is interesting to note that Gate and Richardson (73) have prepared a number of compounds by the reaction between inorganic antimonic acid and polyhydric phenols or α-hydroxy acids. The structures they suggest for several of these are similar to those suggested by Nakai. With tartaric acid, however, a 1:2 tartaric acid–antimonic acid complex is obtained. There is little evidence that any of the proposed structures are correct.

Many salts of stibonic acids have been reported. The work of Fargher and Gray (61) and of Gray and Lamb (62) has already been mentioned. These authors dissolved *m*-acetamidobenzenestibonic acid in aqueous sodium hydroxide to give a neutral solution (presumably pH 7). When this solution was saturated with carbon dioxide and then sodium chloride was added, a sodium salt precipitated which, after extraction with methanol, gave a residue for which the analytical values (after correction for occluded sodium chloride) indicated a Na:Sb ratio of 1:2.9. The *p*-acetamidobenzenestibonic acid when similarly treated gave a sodium salt with a Na:Sb ratio of 1:2.8. Sodium salts of *p*-bromobenzenestibonic acid, *p*-ethoxybenzenestibonic acid, and benzenestibonic acid were also obtained by evaporating to dryness "neutral" solutions of these acids in sodium hydroxide solution; the Na:Sb ratio in these cases was 1:1.35, 1:1.27, and 1:2.06, respectively. With *p*-aminobenzenestibonic acid the salt that precipitated from alkaline solution gave analytical values in agreement with the formula $H_2NC_6H_4SbO_3HNa{\cdot}2H_2O$. This salt could be recrystallized from aqueous acetone and did not lose water below 160°, at which temperature decomposition occurred. This compound probably possesses the structure $[H_2NC_6H_4Sb(OH)_5]^-Na^+$.

The structures of the alkali metal salts of arylstibonic and diaryl- and dialkylstibinic acids probably are closely related to the structure of the salts of antimonic acid, which have been investigated by numerous workers. The structure $[Sb(OH)_6]^-$ for the antimonate ion was originally proposed by Pauling (74) on theoretical grounds and has since been clearly established not only by X-ray diffraction studies (75, 76) but by a number of other physical measurements (77–79). When aqueous solutions of sodium or potassium antimonate are progressively acidified, polymerization occurs. The nature of this polymerization has been the subject of several investigations. Ricca and co-workers (80) have claimed that aggregation occurs to first form $[HSb_2O_6]^-$, followed by formation of $[HSb_6O_{17}]^{3-}$ at lower pH values. Souchay and Peschanski (78) also arrived at the structure $[HSb_6O_{17}]^{3-}$. Recently Gate and Richardson (81) have prepared a dilute aqueous solution of the polymerized acid free from metal ions by passing an aqueous solution of the potassium salt through an ion exchange resin. The resulting solution was very sensitive to the presence of electrolytes as shown by the turbidity produced by the addition of KCl. Conductometric and potentiometric titrations of the acid solutions with 0.1*N* NaOH, both alone and in the presence of mannitol, and ultraviolet absorption measurements at various dilutions suggest that various polymeric species are present in solution. For a 0.01*M* solution the average formula was judged to be $(H_4Sb_6O_{17})_n$, but a 0.001*M* solution or a 0.01*M* solution in the presence of mannitol was approximately $H_n[Sb(OH)_6]_n$.

In the light of these results with antimonic acid, it seems probable that the benzenestibonate anion possesses the structure $[C_6H_5Sb(OH)_5]^-$, and that progressive neutralization of the aqueous solutions of this anion produces salts of various polymerized acids until eventually a polymerized arylstibonic acid is precipitated from solution.

In addition to alkali metal salts, various other salts of arylstibonic acids have been reported. Niyogy (82) has prepared a number of amine salts of *p*-aminobenzenestibonic acid, which were formulated as trimers with a ratio of amine : Sb of 1 : 3. Various heavy metal salts of arylstibonic acids have been reported (36, 83), but their constitution is unknown. O'Donnell (14) has reported the preparation of a thorium salt of benzenestibonic acid. No preparations of salts of dialkyl- or diarylstibinic acids have been reported.

Although aromatic stibonic acids lose water on heating, in only a few cases have oxides corresponding to the empirical formula $ArSbO_2$ been isolated and characterized. Although Schmidt (2) determined the weight loss of benzenestibonic acid when heated to varying temperatures, in no case did the results agree with the theoretical weight loss for the formation of $C_6H_5SbO_2$. Nakai and co-workers (65) heated benzenestibonic acid to 100° and obtained a weight loss corresponding to the formation of $(C_6H_5SbO_2)_2$-H_2O. In contrast to these results, Doak (64) obtained weight loss and analytical values for antimony in agreement with the formula $ArSbO_2$, when benzenestibonic and *o*-toluenestibonic acids were heated to 125° for one week. With other stibonic acids, however, the weight losses at 125° and the analytical values indicate that decomposition occurs, presumably with rupture of the carbon–antimony bond. Diphenylstibinic acid, when heated to 125–130°, loses the theoretical weight consistent with formation of $[(C_6H_5)_2SbO]_2O$ (2). A similar type of compound, $[(C_6H_5CH_2)_2SbO]_2O$, has been reported to be formed by the hydrolysis of dibenzylantimony trichloride in aqueous sodium carbonate solution (84).

Diphenylstibinic acid reacts with acetic acid to form the mixed anhydride $(C_6H_5)_2Sb(O)OCOCH_3$ (58). A similar anhydride $(o\text{-}ClC_6H_4)_2Sb(O)OCOCH_3$ is obtained when bis(*o*-chlorophenyl)stibinic acid is treated with aqueous acetic acid (85). The exact structure of these compounds is not known.

The existence of acid halides derived from stibonic or stibinic acids is uncertain. Reutov and Ptitsyna (46) have described the isolation of $p\text{-}CH_3C_6H_4SbOCl_2$ from the reaction of potassium *p*-tolylazoformate and antimony trichloride, but their formulation is based only on antimony analyses. Reutov and Kondratyeva (86) have described the preparation of

$(p\text{-}IC_6H_4)_2SbOCl$ from the corresponding diazonium hexachloroantimonate, but this formulation is also based only on antimony analyses.

(4) *Other reactions*

Aromatic stibonic and diarylstibinic acids react reversibly with hydrochloric acid to give arylantimony tetrachlorides and diarylantimony trichlorides, respectively:

$$ArSbO_3H_2 + 4HCl \rightleftharpoons ArSbCl_4 + 3H_2O$$

$$Ar_2SbO_2H + 3HCl \rightleftharpoons Ar_2SbCl_3 + 2H_2O$$

Since the arylantimony tetrachlorides are unstable, hygroscopic, often low-melting solids (27), they are seldom isolated. On the other hand the diarylantimony trichlorides are more stable; and, since they usually give satisfactory melting points, they are frequently used to characterize diarylstibinic acids (86–88). The reaction of both aliphatic and aromatic stibonic acids with sulfur tetrafluoride to give either the tetrafluorides, $RSbF_4$, or the difluorostibine oxides, $RSbOF_2$, has been claimed in a patent (1). Since aliphatic stibonic acids are otherwise unknown and their source is not given in the patent, the validity of these claims is difficult to judge. Kawasaki and Okawara (89) have reported that the interaction of benzenestibonic acid and acetylacetone in concentrated hydrochloric acid yields the following hexacoordinate derivative of antimony:

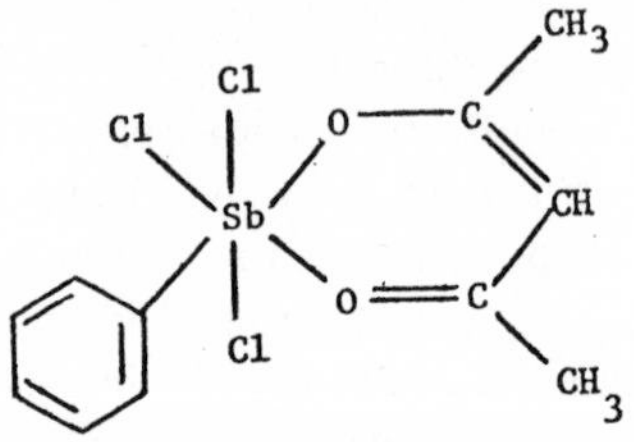

Aromatic stibonic and diarylstibinic acids are readily reduced by sulfur dioxide and hydriodic acid. The reaction is usually carried out in hydrochloric acid solution, and hence is actually the reduction of the arylantimony tetrachloride or diarylantimony trichloride, and the products are the aryldichlorostibine (13) or the diarylchlorostibine (48). The reduction of aromatic stibonic acids can also be carried out very effectively with stannous chloride in hydrochloric acid solution (17, 90). In contrast to arsonic acids, the reduction of aromatic stibonic acids or diarylstibinic acids to the corresponding stibines, $ArSbH_2$ or Ar_2SbH, has not been reported; the reduction of stibonic acids to compounds containing the Sb—Sb bond has been reported but is questionable (cf. VIII-5).

Many reactions of aromatic stibonic acids or diarylstibinic acids involving the aryl group have been carried out. However, the carbon–antimony bond is relatively weak, and under vigorous reaction conditions this bond may be split. For this reason the resulting stibonic or stibinic acids frequently contain inorganic antimony compounds as impurities (91). This is particularly true when electron-repelling groups are attached to the ring. *p*-Aminobenzenestibonic (stibanilic) acid is difficult to obtain in a pure state (29, 91). Ida and co-workers (92) have shown that the carbon–antimony bond in *p*-acetamido- and 3-chloro-4-acetamidobenzenestibonic acids is cleaved by warming the acids in alkaline solution. The carbon–antimony bond in *p*-acetamidobenzene stibonic acid is also cleaved in warm dilute mineral acid solution. When electron-attracting groups are present, the acids appear to be quite stable. Biswell and Hamilton (93) reported that the carbon–antimony bond in 2-chloro-5-nitrobenzenestibonic acid is stable to boiling 6*N* potassium hydroxide over a 7 hour period. When this compound is heated with concentrated sulfuric acid, the carbon–antimony bond is cleaved with the formation of *p*-nitrochlorobenzene. Campbell and White (94) have reported the cleavage of the carbon–antimony bond in a diarylstibinic acid by heating to 160° for three minutes in polyphosphoric acid.

Feigl (95) has developed a test for *p*-aminobenzenestibonic and *p*-aminobenzenearsonic acids that takes advantage of the instability of these substances. When an alkaline solution of either acid is warmed with Raney nickel or Devarda's alloy, the carbon–antimony or carbon–arsenic bond is ruptured with the formation of aniline and antimony or arsenic trioxide. Since nitro groups are reduced under these conditions, *p*-nitrobenzenestibonic and arsonic acids also give this test.

Biswell and Hamilton (93) have studied the nucleophilic displacement of chlorine in *o*-chloro-substituted benzenestibonic acids. The chloro group in 2-chloro-4-nitrobenzenestibonic acid is readily replaced by amines on heating with potassium carbonate and copper in amyl alcohol. The chloro group is also replaced by hydroxyl when the stibonic acid is heated to 97° with 6*N* potassium hydroxide for 7 hours. The chloro group in *o*-chlorobenzenestibonic acid is not replaced by amines under the same conditions used for the chloronitro compound. The authors (93) concluded from these results that the stibono group was less activating for nucleophilic displacement than the arsono group. Because of the difference in structure between the two groups, this comparison is perhaps not entirely meaningful.

Relatively little work has been done on electrophilic substitution reactions with arylstibonic acids. Morgan and Micklethwait (96) nitrated benzenestibonic acid with mixed acid at 40–55°. The product after purification was heated with phosphorus pentabromide to give a 70% yield of *m*-nitrobromobenzene. No isomeric nitrobromobenzenes were detected. Later Schmidt (2) nitrated benzenestibonic acid with nitric acid (d. 1.515) in the cold and obtained a nitrobenzenestibonic acid identical with that obtained by nitrating with mixed acid. Thus, benzenestibonic acid is much easier to nitrate than benzenearsonic acid. Although only the *meta* isomer was isolated, it is possible that small amounts of others isomers could be present (cf. II-2-C-(1)). Doak and Steinman (91) have shown that the nitration of *p*-toluenestibonic acid yields 3-nitro-4-toluenestibonic acid.

2. TRIALKYL- AND TRIARYLANTIMONY DIHYDROXIDES, OXIDES, SULFIDES, AND SELENIDES

A. Dihydroxides and oxides

(*1*) *Preparation*

Trimethylantimony dihydroxide is formed by the hydrolysis of a trimethylantimony dihalide. In aqueous solution the reaction is incomplete and reversible (97). The hydrolysis can be brought to completion by passing an aqueous solution of the dichloride through a column containing an anionic exchange resin (98). Other methods for completing the hydrolysis, which have been reported in the older chemical literature, include the reaction of trimethylantimony sulfate with barium hydroxide (99) and the reaction of trimethylantimony diiodide with moist silver oxide (100). Morgan and Yarsley (101) reported the preparation of trimethylantimony dihydroxide by the action of moist silver oxide on trimethylantimony hydroxybromide.

The hydrolysis of tris(trifluoromethyl)antimony dichloride takes an entirely different course (102). When a small amount of water is used, the monohydrate $(CF_3)_3SbCl_2{\cdot}H_2O$ is obtained as a white solid. The dichloride dissolves in excess water, and when the resulting solution is evaporated by "freeze-drying," hygroscopic crystals of a dihydrate are obtained. It has been suggested that this compound should be formulated as $[(CF_3)_3SbCl_2OH]^-H_3O^+$. When an aqueous solution of this acid is treated with excess silver oxide, silver chloride is precipitated and an aqueous solution of $[(CF_3)_3Sb(OH)_3]Ag$ is obtained. This silver salt could not be obtained as a solid, although a benzene adduct $[(CF_3)_3Sb(OH)_3]Ag{\cdot}C_6H_6$ has been isolated. By addition of one equivalent of hydrochloric acid to an aqueous solution of the silver salt and removal of the precipitated silver chloride, an aqueous solution of an acid, formulated as $[(CF_3)_3Sb(OH)_3]H$, is obtained. Electrometric titration of this solution with standard alkali gave an apparent dissociation constant for this acid of $1.4 \pm 0.3 \times 10^{-2}$. If this acid indeed has the structure $[(CF_3)_3Sb(OH)_3]H$, it is surprising to find that it is not a much stronger acid. Unfortunately, few details of the titration are given; in particular the method used for calculating the dissociation constant is omitted. When warmed with $2M$ potassium hydroxide solution, aqueous solutions of the acid $[(CF_3)_3Sb(OH)_3]H$ are decomposed with the formation of fluoroform, but at room temperature in one year only one-half of the theoretical amount of fluoroform is evolved. A number of salts of the acid were prepared, but except for the pyridine salt no analytical figures are given.

When heated *in vacuo* at 110°, trimethylantimony dihydroxide loses weight equivalent to one mole of water, but the resulting compound is extremely hygroscopic (98). The preparation of triethylstibine oxide has been reported in the older chemical literature (103, 104), but no analytical figures are given. With larger alkyl groups, *n*-propyl, *n*-butyl, *n*-amyl (105), isobutyl (34), and *n*-heptyl (106), the preparation of oxides but not dihydroxides has been reported.

Triarylantimony dihydroxides have been prepared by the hydrolysis of triarylantimony dihalides. Thus, triphenylantimony dibromide (107, 108) and diiodide (109) have been hydrolyzed by alcoholic potassium hydroxide solution to give triphenylantimony dihydroxide. It can be purified by recrystallization from benzene-ether (108) or by dissolving the crude product in acetic acid and reprecipitating by the addition of water (107). Wittig and Hellwinkel (110) have prepared $(C_6H_5)_3Sb{=}NSO_2C_6H_4CH_3$-*p* by the action of chloroamine-T on triphenylstibine. Hydrolysis of the stibine imide with water gives triphenylantimony dihydroxide. Senning (111) prepared a similar imide by the reaction of N-sulfinylmethanesulfonamide, CH_3SO_2NSO, with triphenylstibine and hydrolyzed the product with aqueous sodium hydroxide to triphenylantimony dihydroxide. This dihydroxide has also been prepared by the addition of water to an acetic acid solution of triphenylantimony diacetate (112). Hydrolysis of the tri-*p*-biphenylylantimony dichloride, dibromide, or diiodide (51) or tri-*o*-biphenylylantimony dibromide (52) with alcoholic ammonia gives the corresponding tribiphenylylantimony dihydroxides.

It has been reported that when triphenylantimony dihydroxide is heated to its melting point (214–215°) it loses water to form triphenylstibine oxide, mp 295° (108). More recently, Bernstein and co-workers (69) have reported that this same oxide is obtained when triphenylantimony dichloride is hydrolyzed by boiling with water for "several days" and the insoluble residue is dried *in vacuo*. The authors of the present volume have attempted to repeat the experiments of Bernstein and co-workers and have been unable to obtain triphenylstibine oxide by their procedure. In fact, when triphenylantimony dichloride was refluxed with water for as long as one month, titration of the aqueous solution with standard alkali (after removal of the insoluble residue) indicated that only about 10% of the dichloride had been hydrolyzed. Briles and McEwen (113, 114) have shown that the addition of water to an acetone solution of dimethoxytriphenylantimony yields triphenylstibine oxide, mp

221.5–222.0°. These workers also prepared the same oxide by the thermal decomposition of tetraphenylstibonium hydroxide; this reaction is discussed later in this section.

Jensen (115) obtained tri-*p*-tolylstibine oxide (mp 270°) by hydrolysis of the corresponding dibromide and recrystallization of the product from benzene, whereas Michaelis and Genzken (116) reported the preparation of an oxide, mp 220°, by the same procedure. Jensen believes that Michaelis and Genzken were dealing with the dihydroxide, but he did not confirm this point by preparing and characterizing both compounds. These results are somewhat confusing. In general, when triarylantimony dihalides are hydrolyzed in alkaline solution, the analytical values obtained on the products are in better agreement with the dihydroxides rather than with the oxides. Rieche and co-workers (117), however, have reported that the hydrolysis of trialkyl- and triarylantimony peroxides of the type $R_3Sb(OOR')_2$ yields the corresponding oxides.

The oxidation of trialkylstibines has been the subject of several investigations (see also Section VIII-2-B-(2)). Earlier workers found that when trimethylstibine (99) and triethylstibine (103) are oxidized in the air, some cleavage of the carbon–antimony bond occurs, but the exact nature of the products was not determined. Dyke and Jones (105) made a careful study of the air oxidation of tri-*n*-propyl-, tri-*n*-butyl-, and tri-*n*-amylstibines. In each case they obtained compounds which gave analytical values in agreement with the formulation $R_3Sb_3O_4$. They suggested, without proof, that these substances are derivatives of a metaantimonite and have the structure $R_3Sb(SbO_2)_2$. A similar compound, $(C_7H_{15})_3Sb_3O_4$, was obtained by Tseng and Shih (106) through the air oxidation of triheptylstibine. In contrast to the results obtained with air oxidation, a number of trialkylstibine oxides have been obtained in a pure state by oxidation of trialkylstibines with mercuric oxide (34, 105, 106).

The carbon–antimony bond in tribenzylstibine apparently is not cleaved by air oxidation. Although Challenger and Peters (118) originally obtained tribenzylantimony dihydroxide from antimony trichloride and benzylmagnesium chloride, it was later demonstrated that tribenzylstibine is formed when the Grignard reaction is carried out in a carbon dioxide atmosphere and that the stibine is readily oxidized to tribenzylstibine oxide in the air (84). Another stibine which can be oxidized in the air to the corresponding oxide without rupture of the carbon–antimony bond is pentaphenylstibole (119):

```
C6H5—C———C—C6H5
      ‖    ‖
C6H5—C    C—C6H5
      \  /
       Sb—C6H5
```

In contrast to trialkylstibines, triarylstibines are generally quite stable in air. A possible exception to this generalization is the finding of Yusunov and Manulkin (58) that several unsymmetrical stibines of the type $(C_6H_5)_2ArSb$ (where Ar=*o*-$CH_3C_6H_4$, *p*-CH_2=CH—$CH_2C_6H_4$, or *p*-$C_6H_{11}C_6H_4$), all of which are liquids, give rise to some diphenylstibinic acid on standing in the air for one month. The carbon–antimony bond in tri-α-thienylstibine oxide, which has been prepared by the hydrolysis of tri-α-thienylantimony dibromide with alcoholic potassium hydroxide, is cleaved to give inorganic antimony oxides by allowing it to stand in the air for one year (120). The carbon–antimony bond of tri-α-thienylantimony dibromide is cleaved when silver oxide rather than potassium hydroxide is used for the hydrolysis.

Triarylstibines are readily oxidized with hydrogen peroxide under carefully controlled conditions to give triarylantimony dihydroxides (or oxides). Schmidt (55) oxidized triphenylstibine in acetone with 3% hydrogen peroxide and obtained a product which was not characterized but was converted to triphenylantimony diacetate by treatment with acetic acid. When the oxidation with hydrogen peroxide is carried out in 5*N* sodium hydroxide solution, the sole product obtained is diphenylstibinic acid. In a similar manner the oxidation of tris(*p*-acetamidophenyl)stibine gives bis(*p*-acetamidophenyl)-stibinic acid. Monagle (121) has reported that the oxidation of triphenyl-stibine with hydrogen peroxide yields the corresponding oxide, mp 249–251°; by apparently the same method, Venezky (122) obtained an oxide melting sharply at 285°. By contrast with these results Kaufmann (123) has claimed, in a German patent, that the oxidation of both triphenylstibine and tri-*p*-tolylstibine by hydrogen peroxide or sodium peroxide in alkaline solution yields the corresponding triarylantimony dihydroxides; and Yusunov and Manulkin (58) claim to have prepared diphenyl(*p*-bromophenyl)- and diphenyl(*p*-phenetyl)antimony dihydroxides by oxidation of the corresponding stibines in 1*N* potassium hydroxide with 4% hydrogen peroxide. Tri-α-thienylstibine, however, is completely cleaved to inorganic antimony oxides when it is treated with hydrogen peroxide in alkaline solution (124).

The oxidation of triphenylstibine with hydrogen peroxide in aqueous acetone solution has also been investigated by Nerdel and co-workers (125). Under conditions similar to those employed by Schmidt, they obtained a product which decomposed at about 170° and gave analytical results in agreement with the formula $(C_6H_5)_3SbO_{1.4}$. The compound behaved as a peroxide as shown by its reaction with titanium(IV) sulfate. The peroxide content was determined quantitatively by means of two methods illustrated in the following equations:

$$(C_6H_5)_3SbO_{(1+x)} + 2CH_3CO_2H \rightarrow (C_6H_5)_3Sb(OCOCH_3)_2 + (1-x)H_2O + xH_2O_2$$

$$(C_6H_5)_3SbO_x + Br_2 \rightarrow (C_6H_5)_3SbBr_2 + (x/2)O_2$$

In the first reaction the liberated hydrogen peroxide was determined iodimetrically to give a value of $x + 1 = 1.43 \pm 0.01$; in the second reaction the oxygen was measured in a gas buret to give an average value of $x = 1.37 \pm 0.15$.

Nerdel and co-workers regard their oxidation product as a peroxide. It may possibly be a hydrogen peroxide adduct of triphenylstibine oxide. This suggestion is based upon the results of Copley and co-workers (126) who demonstrated that the compound $[(C_6H_5)_3PO]_2H_2O_2$ was obtained by the oxidation of triphenylphosphine with hydrogen peroxide. This product lost hydrogen peroxide on heating *in vacuo* to give triphenylphosphine oxide, but the adduct was regenerated by recrystallizing the oxide from a methanol-hydrogen peroxide mixture.

Carson and Wong (127) have investigated the use of aryl and alkyl thiosulfinates as oxidizing agents for triphenylphosphine, triphenylarsine, and triphenylstibine. With triphenylphosphine the reaction was formulated as follows:

$$RSS(O)R + (C_6H_5)_3P \rightarrow RSSR + (C_6H_5)_3PO$$

Both aromatic and aliphatic thiosulfinates react with triphenylphosphine, but with triphenylarsine and triphenylstibine only the aromatic thiosulfinates react. Triphenylstibine reacts with phenyl benzenethiosulfinate in benzene-methanol solution to give a 92% yield of a crude product, which after recrystallization from benzene-petroleum ether is identical with triphenylantimony dihydroxide obtained by the alkaline hydrolysis of triphenylantimony dibromide.

In connection with experiments designed to investigate the radical cleavage of iodine from the benzene ring, Shubenko (128) irradiated triphenylstibine and *p*-iodoanisole in methanol solution with ultraviolet radiation for 40 hours. He obtained a 43% yield of triphenylantimony dihydroxide, together with some anisole and formaldehyde. Mel'nikov and Rokitskaya (129) have reported that the reaction of triphenylstibine with selenium dioxide yields triphenylstibine oxide (mp 209°) and triphenylstibine selenide.

Briles and McEwen (114) have found that the thermal decomposition of tetraphenylstibonium hydroxide at 70–80° yields benzene and triphenylstibine oxide, mp 221.5–222°. The decomposition of methoxytetraphenylantimony under similar conditions also yields the same oxide (130). It seems possible that the methoxy compound and other alkoxytetraphenylantimony compounds decompose on melting to give triphenylstibine oxide.

It is obvious from the foregoing paragraphs that the reported melting points of triphenylstibine oxide vary over a wide range. Table 7-1 lists these melting points together with the methods used for preparing the oxide.

Just as certain arylstibonic acids will undergo cyclodehydration to form

Table 7-1 Preparation and Melting Points of Triphenylstibine Oxide

Method of Preparation	MP	Literature Reference
Dehydration of $Ph_3Sb(OH)_2$ at its mp	295	108
Dehydration of $Ph_3Sb(OH)_2$ by "drying"	280–285	110
Dehydration of $Ph_3Sb(OH)_2$ in a drying pistol	[a]	111
Hydrolysis of Ph_3SbCl_2 by boiling water	[a]	69
Hydrolysis of $Ph_3Sb(OMe)_2$ by aqueous acetone	221.5–222	113, 114
Hydrolysis of $Ph_3Sb(OOR)_2$ by water	[a]	117
Oxidation of Ph_3Sb with H_2O_2	249–251	121
Oxidation of Ph_3Sb with H_2O_2	285	122
Oxidation of Ph_3Sb with SeO_2	209	129
Thermal decomposition of Ph_4SbOH	221.5–222	114
Thermal decomposition of Ph_4SbOCH_3	221–223	130

[a] Melting point not given.

heterocyclic stibinic acids, it might be expected that diarylstibinic acids would undergo cyclodehydration to give heterocyclic stibine oxides or dihydroxides. This possibility was investigated by Campbell (131). (*p*-Carbethoxyphenyl)-2-biphenylylstibinic acid, prepared by hydrolysis of the corresponding trichloride, was warmed with acetic anhydride and a small amount of sulfuric acid. The principle product obtained was 5-(*p*-carbethoxyphenyl)dibenzostibole 5-oxide (**7**) (named 9-*p*-carbethoxyphenyl-9-stibiafluorene oxide by Campbell).

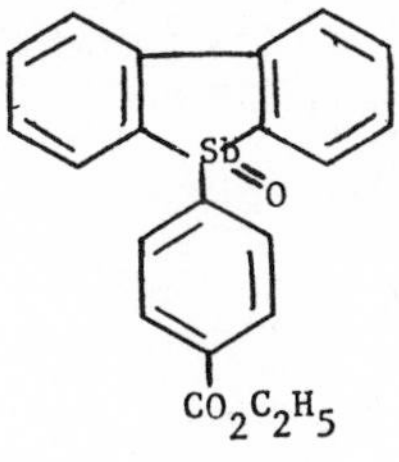

Sb=O
$CO_2C_2H_5$

7

This cyclization procedure was later used for the preparation of a series of heterocyclic antimony compounds; however, the oxides were not characterized but were reduced directly to the heterocyclic stibines (132). The successful preparation of heterocyclic stibine oxides prompted Campbell and White (94) to investigate selected compounds that might be capable of resolution. They attempted first to prepare 3-methoxy-5-(*p*-carbethoxyphenyl)dibenzostibole 5-oxide (**8**) and 3-carbethoxymethoxy-5-*p*-tolyldibenzostibole 5-oxide

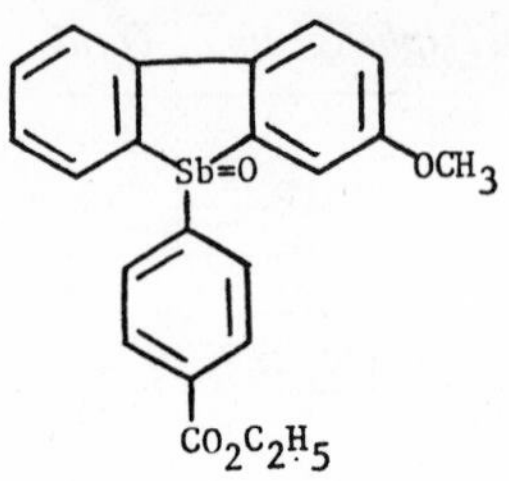

Sb=O OCH_3 $CO_2C_2H_5$

8

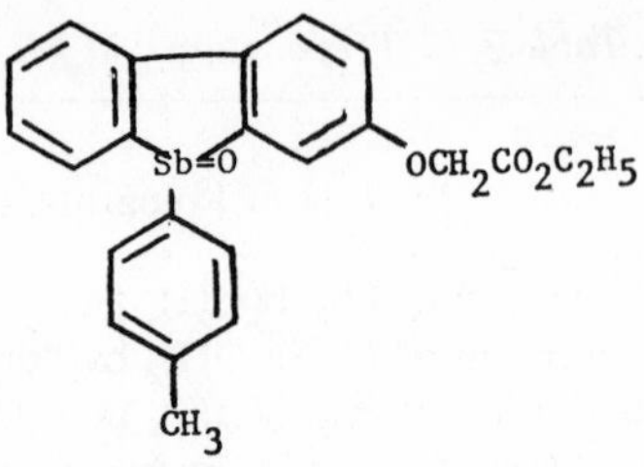

Sb=O $OCH_2CO_2C_2H_5$ CH_3

9

(**9**) by cyclodehydration of the appropriate stibinic acids. Instead of **8** they obtained the two dihydroxides **10** and **11**, and instead of **9** they obtained the

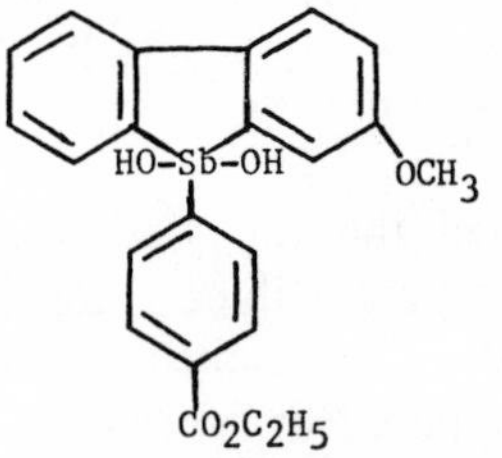

HO–Sb–OH OCH_3 $CO_2C_2H_5$

10

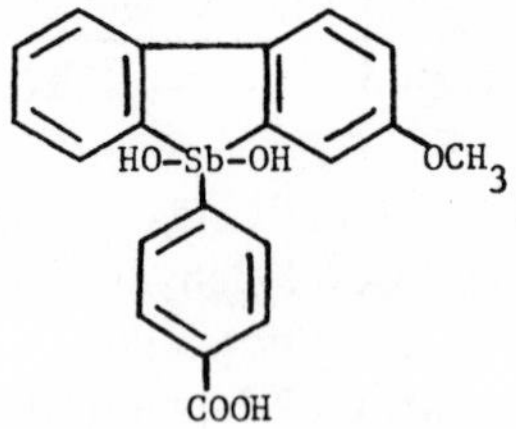

HO–Sb–OH OCH_3 COOH

11

dihydroxides **12** and **13**. In no case did any of these dihydroxides lose water

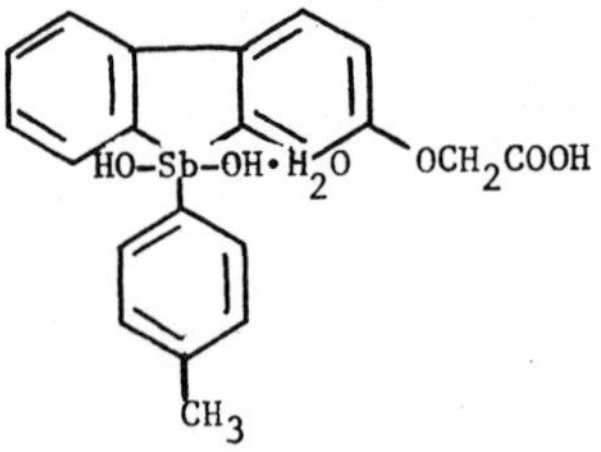

HO–Sb–OH·H_2O OCH_2COOH CH_3

12

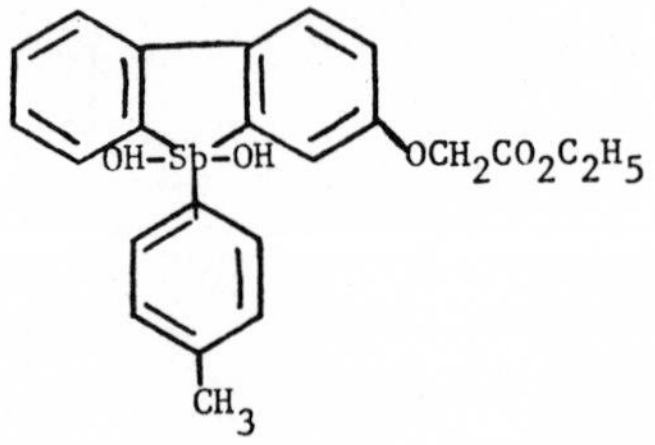

OH–Sb–OH $OCH_2CO_2C_2H_5$ CH_3

13

on prolonged heating at 100° *in vacuo*. An attempt to prepare **9** by cyclodehydration of the stibinic acid with polyphosphoric acid resulted in cleavage of the carbon–antimony bond. In one attempt to prepare cyclic stibine oxides by hydrolysis of cyclic antimony dibromides, the authors (94) obtained the dihydroxides **14** and **15**. The fact that three of the above cyclic antimony dihydroxides were obtained with an additional mole of water

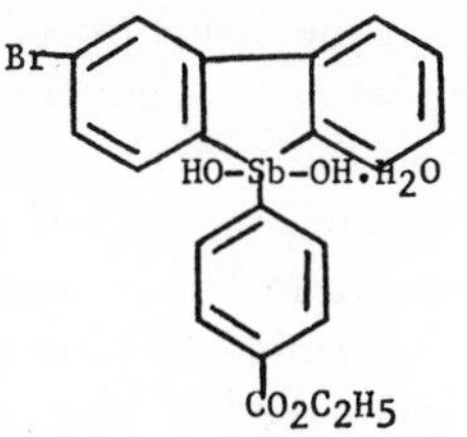

14 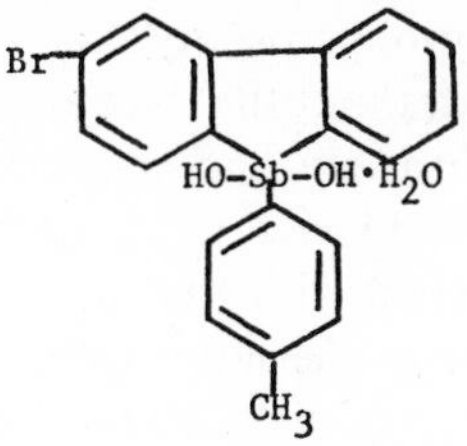 15

prompted Campbell and White to suggest that these compounds are acids with the general formula $[Ar_3Sb(OH)_3]^-H^+$. It seems unlikely, however, that acids containing free protons could be isolated in the solid state.

The fact that all of the heterocyclic stibine oxides except 7 were obtained in a hydrated form, presumably as dihydroxides, caused Campbell and White (94) to abandon this type of compound in their studies of the optical isomerism of organoantimony compounds. They pointed out that, even if they should obtain a triarylstibine oxide of the type abcSbO, the process of salt formation incidental to resolution would undoubtedly involve the dihydroxide as an intermediate and that such an intermediate would possess a plane of symmetry.

(2) *Structure and physical properties*

The structure of these compounds has not been extensively studied. Merck (104), who prepared triethylstibine oxide (or the corresponding dihydroxide) over 100 years ago, reported that an aqueous solution of this material is strongly basic and precipitates metallic hydroxides from solutions of metallic salts. Using a glass electrode, Lowry and Simons (100) titrated an aqueous solution of trimethylantimony dihydroxide (approximately $0.16M$) with standard hydrochloric acid, and obtained a basic dissociation constant of 1.38×10^{-5}; i.e., trimethylantimony dihydroxide is a slightly weaker base than ammonia. Triphenylantimony dihydroxide reacts only with concentrated hydrochloric acid, but does dissolve in weak aqueous alkali. It has been suggested by Campbell and White (94) that the species $[(C_6H_5)_3Sb(OH)_3]^-$ exists in alkaline solution.

No molecular weight data are available on trialkyl- or triarylantimony dihydroxides, and only one infrared spectral study has been reported, in which a peak at 566 cm^{-1} was assigned to the C—Sb asymmetric stretch of trimethylantimony dihydroxide (98). Jensen (108) found that the dipole moment of triphenylantimony dihydroxide in benzene solution is 0 and concluded, therefore, that the Sb—O—H grouping must be linear. This conclusion seems improbable, and another explanation for this finding should be sought.

Several groups of workers have reported on the structure of the triarylstibine oxides, but the data are conflicting and no definite conclusions can be drawn. Bernstein and co-workers (69) studied the hydrolysis of triphenylantimony dichloride to the corresponding oxide in both ordinary water and water enriched with ^{18}O, and found an incorporation of 25 atom % of the ^{18}O isotope in the latter case. An infrared spectrum of the oxide failed to show any bands in the 900–740 cm^{-1} region, where the Sb═O grouping might be expected to absorb. Three bands at 739, 725, and 696 cm^{-1} were almost unaffected by ^{18}O labelling and were assigned to C—H out-of-plane bending and C—C—C deformation. On the basis of these results the authors (69) suggested that $(C_6H_5)_3SbO$ does not contain the Sb═O grouping but is probably polymeric. As stated previously, however, the authors of the present monograph have been unable to prepare triphenylstibine oxide by the procedure of Bernstein and co-workers, whose results seem open to some doubt.

Jensen and Nielsen (68), in an important paper, described the infrared spectrum of triphenylstibine oxide, presumably prepared by the dehydration of triphenylantimony dihydroxide (108). They were unable to find a band in the 700–950 cm^{-1} region which could be associated with the Sb═O stretching frequency, and concluded that triphenylstibine oxide "presumably forms a pseudoionic layer lattice and not individual molecules." In sharp contrast to the results of the previous investigators, Venezky (122) has reported that the infrared spectrum of triphenylstibine oxide exhibits two strong absorption peaks in the 800–600 cm^{-1} region, which were attributed to symmetrical and asymmetrical O—Sb—O stretching modes; both absorptions, however, were absent in the Raman spectrum. Venezky concluded that his results support a polymeric structure for this oxide. Chremos and Zingaro (133) have recently reported that the infrared spectra of trialkylstibine oxides contain a single peak between 650 and 550 cm^{-1} which should be assigned to the fundamental Sb═O stretching frequency. The proton NMR spectrum of a compound designated as $Ph_3SbO{\cdot}0.3H_2O$ has been described, but no information about the source of this substance was given (134).

Jensen has reported that tri-*p*-tolylstibine oxide gives a molecular weight of 528 by cryoscopic measurement in benzene; the calculated value for the monomer is 411 (115). Only a single determination was reported and the depression given was only 0.032°. Jensen also reported that the dipole moment of tri-*p*-tolylstibine oxide is 2.0 D in benzene at 25° and 2.3 D in dioxane at 40°. Several trialkylstibine oxides have been said to occur as syrups or as low-melting solids (34, 105); such properties do not seem consistent with the "ionic layer lattice" structure proposed for triphenylstibine oxide. It seems obvious that further studies are required before the question of the structure of these compounds can be decided unequivocally.

(*3*) *Reactions*

Tertiary stibine oxides and dihydroxides react with both inorganic and organic acids to form compounds of the type R_3SbX_2 (where X^- is the conjugate base of the acid). The acids that have been employed in this reaction include hydrogen chloride (59, 84, 98, 99), hydrogen bromide (84), nitric acid (103, 107, 135), formic acid (98, 135), acetic acid (55, 98, 135, 136), propionic acid (135), butyric acid (135), benzoic acid (135), and acrylic acid (137). Alkyl hydroperoxides react with both trimethylstibine oxide and triphenylantimony dihydroxide to form the diperoxides, $R_3Sb(OOR')_2$ (117). Triarylantimony dichlorides have been obtained by the treatment of triarylstibine oxides with either phosgene or thionyl chloride (138). Trialkylantimony diisocyanates, $R_3Sb(NCO)_2$, have been prepared by the fusion of urea with trialkylstibine oxides or dihydroxides (139, 140). Many of the reactions mentioned in this paragraph are discussed in greater detail in Sections VII-5-A-(3) and VII-5-B.

Oxybis(trimethylantimony) dichloride, $(Me_3SbCl)_2O$, has been prepared by dissolving equimolar quantities of trimethylantimony dichloride and trimethylantimony dihydroxide in water and evaporating the resulting solution to a small volume (98). The analogous dinitrate, $(Me_3SbNO_3)_2O$, has been obtained by the reaction of trimethylantimony dihydroxide in moist acetone with the stoichiometric quantity of nitric acid (135). Trimethylstibine sulfide has been prepared by the interaction of trimethylantimony dihydroxide and hydrogen sulfide in methanol (141). Treatment of triphenylstibine oxide with triphenylantimony dichloride in anhydrous benzene gives a quantitative yield of oxybis(triphenylantimony) dichloride (114). Triphenylstibine oxide has been found to be an effective catalyst for converting phenyl isocyanate to diphenylcarbodiimide at 115–120° (121). By contrast, trialkylstibine oxides at room temperature catalyze the trimerization of alkyl and aryl isocyanates to the corresponding isocyanurates (136). Triphenylstibine oxide in tetrahydrofuran reacts with potassium to give insoluble black products and a red-brown solution (141a).

B. Sulfides and selenides

A number of trialkyl- and triarylstibine sulfides have been reported. Zingaro and Merijanian (142) obtained triethyl-, tri-*n*-propyl-, tri-*n*-butyl-, and tricyclohexylstibine sulfides by refluxing the corresponding stibines with an excess of sulfur in a nitrogen atmosphere. The *n*-butyl compound is a nondistillable liquid; the other compounds are solids. The melting points of the triethyl (118°) and the tri-*n*-propyl (35°) compounds differ significantly from those reported for these same compounds in the earlier literature (103, 105). More recently, Shindo and co-workers (141) have prepared trimethylstibine sulfide by the interaction of trimethylantimony dihydroxide and hydrogen sulfide; they obtained the triethyl- and tricyclohexylstibine sulfides by the reaction of the corresponding dibromides with sodium sulfide. These stibine sulfides were found to give normal molecular weights in chloroform. The infrared spectra of several trialkylstibine sulfides contain a band in the 450–420 cm^{-1} region which has been assigned to the Sb═S stretch (133, 141). The C—Sb asymmetric stretching frequency has been located in the 555–529 cm^{-1} region, while the symmetric stretch occurs between 531 and 506 cm^{-1}. Shindo and co-workers (141, 143) have prepared several complexes of the type $R_2SnX_2 \cdot 2Me_3SbS$, where R is methyl or ethyl and X is chlorine or bromine. These compounds are monomolecular in toluene but dissociate in chloroform or dimethylformamide to the starting materials. The Sb═S stretching frequency occurs near 401 cm^{-1}, about 30 cm^{-1} lower than in trimethylstibine sulfide.

Selenium reacts with triethyl-, tri-*n*-propyl-, and tri-*n*-butylstibines in a similar manner to their reaction with sulfur to give the corresponding trialkylstibine selenides (142). The triethyl compound melts at 124°; the remaining two compounds are liquids. Trimethylstibine was found to react with selenium to give the compound $(CH_3)_3SbSe_2$, which is stable under benzene but decomposes in the air with the formation of elemental selenium. It has been suggested that this compound is a dimer with the following structure:

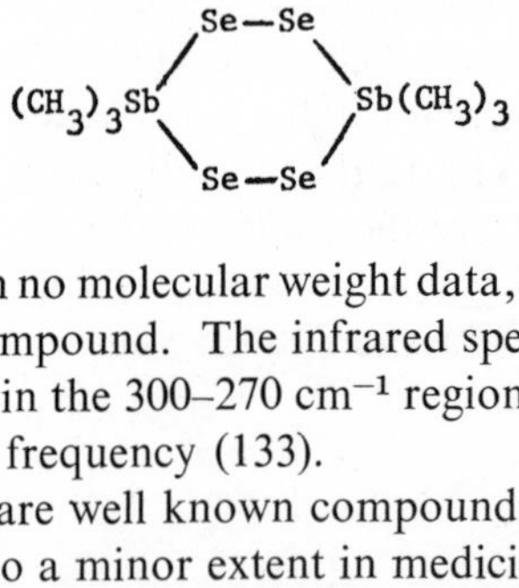

```
          Se—Se
         /     \
(CH3)3Sb        Sb(CH3)3
         \     /
          Se—Se
```

However, there have been no molecular weight data, spectra, or other physical measurements on this compound. The infrared spectra of the trialkylstibine selenides contain a band in the 300–270 cm^{-1} region which has been assigned to the Sb═Se stretching frequency (133).

Triarylstibine sulfides are well known compounds; at one time triphenylstibine sulfide was used to a minor extent in medicine under the name sulfoform (144). They are probably best prepared by the action of hydrogen sulfide on triarylantimony dihalides in alcoholic–ammonia solution (145, 146). Triphenylstibine sulfide has also been obtained by the decomposition of diphenyliodonium sulfide in the presence of finely powdered metallic antimony (147). The reaction involves a free radical mechanism.

The molecular weight of triphenylstibine sulfide has been determined cryoscopically in benzene (found 458.5 and 428.7; calcd. for the monomer,

385) (145). The dipole moment determined in benzene solution is 5.40 D (115). This value is comparable to the values of 5.50 D for $(C_6H_5)_3AsO$ and 4.31 D for $(C_6H_5)_3PO$; both of these compounds are known to be monomolecular. The molecular weight and dipole moment data suggest that triphenylstibine sulfide exists as a monomer. The infrared spectrum of triphenylstibine sulfide has been described, but only phenyl and "X-sensitive" frequencies were identified (68).

It has been reported that triphenylstibine reacts with selenium dioxide to give a mixture of triphenylstibine oxide and triphenylstibine selenide (129). Jensen and Nielsen (68) were unable to obtain the selenide by the reaction between triphenylantimony dichloride and hydrogen selenide, although the corresponding arsenic compound could be prepared. The crystallographic properties of triphenylstibine sulfide, selenide, and a number of related compounds have been investigated (148).

3. COMPOUNDS OF THE TYPE $RSbX_4$ AND $[RSbX_5]^+Y^-$

In the reaction between a diazonium chloride and antimony trichloride (either in aqueous acid solution or in an organic solvent), the principal product is the arylantimony tetrachloride:

$$ArN_2Cl + SbCl_3 \rightarrow ArSbCl_4 + N_2$$

Although a number of these compounds have been prepared in this manner, they are usually hydrolyzed, without isolation, to the aromatic stibonic acids. In a few cases, however, it has proved advantageous to isolate the tetrachloride. For example, a solution of 2-aminothiazole in ethanol–sulfuric acid was diazotized with ethyl nitrite in the presence of antimony trichloride and was then warmed to effect the evolution of nitrogen. Evaporation of the resulting solution gave an almost pure, crystalline 2-thiazolylantimony tetrachloride (148a). It was readily recrystallized from ethanol. The pure compound melts at 133°. By a similar reaction 2-aminopyrimidine leads to 2-pyrimidylantimony tetrachloride, which can be recrystallized from concentrated hydrochloric acid (37).

Arylantimony tetrachlorides can be obtained by treating aromatic stibonic acids with concentrated hydrochloric acid. In this manner Schmidt (2) prepared phenylantimony tetrachloride and several ring-substituted arylantimony tetrachlorides. By the same procedure Biswell and Hamilton (93) prepared both 2-chlorophenyl- and 2-chloro-4-nitrophenylantimony tetrachlorides as pale-yellow crystalline needles.

Finally, arylantimony tetrachlorides have been prepared by chlorination of trivalent monophenylantimony compounds. Thus, Worrall (51) prepared *p*-biphenylylantimony tetrachloride from the corresponding dichloride by chlorination in chloroform solution. This same procedure was used for preparing phenyl- and *p*-tolylantimony tetrachlorides as intermediates in the preparation of the corresponding stibonic acids (149, 150).

No alkylantimony tetrachlorides, tetrabromides, or tetraiodides have been described. Clark (151) treated ethyldiiodostibine with chlorine, with bromine, and with iodine; only inorganic antimony halides were isolated from the reactions. Several alkylantimony tetrafluorides have been reported by the reaction between aliphatic stibonic acids and sulfur tetrafluoride (1). The source of the aliphatic stibonic acids was not mentioned, and the results are highly questionable.

Arylantimony tetrachlorides are generally low-melting, hygroscopic solids, soluble in most organic solvents and decomposed by water. The analytical results obtained on the tetrachlorides that have been recrystallized from hydrochloric acid often indicate that one or more moles of water are present. Thus, 3-nitro-4-chlorophenylantimony tetrachloride has been obtained as a pentahydrate, which readily loses 4 moles of water when kept in a desiccator over a drying agent but which loses the last mole of water very slowly (2). Phenylantimony tetrachloride is an unstable substance in which the carbon–antimony bond is cleaved on standing (2). The decomposition is believed to occur as follows:

$$2C_6H_5SbCl_4 \rightarrow (C_6H_5)_2SbCl_3 + SbCl_3 + Cl_2$$

Both $(C_6H_5)_2SbCl_3$ and chlorine were isolated among the decomposition products.

There has been no experimental work on the structure of the arylantimony tetrachlorides. Antimony pentachloride is a trigonal bipyramid (152, 153), but SbF_3Cl_2 has been shown by means of NMR spectral studies to be an associated molecule with fluorine or chlorine bridges (154).

Arylantimony tetrachlorides react with many amine hydrochlorides to form, in almost quantitative yield, crystalline salts of the type $[RNH_3]^+$-$[ArSbCl_5]^-$. These compounds are readily purified by recrystallization. They often possess sharp, reproducible melting points. Since they can be prepared directly from arylstibonic acids and since they can be readily hydrolyzed to the stibonic acids, these salts have found wide application for both the purification and characterization of arylstibonic acids (2, 13, 28). The insoluble compounds obtained by the reaction of aromatic stibonic acids with methyl violet, crystal violet, or Rhodamine B in concentrated hydrochloric acid are probably salts of this type (155).

If amino groups are present in the arylantimony tetrachloride molecule, zwitterions such as *m*-$(CH_3)_3N^+C_6H_4SbCl_5^-$ can be obtained (32). These generally decompose at high temperatures without melting. It seems probable that many other compounds actually possess this zwitterion structure although they have not been formulated as such. Thus, **16** (22), **17** (37), and *p*-$(CH_3)_2NCH_2C_6H_4SbCl_4\cdot HCl$ (156) probably exist in the zwitterion form.

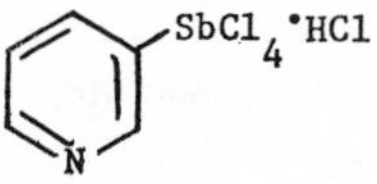

$SbCl_4\cdot HCl$ $SbCl_4\cdot HCl$

16 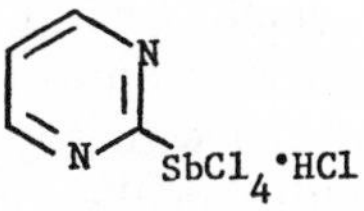 **17**

The reduction of arylantimony tetrachlorides to dichlorostibines is discussed in Section VIII-3-A. The reduction of ammonium phenylpentachloroantimonate to phenylstibine is discussed in Section VIII-1.

4. COMPOUNDS OF THE TYPE R_2SbX_3

A. Preparation

Only a few dialkylantimony trichlorides are known. Morgan and Davies (6) obtained dimethylantimony trichloride by chlorination of dimethylchlorostibine. The tribromide was obtained in a similar manner. Both compounds are unstable in that they lose methyl halide on standing at room temperature; they are hydrolyzed by moisture to dimethylstibinic acid. Morgan and Davies also claimed to have obtained $(CH_3)_2SbOCl$ and $(CH_3)_2SbOBr$ by the air oxidation of dimethylhalostibines. Dimethylantimony trifluoride has been prepared by the interaction of the corresponding trichloride and silver fluoride in aqueous ethanol (157).

Nesmeyanov and Borisov (158) obtained bis(β-chlorovinyl)antimony trichloride in excellent yield by the reaction between antimony pentachloride and acetylene at temperatures below 50°.

Diarylantimony trichlorides are obtained from antimony chlorides by the diazo reaction under a wide variety of reaction conditions; in fact, it is usually difficult to avoid the formation of some of the trichloride even under conditions designed to obtain only the arylantimony tetrachloride. Thus, Doak and co-workers (23) always obtained small amounts of diarylantimony trichlorides in the reaction between a diazonium tetrafluoroborate and antimony trichloride in an organic solvent. It might be expected that diarylantimony trichlorides could be readily prepared from diazonium chlorides and aryldichlorostibines according to the following reaction sequence:

$$[ArN_2]^+Cl^- + Ar'SbCl_2 \rightarrow [Ar'SbCl_3]^-[ArN_2]^+ \rightarrow ArAr'SbCl_3 + N_2$$

Bruker and Makhlis (159) have indeed prepared [*p*-$CH_3C_6H_4SbCl_3$]-[*p*-$CH_3C_6H_4N_2$] and found that when this salt was heated in hydrochloric acid solution, nitrogen was evolved and di-*p*-tolylantimony trichloride was formed. No further attempts to exploit this reaction have apparently been made.

In a series of experiments Reutov and Ptitsyna (46, 160) allowed $[C_6H_5N_2]$-$[SbCl_4]$ to react with zinc powder in ethyl acetate. When this reaction mixture was then treated with additional $[C_6H_5N_2][SbCl_4]$, a vigorous reaction occurred resulting in the formation of excellent yields of diphenylantimony trichloride (which was isolated as $(C_6H_5)_2SbO_2H{\cdot}Sb_2O_3$). It was suggested that the second reaction occurred as follows:

$$C_6H_5SbCl_2 + [C_6H_5N_2][SbCl_4] \rightarrow (C_6H_5)_2SbCl_3 + SbCl_3 + N_2$$

Somewhat similar conclusions were reached by these same authors (46) in their study of the complicated reaction between aryldiazoformates and antimony trichloride in acetone or ethyl acetate. In this case they obtained both an aryldichlorostibine and an aryldiazonium tetrachloroantimonate(III) when the solvent was removed at room temperature. They suggested that the aryldichlorostibines arise by the following reaction sequence:

$$ArN_2CO_2K + SbCl_3 \rightarrow ArN_2CO_2SbCl_2 + KCl$$

$$ArN_2CO_2SbCl_2 \rightarrow ArSbCl_2 + CO_2 + N_2$$

and that the diazonium compounds were obtained by the reaction:

$$2ArN_2CO_2SbCl_2 + 2SbCl_3 + O_2 \rightarrow 2[ArN_2][SbCl_4] + 2CO_2 + 2SbOCl$$

If the solvent is removed, not at room temperature, but at 30–35°, no aryldichlorostibine is obtained, but rather the diarylantimony trichloride (isolated as $Ar_2SbO_2H{\cdot}Sb_2O_3$) is the principal reaction product. This latter result was interpreted as being due to the reaction (initiated by the increased temperature) of the aryldichlorostibine with the diazonium tetrachloroantimonate(III).

The above results suggested that other arylating agents might be used to arylate aryldichlorostibines and diarylchlorostibines. Accordingly, they investigated the reactions of aryldiazoacetates, aryldiazonium tetrachlorozincates, and aryldiazonium chloroantimonates(III) with a variety of arylantimony (and diarylantimony) compounds. In nearly all cases the arylantimony compounds were further arylated under the reaction conditions employed, but there was a wide variation in the yields obtained. Electron-repelling groups attached to the arylating agent retarded arylation; electron-attracting groups accelerated arylation. The diazoacetates were the most active arylating agents; the tetrachlorozincates were the least active. The character of the aryl group originally bonded to the antimony had little effect on the course of the reaction. This method affords an excellent method for preparing both symmetrical and unsymmetrical diarylantimony trichlorides and triarylantimony dichlorides in a pure state and in relatively high yields.

Campbell (131, 132, 161) has used this method for preparing a number of unsymmetrical diarylantimony trichlorides. From *o*-biphenyldiazonium tetrachloroantimonate(III) and *p*-tolyldichlorostibine, *o*-biphenylyl-*p*-tolylantimony trichloride was obtained in 20% yield. A small amount of di-*o*-biphenylyl-*p*-tolylantimony dichloride was also obtained, but was easily separated from the desired product (161). Several other compounds of the type $ArAr'SbCl_3$, where Ar was an *o*-biphenylyl group, were prepared in yields of 20–35% (131–132). Initial attempts to extend the reaction to the preparation of unsymmetrical compounds not containing an *o*-biphenylyl group were unsuccessful (162). Later work (163) showed, however, that this failure was due, in part, to difficulties in isolating the desired products. The best method found for obtaining the diarylantimony trichloride was by first precipitating it from solution as the pyridinium tetrachlorodiphenylantimonate(V). Several unsymmetrical diarylantimony trichlorides could not be obtained due to disproportionation. Thus, from the reaction between *o*-toluenediazonium chloride and *p*-carbethoxyphenyldichlorostibine, only bis(*p*-carbethoxyphenyl)antimony trichloride was obtained (163). Similarly 4-biphenyldiazonium chloride and *p*-carbethoxyphenyldichlorostibine gave di-*p*-biphenylylantimony trichloride (isolated as $(p\text{-}C_6H_4C_6H_4)_2SbCl$) (162).

Diarylantimony trichlorides can be obtained under the conditions of the Nesmeyanov reaction. Nesmeyanov and Kocheshkov (164) obtained principally aryldichlorostibines, together with small amounts of diarylchlorostibines and diarylantimony trichlorides, by the reaction of a diazonium tetrachloroantimonate(III) in an organic solvent with zinc or iron powder. However, when this reaction was carried out at 0° in acetone and iron powder was used to effect the decomposition, diarylantimony trichlorides were usually obtained as the principal and sometimes the sole product of the reaction (85). By-products were aryldichlorostibines and triarylantimony dichlorides. When *p*-anisidine was used as the starting material, tri-*p*-anisylantimony dichloride was the principal product of the reaction. In the preparation of many diarylantimony trichlorides, this reaction appears to be one of the best available.

A modification of the Nesmeyanov reaction involves the decomposition of compounds of the type $[ArN_2]^+[SbCl_6]^-$ by the addition of iron powder to a suspension of the diazonium salt in acetone (86, 165). The reaction proceeds according to the equation:

$$2[ArN_2]^+[SbCl_6]^- + 3Fe \rightarrow Ar_2SbCl_3 + 2N_2 + SbCl_3 + 3FeCl_2$$

Triarylantimony dichlorides may also be formed, and in some cases constitute the sole reaction product. The reaction is of less importance than the reaction employing a diazonium tetrachloroantimonate(III), since the nexachloroantimonate(V) cannot be produced directly from antimony pentachloride and an aryldiazonium chloride but must be prepared from antimony pentachloride and a diazonium tetrachloroferrate(III) (166).

Unsymmetrical diarylantimony trichlorides have been obtained by a similar reaction in which diazonium salts of the type $[Ar'N_2][ArSbCl_5]$ are decomposed in acetone by the addition of iron powder (87). (The diazonium salts were prepared by the reaction between arylantimony tetrachlorides and diazonium tetrachloroferrate(III) compounds in acetone solution (167)). The unsymmetrical diarylantimony trichlorides were usually not isolated from the reaction mixture but were hydrolyzed to the corresponding diarylstibinic acids.

The decomposition of diaryliodonium chloroantimonates under conditions of the Nesmeyanov reaction has also been studied. Antimony trichloride reacts with diaryliodonium chloride to form several different types of compounds (168). Diphenyliodonium chloride forms the compound $[(C_6H_5)_2I]_2$-$[SbCl_5]$ in hydrochloric acid solution but the compound $[(C_6H_5)_2I][SbCl_4]$ in acetone. Di-*p*-tolyl-, bis(*p*-chlorophenyl)-, bis(*p*-bromophenyl)-, and bis(*p*-iodophenyl)iodonium chlorides form complexes of the type $[(p\text{-}XC_6H_4)_2ICl]_3$-$(SbCl_3)_2$ in acetone. This last type of compound may be a mixture of the two types of compounds found in the case of diphenyliodonium chloride. All of the iodonium chloroantimonates react with metallic antimony in hot acetone or ethyl acetate to give mixtures of triarylantimony dichlorides, diarylantimony trichlorides, and diarylchlorostibines. In a later paper (169) the procedure was modified somewhat. Instead of actually isolating the diarylantimony chloroantimonates, it was found that essentially the same results were obtained if a diaryliodonium chloride and antimony trichloride were dissolved in acetone or ethyl acetate and the reaction effected by the addition of antimony metal. Thus, from bis(*p*-chlorophenyl)iodonium chloride under these reaction conditions, the yields of tris(*p*-chlorophenyl)antimony dichloride, bis(*p*-chlorophenyl)antimony trichloride, and bis(*p*-chlorophenyl)chlorostibine were 43, 11, and 15%, respectively.

Diaryliodonium chlorides also react with antimony pentachloride in hydrochloric acid to form compounds of the type $[Ar_2I]^+[SbCl_6]^-$ (88). These decompose sluggishly in acetone suspension upon the addition of powdered iron to give low yields of diarylantimony trichlorides. With powdered antimony, however, reasonably good yields of diarylantimony trichlorides (isolated as diarylstibinic acids) are obtained.

The mechanism of the formation of diarylantimony trichlorides under conditions of the Nesmeyanov reaction has been elucidated by Reutov and co-workers in a series of papers. They first studied the kinetics of the following reaction (170):

$$[p\text{-}XC_6H_4SbCl_5][C_6H_5N_2] + Fe \rightarrow (p\text{-}XC_6H_4)(C_6H_5)SbCl + FeCl_2 + N_2$$

In the above reaction X was C_2H_5O, CH_3, H, Cl, or NO_2. The rate of the reaction was found to decrease with an increase in the electron-attracting power of the substituent X. They also studied the decomposition of compounds of the type $[YC_6H_4N_2][C_6H_5SbCl_5]$ and found that the rate of the reaction decreased with a decrease in the electron-attracting power of the substituent Y (171). These results were interpreted in terms of a heterolytic cleavage of the diazonium cation. They suggested the following mechanism:

$$[ArSbCl_5]^-[Ar'N_2]^+ + Fe \rightarrow [ArSbCl_3]^-[Ar'N_2]^+ + FeCl_2$$

$$[Ar\ddot{S}bCl_3]^-Ar'\text{—}N_2^+ \rightarrow ArAr'SbCl_3 + N_2$$

An excellent review of the mechanism of the Nesmeyanov reaction has been written by Reutov (172).

Mixtures of diarylantimony trichlorides (isolated as $ArSbO_2H$) and arylantimony tetrachlorides (isolated as $ArSbO_3H_2$) have been obtained by Sergeev and Bruker (48) and by Bruker (49, 50) in the reaction between antimony trichloride and arylhydrazines in the presence of cupric chloride. If antimony pentachloride is used in place of antimony trichloride, a diarylantimony trichloride is the sole antimony-containing product isolated (173). The mechanism of the latter reaction is quite similar to that proposed for the reaction with antimony trichloride (cf. VII-1-B-(1)) and involves the air oxidation of the arylhydrazine to form a diazonium cation:

$$ArNHNH_2\cdot HCl + SbCl_5 + O_2 \xrightarrow{CuCl_2} [ArN_2]^+[SbCl_6]^- + 2H_2O$$

$$ArNHNH_2\cdot HCl + [ArN_2]^+[SbCl_6]^- \xrightarrow{CuCl_2} ArSbCl_2 + [ArN_2]Cl + 4HCl + N_2$$

$$ArSbCl_2 + [ArN_2]Cl \longrightarrow [ArN_2][ArSbCl_3] \longrightarrow Ar_2SbCl_3 + N_2$$

In addition to those reactions involving some type of diazo reaction, diarylantimony trichlorides can be readily obtained from diarylstibinic acids by reaction with hydrochloric acid. Since this reaction gives essentially quantitative yields and since diarylantimony trichlorides are readily purified and possess sharp reproducible melting points, this reaction is frequently used for the purification and characterization of diarylstibinic acids. For example, Schmidt (55) prepared $(C_6H_5)_2SbCl_3\cdot H_2O$, mp 175°, from diphenylstibinic acid and dilute hydrochloric acid; Reutov and Kondratyeva (86) prepared a number of diarylantimony trichlorides from diarylstibinic acids and 5N

hydrochloric acid, while Nesmeyanov and co-workers (174) have employed alcoholic hydrochloric acid for the same purpose. Dibenzylantimony tribromide has been obtained from dibenzylstibinic anhydride and hydrobromic acid (84). In a few cases the expected trihalide is not obtained. Thus, Reutov (165) claimed to have obtained $(p\text{-}CH_3C_6H_4)_2SbCl_2OH$ from the corresponding stibinic acid and hot concentrated hydrochloric acid, while Nesmeyanov and co-workers (174) state that $(m\text{-}CH_3C_6H_4)_2SbCl_2OH$, $(o\text{-}ClC_6H_4)_2SbCl_2OH$, and $(p\text{-}IC_6H_4)_2SbCl_2OH$ are obtained when the corresponding acids are treated with alcoholic hydrochloric acid. Finally, Campbell and White (163) have reported the formation of $(4\text{-}C_2H_5O_2CC_6H_4)(3\text{-}C_6H_5C_6H_4)Sb(OH)Cl_2$ when an attempt was made to reduce (4-carbethoxyphenyl)-3-biphenylylantimony trichloride with either stannous chloride or titanous chloride. The hydroxy compound was readily converted to the trichloride upon heating with 4*N* hydrochloric acid.

Diarylantimony trichlorides can also be prepared by chlorination of diarylchlorostibines (27, 51, 175). This method is now seldom used for obtaining this type of compound. Wiberg and Mödritzer (176) have prepared diphenylantimony trichloride by the reaction of chlorine with diphenylstibine. A few other miscellaneous methods for obtaining dialkyl- and diarylantimony trichlorides have been reported. It has been stated that dibenzylantimony trichloride is obtained by the reaction between tribenzylstibine and antimony trichloride (84). The mechanism of this reaction is not obvious. Diphenylantimony trichloride is obtained as a by-product in the preparation of triphenylstibine from chlorobenzene, antimony trichloride, and sodium (107). The reaction has been used for preparative purposes, but the maximum yield of the trichloride was only about 15% (96). Bis(trifluoromethyl)antimony trichloride has been prepared by the chlorination of tetrakis(trifluoromethyl)-distibine in trichlorofluoromethane at $-78°$ (177).

No dialkyl- or diarylantimony triiodides have been reported.

B. Reactions

Dialkylantimony trihalides are relatively unstable in that they decompose with the formation of the alkyl halide on standing at room temperature (6, 177). The diaryl compounds, however, are relatively stable. They react with amine hydrochlorides to form crystalline salts of the type $[R_4N]^+[Ar_2SbCl_4]^-$ (23, 55). The pyridinium salts are readily crystallized from ethanol–hydrochloric acid; and since they can be readily hydrolyzed to diarylstibinic acids, these salts serve as useful derivatives for purifying the acids. The pyridinium salts usually do not possess sharp reproducible melting points and hence have not proved to be of value for characterization purposes (23). Campbell and White (163), however, have reported melting points for several unsymmetrical pyridinium tetrachlorodiarylantimonates. Nesmeyanov, Reutov, and other Russian workers (86, 87, 174) have prepared diazonium tetrachlorodiarylantimonates. These decompose in the range 74–170°, and it is claimed that the decomposition temperatures are sufficiently reproducible for characterization purposes (86). These same diazonium compounds have been obtained by a rather curious procedure, namely, the reaction between a diarylchlorostibine and an aryldiazonium chloride or an aryldiazonium tetrachloroferrate(III) (178). The following mechanism was postulated to account for the results:

$$Ar_2SbCl + 3ArN_2Cl \rightarrow [Ar_2SbCl_4]^-[ArN_2]^+ + 2N_2 + 2Ar\cdot$$

The formation of a diazonium tetrachlorodiarylantimonate(V) in this reaction is in contradiction to earlier work of Bruker and Nikiforova (179) who believed they had obtained the expected diazonium salts, $[Ar_2SbCl_2]^-[Ar'N_2]^+$. Reutov and co-workers (178), however, were unable to obtain such compounds under a variety of reaction conditions, and they doubt if these substances are capable of existence. The decomposition of benzenediazonium or *p*-toluenediazonium tetrachlorodiarylantimonates leads to the formation of the corresponding diarylantimony trichlorides (178).

Diarylantimony trichlorides can be reduced to diarylchlorostibines with stannous chloride in either alcohol or hydrochloric acid (56, 131, 132, 161) or with sulfur dioxide and potassium iodide in hydrochloric acid solution (48, 50, 173). There is one report of failure to reduce certain diarylantimony trichlorides with the above reagents as well as with titanous chloride (163). The reduction of diphenylantimony trichloride to diphenylstibine is discussed in Section VIII-1.

Bis(2-chlorovinyl)antimony trichloride is rapidly hydrolyzed in water but is soluble in dilute hydrochloric acid without decomposition (158). It decomposes with the evolution of acetylene on treatment with cold, aqueous alkali or on heating in alcohol with thiourea or benzyl sulfide. With sulfur dioxide, it forms bis(2-chlorovinyl) sulfone. Reduction with metallic antimony in ether results in the formation of bis(2-chlorovinyl)chlorostibine and (2-chlorovinyl)dichlorostibine.

It has been stated that diphenylantimony trichloride reacts with silver nitrate in alcohol solution to form a basic nitrate (96); the structure of this compound is unknown. Diphenylantimony trichloride reacts with methanol to produce the compound $(C_6H_5)_2Sb(Cl_2)OSb(Cl_2)(C_6H_5)_2$, which is monomolecular in benzene and dissolves in acetonitrile to form a nonconducting solution (180). The formation of $(C_6H_5)_2Sb(OCOCH_3)_3$ in solution has been postulated to explain the fact that diphenylstibinous acetate reacts with lead tetraacetate to form lead diacetate (181). Addition of alcoholic hydrogen chloride to the filtrate, after removal of the lead diacetate, produced diphenylantimony trichloride.

C. Structure and physical properties

Dialkyl- and diarylantimony trihalides occur as crystalline solids with sharp, reproducible melting points. They can be readily purified by recrystallization; the best solvent for this purpose is probably 5*N* hydrochloric acid. Relatively little is known about their structure. Polynova and Porai-Koshits (182) originally described diphenylantimony trichloride as having the configuration of a trigonal bipyramid, but in a later review paper (182a) they formulated the compound as a monohydrate, $Ph_2SbCl_3 \cdot H_2O$, with octahedral geometry. The infrared spectra of dimethylantimony trifluoride and trichloride have been determined (157). The spectrum of the latter compound contains weak bands at 565 and 502 cm^{-1}, which were assigned to the antisymmetric and symmetric C—Sb stretching frequencies, respectively. The spectrum of the trifluoride contains strong bands at 720, 588, and 550 cm^{-1}, which may be associated with Sb—F stretching. Kolditz and co-workers (180) have shown by means of conductivity measurements in acetonitrile that diphenylantimony trichloride is a weak electrolyte.

5. COMPOUNDS OF THE TYPE R_3SbX_2 AND R_3SbXY

A. Preparation

(*1*) *By halogenation of tertiary stibines*

Trialkylantimony dihalides are among the best known organoantimony compounds containing alkyl groups. They were first prepared in the period 1850–1861 (99, 103, 104), and their preparation has been reported a number of times in recent years. With the exception of the fluorides they are best prepared by the halogenation of trialkylstibines (6, 98, 100, 105, 157, 183–187). By this procedure most of the lower trialkylantimony dihalides from methyl to *n*-amyl have been prepared (34, 105). Tri-*n*-heptylantimony diiodide has also been reported (106); an attempt to obtain tricetylantimony dichloride was apparently unsuccessful (14). Other trialkylantimony dihalides prepared by direct halogenation of trialkylstibines include tricyclohexylantimony dibromide (184), trivinylantimony dibromide (188) and diiodide (189–191), tribenzylantimony dichloride and diiodide (84), tris-(chloromethyl)antimony dibromide (9), tris(trimethylsilylmethyl)antimony dibromide (192), *cis*- and *trans*-tripropenylantimony dichlorides, dibromides, and diiodides (193), triisopropenylantimony dichloride, dibromide, and diiodide (188), and tris(trifluoromethyl)antimony dichloride and dibromide (177). Tris(trifluoromethyl)antimony diiodide could not be obtained by this procedure. Hartmann and Kühl (184) reported that when triethylstibine is treated with excess bromine a salt-like compound, which they formulated as $[C_2H_5SbBr]^+[(C_2H_5)_3SbBr_3]^-$, is obtained. This formulation is based upon molecular weight determinations, analysis, and the fact that one-half of the antimony appears to be in the trivalent state.

In addition to direct halogenation, a few trialkylantimony dihalides have been obtained from trialkylstibines by the use of other halogenating agents. Trimethylstibine is such a powerful reducing agent that it will react with hydrochloric acid in a sealed tube to give trimethylantimony dichloride and hydrogen (194). Trimethylstibine also reacts with phosphorus trichloride or phosphorus pentachloride to give trimethylantimony dichloride and elemental phosphorus (195). Antimony trichloride or antimony pentachloride reacts in a similar fashion to give the dichloride and antimony metal. In an attempted preparation of tricyclohexylstibine, Hartmann and Kühl (184) treated the corresponding Grignard reagent with antimony trichloride and subsequently added dilute hydrochloric acid. They obtained, however, tricyclohexylantimony dichloride and suggested the following reaction:

$$(C_6H_{11})_3Sb + 2HCl \rightarrow (C_6H_{11})_3SbCl_2 + H_2$$

Trivinyl- and triisopropenylantimony dichloride have both been obtained from the corresponding stibines and thallic chloride (188).

Triarylantimony dihalides are also readily prepared by halogenation of triarylstibines. The reaction is usually carried out in an organic solvent, whereby the dihalides precipitate from solution in essentially quantitative yield (107, 196, 197). Generally the diiodides are more difficult to obtain in a pure state (197). Worrall (52) was unable to prepare tri-*o*-biphenylylantimony diiodide by direct iodination; Krause and Renwanz (120) obtained only antimony triiodide and α-iodothiophene from tri-α-thienylstibine and iodine, although the dibromides and dichlorides are readily obtained. It has been claimed that use of especially dried solvents and rigid exclusion of moisture is essential in the preparation of the diiodides (196).

Unsymmetrical triarylantimony dihalides of the type $Ar_2Ar'SbX_2$ (58) and mixed arylalkylantimony dihalides of the type Ar_2RSbX_2 and ArR_2SbX_2 (27, 198–200) have also been obtained by direct halogenation of the corresponding stibines. Triarylantimony dihalides have also been obtained in which all three aryl groups are different (94, 162, 163). In these cases, however, it is more difficult to obtain crystalline compounds. Campbell and White (163) failed to obtain crystalline dibromides or dichlorides by halogenation of a series of unsymmetrical triarylstibines of the type $EtO_2CC_6H_4SbArAr'$. The one exception to this finding was with (4-carboethoxyphenyl)(4-methoxy-3-biphenylyl)phenylantimony dichloride, which was obtained as a crystalline solid, mp 168–169°. The difficulty encountered in the crystallization of highly unsymmetrical triarylantimony dihalides is undoubtedly associated with the fact that these groups cannot be readily packed into the crystal lattice (201).

Tri-*p*-biphenylylantimony dibromide and diiodide are obtained in the pure state only by adding slightly less than the theoretical quantity of bromine or iodine, respectively, to a chloroform solution of the stibine (51). Tri-*p*-anisyl- and tri-*p*-phenetylantimony dichlorides cannot be prepared by direct chlorination of the corresponding triarylstibines. Instead, ring chlorination occurs (202). The dichlorides can be obtained, however, by treating the triarylstibines with cupric chloride:

$$Ar_3Sb + 2CuCl_2 \rightarrow Ar_3SbCl_2 + 2CuCl$$

The use of cupric chloride for this purpose has been used with a number of triarylstibines other than the *p*-anisyl and *p*-phenetyl compounds (107, 196). Thallic chloride can also be used for this purpose (203, 204).

Bruker and Nikiforova (179) have stated that triphenylstibine and benzenediazonium chloride react in acetic acid to give triphenylantimony dichloride, biphenyl, and nitrogen. Glushkova and co-workers (146) have obtained not only the expected tri-*m*-tolylstibine in 71.8% yield but also some tri-*m*-tolylantimony dichloride, when they added water to the reaction mixture of *m*-tolyllithium and antimony trichloride.

Only a few mixed halogen compounds of the type R_3SbXY are known. Triphenylantimony bromide iodide has been prepared by the addition of iodine bromide to an acetonitrile solution of triphenylstibine (185). A slightly impure sample of triphenylantimony chloride iodide was obtained by the interaction of the tertiary stibine and iodine monochloride in ether at about $-80°$ (157).

(2) By the diazo reaction

Triarylantimony dihalides have been obtained by means of the diazo reaction under a wide variety of reaction conditions. It might be expected that these dihalides would occur in small amounts as by-products of the preparation of arylantimony tetrahalides (or aromatic stibonic acids) from diazonium salts and antimony trichloride, since Kraft and Rossina (205) found small amounts of triarylarsine oxides from diazonium salts and arsenic oxide under conditions of the Bart reaction. No triarylantimony dihalides or oxides, however, have been found under these reaction conditions. Another obvious method would be the decomposition of an aryldiazonium dichlorodiarylantimonate(III). Bruker and Nikiforova (179) have indeed claimed to have prepared $[C_6H_5N_2]^+[(C_6H_5)_2SbCl_2]^-$, which decomposed in chloroform to yield triphenylantimony dichloride. These results, however, were later challenged by Reutov and co-workers (178) who failed in a number of attempts to obtain the diazonium dichlorodiarylantimonates, but instead obtained compounds of the type $[ArN_2]^+[Ar_2SbCl_4]^-$.

Although diazonium dichlorodiarylantimonates apparently cannot be prepared, it is possible to arylate trivalent diarylantimony compounds by the use of benzenediazoacetates and similar diazo compounds. Thus, diphenylantimony acetate reacts with benzenediazoacetate in acetone solution to give triphenylantimony diacetate (174, 206). Similarly, bis(*o*-chlorophenyl)-chlorostibine reacts with *o*-chlorobenzenediazonium tetrachloroantimonate(III) to give tris(*o*-chlorophenyl)antimony dichloride. The reaction has also been used for preparing unsymmetrical triarylantimony dichlorides, e.g., bis(*o*-chlorophenyl)-*p*-nitrophenylantimony dichloride. In their study of this reaction Nesmeyanov and co-workers (174) have found that the ease with which the reaction occurs depends largely on the nature of the diazo compound; the nature of the trivalent diarylantimony compound is of less importance. Thus, bis(*o*-chlorophenyl)chlorostibine is readily arylated with *o*-chlorobenzenediazonium tetrachloroantimonate(III); no reaction occurs when *o*-chlorobenzenediazonium tetrachlorozincate is employed.

Triarylantimony dichlorides may be formed under the conditions of the Nesmeyanov reaction. Nesmeyanov and co-workers (85, 164) found that an arenediazonium tetrachloroantimonate(III) in organic solvents is decomposed by the addition of either zinc or iron powder with the formation of organoantimony compounds, and that triarylantimony dichlorides were usually, but not always, among the products of the reaction. Triarylantimony dichlorides are also found among the reaction products when diazonium chloroantimonates of the type $[ArN_2]_2^+[SbCl_5]^=$ are decomposed under similar reaction conditions (207). Makin and Waters (175, 208) obtained triarylantimony dichlorides by the decomposition of diazonium compounds in ethyl acetate in the presence of metallic antimony powder and calcium carbonate. From the reaction of benzene-, *p*-chlorobenzene-, or *p*-bromobenzenediazonium chlorides, benzenediazonium tetrachlorozincates, or benzenediazonium tetrachloroantimonates, the corresponding triarylantimony dichlorides were the only organoantimony compounds obtained. With diazonium salts prepared from several other amines, however, triarylstibines, diarylchlorostibines, and aryldichlorostibines were also isolated. Makin and Waters believed that the mechanism of the reaction involves the formation of a covalent diazo compound, produced by intramolecular rearrangement of the diazonium salt in neutral solution. The diazo compound then undergoes homolytic cleavage to form radicals which attack the metallic antimony to yield the various organoantimony compounds formed in the reaction. This work was part of a study of aryl radicals (formed by the homolytic cleavage of diazo compounds), which has been the subject of numerous articles by

Waters and co-workers. Waters' views have been summarized in book form (209).

Makin and Waters (175) also suggest that their mechanism applies to the Nesmeyanov reaction (in which the metallic antimony is presumably formed by reduction of the $SbCl_4^-$ ion by zinc dust). This mechanism of the Nesmeyanov reaction, as well as the mechanism suggested by Nesmeyanov himself (attack of an aryl radical on the $SbCl_4^-$ ion), has been severely criticized by Reutov (172). The latter author has produced considerable evidence to show that the Nesmeyanov reaction proceeds by an ionic mechanism (cf. Section IV-4-F).

In addition to diazonium salts, other onium compounds yield triarylantimony dihalides (among other products) under conditions similar to those used in the Nesmeyanov reaction. Diaryliodonium salts, $[Ar_2I]^+[SbCl_4]^-$ and $[Ar_2I]_2^+[SbCl_5]^=$, when treated with antimony powder in either acetone or ethyl acetate, give mixtures of Ar_3SbCl_2, Ar_2SbCl, and Ar_2SbCl_3 (168, 169). Surprisingly, diarylbromonium and diarylchloronium salts, when treated with antimony powder under similar reaction conditions, give only trivalent compounds, $ArSbCl_2$ and Ar_2SbCl (210).

The isolation of pentacovalent organoantimony compounds in any of the various modifications of the Nesmeyanov reaction, in which a powdered metal is used in excess, is somewhat surprising. This discrepancy was cleared up by Reutov and Ptitsyna (160) who demonstrated that the pentavalent antimony compounds were formed by air oxidation of aryldihalostibines at the boiling point of the solvent. If the solvent is removed at room temperature, or is distilled off *in vacuo* under nitrogen, only trivalent organoantimony compounds are isolated from the reaction.

In addition to onium salts derived from Sb(III), onium salts from Sb(V) may yield triarylantimony dihalides under the conditions of the Nesmeyanov reaction. Benzenediazonium hexachloroantimonate(V), when treated with iron powder in acetone, gave diphenylantimony trichloride (isolated as diphenylstibinic acid), but the corresponding *p*-toluenediazonium salt gave a mixture of tri-*p*-tolylantimony dichloride (29%) and di-*p*-tolylantimony trichloride (58%) (86, 165). A number of other arenediazonium hexachloroantimonates give either the triarylantimony dichlorides, diarylantimony trichlorides, or mixtures of the two types of compounds. Which of the two types of compounds predominates in the reaction mixture is apparently not affected by the electronic nature of the substituent group. Diaryliodonium hexachloroantimonate(V) derivatives react in a similar manner to diazonium salts; but the principal, and often the sole, antimony-containing reaction product is the triarylantimony dichloride (211). This latter reaction has been modified to effect the preparation of unsymmetrical triarylantimony dihalides. Thus, from bis(*p*-bromophenyl)antimony trichloride and di-*p*-tolyliodonium chloride, bis(*p*-bromophenyl)-*p*-tolylantimony dichloride was obtained. Other unsymmetrical dihalides have been similarly prepared, but from di-*p*-tolylantimony trichloride and bis(*p*-chlorobenzene)iodonium chloride, the product was *p*-chlorophenyldi-*p*-tolylstibine.

Reutov (212) has studied the reaction of unsymmetrical diaryliodonium salts with compounds of tin, mercury, bismuth, and antimony. He has written the reaction in general form as follows:

$$xAr_2ICl + MCl_y \rightarrow (Ar_2ICl)_x \cdot MCl_y \xrightarrow{(2x/z)M'} Ar_xMCl_{y-x} + (2x/z)M'Cl_2 + xArI$$

In the above equation M is a metal (with valency y) which is arylated in the reaction and M′ is a metal (with valency z) which acts as a reducing agent. Reutov found that when unsymmetrical iodonium salts were used the aryl group which contained the more electron-withdrawing group served to arylate the metal M. In the case of antimony, this is illustrated by the following two reactions:

$$(C_6H_5)(p\text{-}CH_3OC_6H_4)ICl + (C_6H_5)_2SbCl_3 \xrightarrow{Zn} (C_6H_5)_3SbCl_2 + p\text{-}CH_3OC_6H_4I$$

$$(C_6H_5)(m\text{-}C_2H_5O_2CC_6H_4)ICl + (p\text{-}CH_3C_6H_4)_2SbCl_3 \xrightarrow{Zn} (m\text{-}C_2H_5O_2CC_6H_4)(p\text{-}CH_3C_6H_4)_2SbCl_2 + C_6H_5I$$

Reutov believes that the reaction involves an attack of the reducing agent (metal M′) on the anion of the iodonium salt to form a new and more nucleophilic anion. This step is followed by an S_N2 attack of the anion on the more electron-deficient phenyl group. The reaction was illustrated in the case of mercuric chloride as follows:

$$(p\text{-}CH_3OC_6H_4)(C_6H_5)I^+HgCl_3^- + Hg \longrightarrow [(p\text{-}CH_3OC_6H_4)(C_6H_5)I^+ HgCl^-] \longrightarrow p\text{-}CH_3OC_6H_4I + C_6H_5HgCl$$

(*3*) *Miscellaneous methods*

In addition to the methods described above, trialkyl- and triarylantimony dihalides have been obtained by a few methods of limited importance. Trialkly- and triarylstibine oxides or dihydroxides, when treated with hydrochloric acid, give the dihalides (59, 99). Triarylantimony dichlorides have also been obtained by treating triarylstibine oxides with either phosgene or thionyl chloride (138). Tribenzylantimony dichloride has been obtained, together with other organoantimony compounds, by the action of benzyl chloride on metallic antimony (213). Triphenylantimony diiodide has been obtained by the photolysis of aryl iodides in the presence of triphenylstibine (128, 214). Evidence was presented to show that the reaction involves the homolytic cleavage of the iodoarene and subsequent reaction of atomic iodine with the antimony atom to form the diiodide:

$$(C_6H_5)_3Sb + 2I\cdot \rightarrow (C_6H_5)_3SbI_2$$

Malinovskii and Olifirenko (215) have studied the reaction between triphenylstibine and alkyl halides in the presence of anhydrous aluminum chloride. In nearly all cases the antimony–carbon bond is cleaved, and alkylated benzenes are isolated. In the case of ethyl bromide, however, some ethyldiphenylantimony dibromide is isolated. Triphenylantimony difluoride has been obtained by the metathetical reaction between triphenylantimony dichloride and potassium fluoride (146) or silver fluoride (197). The preparation of tri-*n*-butylantimony dibromide, together with unidentified hydrocarbons, by the reaction between powdered antimony and *n*-butyl bromide in a sealed tube has been claimed by Dyke and Jones (105). No yield data are given.

Trialkylantimony dihalides have been obtained by the elimination of an alkyl halide from pentaalkylantimony compounds. Thus diethyltriisopropenylantimony, when treated with bromine in chloroform solution, gives diethylisopropenylantimony dibromide (216); diethyltrivinylantimony gives diethylvinylantimony dibromide. Similarly, treatment of pentaphenylantimony with excess chlorine yields triphenylantimony dichloride (217). Another type of elimination reaction occurs when 5-phenyl-5,5′-spirobis-(dibenzostibole) is treated with hydrochloric acid (110):

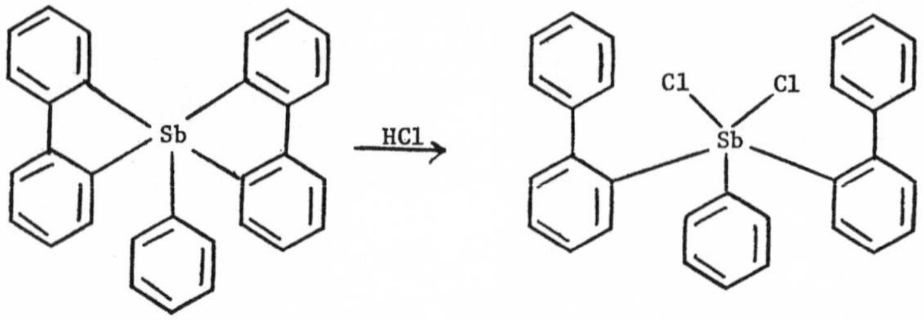

Sb —HCl→ Cl Cl Sb

Treatment of pentaphenylantimony with hot 45% hydrobromic acid gives a 90% yield of triphenylantimony dibromide (217). Stibonium salts have also been converted to antimony dihalides. Thus, Nesmeyanov and co-workers (218) found that the reaction of several tetraalkenylstibonium iodides with bromine at 0° yields trialkenylantimony dibromides. The decomposition of tetraphenylstibonium tribromide at 180° gives triphenylantimony dibromide and bromobenzene (110, 217). The reaction of $Ph_3SbO_{1.4}$ with bromine to yield triphenylantimony dibromide has been mentioned in Section VII-2-A-(1).

Tris(2-chlorovinyl)antimony dichloride has been prepared from antimony pentachloride and acetylene in the presence of mercuric chloride (158). The reaction mixture was fractionated into a major, higher-melting (93–94°) product and a minor, lower-melting (61–62°) product. Considerable evidence was presented (158, 219) to show that the major product is the *trans-trans-trans* isomer, the minor product the *cis-cis-cis* isomer. The structure of the former compound was later confirmed by means of X-ray diffraction studies (220).*

B. Reactions

Both trialkyl- and triarylantimony dihalides are readily reduced to the corresponding stibines. With trialkyl compounds, zinc (98, 101, 221) has usually been used as the reducing agent, but lithium borohydride (222), lithium aluminum hydride (222), and sodium bisulfite (158) have also been used. Triarylantimony dihalides have been reduced with hydrazine (56), with zinc (27), and in a few cases with ammoniacal hydrogen sulfide (161, 162). Usually reaction of a triarylantimony dichloride with hydrogen sulfide leads to the formation of triarylantimony sulfides (145, 146). The electrolytic reduction of triphenylantimony dichloride and trimethylantimony dichloride has been found to be a one-step, two-electron process which involves loss of chlorine as chloride ion and formation of a tertiary stibine (223):

$$R_3SbCl_2 + 2e^- \rightarrow R_3Sb + 2Cl^-$$

Dale and co-workers (177) have reported that the reaction of tris(trifluoromethyl)antimony dichloride and mercury rapidly and quantitatively yields the corresponding tertiary stibine.

Both trialkyl- and triarylantimony dihalides undergo an elimination reaction when heated above their melting points in an inert atmosphere (6, 55, 57, 184, 194, 224):

$$R_3SbX_2 \xrightarrow{\Delta} R_2SbX + RX$$

* More recent work has shown that the lower-melting product has the *cis-trans-trans* configuration; A. N. Nesmeyanov, A. E. Borisov, E. I. Fedin, I. S. Astakhova and Yu. T. Struchkov, *Izv. Akad. Nauk SSSR, Ser. Khim.*, 1977 (1969).

Tris(trifluoromethyl)antimony dibromide decomposes at 20° to yield tris-(trifluoromethyl)stibine, bis(trifluoromethyl)bromostibine, (trifluoromethyl)-dibromostibine, bromotrifluoromethane, and antimony tribromide (177). Several groups of workers have studied the decomposition of unsymmetrical dihalides. Of particular interest is the finding of Steinkopf and co-workers (225) that dimethylphenylantimony dichloride gives a mixture of dimethyl- and diphenylchlorostibine, but with dimethylphenylantimony cyanobromide only methyl bromide is eliminated to give methylphenylcyanostibine. With trimethylantimony cyanobromide, dimethylcyanostibine and methyl bromide are the sole products of the reaction (101). The mechanism of this elimination reaction has not been studied. (For the corresponding reaction with arsenic compounds, cf. Section III-6-A-(9).)

Tris(β-chlorovinyl)antimony dichloride (both the *cis-cis-cis* and *trans-trans-trans* configuration) reacts readily with mercuric chloride to give antimony trichloride and β-chlorovinylmercury chloride; the reaction occurs without change of configuration (219). The carbon–antimony bond in these trivinylantimony compounds is known to be more easily cleaved than in the case of saturated trialkylantimony compounds; e.g., both bromine and iodine react with tris(β-chlorovinyl)antimony dihalides to give *sym*-bromo-chloroethylene and *sym*-chloroiodoethylene, respectively (158).

Tris(trifluoromethyl)antimony dichloride reacts with nitrosyl chloride at $-10°$ to give $NO[(CF_3)_3SbCl_3]$ (102). This compound is probably analogous to the compound $NO[SbCl_6]$ formed from nitrosyl chloride and antimony pentachloride.

The solvolysis of trialkyl- and triarylantimony dihalides is of considerable interest. It was originally proposed by Hantzsch and Hibbert (226) and later by Lowry and Simons (100) that trimethylantimony dihalides are extensively hydrolyzed in aqueous solution:

$$R_3SbX_2 + 2H_2O \rightleftharpoons R_3Sb(OH)X + H_3O^+ + X^-$$

$$R_3Sb(OH)X + 2H_2O \rightleftharpoons R_3Sb(OH)_2 + X^- + H_3O^+$$

Both groups of workers determined the conductivities of these compounds in aqueous solution as a function of concentration. Lowry and Simons also found that the ultraviolet absorption spectrum of trimethylantimony diiodide shows only end absorption (in contrast to $(CH_3)_2TeI_2$, which gives two absorption peaks). On the basis of these results Lowry and Simons suggested that trimethylantimony diiodide might exist in the solid state as $[(CH_3)_3Sb]^{2+}2I^-$. Finally Nylén (97), in work which could serve as a model of careful investigation, demonstrated that the three compounds, $(CH_3)_3AsBr_2$, $(C_2H_5)_3AsBr_2$, and $(CH_3)_3SbBr_2$, are hydrolyzed in aqueous solution according to the equations:

$$R_3MBr_2 + 2H_2O \rightarrow R_3M(OH)Br + H_3O^+ + Br^-$$

$$R_3M(OH)Br \rightleftharpoons R_3MOH^+ + Br^-$$

$$R_3MOH^+ + H_2O \rightleftharpoons R_3MO + H_3O^+$$

Nylén's conclusions were based on determinations of both bromide ion and hydrogen ion. The former was determined by means of a silver–silver bromide electrode; the latter was determined both by the customary electrometric pH determination and also by means of a rate study of a reaction whose rate dependence on pH was known with precision. Long and co-workers (98) have pointed out that the first reaction cannot be completely irreversible in the case of trimethylantimony dibromide, since this compound can be recrystallized from water in essentially quantitative yield. Hartmann and Kühl (184) have found that tricyclohexylantimony dichloride, unlike trimethyl-, triethyl-, and tri-*n*-propylantimony dichlorides, cannot be recrystallized unchanged from aqueous solution. The hydrolysis of tris(trifluoromethyl)-antimony dichloride is discussed in Section VII-2-A-(1).

Another interesting study on the solvolysis of trialkylantimony dihalides has been undertaken by Nefedov and co-workers (221), who demonstrated that $(CH_3)_3Sb$ and $(CD_3)_3SbCl_2$ undergo isotope exchange in alcoholic solution to give some $(CD_3)_3Sb$. The rate of this exchange was determined at 0° and at 10°, and an activation energy of 5 kcal/mole was calculated. A mass spectrum of the exchange products showed the presence of $(CH_3)_3{}^{121}Sb$, $(CH_3)_3{}^{123}Sb$, $(CD_3)_3{}^{121}Sb$, and $(CD_3)_3{}^{123}Sb$, but very little of such mixed products as $(CH_3)_2CD_3{}^{121}Sb$ and $CH_3(CD_3)_2{}^{121}Sb$. From these results the authors (221) concluded that the exchange reaction involves principally an exchange of electrons:

$$(CD_3)_3Sb^{2+} + (CH_3)_3Sb \rightleftharpoons (CD_3)_3Sb + (CH_3)_3Sb^{2+}$$

This exchange may take place through a complex such as

$$[(CH_3)_3Sb{:}Sb(CD_3)_3]^{2+} \quad \text{or} \quad [(CH_3)_3Sb{:}Sb(CD_3)_3Cl]^+$$

Relatively little work has been done on the hydrolysis of triarylantimony dihalides. Bernstein and co-workers (69) claimed to have hydrolyzed triphenylantimony dichloride by boiling with water for several days, but this result is almost certainly in error (cf. Section VII-2-A-(1)). Kolditz, Gitter, and Rösel (180) state that triphenylantimony dichloride cannot be hydrolyzed by water although they concluded from conductivity measurements that solvolysis did occur in methanol solution. It seems probable, however, that the failure to detect hydrolysis is due to the insolubility of triphenylantimony dichloride in water, since it has been shown* that a 0.005*M* solution in 50% ethanol is solvolyzed to approximately 80% according to the equation:

$$(C_6H_5)_3SbCl_2 + 2H_2O \rightleftharpoons (C_6H_5)_3Sb(OH)Cl + H_3O^+ + Cl^-$$

Trialkyl- and triarylantimony dihalides undergo metathetical reactions with metallic salts to give such compounds as $R_3Sb(NO_3)_2$, R_3SbSO_4, etc. (98, 99, 104, 197, 202). Thus trimethylantimony dinitrate (98), trimethylantimony sulfate (98), triethylantimony sulfate (98), tri-*p*-anisylantimony dinitrate (202), and tri-*p*-phenetylantimony dinitrate (202) have been prepared from the corresponding dihalides and the appropriate silver salts. Trimethylantimony dithiocyanate has been obtained from trimethylantimony dibromide and potassium thiocyanate in alcohol solution (226). Triisobutylantimony diisocyanate has been obtained from triisobutylantimony dichloride and silver cyanate in acetonitrile solution (139). It is a liquid, bp 122–124°/0.15 mm. Tri-*n*-butylantimony diisocyanate has been prepared by fusing tri-*n*-butylstibine oxide with urea. Both diisocyanates were characterized by analysis and by the strong absorption at 4.6 μ in their infrared spectra, which was assigned to the isocyanate grouping.

In many cases compounds other than those expected by metathesis are formed. Tris(β-chlorovinyl)antimony dichloride, when treated with silver nitrate, gave the compound $[(ClCH{=}CH)_2Sb]_2AgNO_3$ together with $AgC{\equiv}CAg \cdot AgNO_3$ (158). Oxy compounds of the type $(R_3SbOSbR_3)Y_2$ are also frequently formed. Thus, Long and co-workers (98) obtained $[(CH_3)_3Sb{-}O{-}Sb(CH_3)_3](ClO_4)_2$ when trimethylantimony dibromide was treated with silver perchlorate in either aqueous alcohol or in absolute alcohol. Only $[(C_6H_5)_3SbOSb(C_6H_5)_3](ClO_4)_2$ was obtained from triphenylantimony dichloride and silver perchlorate; alcoholic silver nitrate gave only $[(C_6H_5)_3SbOSb(C_6H_5)_3]_2(NO_3)_2$, and silver sulfate gave only $[(C_6H_5)_3SbOSb(C_6H_5)_3]SO_4$ (197).

There has been considerable confusion in the chemical literature regarding the structure of these oxy compounds, i.e., as to whether they should be regarded as hydroxy compounds, $R_3Sb(OH)Y$, or as oxides, $(R_3SbOSbR_3)Y_2$. Morgan and co-workers (96, 101, 227) prepared a number of compounds which were formulated as hydroxy compounds. Jensen (115) has listed the preparation of $(C_6H_5)_3Sb(OH)Cl$; Wittig and Clauss (217) have reported $(C_6H_5)_3Sb(OH)Br$, and Hartmann and Kühl (184) have reported $(C_6H_{11})_3Sb(OH)Cl$. Many other similar hydroxy compounds have been reported by other workers (169).

As long ago as 1861 Landolt (99) reported that he obtained $(CH_3)_3SbO \cdot (CH_3)_3SbCl_2$ and $(CH_3)_3SbO \cdot (CH_3)_3SbBr_2$ by evaporating an aqueous

* The pH of 0.005*M* triphenylantimony dichloride is 2.7, *not* 1.3 as given in reference 197.

solution of equimolar amounts of trimethylantimony dichloride or dibromide and trimethylantimony dihydroxide. Similar results were also obtained by Hantzsch and Hibbert (226). Lyon, Mann, and Cookson (228) report that they were unable to obtain triphenylantimony dinitrate, as reported by Morgan (227), but invariably obtained the oxide, $[(C_6H_5)_3SbOSb(C_6H_5)_3](NO_3)_2$. Similarly Kolditz, Gitter, and Rösel (180) could not obtain $(C_6H_5)_3Sb(OH)Cl$ but always obtained the oxide. The latter authors established the structure by analyses and molecular weight determinations. In addition, they failed to find any hydroxyl bands in the infrared spectrum. Long and co-workers (98, 197) have made a detailed study of this type of compound. No compounds with analyses corresponding to $R_3Sb(OH)Y$ could be prepared, nor could hydroxyl bands for these compounds be found in their infrared spectra. Accordingly, these authors concluded that compounds of the type $R_3Sb(OH)Y$ probably do not exist in the solid state and that reports of such compounds in the earlier literature are based on analytical errors.

In addition to the preparation of compounds of the type R_3SbY_2 and $(R_3SbY)_2O$ by metathesis with metal salts, these compounds have been obtained by a number of other reactions. Trialkyl- and triarylantimony dihydroxides, or in some cases the corresponding oxides, react with acids to give the expected products (cf. Section VII-2-A-(3)). For example, triphenylantimony diacetate (55, 98), diformate (98), and dinitrate (103, 107) have been obtained in this manner. Triisobutylantimony diacetate has been prepared from the corresponding oxide and acetic acid in benzene solution; removal of the water formed was accomplished with a Dean-Stark trap (136). It is a colorless oil, bp 144–147°/3.3 mm. Triphenylantimony dichloride reacts with concentrated sulfuric acid to give triphenylantimony sulfate (197). This compound has also been obtained by the reaction between triphenylstibine and sulfur trioxide (112); the claim (229) to have prepared it from triphenylstibine and concentrated sulfuric acid appears to be in error (197).

Triphenylantimony dinitrate is obtained when triphenylstibine is treated with hot fuming nitric acid (107); it is best recrystallized from hot nitric acid (197). Treatment of the reaction mixture with water leads to the formation of oxybis(triphenylantimony) dinitrate (96). More recently, triphenylantimony dinitrate was prepared by the reaction between triphenylantimony dichloride and an excess of dinitrogen tetroxide (230). The reaction between the dichloride and silver nitrate in moist acetone yields oxybis(triphenylantimony) dinitrate. In contrast to triphenylstibine, tri-*p*-tolylstibine undergoes ring nitration on treatment with fuming nitric acid; the product is tris(3-nitro-4-methylphenyl)antimony dinitrate (96). Triphenylstibine *is* nitrated, however, when mixed acid is used; nitration occurs in *meta* position.

McKenney and Sisler (134) have prepared compounds of the type $(R_3SbCl)_2NH$ by the reaction of trialkyl- and triphenylstibines with chloramine. These imino compounds are readily hydrolyzed to the oxybis(trialkylantimony) and oxybis(triphenylantimony) dichlorides. The infrared and NMR spectra of the imino and oxybis compounds were shown to be consistent with the proposed structures.

Triphenylstibine is oxidized by Pb(IV) compounds according to the equation (181):

$$Pb(OCOR)_4 + (C_6H_5)_3Sb \rightarrow Pb(OCOR)_2 + (C_6H_5)_3Sb(OCOR)_2$$

Thus, either lead tetraacetate or phenyllead triacetate gives triphenylantimony diacetate. Triphenylantimony dipropionate and triphenylantimony diisobutyrate have been prepared in a similar manner.

Trialkyl- and triarylantimony diperoxides of the type $R_3Sb(OOR')_2$ have been prepared by a variety of methods (117, 231, 232, 233):

$$R_3SbX_2 + 2R'OOH + 2NH_3 \rightarrow R_3Sb(OOR')_2 + 2NH_4X$$

$$R_3SbX_2 + 2R'OONa \rightarrow R_3Sb(OOR')_2 + 2NaX$$

$$[R_3Sb{=}\overset{+}{N}H_2]X^- + 2R'OOH \rightarrow R_3Sb(OOR')_2 + NH_4X$$

$$R_3Sb(OH)_2 + 2R'OOH \rightarrow R_3Sb(OOR')_2 + 2H_2O$$

$$R_3Sb(OR')_2 + 2R''OOH \rightarrow R_3Sb(OOR'')_2 + 2R'OH$$

The trialkyl- or triarylantimony dialkoxides listed in the last equation were obtained from the corresponding dihalides and sodium alkoxides, but were not actually isolated from solution. Reaction of solutions of these dialkoxides with hydrogen peroxide gave polymeric peroxides, and also dihydroperoxides of the type $R_3Sb(OOH)_2$. The trialkyl- and triarylantimony diperoxides are stable in the air and not sensitive to shock but are readily hydrolyzed.

Only a few trialkyl- and triarylantimony dialkoxides have been reported in the chemical literature. Tri-*n*-butylantimony diethoxide was prepared from the corresponding oxide and excess ethanol in refluxing benzene (136). Water was removed by means of a Dean-Stark trap, and the diethoxide was obtained as a colorless oil, bp 104–109°/0.10 mm. Trimethylantimony di-*tert*-butoxide (234) and diethoxide (235) have been prepared in a similar manner. Triphenylantimony dimethoxide has been obtained both by treatment of the dibromide with sodium methoxide and by refluxing methoxytetraphenylantimony with absolute methanol for five weeks (113).

A number of interesting compounds have been prepared by the reaction of the peroxide $(C_6H_5)_3SbO_{1.4}$ (cf. VII-2-A-(1)) with glycols (125). Thus, from pinacol at 120–130°, 3,3,4,4-tetramethyl-1,1,1-triphenyl-2,5-dioxa-2,3,4,5-tetrahydrostibole (**18**) was obtained; at 170°, **19** was obtained. The

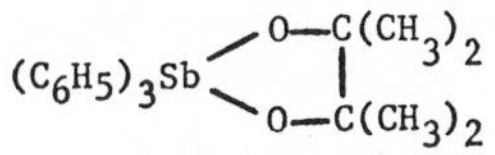

$$(C_6H_5)_3Sb\left<\begin{matrix} O—C(CH_3)_2 \\ | \\ O—C(CH_3)_2 \end{matrix}\right.$$

18

$$\begin{matrix} (C_6H_5)_2Sb\left<\begin{matrix} O—C(CH_3)_2 \\ | \\ O—C(CH_3)_2 \end{matrix}\right. \\ | \\ O \\ | \\ (C_6H_5)_2Sb\left<\begin{matrix} O—C(CH_3)_2 \\ | \\ O—C(CH_3)_2 \end{matrix}\right. \end{matrix}$$

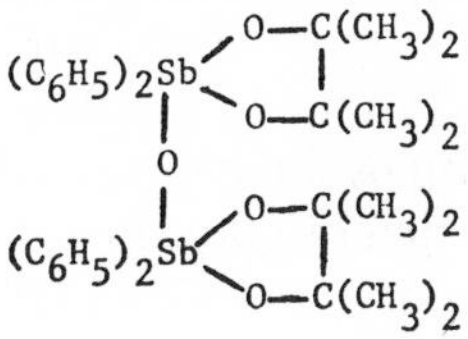

19

compound **20** was obtained when *meso*-hydrobenzoin was heated with the peroxide in dioxane; *meso*-butanediol, without solvent, gave **21**. Action of

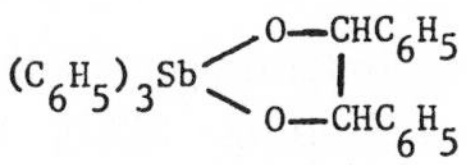

$$(C_6H_5)_3Sb\left<\begin{matrix} O—CHC_6H_5 \\ | \\ O—CHC_6H_5 \end{matrix}\right.$$

20

$$\begin{matrix} (C_6H_5)_2Sb\left<\begin{matrix} O—CH(CH_3) \\ | \\ O—CH(CH_3) \end{matrix}\right. \\ | \\ O \\ | \\ (C_6H_5)_2Sb\left<\begin{matrix} O—CH(CH_3) \\ | \\ O—CH(CH_3) \end{matrix}\right. \end{matrix}$$

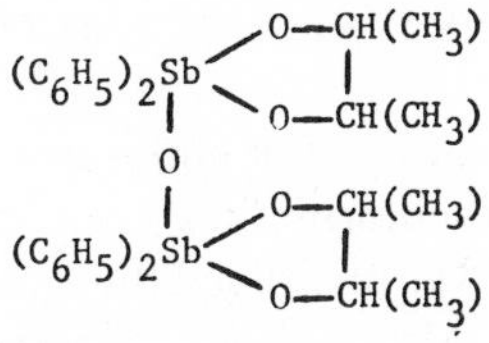

21

bromine on **18** removed a phenyl group with the formation of **22**. Unfortunately no molecular weights or spectral data were given. The structures of these interesting compounds should be further investigated.

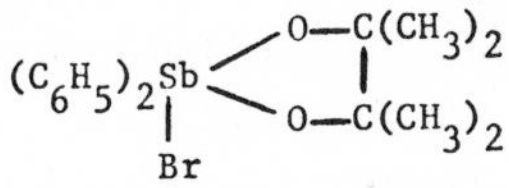

$$\begin{matrix} (C_6H_5)_2Sb\left<\begin{matrix} O—C(CH_3)_2 \\ | \\ O—C(CH_3)_2 \end{matrix}\right. \\ | \\ Br \end{matrix}$$

22

A number of interesting compounds containing the Sb—O—Si linkage have been prepared from trimethyl- or triphenylantimony dichloride and the sodium salt of trimethylsilanol (234). Thus, $Me_3Sb(OSiMe_3)_2$ and $Ph_3Sb(OSiMe_3)_2$ were prepared from the dichlorides and two moles of the sodium salt. When the reaction was carried out with equimolar amounts, the compounds $Me_3Sb(Cl)OSiMe_3$ and $Ph_3Sb(Cl)OSiMe_3$ were obtained. The compound $Ph_3Sb(OSiMe_3)_2$ was also prepared by means of the following reaction (236):

$$Ph_3SbCl_2 + 2Me_4SbOSiMe_3 \rightarrow Ph_3Sb(OSiMe_3)_2 + 2Me_4SbCl$$

Relatively few compounds of the type R_3SbXY are known. Morgan and Yarsley (101) prepared trimethylantimony cyanobromide from trimethylstibine and cyanogen bromide, while Hantzsch and Hibbert (226) prepared triphenylantimony cyanoiodide in a similar manner. Both compounds are readily hydrolyzed; trimethylantimony cyanoiodide was too unstable to be isolated. Dimethylphenylantimony cyanobromide has also been prepared (225). This compound when heated loses methyl bromide to give methylphenylcyanostibine; similarly, trimethylantimony cyanobromide yields dimethylcyanostibine.

Another interesting compound is $Ph_3Sb(Cl)OMe$ which was obtained by Kolditz, Gitter, and Rösel (180) through the reaction of one equivalent of sodium methoxide with triphenylantimony dichloride in methanol solution. Dahlmann and Rieche (233) have prepared peroxybis(trialkylantimony) and peroxybis(triarylantimony) dihalides by the interaction of hydrogen peroxide and trialkyl- or triarylantimony alkoxyhalides:

$$2R_3Sb(OR')X + H_2O_2 \rightarrow R_3Sb(X)OOSb(X)R_3 + 2R'OH$$

The interaction of equimolar quantities of triethylantimony dichloride and aluminum chloride has been found to yield a 1:1 adduct (237). When this material was dissolved in water, the triethylantimony dichloride was recovered. Although the structure of the adduct has not been determined, it may be a tetrachloroaluminate salt, viz., $[Et_3SbCl][AlCl_4]$. The reaction of titanium tetrachloride and triethylantimony dichloride yields ionic complexes, but the structure of these substances has not been elucidated (238). Titanium dichloride is oxidized by triethylantimony dichloride to yield complexes of the type $Et_3Sb \cdot nTiCl_3$. Similar complexes are formed by the interaction of triethylstibine and titanium tetrachloride. Attempts to prepare complexes directly from titanium trichloride were unsuccessful.

The reaction of trialkyl- and triarylantimony dihalides with organometallic reagents can lead to the formation of stibonium salts (Section VII-6-A) or compounds of the R_5Sb type (Section VII-8-A). The conversion of triarylantimony dichlorides to stibine imides is discussed in Section VII-7.

C. Structure and Physical Properties

Triarylantimony dihalides are usually stable, beautifully crystalline solids, which can be readily purified by recrystallization from organic solvents. They are generally insoluble in water. Many of the trialkylantimony dihalides are liquids or low-melting solids. The trimethylantimony dihalides are best recrystallized from water.

The structure of both trialkyl- and triarylantimony dihalides has been the subject of considerable investigation. Wells (239) demonstrated, by means of X-ray diffraction, that trimethylantimony dichloride, dibromide, and diiodide all exist as trigonal bipyramids with the three methyl groups in the equatorial positions and the two halides in apical positions. The antimony-halogen distances, however, are significantly longer than the sum of the covalent radii, a fact that prompted Wells to suggest that these bonds might be intermediate between covalent and ionic bonds. More recently, an X-ray diffraction study of *trans-trans-trans*-tris(2-chlorovinyl)antimony dichloride showed that the three chlorovinyl groups are in a plane with the antimony atom and that each Sb—Cl bond is nearly perpendicular (84°) to this plane (220). The Sb—Cl distance is 2.45 A (compared with 2.49 A in trimethylantimony dichloride), and the Sb—C distance is 2.15 A. Triphenylantimony dichloride has also been shown by means of X-ray diffraction to be a trigonal bipyramid in which the chlorines occupy the apical positions (240, 241). The Sb—Cl distance is 2.48 A, and the Sb—C distance is 2.16 A (241).

Conductivity studies have shown that the bonds in the trialkyl- and triarylantimony dihalides are essentially covalent. Thus, Lowry and Simons (100) in 1930 reported that trimethylantimony dichloride is nonconducting in acetonitrile solution, and Kolditz and co-workers (180) found that tribenzyl- and triphenylantimony dichlorides are essentially nonconductors in acetonitrile solution and in the molten state. More recently, Beveridge and co-workers (185) have made a thorough study of the electrolytic conductance of halogen adducts of the triphenyl derivatives of phosphorus, arsenic, antimony, and bismuth. Very low values were found for the molar conductance of triphenylantimony dichloride and dibromide in acetonitrile. The conductance of acetonitrile solutions of the diiodide was found to increase on standing. Freshly made solutions were pale yellow and were essentially nonconducting. After a short time, however, the solutions became red and were shown to contain triiodide ions; in parallel with these changes the conductance values increased. After standing for 48 hours, brown crystals were deposited. Analysis of this solid suggested the composition Ph_4SbI_2, which can be formulated as $(Ph_4Sb^+)_2(I^-)(I_3^-)$. The migration of a phenyl group under such mild conditions is remarkable. It is reminiscent of the formation of the tetraphenylbismuthonium group by the reaction between triphenylbismuth dihalides and silver perchlorate (242).

Beveridge and co-workers (185) have also attempted to prepare triphenylantimony tetrahalides. Although their attempts were unsuccessful, conductometric titration of triphenylantimony with iodine provided evidence for the formation of $Ph_3SbI^+I_3^-$ and titration of triphenylantimony iodide bromide with iodine provided evidence for $Ph_3SbBr^+I_3^-$. Earlier workers (100) reported that trimethylantimony hexaiodide, Me_3SbI_6, is formed on treating an aqueous solution of the diiodide with iodine.

The trigonal–bipyramidal geometry and the covalent nature of the trialkyl- and triarylantimony dihalides are supported by several infrared investigations (68, 98, 135, 186, 197, 243). Thus, Long and co-workers (98) have stated that the presence of only one C—Sb stretching frequency in the infrared spectra of the trimethylantimony dihalides proves that these molecules belong to the highly symmetric D_{3h} point group. It has also been concluded that the infrared spectra of the triphenylantimony dihalides are consistent with trigonal–bipyramidal structures in which the halogen atoms occupy apical positions (197, 243). It should be noted, however, that Borisov and coworkers (244) have reported that the infrared spectra of trivinylantimony dichloride, tri-*n*-butylantimony dichloride, and tri-*n*-butylantimony dibromide exhibit *two* C—Sb vibrations in the 560–460 cm^{-1} region. Further, Jensen and Nielsen (68) have reported that the infrared spectrum of triphenylantimony dibromide contains strong, sharp bands at 238 cm^{-1} and 294 cm^{-1} that should be assigned to Sb—Br vibrations.

Other physical measurements are consistent with the covalent, trigonal-bipyramidal character of the trialkyl- and triarylantimony dihalides. Thus, Jensen (108) found that triphenylantimony dichloride possesses zero dipole moment. A previous finding (245) that this compound has a dipole moment of 1.19 D is apparently in error caused by the extraordinarily large atomic polarization of the compound. Molecular weight determinations show that triphenylantimony dibromide (145) and difluoride (154) are monomeric in benzene; similar results have been obtained with triphenylantimony dichloride, tribenzylantimony dichloride, and triphenylantimony methoxychloride in bromoform (180).

There have been several NMR investigations of trialkyl- and triarylantimony dihalides (134, 154, 246, 247). At $-32°$, the proton NMR spectra of the trimethylantimony dihalides show single, sharp, ringing signals (246). In the case of the difluoride, the methyl proton signal is split by the two equivalent fluorines, and the ^{19}F spectrum also shows evidence of coupling between the fluorine and nine equivalent protons. At room temperature, there appears to be an intermolecular exchange of halogen, since each methyl signal is broadened and the coupling between the methyl protons and the fluorines is no longer observed. Proton NMR spectra of solutions containing any pair of the trimethylantimony dihalides consist of *three* methyl signals, which were shown to be characteristic of the two dihalides and a mixed halide, Me_3SbXY. Thus, when any two of the trimethylantimony dihalides are mixed together in solution, the following chemical reaction results:

$$Me_3SbX_2 + Me_3SbY_2 \rightleftharpoons 2Me_3SbXY$$

The equilibrium constants for these redistribution reactions approach the statistical value of 4. Analogous reactions have been observed with the triphenyl- and tribenzylantimony dihalides (247). Muetterties and coworkers (154) have reported the ^{19}F chemical shift of triphenylantimony chlorofluoride, but the source of this compound was not disclosed.

The structure of a number of compounds of the type R_3SbY_2 (where Y is a negative group other than halogen) has also been studied. The infrared spectra of trimethylantimony carbonate, chromate, oxalate (186), sulfate (98), diformate, diacetate, dipropionate, dibutyrate, and dibenzoate (135) show that these compounds are covalent; i.e., free anions are not present. Triethylantimony sulfate also appears to be covalent, but the exact structure has not been established (98). The trimethylantimony dicarboxylates probably have a trigonal–bipyramidal structure (135). The infrared spectrum of trimethylantimony dinitrate in a potassium bromide pellet shows the presence of nitrate ions and prompted Long and co-workers (98) to conclude that the dinitrate is ionic. Other workers (135, 186), however, demonstrated that trimethylantimony dinitrate reacts readily with solid potassium bromide to form nitrate ions and that the dinitrate in Nujol mulls or in organic solvents gives an infrared spectrum typical of a covalent compound with trigonal–bipyramidal geometry. The ultraviolet absorption spectrum in chloroform also shows the presence of covalently bonded nitrato groups (186). The Raman spectrum of solid, anhydrous trimethylantimony dinitrate is also consistent with the conclusion that this compound exists as a trigonal–bipyramidal molecule with covalently bonded unidentate nitrato groups in apical positions (248). By contrast, the Raman spectra of aqueous solutions of trimethylantimony dinitrate and perchlorate clearly show that these compounds are ionized in water. The general simplicity of these spectra suggests that the cation is the planar Me_3Sb^{2+} ion. The infrared spectra of solid triphenylantimony dinitrate (197, 230), diformate (197), diacetate (197), and sulfate (197) indicate that these groups are covalently bonded to the antimony atom.

An X-ray diffraction study of triphenylantimony dimethoxide shows that this compound is monomeric and contains pentacoordinated antimony (249). The molecular structure is a trigonal bipyramid in which the two oxygen atoms occupy apical positions. The proton NMR spectrum (113) in carbon tetrachloride and in methanol is consistent with the molecular structure in the solid state.

Comparatively little is known about the structure of compounds of the type $(R_3SbY)_2O$. The two compounds in which Y is Cl or ClO_4 and R is CH_3 have essentially identical infrared spectra (except for bands associated with the perchlorate ion) (98). Since the spectrum of the perchlorate is consistent with the presence of a perchlorate ion, and since the chloride does not contain the 277 cm^{-1} band previously found in trimethylantimony dichloride and assigned to the Sb—Cl asymmetric stretch, it was concluded that both compounds are ionic with the structure $[Me_3SbOSbMe_3]^{2+}2Y^-$. The infrared spectrum of oxybis(trimethylantimony) dinitrate suggests that the bond between the antimony atom and the ONO_2 group is intermediate in character between a covalent and an ionic bond (135). The spectrum of the corresponding dibromide is very similar to that of the dinitrate. The spectra of the aryl compounds (except in the case where Y is perchlorate) are best interpreted in terms of pentacovalent antimony (197). In agreement with this spectral data it has been found that oxybis(triphenylantimony) dichloride (180) and dinitrate (230) are weak electrolytes in acetonitrile and that the dichloride is monomolecular in bromoform (180). McKenney and Sisler (134) have reported the infrared and proton NMR spectra of a number of oxybis-(tertiary antimony) derivatives.

The question of p_π-d_π resonance between the benzene ring and the vacant *d*-orbitals of the antimony atom has been considered by Jaffé (250) and by Rao and co-workers (251). Jaffé concluded that there is only weak conjugation in the case of triphenylantimony dichloride. This conclusion was based on the ultraviolet absorption spectrum which shows well-defined vibrational bands in the 250–270 mμ region, which are characteristic of unperturbed or weakly-perturbed benzene rings. Rao and co-workers came to a similar conclusion.

The diamagnetic susceptibilities of tribenzylantimony dichloride and a number of triarylantimony dihalides have been determined by Parab and Desai (252). The average molar susceptibility of the pentavalent antimony atom was found to be -15.50×10^{-6} cgs unit.

6. STIBONIUM SALTS

A. Preparation and reactions

Tetraalkylstibonium halides were first prepared and studied over 100 years ago (253–255). The method of preparation involved the quaternization of a trialkylstibine with an alkyl halide, and this is still the principal method used for preparing these compounds. Thus, tetramethylstibonium iodide crystallizes from solution when trimethylstibine and methyl iodide are allowed to react in ether solution for one week (256, 257). Other stibonium salts prepared by this technique include trimethylbenzylstibonium chloride (258), trivinylmethylstibonium iodide (190), and tris(trimethylsilylmethyl)methylstibonium iodide (192). The rate of quaternization decreases with increasing size of the alkyl group of the alkylating agent. Quaternization of tripropylstibine with methyl iodide required one week, but quaternization of this same stibine with ethyl iodide required one month (105). Quaternary salts containing three different groups have also been prepared (198). Apparently no attempt has been made to prepare and resolve quaternary stibonium salts containing four different groups.

Neither triarylstibines (259) nor diarylalkylstibines (198) can be quaternized by treatment with alkyl halides. Triphenylstibine, however, has been quaternized by use of the powerful methylating agent, trimethyloxonium fluoborate in liquid sulfur dioxide; triphenylmethylstibonium fluoborate is the product (259). Aryldialkylstibines are readily quaternized with alkyl halides (198).

Both tetraalkyl- and tetraarylstibonium halides have been prepared by the action of halogens on pentaalkylantimony or pentaarylantimony compounds. Thus, pentamethylantimony and bromine give tetramethylstibonium bromide (217). A number of stibonium salts containing unsaturated groups, e.g., tetravinylstibonium bromide (191) and iodide (188), tetrapropenylstibonium bromide (193), and tetraisopropenylstibonium bromide and triiodide (218), have been prepared in this manner. The *cis* and *trans* isomers of diethyltripropenylantimony react with bromine at low temperatures to form the isomeric diethyldipropenylstibonium bromides (260). When pentaalkenylantimony compounds are treated with two molar equivalents of bromine, compounds of the type R_3SbBr_2 may be obtained (218). When pentaphenylantimony is treated with one equivalent of bromine or iodine, tetraphenylstibonium bromide or iodide is formed (217, 257). With excess of the halogen, however, the stibonium trihalide, $(C_6H_5)_4SbBr_3$ or $(C_6H_5)_4SbI_3$, is formed. With chlorine, even in excess, tetraphenylstibonium chloride is the sole product. One phenyl group is cleaved from pentaphenylantimony when it is treated with triphenylboron; the product is $[(C_6H_5)_4Sb]^+[(C_6H_5)_4B]^-$ (217). Pentaphenylantimony can also be converted to tetraphenylstibonium salts by means of hydrogen bromide (217), acetic acid (217), and mercuric chloride (261).

Pentaisopropenylantimony and the *cis* and *trans* isomers of pentapropenylantimony react with thallium(III) bromide to give thallium (I) bromide and stibonium compounds of the type $[R_4Sb][TlBr_4]$ (218). It was suggested that an alkenylthallium dibromide is formed and decomposes to an alkenyl bromide and thallium(I) bromide. Pentaethylantimony reacts with triethylaluminum, diethylaluminum chloride, ethylaluminum dichloride, and aluminum chloride to yield tetraethylstibonium salts of Et_4Al^-, Et_3AlCl^-, $Et_2AlCl_2^-$, and $AlCl_4^-$, respectively (262).

Probably the best method for preparing tetraphenylstibonium halides involves the reaction of triphenylstibine, bromobenzene, and anhydrous aluminum chloride (263). The method is similar to that used for the preparation of tetraphenylarsonium salts (cf. V-9-A). The reactants are heated to 230° after which the mixture is poured into water and the solution boiled and filtered. Addition of potassium bromide to the filtrate precipitates tetraphenylstibonium bromide in excellent yield. It has recently been found, however, that the material prepared in this way is a mixture of tetraphenylstibonium bromide and chloride (130). The pure chloride can be obtained by using chlorobenzene rather than bromobenzene and subsequently precipitating tetraphenylstibonium chloride by the addition of sodium chloride. Another good method for preparing tetraarylstibonium halides involves the reaction of triarylantimony dichlorides with aryl Grignard reagents (264, 265). Tetraallylstibonium chloride has been prepared by the interaction of triallylantimony dichloride and allylmagnesium chloride in ether–tetrahydrofuran (266), and several tetraalkyl tetrahaloaluminates have been obtained by reaction of a dialkylaluminum halide with an antimony trihalide (267). It has been stated that tetraphenylstibonium bromide can be prepared directly from antimony pentachloride and the phenyl Grignard reagent, but no details of this reaction are available (268).

A few other procedures for preparing stibonium salts are available. Wittig and Schwarzenbach (269) investigated use of the reagent $Zn(CH_2X)_2$ (where X is Cl or I) formed by the action of diazomethane on the zinc halide. The reagent (X = Cl) reacted with triphenylphosphine or triphenylstibine to give, after hydrolysis, methyltriphenylphosphonium or methyltriphenylstibonium chlorides in excellent yields. Nesmeyanov and Makarova (270, 271), in their study of diaryliodonium salts as arylating agents, converted triphenylstibine to tetraphenylstibonium fluoborate in 69% yield by the use of diphenyliodonium fluoborate. The conversion of an ylid to a stibonium perchlorate is discussed in Section VII-7.

A number of stibonium salts have been prepared from the halides or the fluoborates by metathesis. It is also possible to convert a stibonium salt to a stibonium hydroxide which can then react with an acid to form a different stibonium salt (130, 268).

Several groups of workers have considered tetraphenylstibonium sulfate as a reagent for precipitating various anions. A large number of anions of organic acids can be precipitated as stibonium salts. The latter usually possess sharp melting points and can be readily purified, so that it has been suggested that tetraphenylstibonium salts can be used for characterizing organic acids (268). In general the solubility of onium salts decreases in the order—phosphonium > arsonium > stibonium (272). Permanganate (272), perrhenate (273), and molybdate (273) can be readily determined as stibonium salts; fluoride can be determined as a stibonium salt by the use of a solvent extraction technique (274, 275). Tetraphenylstibonium iodide can be prepared by adding potassium iodide to an aqueous solution of tetraphenylstibonium bromide; tetraphenylstibonium chloride can be obtained from the bromide by use of an anionic ion-exchange resin in the chloride form (264).

Tetraalkylstibonium halides react readily with heavy metal halides such as $PtCl_4$, $AuCl_3$, or $HgCl_2$ to form stibonium salts of the complex, heavy metal anion. Most of the work on this type of compound was done around the middle of the nineteenth century; there has been little modern work on this subject. Dyke and Jones (105) in 1930 obtained salts of the type $[R_4Sb]_2$-$[PtCl_6]$ in agreement with older work (254). Compounds of the type $[R_4Sb]^+[AuCl_4]^-$ have also been reported (276). The formulas of the mercuric halide and bismuth trichloride adducts, reported in the older chemical literature, have not been explained in terms of modern valency theories. For example, $4(C_2H_5)_4SbCl \cdot 3HgCl_2$, $2(C_2H_5)_4SbCl \cdot 3HgCl_2$ (277), and $(C_2H_5)_4SbCl \cdot 4HgCl_2$ (276) have been reported. More recently, tetraethylstibonium salts of chlorotitanates (238) and chloroaluminates (237) have been reported.

Tetraalkylstibonium halides react with moist silver oxide to give tetraalkylstibonium hydroxides (105, 253, 255). These are strong bases, which react with acids to form stibonium salts and absorb carbon dioxide from the atmosphere. Tetraphenylstibonium hydroxide, which is virtually insoluble in water, has been obtained by treating an aqueous solution of the bromide or sulfate with aqueous ammonia (268, 274).

Several alkoxytetraphenylantimony compounds have been prepared by the reaction between tetraphenylstibonium bromide (or chloride) and the appropriate sodium alkoxides (113, 130). The alkoxy compounds are unstable and decompose rapidly at about 130° (cf. Section VII-2-A-(1)). The trimethylsiloxy compounds, $Me_4SbOSiMe_3$ and $Ph_4SbOSiMe_3$, have been obtained by the reaction between the sodium salt of trimethylsilanol and tetramethylstibonium chloride or tetraphenylstibonium bromide, respectively (234). The preparation of an ethoxytetraarylantimony compound from a pentaarylantimony compound is mentioned in Section VII-8-B. The conversion of methoxytetraphenylantimony to triphenylantimony dimethoxide has been discussed in Section VII-5-B.

Ingold, Shaw, and Wilson (278) nitrated trimethylphenylstibonium salts and found only 86.3% *meta* nitration compared with 98.2% for the corresponding arsonium salts and 100% for the phosphonium salts. This decrease in *meta*-directing power was attributed to the increased shielding effect of the electrons as the atomic number of the central atom increases. The increased shielding causes the positively-charged antimony atom to have less effect in lowering the electron density on the phenyl groups.

The reaction between stibonium salts and metal hydrides was first investigated by Wiberg and Mödritzer (279) in an attempt to prepare $(CH_3)_4SbH$. Tetramethylstibonium iodide was treated with both lithium aluminum hydride and lithium borohydride; the products were trimethylstibine and hydrogen. A similar reaction with tetraphenylstibonium bromide and lithium borohydride or lithium aluminum hydride was reported to give hydrogen, triphenylstibine, and biphenyl (280). In a later investigation Sauers (281) failed to obtain hydrogen or biphenyl when tetraphenylstibonium bromide was treated with lithium aluminum hydride but obtained only triphenylstibine and benzene. He suggests an intramolecular decomposition reaction as follows:

$$(C_6H_5)_3Sb\begin{matrix} \nearrow H \\ \nwarrow C_6H_5 \end{matrix} \longrightarrow (C_6H_5)_3Sb + C_6H_6$$

He further suggests that the formation of hydrogen found by the earlier workers could be attributed to the use of moist reactants. Sauers' results have been confirmed by Nesmeyanov and co-workers (282).

When tetraphenylstibonium hydroxide is heated to 100°, the following dehydration occurs (268, 274, 283):

$$2Ph_4SbOH \rightarrow (Ph_4Sb)_2O + H_2O$$

It has been recently reported (283) that recrystallization of the hydroxide from hot aqueous methanol yields the same oxide. Earlier workers (268) had stated that the hydroxide could be recrystallized from aqueous alcohol but made no mention of any change in composition. The thermal decomposition of methyltrivinylstibonium iodide gives methyl iodide and an 80% yield of trivinylstibine (190). The thermal decomposition of other stibonium salts has apparently not been investigated.

The conversion of stibonium salts to ylids and quinquenary derivatives is discussed in Sections VII-7 and VII-8, respectively.

B. Structure

There has been relatively little work done on the structure of stibonium salts. Wheatley (284) investigated the structure of the complex stibonium salt $[(CH_3)_4Sb]^+Al[OSi(CH_3)_3]_4^-$ by means of two-dimensional X-ray diffraction analysis. The compound is ionic; the closest approach between the anion and cation is 3.30 A; the Sb—C distance is 2.20 A. A proton NMR spectrum of this salt gives two sharp single peaks with an integrated intensity of 3:1, corresponding to the methyl groups on the silicon and on the antimony, respectively (285, 286). The infrared spectrum appears to be consistent with the ionic structure of the compound, but the cryoscopic molecular weight shows little or no dissociation (286).

The Raman spectrum of tetramethylstibonium chloride has been determined in aqueous solution (287). The results are in accord with an antimony atom with four tetrahedrally arranged methyl groups. It was also found that the C—Sb force constant in this compound as well as the C—As force constant in tetramethylarsonium chloride increases by about 20% over the corresponding force constants in trimethylstibine and trimethylarsine. By contrast the C—N force constant decreases by about 9% in going from trimethylamine to tetramethylammonium ion. Although the author of the paper offers no explanation for this result, it can possibly be attributed to the fact that both antimony and arsenic utilize principally p orbitals in the stibines and arsines but sp^3 orbitals in stibonium and arsonium salts. The result is a stronger bond in these onium compounds. By contrast nitrogen uses sp^3 hybrid orbitals in both the amine and in ammonium salts.

The infrared spectrum of tetramethylstibonium triiodomercurate(II) has been investigated by Cullen and co-workers (288). A strong band observed at 572 cm^{-1} was assigned to the C—Sb asymmetric stretch, and a very weak band at 529 cm^{-1} to the (forbidden) C—Sb symmetric stretch. These values are in good agreement with the Raman data discussed in the above paragraph. The C—H frequencies observed in the infrared and Raman spectra are abnormally high and suggest that onium ion formation leads to stronger carbon–hydrogen bonding.

Recently, Shindō and Okawara (135) have described the infrared spectra of tetramethylstibonium iodide and nitrate as mulls in Nujol or hexachlorobutadiene. Their results are also consistent with T_d symmetry for the tetramethylstibonium group. The infrared spectra of tetraphenylstibonium chloride (243), bromide (243), and perchlorate (130) have likewise been interpreted in terms of ionic structures for these compounds.

The question as to whether there can be d-orbital resonance of the type $(CH_3)_4Sb^+ \leftrightarrow H^+CH_2{=}Sb(CH_3)_3$ has been considered by Doering and Hoffmann (256). These authors compared the rate of OD^- catalyzed deuterium exchange with tetramethylammonium, phosphonium, arsonium, and stibonium ions and found that the rate was much slower with ammonium ions, which cannot be involved in d-orbital resonance. On this basis they concluded that d-orbital resonance probably contributes to the transition state in the deuterium exchange reaction of onium compounds other than ammonium.

Several investigators (130, 251, 289) have examined the ultraviolet absorption spectra of tetraphenylstibonium halides. The fine structure in the 250–270 mμ region is characteristic of unperturbed benzene rings and indicates that there is comparatively little p_π-d_π resonance between π-orbitals of the benzene ring and the vacant d-orbitals of antimony. These results are not necessarily contradictory to those of Doering and Hoffmann since the spectral results do not rule out some conjugation, and even slight p_π-d_π resonance might markedly affect the rate of deuterium exchange.

In sharp contrast to the stibonium salts discussed above, tetraphenylstibonium hydroxide (283) and methoxytetraphenylantimony (249) have been shown by means of X-ray diffraction to be covalent compounds with trigonal–bipyramidal geometry. In both compounds the oxygen atom occupies an apical position, and the axial C—Sb distance is significantly longer than the mean equatorial C—Sb distance. These results are consistent with the generally accepted rule that the more electronegative substituents occupy apical positions (290) and that axial bond lengths are greater than equatorial ones (291). Although these generalizations have been the subject of considerable speculation (291–294), the various explanations must still be considered controversial.

Doak and co-workers (130) have reported that the infrared spectra of tetraphenylstibonium hydroxide and of several alkoxytetraphenylantimony compounds contain a weak or medium intensity band between 335 and 320 cm^{-1} that is probably due to the Sb—O stretching vibration. According to Beauchamp and co-workers (283), however, the spectrum of the hydroxide contains peaks at 528 and 795 cm^{-1} that may be due to Sb—O stretching and Sb—O—H bending modes, respectively. The latter workers also noted two bands in the OH stretching region—a broad band at about 3320 cm^{-1}, which was assigned to the hydrogen–bonded OH group, and a sharp peak at 3620 cm^{-1}, which corresponds to the free OH group. The spectrum of the anhydride $(Ph_4Sb)_2O$ does not contain the OH bands nor the bands at 795 and 528 cm^{-1} but does exhibit a strong doublet at 660 and 652 cm^{-1}, which may be due to the Sb—O—Sb asymmetric stretching mode (283). The infrared spectrum of trimethylsiloxytetramethylantimony $Me_4SbOSiMe_3$ exhibits a strong band at 969 cm^{-1}, which has been assigned to a Si—O—Sb stretching mode (234).

The alkoxytetraphenylantimony compounds give the expected proton NMR spectra in deuteriochloroform or carbon tetrachloride (113, 130). The methyl peak of methoxytetraphenylantimony is, however, not observed when the spectrum is determined in methanol (113). The absence of the methyl peak has been attributed to a rapid exchange of methoxyl groups between the solvent and the antimony compound. The NMR spectrum of methoxytetra-*p*-tolylantimony in a mixture of deuteriochloroform and carbon tetrachloride shows only a single *p*-methyl peak even at $-60°$ (249). This result has been ascribed either to the existence of a rapidly established equilibrium mixture of trigonal–bipyramidal and square–pyramidal conformations or to the exclusive presence of the latter conformation.

The ultraviolet absorption spectra of tetraphenylstibonium hydroxide and the alkoxytetraphenylantimony compounds are virtually identical to one another (130). There are no clearly defined maxima or minima in the readily accessible region of the ultraviolet. The spectra exhibit a very broad, intense shoulder near 220 mμ and several inflection points in the benzenoid fine-structure region. They are quite different from the spectra of other compounds in which several phenyl groups are bonded to a single central atom

(250). Although it is not easy to explain these spectra, it is obvious that there is considerable perturbation of the benzene spectrum.

7. COMPOUNDS OF THE TYPE $R_3Sb{=}CR_2'$ and $R_3Sb{=}NR'$

Relatively little work has been done on the reactions of triarylalkylstibonium compounds with phenyllithium to form antimony ylids (295). Henry and Wittig (259) heated a solution of triphenylmethylstibonium iodide with phenyllithium and then treated the reaction mixture with benzophenone. Only trace amounts of triphenylstibine oxide and diphenylethylene were produced; the principal products were triphenylstibine and diphenylacetaldehyde. These results suggest that the ylid, $(C_6H_5)_3Sb{=}CH_2 \leftrightarrow (C_6H_5)_3\overset{+}{Sb}{-}\overset{-}{C}H_2$, is formed but that this undergoes the normal Wittig reaction to an even smaller extent than does the corresponding arsenic ylid. Thus, the success of the Wittig reaction in the formation of alkenes is in the order phosphorus > arsenic > antimony. Wittig and Laib (296) had earlier attempted to prepare other antimony ylids. Trimethylstibine and 9-bromofluorene failed to form a stibonium salt. Dimethylbenzylstibine reacted with benzyl bromide to give dimethyldibenzylstibonium bromide; but when this salt was treated with phenyllithium it underwent the Stephens rearrangement to give dimethyl(1,2-diphenylethyl)stibine:

550

$$(C_6H_5CH_2)_2\overset{+}{Sb}(CH_3)_2 + C_6H_5Li \longrightarrow (C_6H_5\overset{-}{C}H)(C_6H_5CH_2)\overset{+}{Sb}(CH_3)_2 + C_6H_6$$

$$(C_6H_5\overset{-}{C}H)(C_6H_5CH_2)\overset{+}{Sb}(CH_3)_2 \longrightarrow C_6H_5CH(CH_2C_6H_5){-}Sb(CH_3)_2$$

Lloyd and Singer (297) have succeeded in preparing triphenylstibonium tetraphenylcyclopentadienylide by heating diazotetraphenylcyclopentadiene in molten triphenylstibine. The ylid was isolated as ochre-colored crystals, which melt at 196–198° and are stable in air. Addition of ether to a perchloric acid solution of the ylid gives a stibonium perchlorate. The ylid appears to be less stable than the corresponding phosphorus or arsenic analogs and decomposes on heating in ethanol or nitromethane to yield tetraphenylcyclopentadiene. It does not undergo the Wittig reaction with benzaldehyde but reacts with *p*-nitrobenzaldehyde to give a fulvene in high yield. The ultraviolet spectrum of the ylid (λ_{max}, 349 mμ; ε, 41,500) exhibits a bathochromic shift compared to the spectra of the corresponding phosphorus, arsenic, and sulfur ylids.

Closely related to the ylids are compounds of the type $Ar_3Sb{=}NR'$. The first of these, $(C_6H_5)_3Sb{=}NCOCH_3$, was prepared by Petrenko (298) from triphenylstibine and the sodium salt of N-bromoacetamide. The product is a solid, mp 157–159°, readily recrystallized from ether. It forms 1:1 adducts with both cupric chloride and mercuric chloride. The reaction was later extended to the preparation of a number of other compounds of the type $Ar_3Sb{=}NCOCH_2X$, where Ar is *o*-, *m*-, and *p*-tolyl as well as phenyl, and X is Cl, Br, and I as well as H (299, 300). In a similar manner, Petrenko (301) used N-chlorosulfonamides to obtain the compounds $(C_6H_5)_3Sb{=}NSO_2C_6H_5$ and $(C_6H_5)_3Sb{=}NSO_2C_6H_4CH_3$-*p*. These two compounds also form 1:1 adducts with cupric chloride and with mercuric chloride. The tosyl compound was described independently by Wittig and Hellwinkel (110), who apparently were unaware of Petrenko's work. They found that it is readily hydrolyzed to triphenylantimony dihydroxide. Conversion of stibine imides to quinquenary antimony derivatives is discussed in Section VII-8-A.

Appel (302) has reported that the interaction of triarylantimony dichlorides and sodium amide results in the formation of triarylstibine imides:

$$Ar_3SbCl_2 + 2NaNH_2 \rightarrow Ar_3Sb{=}NH + 2NaCl + NH_3$$

These compounds are less stable than the corresponding phosphorus or arsenic derivatives and readily lose ammonia to form compounds of higher molecular weight. In the case of triphenylstibine imide, the main product has the following structure:

$$Ph_3Sb{\equiv}N{\equiv}Sb(Ph)_3{\equiv}N{\equiv}SbPh_3$$

8. COMPOUNDS OF THE TYPE R_5Sb

A. Preparation

Pentamethylantimony has been prepared by Wittig and Torssell (257) by the reaction between trimethylantimony dibromide and methyllithium in ether solution; the yield was 63%. It can also be prepared from tetramethylstibonium bromide and methyllithium, but the yield is somewhat lower. Pentamethylantimony is a liquid, bp 126–127°, mp −16 to −18°. It is decomposed by water but does not ignite in air. By contrast pentamethylarsenic cannot be made either from trimethylarsenic dibromide or tetramethylarsonium bromide. The reason for this difference between the two elements can hardly be due to a difference in size, since pentaphenyl derivatives of both elements are known. Recently, Takashi (262) obtained pentaethylantimony by the reaction of tetraethylstibonium chloride with ethyllithium or diethylzinc and by the reaction of triethylantimony dichloride with diethylmagnesium.

A number of pentaalkyl compounds containing unsaturated organic groups have been prepared by Nesmeyanov and co-workers (188, 191, 193, 216, 303, 304). Thus, pentavinylantimony was prepared from trivinylantimony dibromide and vinylmagnesium bromide (191). *Cis*- and *trans*-pentapropenylantimony and pentaisopropenylantimony were also prepared by similar procedures, except that lithium reagents rather than Grignard reagents were used (193, 303, 304). The pentaalkenyl compounds were also prepared by the reaction between antimony pentachloride and the appropriate lithium reagent (218). When heated to 180°, both pentaisopropenylantimony and pentavinylantimony undergo cleavage of the C—Sb bond with the formation of triisopropenylstibine and trivinylstibine, respectively (188).

In addition to the above symmetrical pentaalkylantimony compounds, Nesmeyanov and co-workers prepared diethyltriisopropenylantimony from triisopropenylantimony dibromide and ethylmagnesium bromide in ether-tetrahydrofuran solution (216). Diethyltrivinylantimony and both *cis*- and *trans*-diethyltripropenylantimony were prepared by similar reactions. When treated with bromine, the vinyl and isopropenyl compounds react with elimination of two moles of alkenyl bromide to give diethylvinyl- and diethylisopropenylantimony dibromides.

Pentaphenylantimony is obtained when either triphenylantimony dibromide or tetraphenylstibonium bromide is treated with phenyllithium (217). The yield in the latter case is 90%. Antimony pentachloride and phenyllithium give pentaphenylantimony in 45% yield (217). It can also be obtained in small yield by the reaction between tetraphenylstibonium bromide and methyllithium (257). It has been suggested that the reaction proceeds as follows:

$$[(C_6H_5)_4Sb]Br + CH_3Li \rightarrow [(C_6H_5)_4Sb(CH_3)_2]Li$$

$$\rightleftharpoons (C_6H_5)_3Sb(CH_3)_2 + C_6H_5Li$$

$$(C_6H_5)_4SbBr + C_6H_5Li \rightarrow (C_6H_5)_5Sb$$

In a later paper Wittig and Hellwinkel (110) demonstrated that many compounds of the type Ph_3SbY_2 react with phenyllithium to give pentaphenylantimony. For example, both the stibine imide, $(C_6H_5)_3Sb{=}NSO_2C_6H_4CH_3$-*p*, and $(C_6H_5)_3Sb(OCOCH_3)_2$ react to give pentaphenylantimony in 85 and 80% yields, respectively.

This latter method was then extended to the preparation of the interesting spirocyclic compound 5-phenyl-5,5′-spirobis(dibenzostibole) (110):

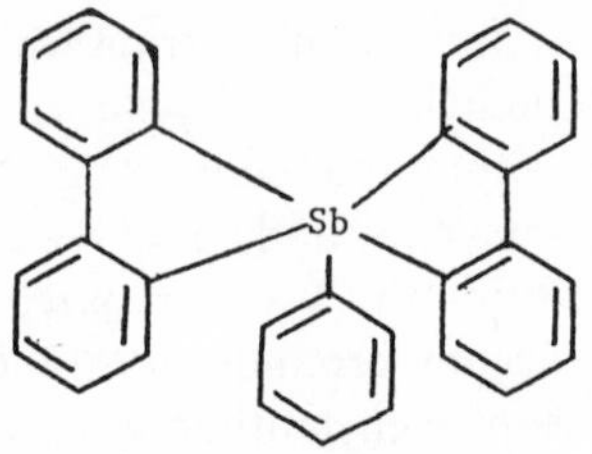

In this case 9-phenylbenzostibole was treated with chloramine-T to give the following stibine imide:

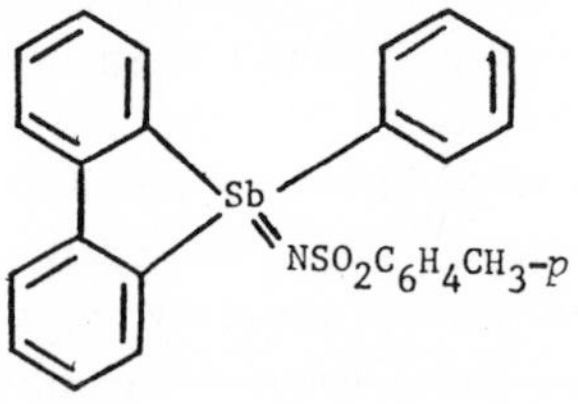

Reaction of this compound with 2,2′-dilithiobiphenyl then led to the desired spiro compound.

Nesmeyanov and co-workers (305) have obtained pentaphenylantimony in low yield by the reaction between triphenylantimony dibromide and cyclopentadienylmagnesium bromide. The mechanism of this unusual reaction has not been elucidated.

B. Reactions

Halogens cleave one phenyl group from pentaphenylantimony with the formation of such stibonium salts as $(C_6H_5)_4SbBr_3$, $(C_6H_5)_4SbBr$, $(C_6H_5)_4SbI_3$, $(C_6H_5)_4SbI$, and $(C_6H_5)_4SbCl$ (217). When heated with 40% hydrobromic acid, pentaphenylantimony yields a mixture of tetraphenylstibonium bromide and oxybis(triphenylantimony) dibromide (which Wittig and Clauss (217) formulated as triphenylantimony hydroxybromide). Warming with acetic acid cleaves one phenyl group with the formation of tetraphenylstibonium acetate (identified as the iodide). Heating pentaphenylantimony with 45% hydrobromic acid gives a 90% yield of triphenylantimony dibromide. In a similar manner 5-phenyl-5,5′-spirobis(dibenzostibole), when heated with concentrated hydrochloric acid, gives bis(2-biphenylyl)phenylantimony dichloride (110). Since tetraphenylstibonium bromide yields triphenylantimony dibromide on heating with 45% hydrobromic acid, the cleavage of two phenyl groups from pentaphenylantimony and similar compounds by the action of concentrated hydrohalic acids is not surprising.

Pentaphenylantimony reacts with triphenylboron to give tetraphenylstibonium tetraphenylborate. The mechanism of the reaction whereby a phenyl group is cleaved from pentaphenylantimony by such mild reagents as triphenylboron or acetic acid is unknown. It is possible, however, that the reaction involves a homolytic cleavage of a C—Sb bond. Thus, Razuvaev and co-workers (306) demonstrated that pentaphenylantimony reacts with chloroform in a sealed tube at 100° to give tetraphenylstibonium bromide, benzene, and biphenyl. If α,α'-diphenyl-β-trinitrohydrazyl (a reagent which is known to give a color change in the presence of free radicals) is added to the initial reaction mixture, a color change from violet to yellow is produced. Carbon tetrachloride also reacts with pentaphenylantimony in the presence of mercury to give tetraphenylstibonium chloride and phenylmercuric chloride. When pentaphenylantimony is heated with labelled benzene, unlabelled triphenylstibine and unlabelled biphenyl are produced. It was also shown that when pentaphenylantimony is heated in a sealed tube with either methanol or ethanol, a mixture of compounds, in which one or two phenyl groups have been cleaved from the antimony, is obtained. Thus, treatment of the reaction mixture with hydrochloric acid gives 80% triphenylantimony dichloride and 20% tetraphenylstibonium chloride. Benzene (identified as nitrobenzene) was the only other reaction product isolated. When pentaphenylantimony is heated with methyl iodide in a sealed tube at 100°, triphenylstibine is obtained in 95% yield, while tetraphenylstibonium iodide is formed in only 3.5% yield (307). In contrast, the analogous reaction with pentaphenylphosphorus gives a quantitative yield of tetraphenylphosphonium iodide. Razuvaev and co-workers (307) have suggested that the reaction with pentaphenylantimony first forms tetraphenylstibonium iodide, which then decomposes into triphenylstibine and iodobenzene. They showed that the stibonium iodide does decompose in this manner under the reaction conditions used.

The spiro compound 5-phenyl-5,5′-spirobis(dibenzostibole) can be recrystallized from alcohol; when heated with this solvent for several hours, however, one C—Sb bond is cleaved to yield 5-(2-biphenylyl)-5-ethoxy-5-phenyldibenzostibole (110). Pentaethylantimony reacts vigorously with methanol at room temperature to give methoxytetraethylantimony (262).

Pentaphenylantimony reacts with phenyllithium in ether solution to form $Li^+[(C_6H_5)_6Sb]^-$ (217). It is a white powder which melts unsharply at about

185°. When crystallized from tetrahydrofuran, it is obtained as $Li[(C_6H_5)_6Sb]\cdot 4C_4H_8O$; the solvent of crystallization is lost at 175°.

The reaction of pentaethylantimony with various aluminum compounds to yield stibonium salts has been discussed in Section VII-6-A.

C. Structure and properties

Pentaphenylantimony is a crystalline solid, mp 169–170°; it decomposes to triphenylstibine and biphenyl when heated to 200° (217). It is soluble in organic solvents, insoluble in water, and possesses a dipole moment of 1.59 D in benzene solution (308). Two-dimensional (308, 309) and three-dimensional (310) X-ray diffraction analysis shows that the molecule is a slightly-distorted square pyramid in which the antimony atom is about 0.5 A above the basal plane. The Sb—C (axial) distance is 2.115 A, while the Sb—C (basal) bonds have an average length of 2. 216A (310). Ideal C_{4v} symmetry is not observed, since the C (axial)—Sb—C (basal) angles deviate alternatively by +4 and −4° from their mean. If only the antimony atom and the five carbons bonded to it are considered, the molecule has virtual C_{2v} symmetry. The X-ray results are surprising, since square pyramidal geometry is usually associated with pentacovalent molecules which contain a nonbonding electron pair (291), such as the SbF_5^{2-} ion (311). Both pentaphenylphosphorus and pentaphenylarsenic are trigonal bipyramids (312). It has been suggested (without confirmation) that crystal packing factors in the solid state cause pentaphenylantimony to exhibit square pyramidal geometry (313).

The infrared and Raman absorption of pentamethylantimony has been recently investigated (314). The results leave little doubt that the molecule has a trigonal-bipyramidal skeleton (point group D_{3h}). The C—Sb equatorial stretching frequencies are located at 493 and 515 cm^{-1} and the axial stretches at 414 and 456 cm^{-1}. In accordance with theory, the C—Sb symmetric stretches at 493 and 414 are absent from the infrared spectrum of liquid pentamethylantimony; they are present, however, as weak bands in the infrared spectrum of the solid compound. This apparent breakdown of the D_{3h} selection rules was attributed to intermolecular interactions in the crystalline state and not to any gross change in molecular stereochemistry.

The infrared spectra of the *cis* and *trans* isomers of pentapropenylantimony (193, 315), diethyltripropenylantimony (260), and diphenyltripropenylantimony (260) have been described. The spectra of the *trans* isomers contain a strong absorption band near 970 cm^{-1} which is absent or weak in the spectra of the *cis* isomers. A number of generalizations concerning the spectra and stereochemistry of various propenyl-substituted antimony compounds are discussed in Section VIII-2-C.

A proton NMR study of pentamethylantimony shows only a single sharp proton resonance at 9.3 τ with a half-height line width of 1.5 cps, and this single resonance is maintained as low as $\sim -100°$ in carbon disulfide solution (154). Since it seems unlikely that the five methyl groups are geometrically equivalent, the NMR results suggest that there is either a rapid intermolecular exchange of methyl groups or some type of intramolecular exchange such as has been suggested for the fluorine groups in phosphorus pentafluoride (316).

Preliminary measurements of the electronic absorption spectrum of cyclohexane solutions of pentamethylantimony reveal at least two absorption bands at 238 mμ (ε, 5×10^3) and at about 250 mμ (ε, 10^3) (314).

The infrared spectra of pentaphenylantimony (243) and pentaethylantimony (262) have been recently described.

REFERENCES

1. W. C. Smith (to E. I. duPont de Nemours and Co.), U.S. Pat. 2,950,306 (Aug. 23, 1960); *C.A.*, **55**, 2569 (1961).
2. H. Schmidt, *Justus Liebigs Ann. Chem.*, **421**, 174 (1920).
3. B. A. Arbuzov and O. D. Samoilova, *Dokl. Akad. Nauk SSSR*, **91**, 531 (1953); through *C.A.*, **48**, 10540 (1954).
4. F. G. Mann, *The Heterocyclic Derivatives of Phosphorus, Arsenic, Antimony, Bismuth, and Silicon*, Interscience, New York, 1950, p. 5.
5. G. Long, G. O. Doak, and L. Madamba, unpublished results.
6. G. T. Morgan and G. R. Davies, *Proc. Roy. Soc., Ser. A*, **110**, 523 (1926).
7. F. A. Paneth and H. Loleit, *J. Chem. Soc.*, 366 (1935).
8. A. Ya. Yakubovich, V. A. Ginsburg, and S. P. Makarov, *Dokl. Akad. Nauk SSSR*, **71**, 303 (1950); through *C.A.*, **44**, 8319 (1950).
9. A. Ya. Yakubovich and S. P. Makarov, *Zh. Obshch. Khim.*, **22**, 1528 (1952).
10. Chem. Fabrik von Heyden Akt.-Ges., Ger. Pat. 254,421 (Aug. 3, 1911); *C.A.*, **7**, 1266 (1913).
11. F. Dunning and E. E. Reid, *J. Amer. Chem. Soc.*, **49**, 2869 (1927).
12. G. M. Dyson, *Rec. Trav. Chim. Pays-Bas*, **57**, 1016 (1938).
13. P. Pfeiffer and P. Schmidt, *J. Prakt. Chem.*, **152**, 27 (1939).
14. G. J. O'Donnell, *Iowa State Coll. J. Sci.*, **20**, 34 (1945).
15. Chem. Fabrik von Heyden Akt.-Ges., Brit. Pat. 244,746 (Dec. 18, 1924); *C.A.*, **21**, 248 (1927).
16. G. J. Burrows and E. Ritchie, *J. Proc. Roy. Soc. N.S. Wales*, **72**, 118 (1939).
17. I. G. M. Campbell, *J. Chem. Soc.*, 4 (1947).
18. C. S. Gibson and R. Kingam, Brit. Pat. 569,037 (May 2, 1945); *C.A.*, **42**, 217 (1948).
19. E. Eimers (to Farbenfabriken Bayer Akt.-Ges.), Ger. Pat. 1,064,948 (Sept. 10, 1959); through *C.A.*, **55**, 19862 (1961).
20. C. T. Chou and J. Y. Chi, *Hua Hsueh Hsueh Pao*, **25**, 38 (1959); through *C.A.*, **54**, 22438 (1960).
21. T. C. Sun and J. Y. Chi, *Yao Hsueh Hsueh Pao*, **8**, 166 (1960); through *C.A.*, **58**, 5721 (1963).
22. A. Binz and O. v. Schickh, *Ber.*, **69**, 1527 (1936).
23. G. O. Doak, L. D. Freedman, and S. M. Efland, *J. Amer. Chem. Soc.*, **74**, 830 (1952).
24. R. D. Englert and O. J. Sweeting, *J. Amer. Chem. Soc.*, **70**, 2977 (1948).

25. H. Tomono, Y. Yamakawa, and R. Nakai, *Bull. Inst. Chem. Res., Kyoto Univ.*, **26,** 99 (1951).
26. E. Eimers (to Farbenfabriken Bayer Akt.-Ges.), Ger. Pat. 1,069,149 (Nov. 19, 1959); through *C.A.*, **55,** 23446 (1961).
27. G. T. Morgan and G. R. Davies, *Proc. Roy. Soc., Ser. A*, **127,** 1 (1930).
28. G. O. Doak and H. G. Steinman, *J. Amer. Chem. Soc.*, **68,** 1987 (1946).
29. C. S. Hamilton and R. E. Etzelmiller, *J. Amer. Chem. Soc.*, **50,** 3360 (1928).
30. W. Riddell and S. Basterfield, *Trans. Roy. Soc. Can.*, **23,** Sect. 3, 45 (1929).
31. I. E. Balaban, *J. Chem. Soc.*, 1685 (1930).
32. P. Pfeiffer and K. Schneider, *Ber.*, **68,** 50 (1935).
33. P. Pfeiffer and H. Böttcher, *Ber.*, **70,** 74 (1937).
34. M. E. Brinnand, W. J. C. Dyke, W. H. Jones, and W. J. Jones, *J. Chem. Soc.*, 1815 (1932).
35. G. T. Morgan and J. W. Cook, *J. Chem. Soc.*, 737 (1930).
36. M. M. Barnett, A. H. C. P. Gillieson, and W. O. Kermack, *J. Chem. Soc.*, 433 (1934).
37. K. Kinoshita, *Yakugaku Zasshi*, **78,** 41 (1958); through *C.A.*, **52,** 11079 (1958).
38. R. Nakai, H. Tomono, and T. Azuma, *Bull. Inst. Chem. Res., Kyoto Univ.*, **22,** 92 (1950).
39. S. M. Mistry and P. C. Guha, *J. Indian Inst. Sci.*, **15A,** 25 (1932); *C.A.*, **26,** 4589 (1932).
40. A. B. Bruker, *Zh. Obshch. Khim.*, **18,** 1297 (1948); *C.A.*, **43,** 4647 (1949).
41. H. Tomono, *Bull. Inst. Chem. Res., Kyoto Univ.*, **21,** 41 (1950).
42. H. Tomono, *Bull. Inst. Chem. Res., Kyoto Univ.*, **22,** 49 (1950).
43. H. Tomono, *Bull. Inst. Chem. Res., Kyoto Univ.*, **23,** 45 (1950).
44. H. Tomono, *Bull. Inst. Chem. Res., Kyoto Univ.*, **24,** 54 (1951).
45. R. Nakai, H. Tomono, and T. Azuma, *Bull. Inst. Chem. Res., Kyoto Univ.*, **25,** 72 (1951).
46. O. A. Reutov and O. A. Ptitsyna, *Izv. Akad. Nauk SSSR, Otd. Khim. Nauk*, 93 (1952).
47. A. N. Nesmeyanov, O. A. Reutov, and P. G. Knol, *Izv. Akad. Nauk SSSR, Otd. Khim. Nauk*, 410 (1954).
48. P. G. Sergeev and A. B. Bruker, *Zh. Obshch. Khim.*, **27,** 2220 (1957).
49. A. B. Bruker, *Zh. Obshch. Khim.*, **27,** 2593 (1957).
50. A. B. Bruker, *Zh. Obshch. Khim.*, **27,** 2223 (1957).
51. D. E. Worrall, *J. Amer. Chem. Soc.*, **52,** 2046 (1930).
52. D. E. Worrall, *J. Amer. Chem. Soc.*, **62,** 2514 (1940).
53. A. Ya. Yakubovich and G. V. Motsarev, *Zh. Obshch. Khim.*, **23,** 1414 (1953).
54. A. F. Voigt, *Acta Chem. Scand.*, **1,** 118 (1947).
55. H. Schmidt, *Justus Liebigs Ann. Chem.*, **429,** 123 (1922).
56. L. A. Woods, *Iowa State Coll. J. Sci.*, **19,** 61 (1944).
57. A. E. Goddard and V. E. Yarsley, *J. Chem. Soc.*, 719 (1928).
58. F. Yu. Yusunov and Z. M. Manulkin, *Zh. Obshch. Khim.*, **31,** 3757 (1961).
59. G. T. Morgan and G. R. Davies, *Proc. Roy. Soc., Ser. A*, **143,** 38 (1933).
60. H. Schmidt, *Ber.*, **55,** 697 (1922).
61. R. G. Fargher and W. H. Gray, *J. Pharmacol. Exp. Ther.*, **18,** 341 (1921).
62. W. H. Gray and I. D. Lamb, *J. Chem. Soc.*, 401 (1938).
63. A. D. Macallum, *J. Soc. Chem. Ind.*, **42,** 468T (1923).
64. G. O. Doak, *J. Amer. Chem. Soc.*, **68,** 1991 (1946).
65. R. Nakai, R. Toyoda, and H. Tomono, *Repts. Inst. Chem. Res., Kyoto Univ.*, **18,** 22 (1949).
66. M. Ida, *Yakugaku Zasshi*, **69,** 182 (1949).

67. K. Dehnicke, *Z. Anorg. Allg. Chem.*, **312,** 237 (1961).
68. K. A. Jensen and P. H. Nielsen, *Acta Chem. Scand.*, **17,** 1875 (1963).
69. J. Bernstein, M. Halmann, S. Pinchas, and D. Samuel, *J. Chem. Soc.*, 821 (1964).
70. M. Edstrand, *Ark. Kemi*, **6,** 89 (1953).
71. L. S. P. Madamba, M.S. Thesis, North Carolina State University, 1964.
72. R. Nakai, R. Toyoda, and H. Tomono, *Repts. Inst. Chem. Res., Kyoto Univ.*, **19,** 71 (1949).
73. S. H. Gate and E. Richardson, *J. Inorg. Nucl. Chem.*, **23,** 265 (1961).
74. L. Pauling, *J. Amer. Chem. Soc.*, **55,** 1895 (1933).
75. J. Beintema, *Rec. Trav. Chim. Pays-Bas*, **56,** 931 (1937).
76. N. Schrewelius, *Z. Anorg. Allg. Chem.*, **238,** 241 (1938).
77. H. Brintzinger, *Z. Anorg. Allg. Chem.*, **256,** 98 (1948).
78. P. Souchay and D. Peschanski, *Bull. Soc. Chim. Fr.*, **15,** 439 (1948).
79. B. Ricca and G. D'Amore, *Ann. Chim.* (Rome), **46,** 483 (1956).
80. B. Ricca, G. D'Amore, and A. Bellomo, *Ann. Chim.* (Rome), **46,** 491 (1956).
81. S. H. Gate and E. Richardson, *J. Inorg. Nucl. Chem.*, **23,** 257 (1961).
82. S. C. Niyogy, *J. Indian Chem. Soc.*, **5,** 285 (1928).
83. Q. Mingoia, *Gazz. Chim. Ital.*, **62,** 343 (1932).
84. I. P. Tsukervanik and D. Smirnov, *Zh. Obshch. Khim.*, **7,** 1527 (1937); through *C.A.*, **31,** 8518 (1937).
85. A. N. Nesmeyanov, O. A. Reutov, and O. A. Ptitsyna, *Dokl. Akad. Nauk SSSR*, **91,** 1341 (1953); *C.A.*, **48,** 11375 (1954).
86. O. A. Reutov and V. V. Kondratyeva, *Zh. Obshch. Khim.*, **24,** 1259 (1954).
87. O. A. Reutov and A. G. Markovskaya, *Dokl. Akad. Nauk SSSR*, **99,** 543 (1954); *C.A.*, **49,** 15767 (1955).
88. O. A. Ptitsyna, A. N. Kozlova, and O. A. Reutov, *Izv. Akad. Nauk SSSR, Otd. Khim. Nauk*, 634 (1962).
89. Y. Kawasaki and R. Okawara, *Bull. Chem. Soc. Jap.*, **40,** 428 (1967).
90. G. O. Doak and H. H. Jaffé, *J. Amer. Chem. Soc.*, **72,** 3025 (1950).
91. G. O. Doak and H. G. Steinman, *J. Amer. Chem. Soc.*, **68,** 1989 (1946).
92. M. Ida, Z. Toyoshima, and T. Nakamura, *Yakugaku Zasshi*, **69,** 178 (1949); *C.A.*, **44,** 3929 (1950).
93. C. B. Biswell and C. S. Hamilton, *J. Amer. Chem. Soc.*, **57,** 913 (1935).
94. I. G. M. Campbell and A. W. White, *J. Chem. Soc.*, 1491 (1959).
95. F. Feigl, *Anal. Chem.*, **33,** 1118 (1961).
96. G. T. Morgan and F. M. G. Micklethwait, *J. Chem. Soc.*, **99,** 2286 (1911).
97. P. Nylén, *Z. Anorg. Allg. Chem.*, **246,** 227 (1941).
98. G. G. Long, G. O. Doak, and L. D. Freedman, *J. Amer. Chem. Soc.*, **86,** 209 (1964).
99. H. Landolt, *J. Prakt. Chem.*, **84,** 328 (1861).
100. T. M. Lowry and J. H. Simons, *Ber.*, **63,** 1595 (1930).
101. G. T. Morgan and V. E. Yarsley, *Proc. Roy. Soc., Ser. A*, **110,** 534 (1926).
102. H. J. Emeléus and J. H. Moss, *Z. Anorg. Allg. Chem.*, **282,** 24 (1955).
103. C. Löwig and E. Schweizer, *Justus Liebigs Ann. Chem.*, **75,** 315 (1850).
104. W. Merck, *Justus Liebigs Ann. Chem.*, **97,** 329 (1856).
105. W. J. C. Dyke and W. J. Jones, *J. Chem. Soc.*, 1921 (1930).
106. C. L. Tseng and W. Y. Shih, *J. Chinese Chem. Soc.*, **4,** 183 (1936); through *C.A.*, **31,** 669 (1937).
107. A. Michaelis and A. Reese, *Justus Liebigs Ann. Chem.*, **233,** 39 (1886).
108. K. A. Jensen, *Z. Anorg. Allg. Chem.*, **250,** 257 (1943).
109. G. A. Razuvaev and M. A. Shubenko, *Zh. Obshch. Khim.*, **21,** 1974 (1951).
110. G. Wittig and D. Hellwinkel, *Chem. Ber.*, **97,** 789 (1964).

111. A. Senning, *Acta Chem. Scand.*, **19,** 1755 (1965).
112. M. Becke-Goehring and H. Thielemann, *Z. Anorg. Allg. Chem.*, **308,** 33 (1961).
113. G. H. Briles and W. E. McEwen, *Tetrahedron Lett.*, 5191 (1966).
114. G. H. Briles and W. E. McEwen, *Tetrahedron Lett.*, 5299 (1966).
115. K. A. Jensen, *Z. Anorg. Allg. Chem.*, **250,** 268 (1943).
116. A. Michaelis and U. Genzken, *Justus Liebigs Ann. Chem.*, **242,** 164 (1887).
117. A. Rieche, J. Dahlmann, and D. List, *Justus Liebigs Ann. Chem.*, **678,** 167 (1964).
118. F. Challenger and A. T. Peters, *J. Chem. Soc.*, 2610 (1929).
119. E. H. Braye, W. Hübel, and I. Caplier, *J. Amer. Chem. Soc.*, **83,** 4406 (1961).
120. E. Krause and G. Renwanz, *Ber.*, **65,** 777 (1932).
121. J. J. Monagle, *J. Org. Chem.*, **27,** 3851 (1962).
122. D. L. Venezky, Abstracts of papers of the 156th meeting of the American Chemical Society, Sept. 1968, Inor 73.
123. L. Kaufmann, Ger. Pat. 360,973 (May 17, 1918); through *C.A.*, **18,** 841 (1924).
124. A. Étienne, *Bull. Soc. Chim. Fr.*, **14,** 50 (1947).
125. F. Nerdel, J. Buddrus, and K. Höher, *Chem. Ber.*, **97,** 124 (1964).
126. D. B. Copley, F. Fairbrother, J. R. Miller, and A. Thompson, *Proc. Chem. Soc.*, 300 (1964).
127. J. F. Carson and F. F. Wong, *J. Org. Chem.*, **26,** 1467 (1961).
128. M. A. Shubenko, *Sb. Statei Obshch. Khim.*, **2,** 1043 (1953); through *C.A.*, **49,** 6856 (1955).
129. N. N. Mel'nikov and M. S. Rokitskaya, *Zh. Obshch. Khim.*, **8,** 834 (1938); through *C.A.*, **33,** 1267 (1939).
130. G. O. Doak, G. G. Long, and L. D. Freedman, *J. Organometal. Chem.*, **12,** 443 (1968).
131. I. G. M. Campbell, *J. Chem. Soc.*, 4448 (1952).
132. I. G. M. Campbell and D. J. Morrill, *J. Chem. Soc.*, 1662 (1955).
133. G. N. Chremos and R. A. Zingaro, Abstracts of papers of the 156th meeting of the American Chemical Society, September 1968, Inor 74.
134. R. L. McKenney and H. H. Sisler, *Inorg. Chem.*, **6,** 1178 (1967).
135. M. Shindō and R. Okawara, *J. Organometal. Chem.*, **5,** 537 (1966).
136. S. Herbstman, *J. Org. Chem.*, **30,** 1259 (1965).
137. J. R. Leebrick (to M & T Chemicals Inc.), U.S. Pat. 3,287,210 (Nov. 22, 1966); through *C.A.*, **66,** 7948 (1967).
138. R. Appel and W. Heinzelmann (Badische Anilin- & Soda-Fabrik A.-G.), Ger. Pat. 1,192,205 (May 6, 1965); through *C.A.*, **63,** 8405 (1965).
139. W. Stamm, *J. Org. Chem.*, **30,** 693 (1965).
140. W. Stamm (to Stauffer Chemical Co.), Ger. Pat. 1,229,529 (Dec. 1, 1966); through *C.A.*, **66,** 2760 (1967).
141. M. Shindo, Y. Matsumura, and R. Okawara, *J. Organometal. Chem.*, **11,** 299 (1968).
141a. F. Hein and H. Hecker, *Z. Naturforsch.*, *B*, **11,** 677 (1956).
142. R. A. Zingaro and A. Merijanian, *J. Organometal. Chem.*, **1,** 369 (1964).
143. M. Shindo and R. Okawara, *Inorg. Nucl. Chem. Lett.*, **3,** 75 (1967).
144. G. M. Dyson, *Pharm. J.*, **121,** 596 (1928); through *C.A.*, **23,** 4534 (1929).
145. W. J. Lile and R. C. Menzies, *J. Chem. Soc.*, 617 (1950).
146. V. P. Glushkova, T. V. Talalaeva, Z. P. Razmanova, G. S. Zhdanov, and K. A. Kocheshkov, *Sb. Statei Obshch. Khim.*, **2,** 992 (1953); through *C.A.*, **49,** 6859 (1955).
147. R. B. Sandin, F. T. McClure, and F. Irwin, *J. Amer. Chem. Soc.*, **61,** 2944 (1939).
148. G. S. Zhdanov, V. A. Pospelov, M. M. Umanskii, and V. P. Glushkova, *Dokl. Akad. Nauk SSSR*, **92,** 983 (1953); through *C.A.*, **49,** 12075 (1955).
148a. D. B. Meyers and J. W. Jones, *J. Amer. Pharm. Ass.*, **39,** 401 (1950).
149. J. Hasenbäumer, *Ber.*, **31,** 2910 (1898).

150. H. Lecoq, *J. Pharm. Belg.*, **19,** 173 (1937).
151. R. E. D. Clark, *J. Chem. Soc.*, 2737 (1930).
152. H. Moureu, M. Magat, and G. Wetroff, *Proc. Indian Acad. Sci.*, *A*, **8,** 356 (1938).
153. S. M. Ohlberg, *J. Amer. Chem. Soc.*, **81,** 811 (1959).
154. E. L. Muetterties, W. Mahler, K. J. Packer, and R. Schmutzler, *Inorg. Chem.*, **3,** 1298 (1964).
155. L. M. Kul'berg and V. F. Barkovskii, *Dokl. Akad. Nauk SSSR*, **85,** 335 (1952); through *C.A.*, **47,** 12280 (1953).
156. J. Y. Chi, *Hua Hseuh Hsueh Pao*, **25,** 189 (1959); through *C.A.*, **54,** 4451 (1960).
157. G. O. Doak and G. G. Long, *Trans. N.Y. Acad. Sci.*, **28,** 402 (1966).
158. A. N. Nesmeyanov and A. E. Borisov, *Izv. Akad. Nauk SSSR, Otd. Khim. Nauk*, 251 (1945); through *C.A.*, **40,** 2123 (1946).
159. A. B. Bruker and E. S. Makhlis, *Zh. Obshch. Khim.*, **7,** 1880 (1937); through *C.A.*, **32,** 72 (1938).
160. O. A. Reutov and O. A. Ptitsyna, *Dokl. Akad. Nauk SSSR*, **79,** 819 (1951).
161. I. G. M. Campbell, *J. Chem. Soc.*, 3109 (1950).
162. I. G. M. Campbell, *J. Chem. Soc.*, 3116 (1955).
163. I. G. M. Campbell and A. W. White, *J. Chem. Soc.*, 1184 (1958).
164. A. N. Nesmeyanov and K. A. Kocheshkov, *Izv. Akad. Nauk SSSR, Otd. Khim. Nauk*, 416 (1944); *C.A.*, **39,** 4320 (1945).
165. O. A. Reutov, *Dokl. Akad. Nauk SSSR*, **87,** 991 (1952); *C.A.*, **48,** 143 (1954).
166. O. A. Reutov, *Dokl. Akad. Nauk SSSR*, **87,** 73 (1952).
167. O. A. Reutov and A. G. Markovskaya, *Dokl. Akad. Nauk SSSR*, **98,** 979 (1954); *C.A.*, **49,** 2926 (1955).
168. O. A. Reutov, O. A. Ptitsyna, and G. Ertel, *Chem. Tech.* (Berlin), **10,** 201 (1958).
169. O. A. Ptitsyna, O. A. Reutov, and G. Ertel, *Izv. Akad. Nauk SSSR, Otd. Khim. Nauk*, 265 (1961).
170. O. A. Reutov, A. G. Markovskaya, and R. E. Mardaleishvili, *Dokl. Akad. Nauk SSSR*, **104,** 253 (1955); through *C.A.*, **50,** 6160 (1956).
171. O. A. Reutov, A. G. Markovskaya, and R. E. Mardaleishvili, *Zh. Fiz. Khim.*, **30,** 2533 (1956); through *C.A.*, **51,** 9511 (1957).
172. O. A. Reutow, *Tetrahedron*, **1,** 67 (1957).
173. A. B. Bruker, *Zh. Obshch. Khim.*, **27,** 2700 (1957).
174. A. N. Nesmeyanov, O. A. Reutov, O. A. Ptitsyna, and P. A. Tsurkan, *Izv. Akad. Nauk SSSR, Otd. Khim. Nauk*, 1435 (1958).
175. F. B. Makin and W. A. Waters, *J. Chem. Soc.*, 843 (1938).
176. E. Wiberg and K. Mödritzer, *Z. Naturforsch.*, *B*, **12,** 131 (1957).
177. J. W. Dale, H. J. Emeléus, R. N. Haszeldine, and J. H. Moss, *J. Chem. Soc.*, 3708 (1957).
178. O. A. Reutov, O. A. Ptitsyna, A. N. Lovtsova, and V. F. Petrova, *Zh. Obshch. Khim.*, **29,** 3857 (1959).
179. A. B. Bruker and N. M. Nikiforova, *Zh. Obshch. Khim.*, **18,** 1133 (1948); through *C.A.*, **43,** 1737 (1949).
180. L. Kolditz, M. Gitter, and E. Rösel, *Z. Anorg. Allg. Chem.*, **316,** 270 (1962).
181. V. I. Lodochnikova, E. M. Panov, and K. A. Kocheshkov, *Zh. Obshch. Khim.*, **34,** 946 (1964).
182. T. N. Polynova and M. A. Porai-Koshits, *Zh. Strukt. Khim.*, **2,** 477 (1961).
182a. T. N. Polynova and M. A. Porai-Koshits, *Zh. Strukt. Khim.*, **7,** 642 (1966).
183. H. Hibbert, *Ber.*, **39,** 160 (1906).
184. H. Hartmann and G. Kühl, *Z. Anorg. Allg. Chem.*, **312,** 186 (1961).
185. A. D. Beveridge, G. S. Harris, and F. Inglis, *J. Chem. Soc.*, *A*, 520 (1966).

186. H. C. Clark and R. G. Goel, *Inorg. Chem.*, **5**, 998 (1966).
187. G. O. Doak, G. G. Long, and M. E. Key, *Inorg. Syn.*, **9**, 92 (1967).
188. A. N. Nesmeyanov, A. E. Borisov, and N. V. Novikova, *Izv. Akad. Nauk SSSR, Otd. Khim. Nauk*, 1578 (1961).
189. L. Maier, D. Seyferth, F. G. A. Stone, and E. G. Rochow, *Z. Naturforsch., B*, **12,** 263 (1957).
190. L. Maier, D. Seyferth, F. G. A. Stone, and E. G. Rochow, *J. Amer. Chem. Soc.*, **79,** 5884 (1957).
191. A. N. Nesmeyanov, A. E. Borisov, and N. V. Novikova, *Izv. Akad. Nauk SSSR, Otd. Khim. Nauk*, 952 (1960).
192. D. Seyferth, *J. Amer. Chem. Soc.*, **80,** 1336 (1958).
193. A. N. Nesmeyanov, A. E. Borisov, and N. V. Novikova, *Izv. Akad. Nauk SSSR, Otd. Khim. Nauk*, 612 (1961).
194. A. B. Burg and L. R. Grant, *J. Amer. Chem. Soc.*, **81,** 1 (1959).
195. R. R. Holmes and E. F. Bertaut, *J. Amer. Chem. Soc.*, **80,** 2983 (1958).
196. J. I. Harris, S. T. Bowden, and W. J. Jones, *J. Chem. Soc.*, 1568 (1947).
197. G. O. Doak, G. G. Long, and L. D. Freedman, *J. Organometal. Chem.*, **4,** 82 (1965).
198. G. Grüttner and M. Wiernik, *Ber.*, **48,** 1759 (1915).
199. A. N. Tatarenko, *Dokl. Akad. Nauk Uzb. SSR*, 35 (1955); through *C.A.*, **52,** 20005 (1958).
200. A. N. Tatarenko and M. A. Edel'man, *Tr. Tashkentsk. Farmatsevt. Inst.*, **2,** 349 (1960); through *C.A.*, **56,** 7350 (1962).
201. F. G. Mann and J. Watson, *J. Chem. Soc.*, 505 (1947).
202. C. Löloff, *Ber.*, **30,** 2834 (1897).
203. A. E. Goddard, *J. Chem. Soc.*, **121,** 36 (1922).
204. A. E. Goddard, *J. Chem. Soc.*, **123,** 1161 (1923).
205. M. Ya. Kraft and S. A. Rossina, *Dokl. Akad. Nauk SSSR*, **55,** 821 (1947).
206. O. A. Reutov and O. A. Ptitsyna, *Dokl. Akad. Nauk SSSR*, **89,** 877 (153); through *C.A.*, **48,** 5135 (1954).
207. A. N. Nesmeyanov, H. K. Gipp, L. G. Makarova, and K. K. Mozgova, *Izv. Akad. Nauk SSSR, Otd. Khim. Nauk*, 298 (1953).
208. W. A. Waters, *J. Chem. Soc.*, 2007 (1937).
209. W. A. Waters, *The Chemistry of Free Radicals*, 2nd ed., Clarendon Press, Oxford, 1948, pp. 146–165.
210. A. N. Nesmeyanov, O. A. Reutov, T. P. Tolstaya, O. A. Ptitsyna, L. S. Isaeva, M. F. Turchinskii, and G. P. Bochkareva, *Dokl. Akad. Nauk SSSR*, **125,** 1265 (1959).
211. O. A. Reutov and A. N. Lovtsova, *Vestn. Mosk. Univ., Ser. Mat., Mekh., Astron., Fiz., Khim.*, **13,** 191 through *C.A.*, (1958); **53,** 11283 (1959).
212. O. A. Reutow, *Theoret. Org. Chem., Papers Kekule Symposium, London*, 176 (1958).
213. H. Grohn, H. Friedrich, and R. Paudert, *Z.Chem.*, **2,** 24 (1962).
214. G. A. Razuvaev and M. A. Shubenko, *Dokl. Akad. Nauk SSSR*, **67,** 1049 (1949); through *C.A.*, **44,** 1435 (1950).
215. M. S. Malinovskii and S. P. Olifirenko, *Zh. Obshch. Khim.*, **25,** 2437 (1955).
216. A. N. Nesmeyanov, A. E. Borisov, and N. V. Novikova, *Izv. Akad. Nauk SSSR, Otd. Khim. Nauk*, 730 (1961).
217. G. Wittig and K. Clauss, *Justus Liebigs Ann. Chem.*, **577,** 26 (1952).
218. A. N. Nesmeyanov, A. E. Borisov, and N. V. Novikova, *Izv. Akad. Nauk SSSR, Ser. Khim.*, 1202 (1964).
219. A. N. Nesmeyanov, *Izv. Akad. Nauk SSSR, Otd. Khim. Nauk*, 239 (1945); through *C.A.*, **40,** 2122 (1946).

220. Yu. T. Struchkov and T. L. Khotsyanova, *Dokl. Akad. Nauk SSSR*, **91**, 565 (1953); through *C.A.*, **48**, 422 (1954).
221. V. N. Nefedov, I. S. Kirin, V. M. Zaitsev, G. A. Semenov, and B. E. Dzevitskii, *Zh. Obshch. Khim.*, **33**, 2407 (1963).
222. E. Wiberg and K. Mödritzer, *Z. Naturforsch.*, *B*, **11**, 747 (1956).
223. R. E. Dessy, T. Chivers, and W. Kitching, *J. Amer. Chem. Soc.*, **88**, 467 (1966).
224. S. Herbstman, *J. Org. Chem.*, **29**, 986 (1964).
225. W. Steinkopf, I. Schubart, and J. Roch, *Ber.*, **65**, 409 (1932).
226. A. Hantzsch and H. Hibbert, *Ber.*, **40**, 1508 (1907).
227. G. T. Morgan, F. M. G. Micklethwait, and G. S. Whitby, *J. Chem. Soc.*, **97**, 34 (1910).
228. D. R. Lyon, F. G. Mann, and G. H. Cookson, *J. Chem. Soc.*, 662 (1947).
229. P. May, *J. Chem. Soc.*, **97**, 1956 (1910).
230. G. C. Tranter, C. C. Addison, and D. B. Sowerby, *J. Organometal. Chem.*, **12**, 369 (1968).
231. A. Rieche, J. Dahlmann, and D. List, *Angew. Chem.*, **73**, 494 (1961).
232. A. Rieche and J. Dahlmann (to Deutsche Akademie der Wissenschaften zu Berlin), Ger. Pat. 1,158,975 (Dec. 12, 1963); *C.A.*, **60**, 9313 (1964).
233. J. Dahlmann and A. Rieche, *Chem. Ber.*, **100**, 1544 (1967).
234. H. Schmidbaur, H. S. Arnold, and E. Beinhofer, *Chem. Ber.*, **97**, 449 (1964).
235. Y. Matsumura, M. Shindo, and R. Okawara, *Inorg. Nucl. Chem. Lett.*, **3**, 219 (1967).
236. H. Schmidbaur and M. Schmidt, *Angew. Chem.*, **73**, 655 (1961).
237. Y. Takashi and I. Aishima, *J. Organometal. Chem.*, **8**, 209 (1967).
238. Y. Takashi, *Bull. Chem. Soc., Jap.* **40**, 1194 (1967).
239. A. F. Wells, *Z. Kristallogr.*, **99**, 367 (1938).
240. T. N. Polynova and M. A. Porai-Koshits, *Zh. Strukt. Khim.*, **1**, 159 (1960).
241. T. N. Polynova and M. A. Porai-Koshits, *Zh. Strukt. Khim.*, **7**, 742 (1966).
242. G. O. Doak, G. G. Long, S. K. Kakar, and L. D. Freedman, *J. Amer. Chem. Soc.*, **88**, 2342 (1966).
243. K. M. Mackay, D. B. Sowerby, and W. C. Young, *Spectrochim. Acta*, *A*, **24**, 611 (1968).
244. A. E. Borisov, N. V. Novikova, N. A. Chumaevskii, and E. B. Shkirtil, *Dokl. Akad. Nauk SSSR*, **173**, 855 (1967).
245. P. F. Oesper and C. P. Smyth, *J. Amer. Chem. Soc.*, **64**, 173 (1942).
246. G. G. Long, C. G. Moreland, G. O. Doak, and M. Miller, *Inorg. Chem.*, **5**, 1358 (1966).
247. C. G. Moreland, M. H. O'Brien, C. E. Douthit, and G. G. Long, *Inorg. Chem.*, **7**, 834 (1968).
248. A. J. Downs and I. A. Steer, *J. Organometal. Chem.*, **8**, P21 (1967).
249. K. Shen, W. E. McEwen, S. J. LaPlaca, W. C. Hamilton, and A. P. Wolf, *J. Amer. Chem. Soc.*, **90**, 1718 (1968).
250. H. H. Jaffé, *J. Chem. Phys.*, **22**, 1430 (1954).
251. C. N. R. Rao, J. Ramachandran, and A. Balasubramanian, *Can. J. Chem.*, **39**, 171 (1961).
252. N. K. Parab and D. M. Desai, *J. Indian Chem. Soc.*, **35**, 569 (1958).
253. H. Landolt, *Justus Liebigs Ann. Chem.*, **78**, 91 (1851).
254. H. Landolt, *Justus Liebigs Ann. Chem.*, **84**, 44 (1852).
255. S. Friedländer, *J. Prakt. Chem.*, **70**, 449 (1857).
256. W. von E. Doering and A. K. Hoffmann, *J. Amer. Chem. Soc.*, **77**, 521 (1955).
257. G. Wittig and K. Torssell, *Acta Chem. Scand.*, **7**, 1293 (1955).
258. F. Challenger and L. Ellis, *J. Chem. Soc.*, 396 (1935).

259. M. C. Henry and G. Wittig, *J. Amer. Chem. Soc.*, **82,** 563 (1960).
260. A. N. Nesmeyanov, A. E. Borisov, and N. V. Novikova, *Izv. Akad. Nauk SSSR, Ser. Khim.*, 1197 (1964).
261. G. Wittig and D. Hellwinkel, *Chem. Ber.*, **97,** 769 (1964).
262. Y. Takashi, *J. Organometal. Chem.*, **8,** 225 (1967).
263. J. Chatt and F. G. Mann, *J. Chem. Soc.*, 1192 (1940).
264. H. H. Willard, L. R. Perkins, and F. F. Blicke, *J. Amer. Chem. Soc.*, **70,** 737 (1948).
265. H. E. Affsprung and A. B. Gainer, *Anal. Chim. Acta*, **27,** 578 (1962).
266. A. E. Borisov, N. V. Novikova, and A. N. Nesmeyanov, *Izv. Akad. Nauk SSSR, Ser. Khim.*, 1506 (1963).
267. Y. Takashi, I. Aijima, Y. Kobayashi, and Y. Sumita (to Asahi Chemical Industry Co., Ltd.), Jap. Pat. 22,571 (Oct. 24, 1963); through *C.A.*, **60,** 3004 (1964).
268. H. E. Affsprung and H. E. May, *Anal. Chem.*, **32,** 1164 (1960).
269. G. Wittig and K. Schwarzenbach, *Justus Liebigs Ann. Chem.*, **650,** 1 (1961).
270. L. G. Makarova and A. N. Nesmeyanov, *Izv. Akad. Nauk SSSR, Otd. Khim. Nauk*, 617 (1945); through *C.A.*, **40,** 4686 (1945).
271. A. N. Nesmeyanov and L. G. Makarova, in A. N. Nesmeyanov, *Selected Works in Organic Chemistry*, Macmillan, New York, 1963, p. 748.
272. H. H. Willard and L. R. Perkins, *Anal. Chem.*, **25,** 1634 (1953).
273. M. Ziegler and M. Gindl, *Naturwissenschaften*, **54,** 19 (1967).
274. K. D. Moffett, J. R. Simmler, and H. A. Potratz, *Anal. Chem.*, **28,** 1356 (1956).
275. J. B. Orenberg and M. D. Morris, *Anal. Chem.*, **39,** 1776 (1967).
276. A. Partheil and E. Mannheim, *Arch. Pharm.*, **238,** 166 (1900).
277. R. Löwig, *Justus Liebigs Ann. Chem.*, **97,** 322 (1856).
278. C. K. Ingold, F. R. Shaw, and I. S. Wilson, *J. Chem. Soc.*, 1280 (1928).
279. E. Wiberg and K. Mödritzer, *Z. Naturforsch.*, *B*, **11,** 750 (1956).
280. E. Wiberg and K. Mödritzer, *Z. Naturforsch.*, *B*, **11,** 753 (1956).
281. R. R. Sauers, *Chem. Ind.* (London), 717 (1960).
282. A. N. Nesmeyanov, A. E. Borisov, and N. V. Novikova, *Izv. Akad. Nauk SSSR, Ser. Khim.*, 815 (1967).
283. A. L. Beauchamp, M. J. Bennett, and F. A. Cotton, *J. Amer. Chem. Soc.*, **91,** 297 (1969).
284. P. J. Wheatley, *J. Chem. Soc.*, 3200 (1963).
285. H. Schmidbaur, *Angew. Chem.*, **75,** 137 (1963).
286. H. Schmidbaur, *Chem. Ber.*, **97,** 842 (1964).
287. H. Siebert, *Z. Anorg. Allg. Chem.*, **273,** 161 (1953).
288. W. R. Cullen, G. B. Deacon, and J. H. S. Green, *Can. J. Chem.*, **43,** 3193 (1965).
289. H. Schindlbauer, *Spectrochim. Acta*, **20,** 1143 (1964).
290. E. L. Muetterties and R. A. Schunn, *Quart. Rev.* (London), **20,** 245 (1966).
291. R. J. Gillespie, *J. Chem. Soc.*, 4672 (1963).
292. F. A. Cotton, *J. Chem. Phys.*, **35,** 228 (1961).
293. R. E. Rundel, *Rec. Chem. Progr.*, **23,** 195 (1962).
294. R. R. Holmes, R. P. Carter, Jr., and G. E. Peterson, *Inorg. Chem.*, **3,** 1748 (1964).
295. A. W. Johnson, *Ylid Chemistry*, Academic Press, New York, 1966, pp. 301–302.
296. G. Wittig and H. Laib, *Justus Liebigs Ann. Chem.*, **580,** 57 (1953).
297. D. Lloyd and M. I. C. Singer, *Chem. Ind.* (London), 787 (1967).
298. L. P. Petrenko, *Zh. Obshch. Khim.*, **24,** 520 (1954).
299. L. P. Petrenko, *Tr. Voronezhsk. Gos. Univ.*, **49,** 19 (1958); through *C.A.*, **56,** 2470 (1962).
300. L. P. Petrenko, *Tr. Voronezhsk. Gos. Univ.*, **49,** 25 (1958); through *C.A.*, **56,** 2471 (1962).

301. L. P. Petrenko, *Tr. Voronezhsk. Gos. Univ.*, **57**, 145 (1959); through *C.A.*, **55**, 6425 (1961).
302. R. Appel, Abstracts of papers, 142nd Meeting of the American Chemical Society, Atlantic City, September 1962, p. 38N.
303. A. N. Nesmeyanov, A. E. Borisov, and N. V. Novikova, *Izv. Akad. Nauk SSSR, Otd. Khim. Nauk*, 147 (1960).
304. A. N. Nesmeyanov, A. E. Borisov, and N. V. Novikova, *Tetrahedron Lett.*, No. 8, 23 (1960).
305. N. A. Nesmeyanov, V. V. Pravdina, and O. A. Reutov, *Dokl. Akad. Nauk SSSR*, **155**, 1364 (1964).
306. G. A. Razuvaev, N. A. Osanova, N. P. Shulaev, and B. M. Tsigin, *Zh. Obshch. Khim.*, **30**, 3234 (1960).
307. G. A. Razuvaev, N. A. Osanova, and Yu. A. Sangalov, *Zh. Obshch. Khim.*, **37**, 216 (1967).
308. P. J. Wheatley and G. Wittig, *Proc. Chem. Soc.*, 251 (1962).
309. P. J. Wheatley, *J. Chem. Soc.*, 3718 (1964).
310. A. L. Beauchamp, M. J. Bennett, and F. A. Cotton, *J. Amer. Chem. Soc.*, **90**, 6675 (1968).
311. A. Byström and K. A. Wilhelmi, *Ark. Kemi*, **3**, 461 (1951).
312. P. J. Wheatley, *J. Chem. Soc.*, 2206 (1964).
313. E. L. Muetterties, W. Mahler, and R. Schmutzler, *Inorg. Chem.*, **2**, 613 (1963).
314. A. J. Downs, R. Schmutzler, and I. A. Steer, *Chem. Commun.*, 221 (1966).
315. A. E. Borisov, N. V. Novikova, and N. A. Chumaevskii, *Dokl. Akad. Nauk SSSR*, **136**, 129 (1961).
316. R. S. Berry, *J. Chem. Phys.*, **32**, 933 (1960).

VIII Trivalent organoantimony compounds

1. PRIMARY AND SECONDARY STIBINES

Only a few primary and secondary stibines are known. Burg and Grant (1) treated dimethylbromostibine with sodium borohydride in diglyme at $-78°$. The reaction products were removed by distillation in high vacuum, condensed in cold traps, and shown to be CH_3SbH_2, $(CH_3)_2SbH$, and $(CH_3)_2SbBH_2$. The methylstibine was separated and purified by distillation and fractional condensation. It proved to be stable at $-78°$ but decomposed slowly above this temperature. No elemental analyses were performed, but a molecular weight of 135.6 (theoretical, 138.8) was calculated from vapor density measurements. A boiling point of 41° was calculated by extrapolation from vapor tension measurements made at lower temperatures; the Trouton's constant was calculated to be 20.9 cal/°mole. Above $-78°$ methylstibine decomposed to give hydrogen and a black solid of unknown composition.

Dimethylstibine (as formed in the above reaction sequence) could not be separated satisfactorily from $(CH_3)_2SbBH_2$ but could be obtained from dimethylbromostibine by reduction with $LiHB(OCH_3)_3$ at temperatures below $-40°$. The yield was 35%, and a small amount of methylstibine was also formed in the reaction. No elemental analyses were performed, but a molecular weight of 152.2 (theoretical, 152.8) was found from vapor density measurements. The boiling point was calculated to be 60.7°, and the value of Trouton's constant was 22.1 cal/°mole.

Dimethylstibine slowly decomposes at room temperature with the formation of hydrogen and tetramethyldistibine. The decomposition is accelerated by mercury and by diborane. Dimethylstibine reacts with HCl to form dimethylchlorostibine and hydrogen, and with bromodiborane to form dimethylbromostibine and diborane. There is no reaction, however, with hydrogen sulfide. Dimethylstibine does not react with liquid ammonia, but in the presence of sodium, hydrogen is evolved; this result the authors ascribed to the following reaction:

$$(CH_3)_2SbH + Na + 2NH_3 \rightarrow (CH_3)_2SbNH_2 + NaNH_2 + 1.5H_2$$

Some tetramethyldistibine was also produced in the reaction.

In contrast to the difficulties encountered with methyl- and dimethylstibine, Issleib and Hamann (2) readily obtained dicyclohexylstibine in 93% yield by the reduction of dicyclohexylchlorostibine with lithium aluminum hydride. The compound was isolated as an orange-yellow oil, which gave satisfactory analytical values for antimony. It was found to react with phenyllithium in ether solution to give lithium dicyclohexylstibide. Issleib and co-workers have also prepared diethyl- (3) and di-*tert*-butylstibine (4) by the reduction of the corresponding halostibines with lithium aluminum hydride. The diethylstibine was not isolated but was converted to lithium diethylstibide by treatment with phenyllithium. On standing at room temperature, the solution of diethylstibine decomposes to yield hydrogen and tetraethyldistibine. The latter compound on treatment with phenyllithium gives diethylphenylstibine and lithium diethylstibide. Di-*tert*-butylstibine can be isolated by distillation, but only with considerable loss owing to decomposition. The lithium derivative of this stibine has also been prepared and was found to be the most nucleophilic of the dialkylstibides investigated.

The only known primary and secondary arylstibines are phenylstibine and diphenylstibine, both of which were first prepared by Wiberg and Mödritzer (5–7). In a rather naive attempt to prepare $C_6H_5SbH_4$, these authors treated $[C_6H_5SbCl_4]^-NH_4^+$ with lithium borohydride or lithium aluminum hydride in ether solution at $-50°$ (5). In both cases they obtained $C_6H_5SbH_2$. Similarly, when $(C_6H_5)_2SbCl_3$ was dissolved in chloroform-methylene chloride-ether mixture and reduced at $-60°$, $(C_6H_5)_2SbH$ was obtained. The yields of $C_6H_5SbH_2$ varied from 15–20% and of $(C_6H_5)_2SbH$ from 20–60%. In later papers the authors reported improved procedures for preparing both stibines by the use of other substrates for the reduction. Thus, when phenyldiiodostibine was reduced with lithium borohydride at approximately $-50°$, phenylstibine was obtained in 42% yield; lithium aluminum hydride gave the same product but in lower yield (6).

Phenylstibine is a colorless liquid, mp -40 to $-38°$, which possesses a phosphine-like odor. It is readily oxidized in ether solution by oxygen to yield a white solid, the elemental analysis of which suggests that it is either $C_6H_5Sb(OH)_2$ or a mixture of C_6H_5SbO and $C_6H_5SbO_3H_2$. This solid has not been further identified. Phenylstibine reacts with iodine to give phenyldiiodostibine. When an ether solution of the phenylstibine is allowed to stand in the air, or better, when the neat stibine in a sealed tube is allowed to stand for 5–6 days, it was found to decompose with the evolution of hydrogen and the formation of a brownish solid. On the basis of elemental analysis, it was suggested that this is a highly polymerized form of antimonobenzene. This result is somewhat similar to that obtained by the spontaneous decomposition of methylstibine (1). Phenylstibine reacts with phenyldiiodostibine in ether solution with the evolution of hydrogen and the formation of hydrogen iodide. A third product of the reaction is a dark-colored insoluble product, similar to that obtained by the decomposition of phenylstibine. The following equations were suggested to explain these results:

$$C_6H_5SbH_2 \rightarrow (C_6H_5Sb)_x + H_2$$

$$C_6H_5SbH_2 + C_6H_5SbI_2 \rightarrow 2(C_6H_5Sb)_x + 2HI$$

Reduction of diphenylchlorostibine with lithium borohydride at temperatures below $-50°$ gives diphenylstibine in a 65% yield (7). The use of lithium aluminum hydride gives the same product but in lower yield (7, 8). It can be isolated by distillation *in vacuo* and is obtained as a colorless oil, mp -3 to $-2°$. Diphenylstibine in ether solution is readily oxidized to diphenylstibinic acid by a stream of oxygen; chlorine converts the stibine to diphenylantimony trichloride. The reaction with iodine follows a different course. A white precipitate is first formed, but this changes to yellow on the addition of further iodine. The white precipitate was not identified, but the yellow precipitate was shown to be diphenyliodostibine. The following equations were suggested (7) to explain the results:

$$2(C_6H_5)_2SbH + I_2 \rightarrow [(C_6H_5)_2Sb]_2 + 2HI$$

$$[(C_6H_5)_2Sb]_2 + I_2 \rightarrow 2(C_6H_5)_2SbI$$

Nesmeyanov and co-workers (9, 10) have investigated the addition of diphenylstibine to a number of alkynes. Thus, this stibine reacts with phenylacetylene as follows (9):

$$Ph_2SbH + PhC{\equiv}CH \rightarrow Ph_2SbCH{=}CHPh$$

The infrared spectrum of the resulting tertiary stibine indicates that it has the *trans* configuration. Diphenylstibine adds to both diphenylethynylstibine and dibutylethynylstibine in a similar manner (10). (Dibutylstibine also adds to dibutylethynylstibine (11), but the source of the secondary stibine has not been given.) In contrast, the reaction of diphenylstibine and triethylethynylsilane leads to the formation of tetraphenyldistibine and triethylvinylsilane. Nesmeyanov and co-workers (10) have suggested a radical mechanism for this reaction.

Diphenylstibine is a powerful reducing agent. Thus, it reacts with acids such as hydrochloric, acetic, and benzoic acids to liberate hydrogen (9):

$$Ph_2SbH + HY \rightarrow H_2 + Ph_2SbY$$

With carbon tetrachloride, diphenylstibine gives diphenylchlorostibine, hydrogen chloride, benzene, and a solid which was believed to be antimonobenzene. Diphenylstibine reacts with benzotrichloride to yield diphenylchlorostibine and benzylidene chloride.

Except for a few melting and boiling points, little is known about the physical properties of primary and secondary stibines. The infrared spectra of dibutylstibine and diphenylstibine have been reported (12). The Sb—H stretching frequency occurs as an intense band at 1847 cm^{-1} in the spectrum of the butyl compound, while the phenyl compound exhibits a similar absorption at 1835 cm^{-1}. Bands attributed to the Sb—H stretch have been reported at 1842–1845 cm^{-1} and 1850 cm^{-1} for dicyclohexylstibine (2) and di-*tert*-butylstibine (4), respectively.

2. TERTIARY STIBINES

A. Preparation

(*1*) *From organometallic compounds*

The reaction between alkyl Grignard reagents and antimony trichloride has been the most popular method used for the preparation of trialkylstibines. Since they are all readily oxidized and the lower members are spontaneously inflammable in air, the preparation and isolation are carried out in an inert atmosphere. If a very pure sample is required, fractionation is best done by vacuum line techniques (13). Otherwise the product can be isolated by stripping off the solvent and distilling the stibine *in vacuo* in an inert atmosphere. Trimethyl- and triethylstibines are difficult to separate from ether when prepared in this solvent. An attempt to avoid this difficulty with trimethylstibine by the use of di-*n*-butyl ether as the solvent was unsuccessful in that the stibine and the ether formed a constant boiling mixture (14). These stibines have usually been isolated by converting them to the corresponding trimethyl- and triethylantimony dihalides, which can subsequently be reduced to the desired stibines (15–19). Other trialkylstibines, R_3Sb, where R is propyl, butyl, isobutyl, 2-methylbutyl, pentyl, isopentyl, hexyl (18), heptyl (20), and cetyl (21), have been obtained by distillation. A particularly good description of the techniques employed in the preparation of these compounds has been given by Dyke, Davies, and Jones (18); many later workers have followed their procedure. Other trialkylstibines prepared by means of the Grignard reaction include trivinylstibine (22–24) and tris(trimethylsilylmethyl)stibine (25) (both of which were prepared in THF solution), tris(perfluorovinyl)stibine (26), tribenzylstibine (27–29), triethynylstibine (30), tris(2-phenyl-1-ethynyl)-stibine (31), and tricyclohexylstibine (2, 19, 32). The preparation of tri-1-butenylstibine (33) and tris[(dimethyl-*p*-tolylsilyl)methyl]stibine (34) has been described only in the patent literature.

The reaction between *tert*-butylmagnesium chloride and antimony trichloride does not give the expected tri-*tert*-butylstibine but rather di-*tert*-butylchlorostibine (19). Triisopropylstibine apparently cannot be distilled without decomposition (18). However, triisopropylantimony dibromide can be isolated by adding bromine to the solution obtained by the interaction of isopropylmagnesium bromide and antimony trichloride in ether (19).

A few mixed tertiary alkylstibines have been obtained by utilizing alkyl-dichloro- or dialkylchlorostibines in place of antimony trichloride. Thus, Steinkopf and co-workers (35) prepared 1-methylhexahydroantimonin (**1**) from 1,5-cyclopentylbis(magnesium chloride) and methyldichlorostibine.

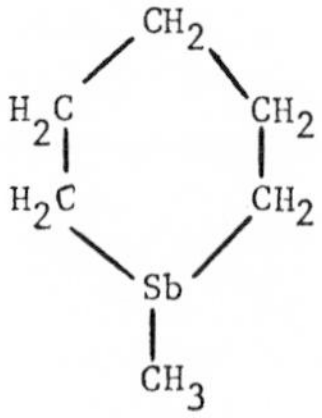

CH_2
H_2C CH_2
H_2C CH_2
Sb
CH_3

1

Ramsden (33) has reported the preparation of ethyldivinyl-, cyclopentyl-divinyl-, dimethylvinyl-, and dilaurylvinylstibines; Brinnand and co-workers (36) obtained dibutylethyl-, diamylethyl-, and dicyclohexylethylstibines from ethyldiiodostibine and the appropriate Grignard reagent. Diethylbromo-stibine reacts with ethynylenebis(magnesium bromide) to give ethynylene-bis(diethylstibine) (19); dibutylbromostibine and allylmagnesium bromide give dibutylallylstibine (37).

Triarylstibines are also readily prepared by means of the Grignard reaction. Triphenylstibine is the subject of an *Organic Syntheses* preparation (38). Other triarylstibines prepared by the Grignard reaction include tri-α-naphthyl- (39), tri-β-naphthyl- (40), tris(*m*-trifluoromethylphenyl)- (21), tris(pentafluorophenyl)- (41), tri-α-thienyl- (42, 43), and several tris(*p*-alkenylphenyl)stibines (44). Usually the yields are quite satisfactory. All three halogens of the antimony trihalide used in the reaction are replaced by the aryl group except in the case of *o*-anisylmagnesium bromide where some di-*o*-anisylchlorostibine was also isolated (40). An attempt to prepare tris(*p*-dimethylaminophenyl)stibine from the corresponding Grignard reagent was apparently unsuccessful (45), although the desired stibine can be obtained from the lithium derivative (46).

The preparation of triarylstibines containing either two or three different aryl groups and the preparation of mixed alkylarylstibines have been readily achieved by the reaction between the appropriate Grignard reagent and a substituted antimony halide, $RSbCl_2$ or R_2SbCl. Using diphenylchlorostibine and a series of aryl Grignard reagents, Yusunov and Manulkin prepared various stibines of the type $(C_6H_5)_2SbC_6H_4Y$, where Y was *p*-OCH_3, *p*-OC_2H_5, *p*-OC_6H_5, *o*-CH_3, *p*-Br, *p*-C_6H_{11}, 2,4,6-$(CH_3)_3$ (47), *p*-CH_2—CH=CH_2, and *p*-CH=$CHCH_3$ (48). In a similar reaction with alkyl Grignard reagents, a number of diphenylalkylstibines, $(C_6H_5)_2SbR$, were obtained (32, 49). From ethyldiiodostibine and aryl Grignard reagents, mixed ethyldiarylstibines have been obtained (36); from 5-chlorodibenzostibole and methylmagnesium iodide, Morgan and Davies (50) obtained 5-methyldibenzostibole.

Campbell (51) has achieved ring closure by the reaction between an aryldichlorostibine and a di-Grignard reagent. Thus, from the Grignard reagent obtained from 2,2′-dibromo-4-methyldiphenyl ether and various aryldichlorostibines, derivatives of 10-phenylphenoxantimonin (**2**) were obtained.

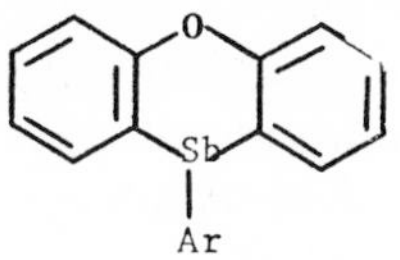

2

The preparation of triarylstibines with three different aryl groups is also due to the work of Campbell. From 2-biphenylyl-*p*-tolylchlorostibine and phenylmagnesium chloride, 2-biphenylylphenyl-*p*-tolylstibine was obtained as a non-crystallizable syrup (52). From phenylmagnesium bromide and various diarylchlorostibines of the type (*p*-$C_2H_5O_2CC_6H_4$)ArSbCl, a number of *p*-carbethoxyphenyl-substituted tertiary stibines were obtained (52, 53). The majority of these were syrups which, without any attempt at crystallization, were hydrolyzed to yield crystalline tertiary stibines containing a *p*-carboxyphenyl group.

The preparation of tertiary stibines from organolithium compounds has also been extensively investigated. The yields are excellent, often exceeding those obtained when Grignard reagents are used. Thus, Talalaeva and Kocheshkov (54) obtained the following stibines (yields in parentheses): triphenyl- (96–97%), tri-*p*-tolyl- (95.3%), tri-*p*-phenetyl- (85.5%), tris(*p*-dimethylaminophenyl)- (78.1%), tri-*o*-anisyl- (31%), tri-*p*-biphenylyl- (51%), and tri-α-naphthyl- (17%). The reaction generally leads to the formation of the tertiary stibine as the only organoantimony compound, although it has been reported that *m*-tolyllithium gives a mixture of tri-*m*-tolylstibine and tri-*m*-tolylantimony dichloride (39). Among the few failures with lithium reagents is that of O'Donnell (21) who was unable to prepare tri-α-pyridylstibine from the lithium derivative. The organolithium reaction works well for the preparation of trialkylstibines. Nesmeyanov and co-workers employed lithium reagents for the preparation of triisopropenylstibine (55) and of both *cis* and *trans* isomers of tripropenylstibine (56–58). O'Donnell (21) has prepared tris(4-dibenzofuryl)stibine from the lithium reagent and antimony trichloride, as well as diphenyl-4-dibenzofurylstibine from the same lithium reagent and diphenylchlorostibine. Zakharkin and co-workers (59) obtained compound

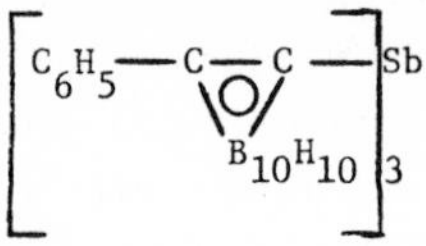

$$[C_6H_5-C\overset{\displaystyle\frown}{(B_{10}H_{10})}C-]_3Sb$$

3

3, which is termed tris(phenylbarenyl)stibine, from the corresponding lithium derivative. Phenyldiiodostibine and 2-biphenylyllithium react in ether to give bis(2-biphenylyl)phenylstibine (60).

Several interesting heterocyclic tertiary stibines have been obtained by the reaction between dilithio reagents and substituted antimony halides. From phenyldiiodostibine and 2,2′-dilithiobiphenyl, Heinekey and Millar (61) obtained 5-phenyldibenzostibole (**4**). This same compound was later pre-

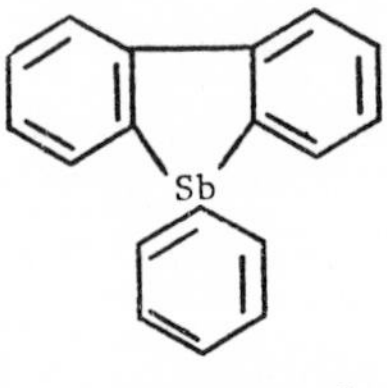

Sb

4

pared from 5-chlorodibenzostibole and phenyllithium (60).

The reaction of 1,4-dilithio-1,2,3,4-tetraphenylbutadiene with phenyldichlorostibine leads to the formation of pentaphenylstibole (**5**). This

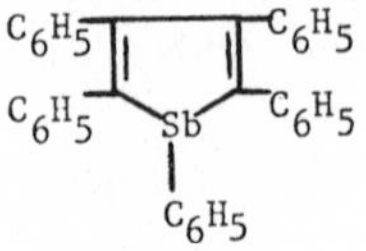

C_6H_5 C_6H_5
C_6H_5 Sb C_6H_5
C_6H_5

5

reaction was first investigated by Leavitt and co-workers (62, 63) who found that the dilithio reagent would react with halides of S, P, As, Sb, C, Ge, and Sn to form heterocyclic compounds. This reaction was later duplicated and extended by Braye and co-workers (64). The necessary dilithio reagent was readily obtained from diphenylacetylene and lithium.

In a few cases organosodium compounds have been used in the preparation of tertiary stibines. Hartmann and Kühl (19, 65) prepared ethynylenebis-(dimethylstibine) by the reaction between dimethylbromostibine and ethynylsodium. In order to account for their failure to obtain the expected dimethylethynylstibine, the authors postulated the following mechanism:

$$(CH_3)_2SbBr + NaC{\equiv}CH \rightarrow (CH_3)_2SbC{\equiv}CH + NaBr$$

$$(CH_3)_2SbC{\equiv}CH + NaC{\equiv}CH \rightarrow [(CH_3)_2Sb(C{\equiv}CH)_2]Na$$

$$[(CH_3)_2Sb(C{\equiv}CH)_2]Na + (CH_3)_2SbBr \rightarrow (CH_3)_2SbC{\equiv}CSb(CH_3)_2 + NaBr + C_2H_2$$

Diisopropylbromostibine and ethynylsodium give both ethynylenebis(diisopropylstibine) and ethynyldiisopropylstibine in yields of 30–40% and 60–70%, respectively. Di-*p*-tolylchlorostibine and bis(*p*-chlorophenyl)chlorostibine react with ethynylsodium in liquid ammonia to give the corresponding ethynylenebisstibines; *p*-tolyldichlorostibine with the same sodium reagent gives diethynyl-*p*-tolylstibine (66). Another interesting compound prepared from an organosodium reagent is tricyclopentadienylstibine (67). The reaction is run in THF at 0°; at higher temperatures tetracyclopentadienyldistibine is formed. Presumably the C—Sb bonding involves σ bonds, but no structural data have been reported.

The interaction of an aryl halide, an antimony trihalide, and sodium leads to the formation of a triarylstibine. This reaction, which has been considered a variant of the Wurtz-Fittig reaction, probably proceeds through an arylsodium intermediate. This method was first used by Michaelis and Reese (68) for the preparation of triphenylstibine and was later extended to the isomeric tritolylstibines (69). More recently, Olifirenko (70) has stated that triphenylstibine is best prepared by the interaction of bromobenzene, antimony trichloride, and sodium in a toluene–benzene mixture; in this way he obtained a 72% yield. In a similar manner Worrall prepared tri-*p*-biphenylyl- (71) and tri-*o*-biphenylylstibines (72) in 86% and 64% yields, respectively, while Étienne (73) obtained tri-α-furylstibine in 70% yield. An attempt to prepare 2-phenylisostibindoline from α,α′-dibromo-*o*-xylene, phenyldichlorostibine, and sodium gave only triphenylstibine, although the analogous reaction with phenyldichloroarsine was successful (74).

In recent years the use of organoaluminum compounds for the preparation of trialkylstibines has been investigated. The descriptions of the reaction conditions used are fragmentary in some cases (75, 76), but the preparation of triisobutylstibine in 77% yield (compared to a 22–23% yield from the Grignard reagent) from triisobutylaluminum etherate and antimony trifluoride has been adequately described (77). In later communications without experimental details, Stamm and Breindel (78, 79) have stated that trialkylaluminum compounds in excess react readily with antimony trioxide to give trialkylstibines. In view of the large yields claimed in this procedure and the ready availability and low cost of antimony trioxide, this method may prove to be one of the best methods for preparing trialkylstibines, provided that the necessary equipment for handling the spontaneously-inflammable trialkylaluminums is available.

The interaction of diphenylcadmium and antimony trichloride in benzene solution has been found to give a 91% yield of triphenylstibine (80).

In all of the reactions noted above a reactive organometallic compound and an antimony halide or antimony trioxide have been allowed to react together. It is also possible to prepare tertiary stibines from antimony metal and a less reactive organometallic compound. Thus, trimethylstibine has been prepared from dimethylmercury and antimony powder in a sealed tube at 150–160° (13), and triphenylstibine has been obtained by heating antimony with triphenylbismuth (81) or tetraphenyltin (82).

(2) By reduction

As mentioned previously, trialkylstibines can be readily prepared from antimony trihalides and Grignard reagents or other active organometallic compounds; the properties of the stibines, however, make it difficult to isolate large amounts of these compounds or to store them for any length of time. For this reason it is common practice to convert the trialkylstibines to trialkylantimony dihalides. The latter are quite stable crystalline solids, which can be readily reduced to the stibines in an inert atmosphere. For example, heating the dihalides with granulated zinc (16, 83) or zinc dust (84) reduces them to tertiary stibines. Trimethylantimony dibromide has been reduced to trimethylstibine with lithium borohydride or lithium aluminum hydride (85), while both *cis-cis-cis* and *trans-trans-trans* tris(β-chlorovinyl)-antimony dichlorides have been reduced with sodium bisulfite in aqueous solution to the corresponding tris(β-chlorovinyl)stibines (86). The use of thiols for the reduction of trimethylstibine oxide or trimethylantimony diethoxide has also been reported (87).

No critical study of these various reducing agents has been made in order to determine the best method. When zinc is used it is difficult to see how contamination with dialkylhalostibines can be avoided, since these substances are produced when trialkylantimony dihalides are heated in an inert atmosphere.

The reduction of triarylantimony dihalides to triarylstibines has also been achieved. Lithium borohydride reduces triphenylantimony dichloride to triphenylstibine (5); hydrazine hydrate has also been used for this purpose (21, 46). Campbell, in her study of the stereochemistry of organoantimony compounds, had occasion to reduce unsymmetrical triarylantimony dihalides. Rather surprisingly, when stannous chloride was used, the reduction was not always successful. Thus, when either 2-biphenylylphenyl-*p*-tolylantimony dibromide or 2-biphenylyl(*p*-carbethoxyphenyl)phenylantimony dibromide was treated with stannous chloride in ethanol–hydrochloric acid solution, the corresponding triarylantimony dichlorides were obtained (52). Both dibromides were readily reduced to the desired tertiary stibines with hydrogen sulfide in alcoholic ammonia. By contrast, stannous chloride has been used successfully for the reduction of several triarylantimony oxides in acetone–hydrochloric acid solution. Thus, 5-*p*-tolyldibenzostibole 5-oxide (obtained by cyclodehydration of the corresponding stibinic acid) was reduced (without isolation) in acetone and 3.5*N* hydrochloric acid with stannous chloride to 5-*p*-tolyldibenzostibole (86); 5-(*p*-carbethoxyphenyl)dibenzostibole 5-oxide was also reduced under similar conditions (88). This reaction was later extended with considerable success to the preparation of a number of other derivatives of 5-phenyldibenzostibole (89). The reduction of triphenylantimony dichloride with sodium diphenyldithiocarbamate has also been reported (90).

In addition to triarylantimony dihalides (or the corresponding oxides), tetraphenylstibonium salts can be reduced with lithium aluminum hydride to triphenylstibine and benzene (91).

(3) *By the diazo reaction*

Although antimony trihalides can be triarylated by reaction with diazonium salts in the presence of a metal such as powdered zinc or iron, in most cases triaryl derivatives of antimony(V) rather than antimony(III) are obtained. There are a few exceptions to this generalization. Nesmeyanov and Kocheshkov (92) obtained tris(*o*-chlorophenyl)stibine, as well as tris(*o*-chlorophenyl) antimony dichloride, bis(*o*-chlorophenyl)chlorostibine, and (after hydrolysis) *o*-chlorostibosobenzene, when *o*-chlorobenzenediazonium tetrachloroantimonate(III) was decomposed in ethyl acetate by means of zinc powder. No triarylstibine was obtained from the corresponding *p*-chloro or *m*-chloro diazonium salt. Nakai and Yamakawa (93) have obtained triphenylstibine, together with trivalent mono- and diarylantimony derivatives, from benzenediazonium fluoborate in acetic anhydride solution by the addition of powdered zinc. The yield of triphenylstibine was only 12%, but when *p*-toluenediazonium fluoborate was used a yield of 42% tri-*p*-tolylstibine was obtained. Makin and Waters (94) prepared several triarylstibines by adding powdered antimony metal to arenediazonium chlorides in acetone kept neutral by the addition of calcium carbonate. The triarylstibines, when formed, were invariably accompanied by other organoantimony compounds. The reaction, which is of more theoretical than practical value, is discussed more fully in Section VII-5-A-(2).

(4) *By the reaction between alkyl or aryl halides and compounds containing an antimony–metal bond*

Tertiary stibines have been prepared by the reaction of alkyl or aryl halides with compounds of the type $(R_2Sb)_nM$, where $n = 1$ and M = sodium or lithium, and $n = 2$ and M = magnesium. These reactive metal derivatives need not be isolated but can be allowed to react in solution to produce the desired tertiary stibines. Herbstman (37) has reported that the addition of dibutylbromostibine to lithium metal in THF produces a deep-red solution. A dark-brown magnesium compound is formed from magnesium metal and dibutylbromostibine under similar conditions. Addition of an alkyı romide to either the lithium or magnesium reagent results in the formation of a dibutylalkylstibine. The yields are considerably larger when the lithium reagent is used. Woods and Gilman (95) reported the preparation of unsymmetrical tertiary stibines by the use of sodium diarylstibides and aryl halides. The necessary sodium compounds were prepared from diaryliodostibines and sodium in liquid ammonia, but details of the reaction procedure were not given.

Compounds of the type R_2SbLi have been prepared from dialkylstibines and phenyllithium in ether solution (2, 3, 4). These compounds (which have been isolated from solution and subjected to elemental analysis) react with symmetrical dibromoalkanes to form ditertiary stibines of the type $R_2Sb(CH_2)_nSbR_2$, where $n = 3, 4, 5$, or 6. With 1,4-dichlorobutane, it is also possible to obtain compounds of the type $R_2Sb(CH_2)_3CH_2Cl$. When 1,2-dichloro- or dibromoethane is used, no tertiary stibines are obtained. Instead ethylene is evolved and tetraalkyldistibines are formed. Methylene dichloride and dibromide also react with the formation of the distibines. The following reactions have been suggested to explain the results with the dihaloethanes:

$$XCH_2CH_2X + LiSbR_2 \rightarrow LiCH_2CH_2X + R_2SbX$$
$$LiCH_2CH_2X \rightarrow C_2H_4 + LiX$$
$$R_2SbX + LiSbR_2 \rightarrow R_2SbSbR_2 + LiX$$

A similar result has been recorded by Hewertson and Watson (96) who obtained tetraphenyldistibine from sodium diphenylstibide and 1,2-dichloroethane. In contrast to these results, sodium diphenylarsenide and 1,2-dichloroethane give $(C_6H_5)_2AsCH_2CH_2As(C_6H_5)_2$.

Trialkyl- and triarylstibines have also been prepared by the interaction of an alkyl or aryl halide, an alloy of antimony, and an alkali metal. The first preparation of a trialkylstibine involved this procedure. In 1850 Löwig and Schweizer (97) heated a potassium–antimony alloy (admixed with sand to moderate the reaction) with ethyl iodide and obtained tetraethylstibonium iodide. This compound on being heated with more potassium–antimony alloy yielded triethylstibine. A few years later Friedländer (98) prepared this stibine in a similar manner, and in 1861 Landolt (99) obtained trimethylstibine from sodium–antimony alloy and methyl iodide. The preparation of the spontaneously inflammable and highly toxic trialkylstibines is a tribute to the skill of these early workers. It has recently been reported that the reaction of ethyl bromide with antimony and sodium or lithium in liquid ammonia yields not only triethylstibine but also tetraethyldistibine (100). Amberger and Salazar (101) have synthesized the four compounds of the type $(Me_3E)_3Sb$ (where E is C, Si, Ge, or Sn) by the following reactions:

$$3Li + Sb \xrightarrow{NH_3} Li_3Sb$$
$$Li_3Sb + 3Me_3EX \longrightarrow (Me_3E)_3Sb + 3LiX$$

Tri-*tert*-butylstibine was obtained in only a 1% yield, but the yields of the other three compounds were 80–85%. Triphenylstibine has been prepared from lithium–antimony alloy and bromobenzene (102), but the yields were inferior to those obtained by the Wurtz-Fittig procedure (70).

Vyazankin and co-workers (103) have investigated the cleavage of Ge—Sb and Sn—Sb bonds by alkyl halides. When $[(C_2H_5)_3Ge]_3Sb$ or $[(C_2H_5)_3Sn]_3Sb$ was treated with benzyl bromide, tribenzylstibine and $(C_2H_5)_3GeBr$ or $(C_2H_5)_3SnBr$ were formed. If bromocyclopentane rather than benzyl bromide was used, tricyclopentylstibine was obtained.

(5) *Miscellaneous procedures*

Diazomethane reacts with antimony trichloride in benzene solution with the formation of either tris(chloromethyl)stibine or chloromethyldichlorostibine (104, 105). If excess diazomethane is used the tertiary stibine is obtained, whereas excess antimony trichloride yields the dichlorostibine. Tris(chloromethyl)stibine is a liquid, bp 105°/3 mm; it is slowly hydrolyzed by water.

Tris(trifluoromethyl)stibine is the principal product obtained when trifluoroiodomethane and antimony metal are heated in a steel cylinder at 165–170° for 7 hours (106). The product mixture also contains bis(trifluoromethyl)iodostibine and trifluoromethyldiiodostibine, which can be separated by fractional distillation. The preparation of tris(trifluoromethyl)stibine appears to be a difficult procedure. Pyrex tubes are unsuitable, and even in steel cylinders the yields are very poor until the cylinders have been used several times.

In addition to the methods already given, tertiary stibines have been obtained by several methods which are not of preparative importance. Pyrolysis of pentavalent antimony compounds of the type $(C_6H_5)_3Sb(OR)_2$ yields triphenylstibine (107). Thus, when **6** is heated to 240–250°, triphenylstibine

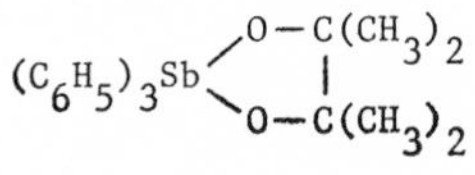

$$(C_6H_5)_3Sb\begin{matrix} \diagup O-C(CH_3)_2 \\ \qquad | \\ \diagdown O-C(CH_3)_2 \end{matrix}$$

6

and acetone are obtained; similarly **7** gives triphenylstibine and benzaldehyde.

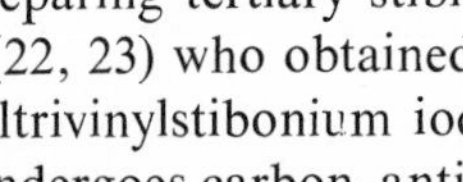

$$(C_6H_5)_3Sb\begin{matrix} \diagup O-CHC_6H_5 \\ \qquad | \\ \diagdown O-CHC_6H_5 \end{matrix}$$

7

In contrast to the studies on the pyrolysis of arsonium salts (cf. IV-4-E), the pyrolysis of stibonium salts has been largely neglected. That such a reaction can be used for preparing tertiary stibines was shown by the work of Maier and co-workers (22, 23) who obtained trivinylstibine in 80% yield by the pyrolysis of methyltrivinylstibonium iodide. Pentaphenylantimony, on being heated to 200°, undergoes carbon–antimony bond cleavage to form triphenylstibine and biphenyl (108).

The interesting tertiary stibines $Sb(CH_2CO_2R)_3$, where R = ethyl, *n*-propyl, or *n*-butyl, have been obtained by passing a stream of ketene through a benzene solution of the corresponding trialkoxystibine (109):

$$3CH_2{=}C{=}O + (RO)_3Sb \rightarrow Sb(CH_2CO_2R)_3$$

A mixture of triphenylstibine, diphenylchlorostibine, and phenyldichlorostibine is obtained from benzene and antimony trichloride when irradiated with neutrons; a similar reaction occurs with benzene and arsenic trichloride and undoubtedly involves the formation of phenyl radicals (110).

The preparation of triphenylstibine from powdered antimony and phenyllithium has been reported by Talalaeva and Kocheshkov (111). The reaction was carried out by refluxing the reactants in ether or xylene in a nitrogen atmosphere; the yield of triphenylstibine was 3.6%. The formation of trimethylstibine and triethylstibine by the reaction of methyl and ethyl radicals, generated by the pyrolysis of tetramethyl- and tetraethyllead (the Paneth technique), was of considerable importance in establishing the existence and properties of radicals (112).

The preparation of tertiary stibines by the addition of secondary stibines to alkynes has been discussed in Section VIII-1.

Finally, one method should be mentioned which fails to produce triarylstibines. In contrast to phosphorus trichloride and arsenic trichloride, both of which react with dimethylaniline to form tris(*p*-dimethylaminophenyl)-phosphine and tris(*p*-dimethylaminophenyl)arsine, respectively, as well as the corresponding halo- and dihalophenylphosphines and -arsines, antimony trichloride fails to react with dimethylaniline (113). These results would indicate that antimony is a poorer electrophile than arsenic or phosphorus.

B. Reactions

(*1*) *Reactions with electrophilic reagents*

Trialkylstibines are powerful reducing agents. Thus, trimethylstibine reduces both phosphorus pentachloride and phosphorus trichloride to elemental phosphorus, and antimony pentachloride and antimony trichloride to antimony (114); in this respect it is a stronger reducing agent than trimethylarsine, which reduces phosphorus pentachloride and antimony pentachloride to the trichlorides. Trialkylstibines reduce mercuric salts to mercury in the cold, auric and silver salts to gold and silver, respectively, in the hot (18), and platinum(IV) to platinum(II) (115). Trimethylstibine reacts with hydrogen chloride in a sealed tube to form trimethylantimony dichloride and hydrogen (1, 99). The expected reaction to form methane and antimony trichloride takes place to only a limited extent (1). Trialkylstibines react readily with chlorine, bromine, or iodine to form the corresponding trialkylantimony dichlorides, dibromides, or diiodides (cf. VII-5-A-(1)), with sulfur to form trialkylstibine sulfides, and with selenium to form trialkylstibine selenides (cf. VII-2-B). An exception to this generalization is found with tris(trifluoromethyl)stibine where the electron-attracting trifluoromethyl groups reduce the nucleophilicity of the antimony sufficiently to prevent the reaction with sulfur (106).

The reaction with halogens is carried out by adding the halogen to a solution of the stibine in an organic solvent, whereupon the dihalide usually crystallizes from solution in essentially quantitative yield (116). When excess bromine (19) or iodine (17) is used, there is the possibility of forming perhalides; apparently this does not occur with chlorine. Cyanogen bromide also reacts with trimethylstibine to form trimethylantimony cyanobromide (83). Trimethylstibine reacts with dimethylchloroborane in a sealed tube at 60° to form trimethylborane and an unidentified antimony compound (1).

McKenney and Sisler (117) have found that trimethyl-, triethyl-, tripropyl-, tributyl-, and triphenylstibines react with either chloramine or a chloramine–ammonia mixture to yield compounds of the type $(R_3SbCl)_2NH$. In some cases compounds of the type R_3SbCl_2 were also formed. The nitrogen-containing compounds have been discussed in Section VII-5-B.

Trialkylstibines react with alkyl halides to form tetraalkylstibonium salts (118, 119). The rate of quaternization is fairly slow (since antimony is not a good nucleophile) and decreases with increasing chain length of the alkyl halide (120). The reaction between trimethylstibine and trifluoroiodomethane takes an entirely different course (121). Dimethyltrifluoromethylstibine and tetramethylstibonium iodide are the principal products of the reaction:

$$2(CH_3)_3Sb + CF_3I \rightarrow CF_3Sb(CH_3)_2 + (CH_3)_4SbI$$

The interaction of 9-bromofluorene and trimethylstibine does not yield the expected stibonium salt but instead gives difluorenyl (122).

Matsumura and Okawara (123) have recently reported that the reaction of trimethylstibine and iodoform in refluxing acetone yields the pentacovalent antimony compound $Me_3Sb(CHI_2)(CH_2I)$. This formulation was based on elemental analysis as well as NMR and infrared spectra. No mechanism for this remarkable transformation has been suggested. Treatment of the pentacovalent compound with hydrobromic acid gives the distibonium salt $[Me_3SbCH_2SbMe_3]Br_2 \cdot 2H_2O$, which undergoes a metathetical reaction with silver nitrate to give the anhydrous distibonium dinitrate.

Trimethylstibine (as well as trimethylamine, trimethylphosphine, and trimethylarsine) reacts with trichloroiodomethane to form a 1:1 adduct (124). Conductometric evidence suggests that the adducts are uncharged species, presumably with the structure $R_3M \rightarrow ICCl_3$. Trimethylstibine and B_2H_6 react at liquid nitrogen temperature to give $(CH_3)_3SbBH_3$; this melts at −35° and dissociates above this temperature (125). Trimethylstibine and trimethylantimony dichloride undergo halogen exchange in alcohol solution (126). The mechanism of this reaction was followed by isotopic labelling and mass spectrometry and has been discussed in a previous section (VII-5-B).

Triarylstibines are also reducing agents but are less reactive in this respect than the trialkyl compounds. With chlorine, bromine, and iodine, they form the corresponding triarylantimony dihalides (68, 127). The dihalides have also been obtained from triarylstibines and cupric (40, 68, 128), ferric (129), or thallic (55, 130, 131) chlorides with the concomitant formation of cuprous, ferrous, and thallous chlorides. Tris(*o*-chlorophenyl)stibine and tri-β-naphthylstibine do not react with cupric chloride (40). Triarylstibines reduce lead tetraacetate to lead diacetate (132). Triphenylstibine reacts with iodine radicals, generated by the photolysis of aryl iodides, to form triphenylantimony diiodide (133, 134).

In contrast to triarylarsines (cf. V-9-A), triarylstibines are not quaternized by treatment with alkyl halides. This result is in accord with the decreased nucleophilicity of antimony as compared with arsenic. However, triphenylstibine and diphenyliodonium fluoborate react to form tetraphenylstibonium fluoborate (135). Triphenylstibine and trimethyloxonium fluoborate give methyltriphenylstibonium fluoborate (136), and triphenylstibine and bis-(chloromethyl)zinc give a product which can be hydrolyzed to methyltriphenylstibonium chloride (137).

Triarylstibines react with compounds containing a nitrogen–halogen bond to form imides of the type $Ar_3Sb{=}NR$. Thus, triphenylstibine reacts with the sodium salt of N-bromoacetamide in the presence of hydrochloric acid to give $(C_6H_5)_3Sb{=}NCOCH_3$ (138), and with the sodium salt of N-chloro-*p*-toluenesulfonamide (chloramine-T) to give $(C_6H_5)_3Sb{=}NSO_2C_6H_4CH_3$-*p* (139, 140). These imides as well as the related ylids have been discussed in Section VII-7.

Triphenylstibine is oxidized by fuming nitric acid to form triphenylantimony dinitrate without nitration of the ring (68, 127), but with tri-*o*-anisylstibine and tri-*o*-phenetylstibine ring nitration occurs (128). It has been reported that triphenylstibine is oxidized with warm sulfuric acid to triphenylantimony sulfate (141), but this finding could not be confirmed (127). Triphenylstibine does react with sulfur trioxide in 1,2-dichloroethane solution to produce triphenylantimony sulfate (142).

The reaction between triphenylstibine and benzenediazonium chloride does not lead to arylation of the antimony atom but rather to the formation of triphenylantimony dichloride (143):

$$(C_6H_5)_3Sb + 2C_6H_5N_2Cl \rightarrow (C_6H_5)_3SbCl_2 + C_6H_5C_6H_5 + N_2$$

(2) *Reactions in which a carbon–antimony bond is cleaved*

Trialkyl- and triarylstibines undergo many reactions in which one or more of the C—Sb bonds are broken. Since the C—Sb bond is somewhat weaker than the C—P or C—As bond in the corresponding phosphines and arsines, reactions with C—M bond cleavage occur with greater ease in the case of the tertiary stibines. Manulkin and co-workers (32, 49) observed that unsymmetrical liquid stibines of the type $(C_6H_5)_2SbR$, where R is an alkyl group, became solid on prolonged standing; this fact was attributed to disproportionation to the symmetrical stibines. Reutov and co-workers (144) refluxed both triphenyl- and tri-*p*-tolylstibines with an excess of ^{124}Sb (0.007 moles to 0.04 moles) suspended in kerosene or tetralin. After having been refluxed for 5 to 10 hours, the resulting stibines were found to have exchanged between 15 and 80% of the antimony. These reactions illustrate the ease of C—Sb bond cleavage; other cleavage reactions are discussed in the following paragraphs.

Trialkylstibines react vigorously with atmospheric oxygen. For example, Seifter (145) noted that trimethylstibine sometimes explodes on contact with air. Dyke, Davies, and Jones (18) reported that trialkylstibines ignite when warmed in the air and burn with the greyish-green flame characteristic of antimony. The affinity of the stibines for oxygen appears to decrease with increasing molecular weight.

Early workers (97, 99) in the field of organoantimony chemistry were aware that some C—Sb bond cleavage occurs when trialkylstibines are oxidized by air, but the exact nature of the products was not determined. Dyke and Jones (120) state that the air oxidation of liquid trialkylstibines results in the formation of two products, viz., $R_3SbO{\cdot}Sb_2O_3$, which is insoluble in excess stibine, and R_3SbO, which is soluble in the stibine. The insoluble material was formulated as a metaantimonite, $R_3Sb(SbO_2)_2$. This formulation was based entirely on analytical results obtained on unrecrystallized products and is open to question. Bamford and Newitt (146) studied the rate of oxidation of trimethyl- and triethylstibines as oxygen was slowly admitted to the neat liquids. No attempt was made to identify the products of the reaction; and other than the finding that the ethyl compound was oxidized at a faster rate than the methyl compound, their results are of little value in determining the mechanistic path of the oxidation.

Unlike other trialkylstibines, tribenzylstibine is oxidized by atmospheric oxygen to tribenzylstibine oxide (27). Triarylstibines are generally stable in air, although Yusunov and Manulkin (48) have noted that several unsymmetrical triarylstibines such as diphenyl-*o*-tolylstibine, on standing in the air for one month, gave diphenylstibinic acid.

Both trialkyl- and triarylstibines are thermally quite stable, although less so than the corresponding arsines or phosphines. Harris and co-workers (40) made a rough qualitative study of the pyrolysis of triphenylstibine and several of its derivatives, both in a nitrogen and in an oxygen atmosphere. Triphenylstibine in nitrogen at 325° gave benzene, biphenyl (10%), and antimony (17%) together with a large amount of an unidentified oil. In an oxygen atmosphere the rate of carbon–antimony bond cleavage was accelerated. Another study was made by Ipatiew and Rasuwajew (147) who compared the decomposition of triphenylphosphine, -arsine, -stibine, and

-bismuth in a hydrogen and in a nitrogen atmosphere at constant pressure. In hydrogen all compounds were reduced to benzene and the element; under nitrogen, except for the phosphine which was stable under these conditions to 350°, biphenyl and the element were obtained. In a hydrogen atmosphere reduction of triphenylbismuth started at 125°; the corresponding stibine, arsine, and phosphine started at 200°, 275°, and 325°, respectively. When the % reduction of the four compounds was plotted as a function of temperature, four roughly parallel lines were obtained. Only a few points were obtained for each compound, and the data do not allow more than qualitative comparisons of the stability of the four compounds. More recently, it has been reported that the vapor-phase catalytic hydrogenation of triphenylstibine yields cyclohexane and antimony (148). The conversion of triethylstibine to ethane by reaction with a hydride of a group IV element will be discussed in Section VIII-6.

Dale, Emeléus, and co-workers (106, 149) have compared the pyrolysis of tris(trifluoromethyl)arsine and the corresponding stibine. The carbon–antimony bond is cleaved in the temperature range 180–220°, whereas the carbon–arsenic bond is stable at these temperatures but is cleaved at 410°. The arsine gives predominantly the saturated fluorocarbons, perfluoropropane and perfluorobutane; tris(trifluoromethyl)stibine gives tetrafluoroethylene, perfluorocyclopropane, and other unsaturated fluorocarbons, as well as large amounts of antimony trifluoride. The authors suggested that in both cases homolytic cleavage results in formation of arsenic or antimony, but in the latter case the lower volatility of the antimony allows the formation of the CF_2 diradical by reaction between the CF_3 radical and antimony. Further reactions of the diradical lead to the formation of the unsaturated compounds.

Cleavage of a carbon–antimony bond by hydrolysis of triarylstibines under both acidic and basic conditions has been studied by various investigators. Triarylstibines, when boiled with alkaline hydrogen peroxide, give diarylstibinic acids (46, 150, 151). Di-*p*-tolylstibinic acid has been obtained in small yield from tri-*p*-tolylstibine by boiling with methanolic-HCl in a carbon dioxide atmosphere (151). Diphenyl-*o*-tolylstibine gives diphenylstibinous acetate when heated with acetic acid (48). The carbon–antimony bond in tri-α-furylstibine is readily cleaved under both acidic and basic conditions (73). With alkaline hydrogen peroxide, furan and antimony trioxide are obtained; with hydrochloric acid, furan and antimony trichloride. Both tris(trifluoromethyl)stibine (106) and dimethyl(trifluoromethyl)stibine (121) are readily cleaved by alkali to give fluoroform. Similarly, tris(perfluorovinyl)stibine gives trifluoroethylene (152). Tris(trifluoromethyl)stibine also gives fluoroform when treated with hydrochloric acid. The formation of fluoroform on hydrolysis is typical of compounds containing the CF_3—M linkage, where M is phosphorus, arsenic, or antimony. The ease of hydrolysis is in the order: phosphorus $<$ arsenic $<$ antimony (121, 153). A similar order is found in the hydrolysis of the perfluorovinyl compounds of these elements (152). In general the difference between the rates of hydrolysis of the carbon–antimony bond and the carbon–arsenic bond is much greater than the corresponding difference between the carbon–arsenic and the carbon–phosphorus bond (153). The cleavage of vinyl- (22, 23) and ethynyl-substituted (19, 65) tertiary stibines has also been investigated.

The reaction between an alkali metal and triphenylstibine results in the cleavage of one carbon–antimony bond and the formation of a product which reacts as if it possesses the structure $(C_6H_5)_2SbM$. This reaction was first observed by Wittenberg and Gilman (154), who treated a solution of triphenylstibine in THF with lithium. The resulting red solution was treated with trimethylchlorosilane and then with water. The products isolated were trimethylphenylsilane and diphenylstibinic acid:

$$(C_6H_5)_3Sb \xrightarrow{Li} C_6H_5Li + (C_6H_5)_2SbLi \xrightarrow{(CH_3)_3SiCl,\ H_2O,\ O_2} (CH_3)_3SiC_6H_5 + (C_6H_5)_2SbO_2H$$

Trimethylsilanol, which might be expected from the hydrolysis of $(C_6H_5)_2$-$SbSi(CH_3)_3$, was not observed, although it was found, together with trimethylphenylsilane, when triphenylphosphine was used rather than triphenylstibine. Cleavage of the carbon–antimony bond in tributylstibine by lithium in boiling THF was unsuccessful (37).

Hewertson and Watson (96) later cleaved one phenyl group from triphenylphosphine, -arsine, and -stibine by treating these compounds in liquid ammonia with sodium. The mechanism of the alkali metal cleavage of such tertiary compounds has been investigated with triphenylphosphine by Britt and Kaiser (155, 156). The first step involves cleavage of a phenyl group and subsequent reaction of the diarylphosphide with the metal to form a radical anion:

$$(C_6H_5)_3P + 2M \rightarrow (C_6H_5)_2PM + C_6H_5M$$

$$(C_6H_5)_2PM + M \rightarrow (C_6H_5)_2PM^- + M^+$$

Detection of the radical anion was done largely by the use of electron spin resonance techniques.

Trialkyl- and triarylstibines undergo exchange reactions with reactive organometallic compounds. Thus, the interaction of tribenzylstibine and ethyllithium in benzene solution gives a precipitate of benzyllithium (28). Woods (46, 95) reported that tertiary stibines react with *n*-butyllithium according to the equation:

$$R_3Sb + n\text{-}C_4H_9Li \rightarrow n\text{-}C_4H_9SbR_2 + RLi$$

The yield of RLi was determined by carbonating the mixture, and the mixed stibines were not isolated. Anhydrous aluminum chloride reacts with triarylstibines with arylation of the aluminum and the formation of antimony trichloride (157). The reaction must occur stepwise, since diphenylchlorostibine was also isolated in the reaction (158). Boron trichloride cleaves the carbon–antimony bond as shown by the reaction between this substance and tris(trichlorovinyl)stibine in benzene at room temperature to give antimony trichloride in 100% yield (159). All three phenyl groups are cleaved from triphenylstibine by treatment with potassium amide in liquid ammonia (160). The reaction is believed to proceed in the following manner:

$$(C_6H_5)_3Sb + 3NH_3 + KNH_2 \rightarrow K[Sb(NH_2)_4] + 3C_6H_6$$

$$K[Sb(NH_2)_4] \rightarrow K[Sb(NH)_2] + 2NH_3$$

The tetraamide was not isolated but presumably was formed and then lost ammonia to form the yellow $K[Sb(NH)_2]$.

(3) *Tertiary stibines as ligands in coordination chemistry*

In a similar manner to tertiary phosphines and arsines, tertiary stibines have found extensive use as ligands in coordination chemistry. In contrast to the phosphines and arsines, however, studies with stibines have been largely restricted to simple trialkyl and triaryl compounds, and few attempts have been made to synthesize stibines specifically for use as ligands. This may be due in part to the greater synthetic difficulties in preparing organoantimony compounds and the decreased stability of the carbon–antimony bond compared with the carbon–phosphorus and carbon–arsenic bonds. Another reason is undoubtedly the decreased stability of coordination compounds formed from tertiary stibines. Chatt (161) has made qualitative comparisons of the stability of the three compounds $[Pr_3P]_2[PtCl_2]_2$, $[Pr_3As]_2[PtCl_2]_2$, and $[Pr_3Sb]_2[PtCl_2]_2$ and found that there is a clear-cut decrease in stability in the order P, As, Sb. He states further that "this order is observed in all of the known complexes of the trialkyls of these elements and may be universally true." This decrease in stability must be associated primarily with the antimony–metal σ bond, since Chatt and Wilkins (162) have demonstrated that the π bond character of the Sb—M, As—M, and P—M bonds, which is due to overlap of filled *d* orbitals of the metal and vacant *d* orbitals of the group V element, remains approximately the same for phosphorus, arsenic, and antimony.

Trimethyl-, triethyl-, tripropyl-, and tributylstibine have been used quite extensively as ligands. Morgan and Yarsley (115) were the first to study coordination compounds of trimethylstibine. More recently, Jensen (163) and Chatt and Wilkins (164) have prepared compounds of the type $(R_3Sb)_2$-PtX_2 by shaking a trialkylstibine with potassium hexachloroplatinate(IV). In contrast to the trialkylphosphines and trialkylarsines where a mixture of *cis* and *trans* isomers is isolated, the trialkylstibines usually give the *cis* isomers exclusively (164). It has been shown that in solution the *trans* isomer is actually formed preferentially but that this is isomerized to the *cis* isomer when attempts are made to isolate the *trans* isomer from solution. A *trans* iodide, $[(C_2H_5)_3Sb]_2PtI_2$, was isolated from benzene solution as a red solid which rapidly isomerized to the yellow *cis* form (165).

Chatt and Venanzi (166) have reported amine complexes of the type $R_3SbPt(NHR)X_2$. These complexes as well as their phosphorus and arsenic analogs are associated in the solid state and in solution, presumably by hydrogen bonding of the type N—H—Cl (167, 168). The association constant can be calculated by measuring the deviation of the N—H stetching absorption from Beer's law. The degree of association was found to be a function of several different factors including the nature of the ligand. Association increased in the order $R_3P < R_3As < R_3Sb$. Thus, for the three compounds $R_3MPt(NHR')X_2$, where R is propyl, M is P, As, or Sb, and NHR′ is piperidine, the association constants are 18, 24, and 38, respectively. The authors attributed these differences to the inductive effect of the ligand on the amine. Thus, the least electronegative ligand, by donating its electrons more completely to the metal atom, causes the metal to release electrons to the nitrogen atom. The association constants, therefore, give a rough measure of the electron-releasing ability of phosphorus, arsenic, and antimony, at least in association with platinum.

Chatt and Venanzi (169) reported a bridged palladium complex, bis(triethylstibine)dichloro-μ,μ'-dichloropalladium, as a dark-red powder which decomposed rapidly in moist air and more slowly in a desiccator. Among compounds of this type, prepared from trialkylphosphines, -arsines, -stibines, dialkyl sulfides, selenides, and tellurides, those prepared from trialkylstibines were the least stable. Benlian and Bigorgne (170) reported the preparation of two coordination compounds of nickel and triethylstibine, $Ni(CO)_3Sb(C_2H_5)_3$ and $Ni(CO)_2[Sb(C_2H_5)_3]_2$, and three molybdenum compounds, $Mo(CO)_5$-$Sb(C_2H_5)_3$, *cis*-$Mo(CO)_4[Sb(C_2H_5)_3]_2$, and *cis*-$Mo(CO)_3[Sb(C_2H_5)_3]_3$. However, only in the case of the last two molybdenum compounds were crystalline materials actually isolated and analyzed. Coates (171) has prepared a series of coordination compounds from trimethylgallium and trimethyl derivatives of nitrogen, phosphorus, arsenic, and antimony. No compound could be prepared from trimethylbismuth. The stability of the compounds was in the order $N > P > As > Sb$.

Triarylstibines have found much greater use as ligands than have the trialkylstibines; the resulting coordination compounds are generally more stable than when trialkylstibines are used. Again the stability appears to be in the order P > As > Sb, but no quantitative studies have been made. Triphenylstibine has been the most widely used, but a few other symmetrical triarylstibines have also been employed.

Coordination compounds of triarylstibines have generally been prepared by displacing another ligand from a coordination compound. Such reactions (172, 173) are illustrated in the following equations:

$$Ir(CO)_2(p\text{-}NH_2C_6H_4CH_3)Br + 2Ph_3Sb \rightarrow Ir(CO)(SbPh_3)_2Br + CO + p\text{-}NH_2C_6H_4CH_3$$

$$V(CO)_6 + 2Ph_3Sb \rightarrow V(CO)_4(Ph_3Sb)_2 + 2CO$$

In addition to iridium and vanadium, coordination compounds containing one or more triarylstibine groups have been obtained from the following elements: gold (174), ruthenium (175), rhodium (176), chromium (177), iron (178), nickel (170, 179), cobalt (180), manganese (181), molybdenum (170, 182), and platinum (163). Most of these coordination compounds are stable crystalline solids, but the vanadium compound $V(CO)_4(SbPh_3)_2$ is pyrophoric (173). The properties of a few representative compounds are given in Table 8-1.

Mercuric chloride does not give a coordination compound with triphenylstibine; instead, an exchange reaction occurs to give phenylmercuric chloride and antimony trichloride (68). The three isomeric tritolylstibines do form compounds of the type $Ar_3Sb \cdot HgCl_2$ (69). The two compounds $(p\text{-}CH_3C_6H_4)_3Sb \cdot HgCl_2$ and $(m\text{-}CH_3C_6H_4)_3Sb \cdot HgCl_2$ are decomposed by boiling alcohol to give antimony trichloride and the corresponding tolylmercuric chloride, but the *o*-isomer is stable in boiling alcohol. Harris and co-workers (40) prepared stable 1:1 mercuric chloride-triarylstibine adducts from tri-*o*-anisyl-, tri-*m*-anisyl-, tri-*o*-phenetyl-, and tris(*p*-phenoxyphenyl)-stibines; these compounds possess sharp melting points. Tris(*p*-bromophenyl)stibine gives a mercuric chloride adduct which decomposes when attempts are made to recrystallize it.

Several groups of workers (182, 186, 187) have determined the C—O stretching frequencies in coordination compounds containing both a carbonyl group and a tertiary phosphine, arsine, or stibine and have found that changing the group V element has little or no effect on this frequency. From these results the various authors concluded that the donor–acceptor properties of the three ligand atoms (phosphorus, arsenic, and antimony) are essentially the same. On the other hand, the H—Ir stretching frequencies of the three compounds $IrHCl_2(Ph_3M)_3$, where M is phosphorus, arsenic, or antimony, were found to be 2200, 2170, and 2100 cm^{-1}, respectively (186).

Table 8-1 Representative Coordination Compounds Containing Triphenylstibine as a Ligand

Compound	Color	mp	Reference
L=$(C_6H_5)_3Sb$			
Co(NO)(CO)L	red	100	183
$Fe(NO)_2(CO)L$	orange-red	105–115	183
$Cr(CO)_5L$	yellow	147–149	177
$IrHCl_2L_2$	yellow-orange	201(d)	184
$Mn(CO)_3ClL_2$	yellow	—	181
$Ni(CO)_3L$	—	96–100	179
$Ru(CO)_2I_2L_2$	yellow	—	175
$(CO)_5Mn–AuL$	—	—	174
$Rh(CO)L_2I_3$	brown	198–200	185
$PtCl_2L_2$	yellow	140(d)	163

Such a difference might be expected in this case where the hydrogen is attached directly to the central atom and any difference in the inductive effect of the ligand would be enhanced.

Solid 1:1 complexes of triphenylphosphine, -arsine, and -stibine with hexamethylbenzene have been studied by Shaw and co-workers (188). Similar complexes of triphenylamine and triphenylbismuth could not be obtained. Naphthalene does not form complexes with any of the triphenyl derivatives. These complexes are believed to be formed by the donation of electrons from the hexamethylbenzene to the vacant *d*-orbitals of phosphorus, arsenic, and antimony.

(4) *Resolution of optically active tertiary stibines*

The theoretical calculations of Weston (189) suggested that tertiary stibines, as well as arsines, should be capable of resolution. The resolution of 2-methyl-10-(*p*-carboxyphenyl)phenoxantimonin (**8**) had been previously

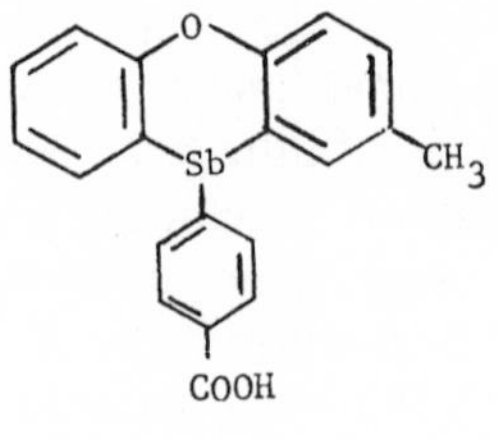

O
Sb
CH_3
COOH

8

accomplished by Campbell (51), but the asymmetry was ascribed to folding about the Sb—O axis rather than to the antimony atom itself. Later studies by Mislow and co-workers (190), however, have demonstrated that such folding does not account for the optical isomerism of the phenoxarsines, and they have attributed the existence of two enantiomeric forms solely to the asymmetry of the arsenic atom (cf. Section IV-5-D). It would seem probable, therefore, that the barrier to folding about the Sb—O axis would also be small and that the optical activity found by Campbell should be ascribed to the asymmetry of the antimony atom. The two enantiomers were separated by means of their strychnine salts and gave rotations of $[\alpha]_D = +77.5°$ and $-77.2°$. They were stable in boiling benzene or chloroform but racemized slowly in boiling alcohol.

Campbell then turned her attention to a series of unsymmetrical 5-phenyldibenzostiboles. The first of these to be prepared was 2-bromo-5-(*p*-carboxyphenyl)dibenzostibole, but the alkaloidal salts of this acid were too insoluble to be used for resolution (86). However, 5-*p*-tolyl-3-dibenzostibolecarboxylic acid was resolved by means of its α-phenylethylamine salts to give two enantiomers, $[\alpha]_D^{20} = \pm 245°$ (pyridine). The enantiomers were optically stable at 20° but racemized slowly at 40°. Racemization was catalyzed by hydrochloric acid; this result led Campbell to suggest that the stibine might be protonated to $[R_3SbH]^+$. However, since sulfuric acid of the same concentration did not catalyze the racemization, this explanation seems unlikely. Campbell (88) also resolved 3-amino-5-*p*-tolyldibenzostibole by means of its hydrogen tartrate, but only one enantiomer, $[\alpha]_D^{22} = +250.5°$, was obtained optically pure. The pure enantiomer racemized in benzene solution. By determining this rate at four different temperatures, the energy of activation was calculated to be approximately 15 kcal/mole. However, there was evidence that the racemization was catalyzed by traces of impurities, and the above figure may therefore be too low. Other derivatives of 5-phenyldibenzostibole which have been resolved include 3-methoxy-5-(*p*-carboxyphenyl)-, 3-carbomethoxy-5-*p*-tolyl-, and 3-methyl-5-(*p*-carboxyphenyl)dibenzostibole (89). With the last compound only one enantiomer was obtained optically pure; the (−) acid (−) ephedrine salt underwent second-order asymmetric transformation which prevented isolation of the (−) enantiomer in a pure form.

In all of the work on the dibenzostiboles, Campbell was concerned that the optical activity might be ascribed to molecular asymmetry in which the two benzene rings of the diphenyl moiety might not be coplanar and hence could assume two configurations. The ultraviolet absorption spectra of both 5-*p*-tolyldibenzostibole and its arsenic analog, however, were found to be consistent with a structure in which both benzene rings were coplanar, and accordingly it was concluded that the optical activity must be ascribed to the asymmetry around the central atom (191) (cf. IV-6-A). This conclusion was substantiated by later work in which non-heterocyclic stibines with three different aryl groups were resolved. Thus, *p*-carboxyphenyl-2-biphenylyl-phenylstibine was resolved into enantiomers, $[\alpha]_D^{20} = +47.2°$ and $-46.9°$ in pyridine (52). In contrast to the optically active derivatives of dibenzostibole, both of these enantiomers were optically stable in solution. Even dry hydrogen chloride in chloroform did not effect racemization. (*p*-Carboxyphenyl)-α-naphthylphenylstibine was also resolved to give the two enantiomers, $[\alpha]_D^{18} = -35.5 \pm 1.5°$ and $+35.1 \pm 1.6°$ in chloroform solution (53).

C. Physical properties

The ultraviolet absorption spectra of dimethylphenylstibine (192), several triarylstibines (52, 191, 193–197), and a number of dibenzostibole derivatives (52, 86, 191) have been reported. All of these spectra are characterized by intense absorption near 250 mμ and by a notable lack of fine structure. Thus, the spectrum of triphenylstibine in ethanol exhibits a broad, structureless band at about 255 mμ with a molar absorbance of about 12,000 (193). The spectrum of dimethylphenylstibine is very similar but less intense (λ_{max} = 250; ε = 3700) (192). 2-Biphenylyl-*p*-tolylphenylstibine exhibits no maxima above 230 mμ, but the spectrum has a well-defined inflection in the 250–255 mμ region (ε = 16,000) (52). The 5-aryldibenzostiboles also absorb strongly near 250 mμ, but in addition they have an intense peak at higher wavelengths (52, 191). For example, 5-*p*-tolyldibenzostibole has a λ_{max} at 287 mμ (ε = 12,400). The spectrum of ethyl 5-*p*-tolyl-3-dibenzostibole-carboxylate exhibits two intense maxima, one at 252.5 mμ (ε = 23,000) and a second at 301.5–302.5 mμ (ε = 21,000) (86). The intensity of absorption and the absence of vibrational structure in the spectra discussed in this paragraph suggest strong interaction of the non-bonded electrons of the antimony atoms and the π-orbitals of the benzene rings (cf. the interpretation of the spectra of tertiary arsines, Section IV-6-A).

Only a few Raman spectra of tertiary stibines have been determined. Rosenbaum and Ashford (198) found that the low frequency spectrum of trimethylstibine consists solely of two, very strong, broad lines at 188 and 513 cm^{-1}. It has been concluded (199) that this result is consistent with the expected trigonal-pyramidal (C_{3v}) symmetry, if one assumes that the angle between an edge of the pyramid and its symmetry axis is 58°. Several investigators (200, 201) have used the Raman data for calculations of the force constants of the C—Sb bond. Relatively recent values obtained by Siebert (201) are 2.18×10^5 dynes/cm for the stretching force constant and 0.145×10^5 dynes/cm for the bending force constant. Nesmeyanov and co-workers (202) have reported the Raman spectra of the *cis* and *trans* isomers of tris(2-chlorovinyl)stibine and have noted that there are no sharp differences between the spectra of the two compounds.

Miller and Lemmon (203) have made a detailed analysis of the infrared (35 to 4000 cm^{-1}) and Raman spectra of triethynylphosphine, -arsine, and -stibine. A strong band at 449.5 cm^{-1} in the vapor-phase infrared spectrum of the stibine was assigned to a C—Sb stretching mode, while a polarized Raman line at 476 cm^{-1} was assigned to the second C—Sb stretch. In all, twelve spectroscopically-active fundamental vibrations of the antimony compound were identified. Hartmann and Kühl (19) have reported that the infrared spectra of dialkylethynylstibines contain bands at 3300 cm^{-1} (characteristic of the C—H stretching band of monosubstituted acetylenes) and 2000 cm^{-1} (assigned to C≡C stretching).

Oswald (204) has investigated the infrared spectra of seven trialkylstibines in the 300–4000 cm^{-1} region. The spectra of five of these compounds (triethyl-, tripropyl-, triisopropyl-, tributyl-, and tri-*sec*-butylstibine) were found to contain a band between 478 and 505 cm^{-1} which was assigned to the C—Sb stretch. This vibration in trimethylstibine was found at a somewhat higher frequency (by 10–20 cm^{-1}), while in triisobutylstibine it occurs at a much higher value, 599 cm^{-1}. Taking 512 cm^{-1} as the stretching frequency of the C—Sb bond in the average trialkylstibine, Oswald calculated a force constant of 2.07×10^5 dynes/cm. There appears to be no obvious reason why the C—Sb bond in the isobutyl compound should have a significantly higher stretching frequency and hence a larger force constant (2.7×10^5 dynes/cm). Borisov and co-workers (12) have also studied the infrared spectra of several trialkylstibines. Their results are in general agreement with those of Oswald. Thus, the Russian workers reported that the C—Sb stretching vibration occurs at 505 cm^{-1} in tributylstibine and at 457 cm^{-1} in triallylstibine. More recently, they have noted (205) that the C—Sb stretching vibration in alkyl, phenyl, alkenyl, and haloalkyl compounds of antimony is in the 455–560 cm^{-1} region.

A medium–strong band near 529 cm^{-1} in the infrared spectra of the *cis* platinum(II) chloride and platinum(II) bromide complexes of triethylstibine has been assigned by Jensen and Nielsen (206) to the C—Sb asymmetric stretch. The spectrum of dibutylallylstibine has been found to exhibit bands characteristic of the allyl group ($\nu_{C=C}$, 1625 cm^{-1}) and of the butyl groups (865, 1140, 1455, and 2880 cm^{-1}) (37). McKenney and Sisler (117) have described the spectrum of tributylstibine, but have not made any specific assignments. The infrared spectra of dimethyl(perfluoroethyl)stibine and diethyl(trifluoromethyl)stibine have been determined in the 700–4000 cm^{-1} region by Pullman and West (207). Almost all of the bands observed were assigned to various C—H and C—F vibrations. The dimethyl compound exhibits a strong band at 925 cm^{-1}, assigned to methyl rocking vibrations, and a medium band at 822 cm^{-1}, assigned to methyl wagging. The infrared spectrum of tris(trifluoromethyl)stibine has been found to contain four strong bands in the C—F stretching vibration region as well as a strong band at 722 cm^{-1} attributed to CF_3 deformation (106).

A study of the infrared spectra of the *cis* and *trans* isomers of tripropenylstibine and of a number of related substances has revealed several interesting generalizations concerning the spectra and the stereochemistry of these compounds (56, 58, 208). All of the *trans* isomers exhibit a characteristic, strong absorption band in the 945–970 cm^{-1} region (assigned to an out-of-plane C—H vibration). Although the *cis* isomers have bands in the nearby 920–940 cm^{-1} region, these bands are considerably weaker than the 945–970 cm^{-1} bands of the *trans* isomers. All of the *trans* isomers (but none of the *cis* isomers) have bands in the 718–726 cm^{-1} region. The spectra of both the *cis* and *trans* isomers exhibit bands in the 655–660 cm^{-1} region, but these bands are 2–2.5 times as intense in the spectra of the *cis* compounds. A relatively strong band at 452 cm^{-1} is present in the spectra of the *cis* tripropenylantimony dihalides and of *cis*-tetrapropenylstibonium bromide; this band is absent in all the *trans* isomers as well as in *cis*-tripropenylstibine and *cis*-pentapropenylantimony. The ditertiary stibine, vinylenebis(dibutylstibine), has a strong absorption band at 970 cm^{-1} and has been assigned the *trans* configuration (11). *Trans*-diphenylstyrylstibine has been found to have strong absorption bands at 980 and 990 cm^{-1}. The spectrum of the isomeric diphenyl(1-phenylvinyl)stibine does not contain these frequencies, but it exhibits an absorption at 1685 cm^{-1} which is absent in the spectrum of the diphenylstyrylstibine (9).

Benlian and Bigorgne (170) have used infrared spectroscopy to study the C—O stretching frequency in complexes of the types $Ni(CO)_{4-n}L_n$ and $Mo(CO)_{6-n}L_n$, where L is triethyl- or triphenylstibine and n is 1, 2, or 3. In a later paper Bouquet and co-workers (209) reported a systematic study of the Raman and infrared spectra of the nickel complexes and were able to assign the Ni—C and Ni—C—O vibrations.

Sterlin and Dubov (210) have studied the infrared spectra of a series of perfluorovinyl compounds of the type $(CF_2{=}CF)_xM$, where M is B, P, As, Sb, Bi, Si, Ge, Sn, Pb, or Hg. With the exception of the boron derivative, the frequency of the C=C bond was always found in 1720–1730 cm^{-1} region.

The infrared spectrum of triphenylstibine has been described by a large number of investigators (127, 194, 195, 206, 211–214). In a recent paper Mackay and co-workers (214) have discussed the spectrum of this stibine in terms of the notation used by Whiffin in his analysis of the spectra of the halobenzenes (cf. Section V-7). The vibrations q and t (which contain contributions from Ph—Sb stretching) were assigned to strong, relatively broad bands at 1063 and 268 cm^{-1}, respectively, and the vibration *y* (which involves Ph—Sb bending) was assigned to a complex, very strong absorption centered at 453 cm^{-1}.

The infrared spectrum of 1,4-phenylenebis(diphenylstibine) contains strong bands at 480, 454, and 333 cm^{-1}, which have been attributed to C=C deformation or C—Sb stretching (197).

Only a few nuclear magnetic resonance spectra of tertiary stibines have been published. Several groups of workers (215–217) have included trimethylstibine in a study of the proton NMR spectra of the trimethyl derivatives of the group V elements. It was concluded that the proton chemical shift could be correlated with the electronegativity of the atom bonded to the methyl group. The difference between the proton chemical shifts of the CH_3 and CH_2 protons in various ethyl compounds has also been correlated with the electronegativity of the central atom (218, 219). Thus, this difference in triethylstibine is very small (0.1 ppm) and is in line with the low electronegativity of antimony. The proton NMR spectrum of trivinylstibine (220) has been reported as part of an investigation of a large number of vinyl compounds, while the proton spectrum of triphenylstibine (117) has been determined in connection with a study of the chloramination of tertiary stibines. Nuclear quadrupole resonances of ^{121}Sb and ^{123}Sb in triphenylstibine (221) and the ^{19}F NMR spectrum of tris(perfluorovinyl)stibine (222) have also been described.

Fritz and Schwarzhans (223) have investigated the effect of complex formation on the proton NMR spectra of alkyl derivatives of group V and group VI elements. Included in this study were two platinum(II) complexes of trimethylstibine. It was found that the proton signal of the methyl groups is shifted to lower fields when the stibine is bound to platinum.

The Mössbauer isomer shifts of the ^{121}Sb nucleus in seven organoantimony compounds of the types Ar_3Sb, Ar_4SbX, and Ar_3SbX_2 have been reported (224). The 37.2 KeV ^{121}Sb gamma ray used was derived from ^{121}Sn (which decays to ^{121}Sb) in the form of stannic oxide. The spectra of the organoantimony compounds consist mainly of broadened single lines with appreciable asymmetry. The tertiary stibines, namely tris(*p*-chlorophenyl)- and tris(*p*-methoxyphenyl)stibine, have isomer shifts of -9.3 and -9.0 mm/sec, respectively, while the isomer shifts of the other compounds range from -5.5 to -6.9 mm/sec.

Very little is known about interatomic distances in tertiary stibines. The structure of tris(trifluoromethyl)stibine has been determined in the gas phase by means of electron diffraction measurements (225). The observed C—Sb distance of 2.202 ± 0.016 A is only slightly longer than the 2.18 A calculated from the sum of the covalent radii. The C—Sb—C angle was found to be $100.0° \pm 3.5°$, a value not significantly different from the C—P—C and

C—As—C angles in tris(trifluoromethyl)phosphine and tris(trifluoromethyl)-arsine, respectively. A number of investigators (39, 226–228) have used X-ray spectroscopy to study the crystal structure of triphenylstibine, but no information about bond distances or bond angles in this compound is available. Mootz and co-workers (229) have reported that the tris(phenylethynyl) compounds of phosphorus, arsenic, and antimony have three molecules in the unit cell and exhibit the expected pyramidal structure (C_3 symmetry) with the lone pair presumably occupying the fourth coordination position around the heteroatom; the arrangement of the three phenyl groups is described as propeller-like.

Only a few thermochemical measurements have been made on tertiary stibines. Long and Sackman (230) have investigated the combustion of trimethylstibine in a static bomb calorimeter and have reported that the main solid product is antimony tetroxide, Sb_2O_4, although significant amounts of antimony trioxide, antimony metal, and soot are also formed. By analyzing the products and applying the necessary corrections, they were able to obtain −698.0 kcal/mole for the heat of combustion of liquid trimethylstibine to antimony tetroxide, carbon dioxide, and liquid water. This value corresponds to a heat of formation for trimethylstibine at 25° of 1.4 kcal/mole; i.e., the reaction between the elements to form trimethylstibine is slightly endothermic. The heat of formation thus obtained was used in the usual way to derive the mean dissociation energy of the methyl–antimony bond as 49.7 kcal/mole. It has been suggested that this figure should be regarded as a minimum value; using the same heat of combustion data, Skinner (231) obtained 51.5 kcal/mole for the methyl–antimony bond energy. Kinetic studies of the thermal decomposition of trimethylstibine have led to 57.0 kcal/mole for the energy required to cleave the *first* C—Sb bond in this compound (232). The validity of this result is not certain, since the pyrolysis was quite complex and the interpretation of the data was not straightforward.

Lautsch and co-workers (233, 234) have employed a Berthelot bomb calorimeter for the determination of the heats of combustion of the triethyl derivatives of phosphorus, arsenic, antimony, and bismuth, and have used their data to calculate the mean dissociation energies of the ethyl-heteroatom bonds. They noted that the calculated bond energies decrease with increasing atomic number of the heteroatom and are thus in accord with the chemical properties of these compounds. Triethylstibine was found to have a heat of combustion of −1162.4 kcal/mole, a value which leads to 29.2 kcal/mole for the ethyl–antimony bond energy.

Birr (235, 236) has measured the heat of combustion of triphenylstibine by means of static bomb calorimetry, and has used this data to obtain a value of 61.3 kcal/mole for the mean dissociation energy of the phenyl–antimony bond. There are errors in Birr's calculations, however, and Skinner (231) has recalculated the value of the dissociation energy to be 58.3 kcal/mole. The fact that the triphenyl compound has a higher bond energy than the trimethyl compound has been attributed by Birr (236) to the existence of π-bonding between the antimony atom and the benzene rings.

Winters and Kiser (237) have investigated the ionization and subsequent fragmentation of trimethylstibine in a time-of-flight mass spectrometer. The major ion was found to be the dimethylantimony ion; however, the parent molecule-ion was also present in high abundance. In addition, the following ions were observed with relative abundances decreasing in the order given: Sb^+, CH_3Sb^+, CH_2Sb^+, $(CH_2)_2Sb^+$, SbH^+, $CHSb^+$, SbH_2^+, and CH_3SbH^+. Since n-heptane was used as a solvent for the trimethylstibine sample, it was not possible to study the hydrocarbon ions that may have been formed from the antimony compound. The ionization potential for trimethylstibine was found to be 8.04 ± 0.16 eV. Appearance potentials for all the other positive ions produced were also determined, and the corresponding heats of formation were calculated.

The mass spectra of triphenylstibine, trivinylstibine, and tris(perfluorovinyl)stibine are discussed in Section IV-6-C. Vilesov and Zaitsev (238) have found by means of a photoionization method that the first ionization potential of triphenylstibine is 7.3 eV. They concluded that the positive ion resulted from the removal of an electron from the lone pair on the antimony atom.

Long and Sackman (13) have measured the vapor pressure, melting point, and liquid density of a carefully purified sample of trimethylstibine. The vapor pressure results were obtained over the range $-25°$ to $+15°$ and lead to an extrapolated normal boiling point of 79.4°, a heat of vaporization of 7.82 kcal/mole, and a Trouton's constant of 22.1. These data are in good agreement with some earlier results of Rosenbaum and Sandberg (239). The freezing point ($-87.6°$) observed by Long and Sackman is, however, about 25° lower than an earlier value ($-62.0°$) reported by Bamford and coworkers (84). The latter group of investigators reported that triethylstibine has a freezing point of $-98.0°$, an extrapolated normal boiling point of 161.4°, a heat of vaporization of 9.993 kcal/mole, and a Trouton's constant of 23.0. Trivinylstibine has a much lower freezing point ($-157.0°$), but its other physical properties are not unexpected: an extrapolated normal boiling point of 149.9°, a heat of vaporization of 9.21 kcal/mole, and a Trouton's constant of 21.8 (22, 23). Strangely enough, the boiling points of the tris(trifluoromethyl) derivatives of the group V elements are lower than the boiling points of the corresponding trimethyl compounds (106). Thus, tris(trifluoromethyl)stibine has a boiling point of 72° and a comparatively high melting point of $-58°$. The Trouton's constant, derived in the usual way from vapor pressure measurements, is 24.0.

Forward, Bowden, and Jones (240) have measured the densities, vapor pressures, and surface tensions of the triphenyl derivatives of phosphorus, arsenic, antimony, and bismuth at temperatures above the melting points of the compounds. The density of triphenylstibine (mp 78.5°) between 80° and 260° was found to obey the following equation:

$$D_{4^\circ}^{t^\circ} = 1.4564 - 0.0009426t$$

Vapor pressure measurements between 230° and 280° led to an extrapolated normal boiling point of 377° and a heat of vaporization of 19.88 kcal/mole. The surface tension data in conjunction with the liquid densities yielded a parachor of 635.4. Since this value was found to be essentially independent of temperature, it was concluded that liquid triphenylstibine is not associated. If Sugden's value of 190.9 for the parachor of the phenyl group is used, the atomic parachor of antimony turns out to be 65.4. This is in reasonable agreement with other values found in the literature. Forward and his co-workers also studied the phase equilibria between triphenylstibine and each of the following compounds: triphenylamine, triphenylphosphine, triphenylbismuth, triphenylmethane, and tetraphenyltin.

The dipole moment of triphenylstibine has been determined in several laboratories (241–244). The results, which vary from 0.57 D to 0.84 D, are consistent with the expected pyramidal structure of this compound. The dipole moment of 1,4-phenylenebis(diphenylstibine) has been found to be 1.14 D (197). In both the Ph_3M and the p-$Ph_2MC_6H_4MPh_2$ series (where M is P, As, Sb, or Bi), the dipole moment decreases in the order P > As > Sb > Bi. In sharp contrast to these results, the dipole moments of the tris(perfluorovinyl) derivatives of the group V elements are in the order Bi > Sb > As > P (152). The dipole moment of tris(perfluorovinyl)stibine has the relatively large value of 2.84 D.

The molar diamagnetic susceptibility of triphenylstibine was found by Pascal (245) to be -182.2×10^{-6} cgs unit. Parab and Desai (246, 247) obtained a similar value for this compound and also determined the susceptibilities of four other tertiary stibines. Their experimental results led to a value of -24.70×10^{-6} cgs unit for the molar diamagnetic susceptibility of trivalent antimony. Fischer and co-workers (248) have reported that the molar susceptibility of tricyclopentadienylstibine is $-178 \pm 40 \times 10^{-6}$ cgs unit.

Bothorel (249, 250) has obtained Rayleigh scattering data on cyclohexane solutions of triphenylamine, -phosphine, -stibine, and -bismuth and has also determined the magnetic susceptibility of single crystals of these compounds. His results enabled him to calculate the angles between the benzene rings and to conclude that the lone pairs of electrons of the heteroatoms were delocalized to a large extent. Other workers (244) have used Kerr constants in conjunction with dipole moment and refractivity data to deduce the conformation of the triphenyl derivatives in benzene solution.

Hartmann and Kühl (19) have reported normal molecular weights for tertiary stibines of the types $R_2SbC{\equiv}CH$ and $R_2SbC{\equiv}CSbR_2$, where R is Me, Et, *i*-Pr, or *t*-Bu. The cryoscopic molecular weight of tri-*tert*-butylstibine in benzene has also been found to be normal (101).

3. HALO- AND DIHALOSTIBINES

A. Preparation

The reaction between an antimony trihalide and a Grignard reagent normally leads to the formation of the tertiary stibine. With *tert*-butylmagnesium chloride, however, only two halogens can be replaced, and di-*tert*-butylchlorostibine is the sole antimony-containing product of the reaction (19). Harris and co-workers (40) obtained a mixture of tri-*o*-tolylstibine and di-*o*-tolylchlorostibine from *o*-tolylmagnesium chloride and antimony trichloride. Other examples of the preparation of halo- and dihalostibines by means of the Grignard reaction have been reported in the patent literature (33).

Organometallic compounds less active than Grignard reagents have been used successfully for replacing one or two halogens of an antimony trihalide. Goddard and co-workers (251) first reported the use of this method in which diphenylchlorostibine was obtained from antimony trichloride and tetraphenyllead. Later workers (32) reported that the product was difficult to purify when prepared by this procedure and that better results were obtained when tetraphenyltin was used. Diphenylchlorostibine has been obtained in low yield by the interaction of triphenylbismuth and antimony trichloride in refluxing chloroform (252). Kharasch and co-workers (253) used tetraethyllead for the preparation of ethyldichlorostibine, and Maier (254) prepared vinyldichlorostibine in a similar manner from tetravinyllead. Müller and Dathe (255–258) have used organosilicon derivatives to prepare the previously unknown diphenylfluorostibine. Thus, they found that antimony trifluoride can be di-arylated with an arylfluorosilicate in aqueous solution:

$$2PhSiF_5^{2-} + SbF_3 \rightarrow Ph_2SbF + 2SiF_6^{2-}$$

The secondary fluorostibine can also be obtained by the addition of phenyltrifluorosilane to an aqueous solution of ammonium pentafluoroantimonate(III) and ammonium fluoride:

$$2PhSiF_3 + SbF_5^{2-} + 2F^- \rightarrow Ph_2SbF + 2SiF_6^{2-}$$

The preparation of phenyldifluorostibine via organosilicon compounds was, however, not successful.

Both dialkyl- and diarylhalostibines have been prepared by the pyrolysis of the pentavalent antimony dihalides:

$$R_3SbX_2 \xrightarrow{\Delta} R_2SbX + RX$$

Since the dialkylhalostibines are spontaneously inflammable liquids, the pyrolysis and subsequent distillation must be carried out in an inert atmosphere. The procedure can be used for preparing both chloro- and bromostibines (16). Excellent yields of dimethyl-, diethyl-, and diisopropylchlorostibines (19) as well as dipropyl- (259), dibutyl-, and diisobutylbromostibines (37) have been obtained by this method. The results obtained with triarylantimony dihalides are less satisfactory. Goddard and Yarsley (151) claimed to have prepared di-*p*-tolylchlorostibine and the corresponding bromo- and iodostibines by pyrolysis of the corresponding tri-*p*-tolylantimony dihalides. Blicke and co-workers (260), however, questioned the reliability of the results, at least as far as the iodo compound obtained, since the di-*p*-tolyliodostibine they obtained from di-*p*-tolystibinous acetate and potassium iodide melted at 76–78°, whereas Goddard and Yarsley reported a melting point of 233°. Other workers have reported that only trace amounts of diarylhalostibines are obtained by the pyrolysis of triarylantimony dihalides (71, 150).

The pyrolysis of dimethylantimony trichloride leads to the formation of methyldichlorostibine (16). The present authors (261) have prepared this compound in much better yield from tetramethyllead and antimony trichloride. Methyldibromo- and methyldiiodostibine are formed by the spontaneous loss of methyl bromide or iodide when the corresponding dimethylantimony tribromide or triiodide is stored in a desiccator (16).

A redistribution reaction between trialkyl- or triarylstibines and an antimony trihalide leads to the formation of a mixture of antimony trihalides and halo- and dihalostibines. Generally the mixture is difficult to separate, and the procedure has found relatively little use for preparative purposes. Thus, although Maier and co-workers (22, 23) were able to separate analytically pure divinylbromostibine from the mixture of divinylbromostibine, vinyldibromostibine, and antimony tribromide, they were unable to separate the other two components. When antimony trichloride was used, the mixture of vinyldichloro-, divinylchlorostibines, and antimony trichloride was not separated. Worrall (71) succeeded in separating the mixture of *p*-biphenylyldichlorostibine and di-*p*-biphenylylchlorostibine (obtained from tri-*p*-biphenylylstibine and antimony trichloride) only by a tedious fractional crystallization procedure. When tri-*o*-biphenylylstibine and antimony trichloride were heated with xylene in a sealed tube and the product was repeatedly recrystallized from benzene, a substance was obtained which was formulated as *o*-$C_6H_5C_6H_4Sb(Cl)OH$ on the basis of analytical results (72).

Dale and co-workers (106) heated tris(trifluoromethyl)stibine and antimony triiodide in a sealed tube for seven days and obtained some bis(trifluoromethyl)iodostibine admixed with starting materials; the products were identified spectroscopically and were not separated. In the patent literature the claim has been made that the redistribution reaction is quite successful if a solvent is used (262, 263). Thus, triphenylstibine and antimony trichloride, when heated in a 1:2 ratio in methylene chloride for 64 hours, gave phenyldichlorostibine in 95% yield (262). If the ratio is reversed, a 91.6% yield of diphenylchlorostibine was reported. More recently, Besolova and co-workers (264) have reported that the addition of antimony trichloride to a cold benzene solution of tris(carbethoxymethyl)stibine yields a dichlorostibine:

$$(EtO_2CCH_2)_3Sb + 2SbCl_3 \rightarrow 3EtO_2CCH_2SbCl_2$$

One of the best methods for preparing aryldichloro- or diarylchlorostibines is by reduction of the corresponding arylantimony tetrachlorides or diarylantimony trichlorides, respectively. Since the readily available arylstibonic and diarylstibinic acids are converted to these chlorides with hydrochloric acid, the best procedure is to reduce a solution of the stibonic or stibinic acid in hydrochloric acid without isolation of the intermediate chlorides. The usual reducing agents are either sulfur dioxide (often with the addition of a trace of hydriodic acid) or stannous chloride. There are a number of literature descriptions of both procedures. Thus, Sergeev and Bruker (265) have described the preparation of both phenyldichloro- and diphenylchlorostibines by reduction of benzenestibonic and diphenylstibinic acids, respectively, in hydrochloric acid solution with sulfur dioxide and hydriodic acid. Pfeiffer and Schmidt (266) described the reduction of α-naphthalenestibonic acid to the dichlorostibine in methanol–hydrochloric acid; Morgan and Davies (50) described the preparation of 5-chlorodibenzostibole from the corresponding cyclic acid. An excellent description of the reaction conditions for this reduction has been given by Blicke and Oakdale (267). The yields by this method are usually excellent, and it has been quite widely used.

Meyers and Jones (268) have described the preparation of 2-thiazolyldichlorostibine by reduction of the corresponding tetrachloride. Doak and Jaffé (269) have described a general method for the stannous chloride reduction based on Campbell's preparation of *p*-cyanophenyldichlorostibine (51). Kharasch (270) has also described the preparation of a number of aryldichlorostibines. For example, *p*-hydroxybenzenestibonic acid was reduced with stannous chloride in acetic acid solution; addition of aqueous potassium iodide solution resulted in the precipitation of *p*-hydroxyphenyldiiodostibine. The reduction of unsymmetrical diarylantimony trichlorides with stannous chloride in ethanol-hydrochloric acid was used extensively by Campbell (53,

86, 88, 89) in her studies of the preparation and resolution of unsymmetrical tertiary stibines. Woods (46, 95) has described the preparation of diphenylchlorostibine by a similar procedure. In general, the yields of both aryldichloro- and diarylchlorostibines are excellent by this method. It has been reported that reduction of *p*-carboxymethylthiobenzenestibonic acid with stannous chloride in hydrochloric acid results in cleavage of the carbon–antimony bond, but if the reduction is carried out in an acetic acid-hydrochloric acid mixture at low temperatures, the desired dichlorostibine can be obtained (271). As far as the present authors are aware, the only halides prepared by either of the above two reductive methods have been chloro- or dichlorostibines.

The hydrolysis of aryldihalo- or diarylhalostibines to give the corresponding oxides is reversible, and it is possible to prepare both the mono- and dihalostibines by treating the corresponding oxides with a hydrohalic acid. *o*-Benzylphenyldibromostibine has been obtained from *o*-benzylstibosobenzene and hydrobromic acid (272); phenyldichlorostibine is readily obtained when stibosobenzene in acetic acid is treated with hydrochloric acid (273, 274). Doak and Jaffé (269) have described an improvement on this method. The stiboso compound is suspended in chloroform; the suspension is cooled in dry ice and then treated with dry hydrogen chloride, whereupon the stiboso compound is converted to the crystalline aryldichlorostibine. The reaction between stiboso compounds and hydrohalic acids should be conducted at low temperatures in order to avoid disproportionation to diarylchlorostibines. Nesmeyanov and co-workers (275) have recorded the observation that when *o*-chlorostibosobenzene was treated with hydrochloric acid, presumably at room temperature, bis(*o*-chlorophenyl)chlorostibine was obtained. Diarylhalostibines have also been prepared by the reaction of diarylantimony acetates with hydrohalic acids (260).

Aryldiiodostibines and diaryliodostibines are best prepared by adding hydriodic acid or sodium iodide to a solution of the corresponding aryldichloro- or diarylchlorostibine, respectively (51, 260, 267, 269). The less soluble iodo- or diiodostibine precipitates from solution.

The reaction between a diazonium salt and antimony trichloride in an organic solvent in the presence of a reducing agent, such as zinc, may lead to the formation of aryldichloro- or diarylchlorostibines as the principal products of the reaction (92). The reaction is not straightforward, however, in that triarylstibines and triarylantimony dichlorides may also be isolated from the reaction mixture. Furthermore, with most of the reactions studied, the dichloro- or chlorostibines were not isolated as such but were hydrolyzed to oxides or converted to other derivatives for identification purposes. For this reason this reaction is discussed in a following section dealing with hydrolysis and other products of halo- and dihalostibines.

Aryldichloro- and diarylchlorostibines (or their hydrolysis products) may also be formed (together with other organoantimony compounds) by the decomposition of diaryliodonium (276), bromonium, or chloronium salts (277) in the presence of metallic antimony, by the decomposition of arenediazonium tetrachloroantimonate(III) compounds in ethyl acetate brought about by the addition of zinc (278), and by the decomposition of diazonium salts in acetone in the presence of antimony powder and calcium carbonate (94). None of these methods is probably of synthetic importance.

Chloromethyldichlorostibine has been prepared from diazomethane and antimony trichloride (104, 105). The reaction is run in benzene in a nitrogen atmosphere. No bis(chloromethyl)stibine is obtained, and if excess diazomethane is used, the product is tris(chloromethyl)stibine. On the other hand, when diazoethane is used, no α-chloroethyldichlorostibine is isolated, and bis(α-chloroethyl)chlorostibine is the only product of the reaction.

In their investigations of the direct synthesis of organometallic compounds from alkyl halides and a metal, Maier and co-workers (279) heated metallic antimony in a stream of methyl chloride and hydrogen in the presence of cupric chloride as a catalyst. The methyldichloro- and dimethylchlorostibines which resulted from the reaction were not isolated but were treated with butylmagnesium bromide to give a separable mixture of dibutylmethyl- and butyldimethylstibines. The yields of the two tertiary stibines obtained indicated that 58.2% of the initial reaction product was methyldichlorostibine and 9.7% was dimethylchlorostibine. When methyl bromide was used, a 40% yield of methyldibromostibine and a 38% yield of dimethylbromostibine was indicated.

Cyclic diarylchlorostibines have been prepared by cyclodehydrohalogenation. Thus, 5-chlorodibenzostibole results when 2-biphenylyldichlorostibine is distilled at 100°/25 mm (50); *o*-benzylphenyldichlorostibine fails to undergo ring closure under similar reaction conditions (272).

No difluorostibine and only one fluorostibine (diphenylfluorostibine) has been prepared. Wilkins treated methyldichloro- and dimethylchlorostibines with ammonium fluoride but obtained only trimethylstibine in both cases (280). He attributed this result to disproportionation at the reaction temperature (90°) employed.

B. Reactions

The alkyldihalo- and dialkylhalostibines are highly reactive substances. They are rapidly oxidized in the air, and some are spontaneously inflammable (16). Controlled oxidation of the dialkylhalostibines leads to the formation of compounds which have been formulated as $R_2Sb(O)X$ (16). Alkaline hydrolysis of alkyldihalo- and dialkylhalostibines gives the corresponding oxides,

$(RSbO)_x$ and $(R_2Sb)_2O$, while the action of H_2S leads to the formation of RSbS and $(R_2Sb)_2S$. The aryldihalo- and diarylhalostibines are less susceptible to air oxidation than the corresponding alkyl compounds. They also appear to be thermally more stable, but many of these, such as phenyldichlorostibine, slowly decompose when stored in a desiccator at room temperature. All attempts to recrystallize *m*-anisyldichlorostibine from chloroform resulted in decomposition (269).

Aryldichlorostibines react with amine hydrochlorides to form compounds of the type $[ArSbCl_3]^-[RNH_3]^+$. Pyridine and quinoline have usually been used as the amine, and the salts obtained generally have sharp reproducible melting points. Thus, the following compounds have been reported: $[p\text{-}CH_3C_6H_4SbCl_3]$[H-pyridine], mp 132°; $[C_6H_5SbCl_3]$[H-quinoline], mp 111° (274); and $[\alpha\text{-}C_{10}H_7SbCl_3]$[H-pyridine], mp 90° (266). Aryldichlorostibines react with diazonium salts to form arenediazonium aryltrichloroantimonate(III) compounds. These compounds were first investigated by Bruker and co-workers (281–283). Thus, benzenediazonium chloride and phenyldichlorostibine, both in glacial acetic acid solution, were mixed to yield a precipitate of $[C_6H_5N_2][C_6H_5SbCl_3]$ (281). If an excess of diazonium salt is used, compounds of the type $[ArN_2^+]_2[ArSbCl_4]^=$ may be obtained (282, 283). Diazotization of amino-substituted aryldichlorostibines results in the formation of zwitterions such as $p\text{-}\overset{+}{N}_2C_6H_4\overset{-}{Sb}Cl_3$ (274, 284). It seems probable that many aryldichlorostibines containing a basic nitrogen substituent should be written as zwitterions rather than as hydrohalic salts, e.g. (268):

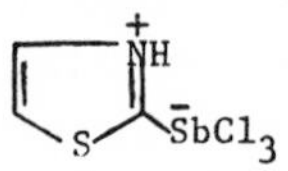

$\overset{+}{N}H$ $\overset{-}{Sb}Cl_3$ S

The reaction between diarylhalostibines and arenediazonium salts does not lead to the expected arenediazonium diaryldichloroantimonate(III) derivatives, as earlier claimed by Bruker and Nikiforova (143), but rather to antimony(V) derivatives $[ArN_2][Ar_2SbCl_4]$ (285).

Aryldihalostibines react with halogens to give arylantimony tetrahalides (cf. VII-3), but there have been no reports of a similar reaction between alkyldihalostibines and halogens. Both dialkyl- and diarylhalostibines react with halogens to give the corresponding dialkyl- and diarylantimony trihalides (cf. VII-4-A). Dihalo- and halostibines can be converted to tertiary stibines by reaction with suitable organometallic reagents (cf. VIII-2-A-(1)). The halogen in a dialkyl- or diarylhalostibine can be replaced by lithium (37), magnesium (37), or sodium (95), but there have been no reports of a reaction between a dihalostibine and a reactive metal. The reduction of dihalo- and halostibines to primary and secondary stibines, respectively, has been discussed in Section VIII-1; the reduction of dihalo- and halostibines to compounds containing antimony–antimony bonds will be described in Section VIII-5.

The antimony–carbon bond in aryldihalo- or diarylhalostibines may be cleaved by such reagents as mercuric chloride (286, 287) or aluminum chloride (288). The carbon–antimony bond in $ClCH_2SbCl_2$ is cleaved with both alkali and water (105). Trifluoromethyldiiodostibine and bis(trifluoromethyl)chlorostibine are cleaved by aqueous potassium hydroxide to form fluoroform (106). In general, however, only the antimony–halogen bond in halo- and dihalostibines is split by water or alkali.

A number of compounds of the types R_2SbY and $RSbY_2$ have been prepared by the reaction of halo- or dihalostibines with ammonium or sodium alkoxides (259), phenoxides (289), mercaptides (271), dithiocarbamates (90, 290–292), azides (293), carboxylates (294–297), or the lithium (298) or sodium (299) salt of trimethylsilanol. The properties of these compounds will be discussed in Section VIII-4. The reaction of diphenylchlorostibine with a sodium dialkyldithiocarbamate does not, however, always yield the expected diphenylantimony dialkyldithiocarbamate; in some cases the phenylantimony bis(dialkyldithiocarbamate) is obtained (292). This result illustrates the ease with which the carbon–antimony bond is sometimes cleaved.

A number of silicon-containing amino derivatives of antimony have been prepared via halo- and dihalostibines and are discussed in Section III-5. The conversion of halo- and dihalostibines to compounds containing antimony–metal or antimony–metalloid bonds will be described in Section VIII-6.

C. Physical properties

Except for melting and boiling point data, very little is known about the physical properties of halo- and dihalostibines. Most of these compounds are colorless liquids or solids, but di-*tert*-butylchlorostibine (19), dicyclohexylchlorostibine (19), and several dialkylbromostibines (37) have been described as yellow liquids while dimethyliodostibine (16) is a yellow solid. Trifluoromethyldiiodostibine has been obtained as a viscous, bright-yellow liquid which freezes to a pale-yellow solid (106). Doak and Jaffé (269) have found by means of cryoscopic measurements in benzene that *p*-tolyl- and *p*-chlorophenyldichlorostibines have normal molecular weights. Saikina (300) has reported that the polarographic reduction of phenyldichlorostibine takes place in 1 *N* sodium hydroxide at a half-wave potential of -1.24 volts (relative to the saturated calomel electrode). Dessy and his co-workers (301) have observed that the electrolytic reduction of diphenyliodostibine gives at first a colorless solution of tetraphenyldistibine, presumably as the result of rapid coupling of diphenylantimony radicals. Further reduction yields an orange solution of the diphenylantimony anion:

$$Ph_2SbSbPh_2 + 2e^- \rightarrow 2Ph_2Sb^-$$

4. OXIDES, SULFIDES, AND RELATED COMPOUNDS

The hydrolysis of alkyl- or aryldihalostibines and dialkyl- or diarylhalostibines gives rise to compounds of the type $(RSbO)_x$ and $R_2SbOSbR_2$, respectively. The former are termed stiboso compounds by analogy with nitroso compounds, although the antimony compounds do not contain an Sb=O linkage but are certainly polymeric. The hydrolysis has been carried out with water, but more often aqueous sodium hydroxide or ammonia is used. Methyldichloro- and methyldibromostibine on hydrolysis with dilute aqueous alkali give an amorphous solid (16). Analytical values for this product are stated to be unsatisfactory for CH_3SbO, but methyldiiodostibine can be obtained by treatment of the solid with hydriodic acid. Hydrolysis of dimethylbromostibine gives bis(dimethylantimony) oxide as a colorless oil, spontaneously inflammable in the air (16). These are the only two aliphatic trivalent antimony oxides which have been reported. Attempts to hydrolyze chloromethyldichlorostibine, either in aqueous alkali or in water, give antimony trioxide (105). The hydrolysis of bis(trifluoromethyl)iodostibine by either hot water or aqueous alkali completely cleaves the carbon–antimony bond (106).

The hydrolysis of aryldihalo- or diarylhalostibines is more satisfactory, although even here the carbon–antimony bond can be cleaved. The hydrolysis of *p*-acetamidophenyldichlorostibine gives *p*-$CH_3CONHC_6H_4SbO{\cdot}H_2O$, but if this product is allowed to stand in aqueous solution acidified with acetic acid, antimony trioxide is formed (150). Morgan and Davies reported that the hydrolysis of 2-biphenylyldichlorostibine (50) and 2-benzylphenyldichlorostibine (272) by aqueous alkali gives rise to antimony trioxide and the corresponding bis(diarylantimony) oxides:

$$4ArSbCl_2 \xrightarrow{OH^-} (Ar_2Sb)_2O + Sb_2O_3$$

Worrall (72), on the other hand, has reported the preparation of 2-stibosobiphenyl by the alkaline hydrolysis of *o*-$C_6H_5C_6H_4Sb(OH)Cl$.

A number of aromatic stiboso compounds and bis(diarylantimony) oxides have been prepared by the alkaline hydrolysis of aryldichloro- and diarylchlorostibines (71, 92, 266, 267). Meyers and Jones (268) have reported the preparation of 2-stibosothiazole by the alkaline hydrolysis of the corresponding dichlorostibine hydrochloride; Chi and Moh (302) have reported the preparation of bis(*p*-stibosophenyl) sulfone in a similar manner. In several papers devoted to the Nesmeyanov reaction in which mixtures of compounds of the type $ArSbCl_2$, Ar_2SbCl, Ar_3Sb, and Ar_3SbCl_2 were formed, the aryldichlorostibines were not isolated as such from the reaction mixture but were hydrolyzed with aqueous ammonia to the stiboso compounds (92, 303). Any bis(diarylantimony) oxide formed by hydrolysis of the diarylchlorostibine was separated from the stiboso compound by extraction with ether. A number of substituted stibosoarenes were thus prepared and characterized. Doak and Jaffé (269) prepared a number of stibosoarenes by the hydrolysis of carefully purified aryldichlorostibines. These authors (304) had previously shown that the presence of impurities has a marked effect on the stability of stiboso compounds. Reutov and Ptitsyna (305) have isolated both stibosoarenes and bis(diarylantimony) oxides after hydrolysis of the reaction mixture obtained by treating antimony trichloride with arylazoformates. Nakai and Yamakawa (93) investigated the reaction of benzenediazonium fluoborate with antimony trichloride and zinc dust in acetic acid solution. Hydrolysis of the reaction mixture gave stibosobenzene in 8% yield; bis(diphenylantimony) oxide and triphenylstibine were also obtained.

The synthesis of several stiboso compounds under conditions where one might expect to obtain arylstibonic acids has been claimed by Mistry and Guha (306). Thus, the preparation of 4,4′-distibosobiphenyl from benzidine and the preparation of 4,4′-distibosodiphenylmethane from 4,4′-diaminodiphenylmethane have been reported. It is difficult to understand why the trivalent antimony compounds are obtained in these cases, since with some other diamines the expected distibonic acids were isolated.

One of the most interesting methods for preparing bis(diaryantimony) oxides is by disproportionation of the corresponding stibosoarenes:

$$4ArSbO \rightarrow [Ar_2Sb]_2O + Sb_2O_3$$

This reaction can proceed further at higher temperatures with the formation of triarylstibines:

$$3ArSbO \rightarrow Ar_3Sb + Sb_2O_3$$

The reaction is of synthetic importance since stibosoarenes are readily available by reduction of arenestibonic acids. Thus, Schmidt (150) prepared tris(*p*-acetamidophenyl)stibine by heating *p*-stibosoacetanilide in an inert atmosphere; Blicke and co-workers (260) have described the synthesis of bis(diphenylantimony) oxide from stibosobenzene. Schmidt (273) has also obtained bis(diphenylantimony) oxide by dissolving stibosobenzene in acetic acid containing tartaric acid. After standing for some time the solution was made alkaline with ammonia to precipitate the oxide.

The mechanism of the formation of bis(diarylantimony) oxides by disproportionation was investigated by Jaffé and Doak (304, 307), who found that the reaction rate was catalyzed by one of the products, the bis(diarylantimony) oxide, and by bases, such as ammonia or sodium hydroxide (small amounts of which were always present in the starting material). From 20 to 100% completion the reaction was shown to obey the rate law $-dx/dt = kx(1 - x)$, where x is the fraction of stibosobenzene remaining at time t. A mechanism was proposed which envisioned a chain reaction initiated by the thermal dissociation of phenyl groups from both stibosobenzene and bis(diphenylantimony) oxide. The rates of disproportionation of seven different aromatic stiboso compounds were determined at 100° and found to follow the Hammett equation with the reaction constant $\rho = -2.9 \pm 0.1$. Taking the relative thermal stability of stibosobenzene as unity, Jaffé and Doak found that the corresponding value for *p*-stibosotoluene was 0.682, for *p*-bromostibosobenzene was 14.68, and for *p*-nitrostibosobenzene was 157.5.

Bis(diarylantimony) oxides have also been obtained by the cleavage of one aryl group from triarylstibines under acid conditions. Schmidt (150) has described the preparation of bis(diphenylantimony) oxide by first refluxing triphenylstibine in methanolic hydrogen chloride in an inert atmosphere and then precipitating the oxide by the addition of base. Yusunov and Manulkin (48) reported that when diphenyl-*o*-tolylstibine was heated with acetic acid, toluene and diphenylantimony acetate were obtained.

Stibosoarenes are amorphous powders, insoluble in water and in most organic solvents. They have not been successfully recrystallized from any solvents, and no molecular weight data are available. Although stibosobenzene is soluble in acetic or fromic acids, cryoscopic measurements indicated reaction with these solvents (304). There seems little doubt, however, that stiboso compounds are polymeric. Various authors (71, 92, 266) have listed melting points for stibosoarenes. However, since all of these melting points lie in the range 100–200° and since it has been shown that there is considerable disproportionation even at 100° (307), it seems probable that most of the melting points listed in the literature are not true melting points but rather decomposition points.

Unlike stibosoarenes, bis(diarylantimony) oxides are readily soluble in and can be recrystallized from a variety of organic solvents. Thus, bis(diphenylantimony) oxide has been recrystallized from both alcohol (260) and ether (307). Bis(diarylantimony) oxides are generally quite low-melting, crystalline solids: $[(C_6H_5)_2Sb]_2O$ melts at 82°; $[(p\text{-}CH_3C_6H_4)_2Sb]_2O$ melts at 107° (307). Apparently no molecular weight data are available. As far as the present authors are aware, there have been no spectral or X-ray diffraction studies on either stibosoarenes or bis(diarylantimony) oxides.

The mercuration of aromatic stiboso compounds has been studied by Hiratsuka (308). In general, the results are quite similar to those obtained with aromatic arsenoso compounds (cf. Section III-2-B). The antimony compounds, however, reacted less readily. Thus, heating was invariably required to bring about reaction between mercuric acetate and stiboso compounds. Once formed, the mercurated stiboso compounds were considerably more stable than their arsenic analogs. For example, Hiratsuka was unable to isolate 2-arsenoso-5-tolylmercuric acetate, but the corresponding antimony compound was readily prepared as follows:

$$CH_3C_6H_4SbO + Hg(OCOCH_3)_2 \longrightarrow H_3C\text{-}C_6H_3(SbO)(HgOCOCH_3)$$

Treatment of the stibosoarylmercuric acetates with hydrochloric acid gave the corresponding dichlorostibinoarylmercuric chlorides:

$$H_3C\text{-}C_6H_3(SbO)(HgOCOCH_3) + 3HCl \longrightarrow H_3C\text{-}C_6H_3(SbCl_2)(HgCl)$$

Aromatic stiboso compounds and bis(diarylantimony) oxides react with hydrochloric acid to give aryldihalo- and diarylhalostibines, respectively (92, 269, 272–274). Doak and Jaffé (269) have given detailed directions for preparing aryldichlorostibines from stibosoarenes and hydrogen chloride in chloroform solution; Nesmeyanov and Kocheshkov (92) have described the preparation of bis(*m*-chlorophenyl)chlorostibine from the corresponding oxide and 5*N* hydrochloric acid.

Bis(diarylantimony) oxides react with carboxylic acids to give compounds of the type Ar_2SbY. The best known of these compounds are those in which Y is an acetate group. Such acetates are generally crystalline compounds with sharp, reproducible melting points, and they have been widely used in the isolation, purification, and characterization of bis(diarylantimony) oxides (92, 150, 260, 267, 303). Schmidt (150) described the preparation of a tartrate, but did not report analytical data. Koton and Florinskii (309) prepared both diphenylantimony acrylate and methacrylate. These compounds polymerized when heated with toluene in the presence of an initiator; they could also be copolymerized with either methyl methacrylate or styrene to form thermoplastic products. Koton and Kiseleva (310) also described diphenylantimony benzoate and diphenylantimony *p*-vinylbenzoate. Several patents have described the preparation of other diarylantimony carboxylates by the reaction between bis(diarylantimony) oxides and carboxylic acids (294, 311, 312). Water can be removed from the reaction mixture by means of a Dean-Stark trap (311). The same products were also obtained from diarylchlorostibines and the sodium carboxylates (cf. Section VIII-3-B) or by refluxing the diarylantimony acetates with the carboxylic acids in toluene solution (312).

There have been no structural studies reported on compounds of the type Ar_2SbY, although their relatively low melting points and their solubilities in organic solvents suggest that the Sb—Y bond may be covalent. Dessy and his co-workers (301) have reported that the electrolytic reduction of diphenylantimony acetate gives a colorless solution of tetraphenyldistibine, presumably as the result of rapid coupling of diphenylantimony radicals. If the solution is further reduced (at −2.45 volts), the distibine takes up two electrons per molecule to give an orange solution of the diphenylantimony anion. Addition of diphenylantimony acetate to the orange solution immediately dissipates the color to give tetraphenyldistibine:

$$Ph_2Sb^- + Ph_2SbO_2CCH_3 \rightarrow Ph_2SbSbPh_2 + CH_3CO_2^-$$

Only a few organoantimony sulfides containing trivalent antimony are known, and apparently no selenides have been reported. Morgan and Davies (16) described the preparation of CH_3SbS by the action of hydrogen sulfide on methyldichlorostibine. It is a yellow powder which melts indefinitely about 70° and is sparingly soluble in carbon disulfide and benzene. No other physical properties are given. In the early literature Hasenbäumer (313) reported the preparation of C_6H_5SbS and *p*-$CH_3C_6H_4SbS$ from the corresponding diarylhalostibines or stiboso compounds by the action of hydrogen sulfide in alcoholic ammonia solution; Michaelis and Günther (314) described the preparation of $[(C_6H_5)_2Sb]_2S$ from the corresponding oxide and hydrogen sulfide in alcohol solution. All of these compounds are reported as low melting solids: C_6H_5SbS, mp 65°; $[(C_6H_5)_2Sb]_2S$, mp 69°. More recently, several new methods for preparing thiostibosobenzene and bis-(diphenylantimony) sulfide from the corresponding oxides have been described. In one method (90) the oxide is dissolved in carbon disulfide, and ammonia is passed through the solution for several hours; the product obtained by removal of the solvent is then recrystallized from alcohol. In another method (292) the sulfides are obtained by reaction of the oxides with benzylammonium benzyldithiocarbamate in chloroform at 25°. Thiostibosobenzene has also been prepared by the thermal decomposition of phenylantimony bis(benzyldithiocarbamate) (292). There have been no structural studies on the thiostiboso compounds or the sulfides, but their relatively low melting points suggest that they are not polymeric.

A number of compounds of the type $RSb(SR')_2$ or $RSbSCR_2'CR_2'S$ (ring-closed) have been reported, particularly in the patent literature (270, 291, 315–319). They have been prepared either by the action of mercaptans on stiboso compounds (315) or aryldichlorostibines (315–317) or by the action of an excess of mercaptan on the sodium salts of stibonic acids (318). Many of these compounds were prepared in attempts to obtain organoantimony compounds of therapeutic value. The compound p-$HO_2CCH_2SC_6H_4Sb(SCH_2CO_2H)_2$ has been obtained from p-$HO_2CCH_2SC_6H_4SbCl_2$ and sodium thioglycollate; when dissolved in aqueous acetone, it is converted to the following anhydride (271):

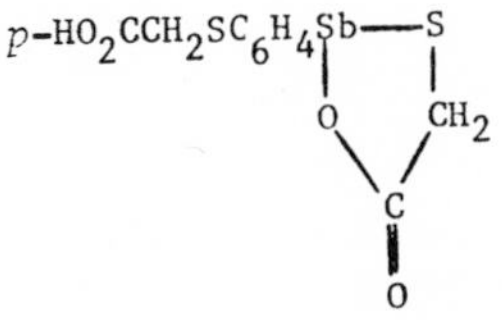

```
p-HO2CCH2SC6H4Sb——S
               |    |
               O   CH2
                \  /
                 C
                 ‖
                 O
```

The reaction of bis(diarylantimony) and bis(dialkylantimony) oxides with mercaptans to give compounds of the type R_2SbSR' has been described in the patent literature (320). Kupchik and McInerney (321) have reported that bis(diphenylantimony) oxide reacts with 6-mercaptopurine to give the expected 6-diphenylantimonymercaptopurine in 78% yield. The infrared spectrum of the product shows a band at 2762 cm^{-1} characteristic of hydrogen-bonded NH and bands at 732 cm^{-1} and 695 cm^{-1} attributable to monosubstituted benzene. The fact that the bonding is between antimony and sulfur was confirmed by showing that bis(diphenylantimony) oxide does not react with purine itself.

The preparation of dithiocarbamates from halo- and dihalostibines has been mentioned in Section VIII-3-B. They have also been obtained by the reaction of a diarylantimony acetate with a salt of a dithiocarbamate (90) and by the reaction of a stiboso compound with benzylamine and carbon disulfide in chloroform (292). Molecular weight determinations on several of these compounds indicate that they are monomolecular (90, 292). The infrared spectra of the phenylantimony bis(dialkyldithiocarbamates) contain a strong band near 1500 cm^{-1}, which has been attributed to the stretching frequency of the partial C=N bond of the following canonical form:

```
          +
PhM(S—C=NR2)2
      |
      S—
```

The ultraviolet absorption spectra show intense absorption near 260 mμ ($\varepsilon > 10^4$); this band is apparently primarily associated not with the phenyl group but with the dialkyldithiocarbamate moieties.

5. COMPOUNDS WITH ANTIMONY–ANTIMONY BONDS

When the arseno compounds (which were believed to have the structure RAs=AsR) were introduced into medicine in 1910 with spectacular success, it was only natural that attempts would be made to prepare the corresponding antimono compounds. Indeed, a number of German patents were soon issued covering not only derivatives of antimonobenzene, but also mixed antimony–arsenic compounds, ArAs=SbAr, known as stibarseno compounds. Antimonobenzene was reported in 1913 as a yellow powder, crystallizable from chloroform, and obtained by the reduction of benzenestibonic acid with sodium tetrathionite (322). In this same patent it was claimed that *m*-aminophenyldichlorostibine could be reduced to the corresponding antimono compound by sodium tetrathionite in alkaline solution in the presence of magnesium chloride. In a second patent (323) the preparation of stibarseno compounds through the reduction of a mixture of arsonic and stibonic acids was claimed. The reduction of a mixture of arsenoso compounds and stibonic acids was also said to yield stibarseno compounds. In yet another patent (324) the preparation of these mixed compounds was claimed to have been achieved through the condensation of primary aromatic arsines and aromatic stiboso compounds:

$$ArAsH_2 + Ar'SbO \rightarrow ArAs{=}SbAr' + H_2O$$

The first description of such compounds in the chemical literature was given by Ehrlich and Karrer (325) in 1913. They described the preparation of ArAs=SbAr type compounds by the condensation of arylarsines with aryldihalostibines. The products were amorphous brown powders for which the analytical results (C, H, As, and Sb for one compound, As and Sb for another) were in only fair agreement with the theoretical values. Somewhat later Schmidt (273) reported the preparation of antimonobenzene by the reduction of stibosobenzene with sodium hypophosphite in glacial acetic acid. It was described as a brown powder which decomposed about 160°, and only antimony analyses were reported. Scmhidt also reported the preparation of *m,m'*-antimonodianiline (for which only analytical values for nitrogen were given) and the unsymmetrical compound $3\text{-}NH_2\text{-}4\text{-}ClC_6H_3Sb{=}AsC_6H_3\text{-}3\text{-}NH_2\text{-}4\text{-}OH$ (without analysis). Riddell and Basterfield (326) have described *p,p'*-antimonodianiline and α-antimononaphthalene; these were obtained by sodium dithionite reduction of *p*-nitrobenzene- and α-naphthalenestibonic acids, respectively, and only antimony analyses were given. Lecoq (327) reduced benzenestibonic acid under conditions of the Clemmensen reaction in a carbon dioxide atmosphere and obtained a brown amorphous powder. Although Lecoq only implied that the product might be antimonobenzene, in a later paper (328) he referred to the preparation of antimonobenzene as given in his earlier paper. No analyses on antimonobenzene were given in either paper.

From all of the above results it is seen that the claims for the preparation of antimonoarenes are not entirely convincing, and indeed the earlier claims were challenged by Klages and Rapp (329) in 1955. All of their attempts to prepare a product which gave analytical values corresponding to C_6H_5Sb met with failure. Reduction of benzenestibonic acid by the method of Lecoq gave a black product containing 78–79% antimony; benzene and stibine were also obtained in the reaction. Reduction of benzenestibonic acid with hypophosphorous acid in refluxing hydrochloric acid solution gave metallic antimony. Reduction of stibosobenzene with sodium hypophosphite in acetone-acetic acid in an inert atmosphere gave a brown, amorphous powder with the approximate composition $C_6H_5O_{0.4}Sb_{1.25}$, while the reduction of phenyldiiodostibine with zinc and hydrochloric acid in ethanol solution gave a product which contained 86% antimony. The authors concluded that the so-called antimonobenzene is a highly polymerized product whose composition lies somewhere between C_6H_5Sb and metallic antimony.

Several other papers on antimonobenzene have been recorded in the modern chemical literature. Wiberg and Mödritzer (6) found that phenylstibine, when allowed to decompose spontaneously in a sealed ampoule, gave hydrogen and a dark-brown solid which they suggest is a highly polymerized form of antimonobenzene. In this case, the values for antimony were only slightly lower than theoretical. Wiberg and Mödritzer also obtained this dark-brown solid by the interaction of phenylstibine and phenyldiiodostibine. It should be recalled that methylstibine loses hydrogen spontaneously to give a black solid (1). When this solid is treated with hydrochloric acid, methane and hydrogen are obtained but in less than the stoichiometric amounts for the formula $(CH_3Sb)_x$. Kuchen and Ecke (330) treated phenyldichlorostibine with diphenylsilane in ether at $-25°$ and obtained a black solid with the composition $C_6H_{5.06}Sb_{1.04}Cl_{0.14}$. Four different preparations of this compound gave essentially the same analyses. An infrared spectrum was identical with that of phenyldichlorostibine. On the basis of these results, the authors suggested that the compound possesses the following structure: $Cl(Ph)Sb(SbPh)_nSb(Ph)Cl$ (where $n = \sim 13$). The compound was not appreciably oxidized by dry air, but in moist air or in moist solvents it was converted to a compound with the approximate composition, C_6H_5SbO. When the original compound was suspended in ether and treated with a stream of oxygen, a product was obtained with the approximate composition, $C_6H_5Sb(OH)_2$. Finally, Nesmeyanov and co-workers (9) have suggested that antimonobenzene is formed as a by-product of the interaction of diphenylstibine and carbon tetrachloride.

In marked contrast to the above results, Issleib and co-workers (4) obtained good elemental analyses and cryoscopic molecular weights on a compound $(t\text{-BuSb})_4$, which they termed tetra-*tert*-butylcyclotetrastibine. It was prepared by ether extraction of the distillation residue from di-*tert*-butylstibine and also by the reaction of lithium di-*tert*-butylstibide with iodine. The compound is described as a red, air-sensitive, amorphous powder.

Although relatively few distibines, $R_2Sb—SbR_2$, have been reported, they have been better characterized than the antimono compounds. Paneth and Loleit (112) obtained both $(CH_3)_2Sb—Sb(CH_3)_2$ and $(C_2H_5)_2Sb—Sb(C_2H_5)_2$, together with trimethyl- and triethylstibines, by the reaction between methyl or ethyl radicals and antimony mirrors. The radicals were obtained by heating the corresponding tetraalkyllead compounds; preparation of the distibines was part of a series of classical experiments performed by Paneth and his co-workers to demonstrate the existence of alkyl radicals. Tetramethyldistibine was produced with both cold and heated antimony mirrors, but tetraethyldistibine required a heated mirror. An attempt to prepare tetrapropyl- and tetrabutyldistibines by the Paneth technique gave only the trialkylstibines (112).

In addition to the above synthesis, tetramethyldistibine has been prepared from dimethylbromostibine and sodium in liquid ammonia in a sealed tube (1). The yield by this method was 75%; use of lithium instead of sodium reduced the yield to 55%. Tetramethyldistibine has also been obtained by the diborane-catalyzed decomposition of dimethylstibine. More recently Shlyk and co-workers (100) have reported that both triethylstibine and tetraethyldistibine can be prepared by the interaction of antimony, sodium (or lithium), and ethyl bromide in liquid ammonia. It was suggested that Na_3Sb (or Li_3Sb) was an intermediate in the formation of the tertiary stibine, while the distibine was formed via $Na_2SbSbNa_2$ (or $Li_2SbSbLi_2$). Tetraethyldistibine has also been obtained by the decomposition of a solution of diethylstibine at room temperature (3).

Tetramethyldistibine occurs as red needles which melt at 17.5° to a yellow oil. On cooling the red crystals to −196°, they become orange in color. The tetraethyl compound undergoes similar color changes on melting and on cooling to low temperatures. Molecular weight data, both cryoscopically in benzene and by the Rast camphor method, are in excellent agreement with the theoretical values for the monomeric form, R_2SbSbR_2 (112, 331). These results suggest that the color changes are not associated with homolytic cleavage to give dialkylantimony radicals. From vapor tension measurements on the methyl compound, an extrapolated boiling point value of 224° and a Trouton's constant of 22.6 were calculated (1).

Both tetramethyl- and tetraethyldistibines react with bromine or iodine to give the corresponding dialkylhalostibines (112). Oxidation with air in benzene solution leads to the formation of dimethyl- and diethylstibinic acids. The antimony–antimony bond is readily cleaved by hydrochloric acid at room temperature according to the equation (1):

$$[(CH_3)_2Sb]_2 + 2HCl \rightarrow 2(CH_3)_2SbCl + H_2$$

At higher temperatures (250° for 15 hours) the following reaction occurs quantitatively:

$$[(CH_3)_2Sb]_2 + 6HCl \rightarrow 2SbCl_3 + 4CH_4 + H_2$$

This result was used as a method for the analysis of the distibine. Tetramethyldistibine is thermally stable at 100°, but at 200° for 20 hours it decomposes almost quantitatively according to the equation:

$$3[(CH_3)_2Sb]_2 \xrightarrow{\Delta} 2Sb + 4(CH_3)_3Sb$$

Tetrakis(trifluoromethyl)distibine is a pale yellow liquid formed by the reaction between bis(trifluoromethyl)iodostibine and excess zinc or mercury (106). It deepens in color on heating but freezes to a colorless solid. It is considerably less stable than the tetramethyl analog. Thus, it slowly decomposes at room temperature (particularly in the light). Bromine at room temperature cleaves the antimony–antimony bond to give bromotrifluoromethane and antimony tribromide, but the carefully controlled addition of chlorine to a solution of the tetrakis compound in trichlorofluoromethane at −78° gives bis(trifluoromethyl)antimony trichloride.

Tetracyclohexyldistibine has been prepared by several different procedures (2). It can be obtained by the reaction between lithium dicyclohexylstibide and 1,2-dichloro- or 1,2-dibromoethane and by the reaction between methylene chloride and this same lithium compound. It is most conveniently prepared, however, by the addition of a sodium dispersion in toluene to dicyclohexylchlorostibine in ether. The mixture is then filtered, and methanol is added to the filtrate to yield the desired distibine as yellow crystals, mp 71–73°, in a 96.2% yield. Tetra-*tert*-butyldistibine has been obtained by the reaction of lithium di-*tert*-butylstibide with 1,4-dichlorobutane at 65° or with di-*tert*-butylchlorostibine (4). Fischer and Schreiner (67) have reported that the reaction of sodium, cyclopentadiene, and antimony trichloride in boiling tetrahydrofuran gives an 83% yield of tetracyclopentadienyldistibine.

Our knowledge of tetraaryldistibines is largely due to the work of Blicke and co-workers at the University of Michigan. Tetraphenyldistibine (260), tetrakis(*p*-bromophenyl)distibine (267), and tetra-*p*-tolyldistibine (267) were prepared by the reduction of the corresponding diaryliodostibines with sodium hypophosphite. The reactions were carried out in "free-radical bulbs," and all subsequent manipulations including melting point determinations were conducted in an inert atmosphere. Tetraphenyldistibine has also been obtained when sodium diphenylstibide in liquid ammonia solution was treated with 1,2-dichloroethane (96). The product was recrystallized from chloroform–methanol solution; the yield was 66%, mp 125°. This same distibine may also be formed by the decomposition of diphenylstibine (7):

$$2(C_6H_5)_2SbH \rightarrow [(C_6H_5)_2Sb]_2 + H_2$$

The product as isolated, however, was somewhat lower in antimony than the theoretical value and contained both triphenylstibine and metallic antimony.

Tetraaryldistibines are pale-yellow to colorless crystalline solids with sharp melting points. A molecular weight determination of tetra-*p*-tolyldistibine by cryoscopic measurements in benzene indicated that it was monomolecular. Tetraphenyldistibine was found to react with iodine to form diphenyliodostibine, and it instantly decolorized permanganate solution. All three tetraaryldistibines, when dissolved in non-polar solvents and allowed to stand, rapidly absorb oxygen in the amount that corresponds to the formation of a peroxide, $Ar_2Sb—O—O—SbAr_2$.

Dessy and his co-workers (301) have reported that the electrolytic reduction of tetraphenyldistibine involves the uptake of two electrons per molecule and yields an orange solution of diphenylantimony anion. The formation of tetraphenyldistibine by the electrolytic reduction of diphenyliodostibine has been previously mentioned (Section VIII-3-C).

6. COMPOUNDS WITH ANTIMONY–METAL OR ANTIMONY–METALLOID BONDS

Except for coordination compounds of tertiary stibines, few compounds containing antimony–metal or antimony–metalloid bonds are known. The preparation of stibides of the types R_2SbM, where M is Li or Na, and $(R_2Sb)_2M$, where M is Mg, has been described in Sections VIII-1 and and VIII-2-A(4). The use of these compounds for the preparation of tertiary stibines and distibines has been discussed in Sections VIII-2-A-(4) and VIII-5, respectively. These stibides have also been used for establishing Sb—Si, Sb—Ge, Sb—Sn, and Sb—Pb bonds. Thus, Campbell and co-workers (332) have prepared compounds of the type Ph_2SbSnR_3 (where R is Et, Pr, or Bu) by means of the following reaction:

$$Ph_2SbNa + R_3SnBr \rightarrow Ph_2SbSnR_3 + NaBr$$

Similarly, Schumann and co-workers (333) have obtained $Ph_2SbSnPh_3$, $(Ph_2Sb)_2SnPh_2$, $(Ph_2Sb)_3SnPh$, and $(Ph_2Sb)_4Sn$ by the reaction between sodium diphenylstibide and the appropriate tin halide:

$$n\mathrm{Ph_2SbNa} + \mathrm{Ph}_{4-n}\mathrm{SnCl}_n \rightarrow (\mathrm{Ph_2Sb})_n\mathrm{SnPh}_{4-n} + n\mathrm{NaCl}$$

where n is 1, 2, 3, or 4. Similar compounds were prepared by the complementary reactions of Ph_3SnLi and an antimony halide:

$$\mathrm{Ph}_{3-n}\mathrm{SbCl}_n + n\mathrm{Ph_3SnLi} \rightarrow \mathrm{Ph}_{3-n}\mathrm{Sb(SnPh_3)}_n + n\mathrm{LiCl}$$

Herbstman (37) prepared $Bu_2SbSiMe_3$ by the reaction between magnesium dibutylstibide and trimethylchlorosilane:

$$\mathrm{(Bu_2Sb)_2Mg + 2Me_3SiCl \rightarrow 2Bu_2SbSiMe_3 + MgCl_2}$$

Finally, Schumann and Schmidt (334) obtained $Ph_2SbGePh_3$ and $Ph_2SbPbPh_3$ in low yields from sodium diphenylstibide:

$$\mathrm{Ph_2SbNa + Ph_3GeCl \rightarrow Ph_2SbGePh_3 + NaCl}$$

$$\mathrm{Ph_2SbNa + Ph_3PbCl \rightarrow Ph_2SbPbPh_3 + NaCl}$$

The preparation of compounds of the type $(Me_3E)_3Sb$, where E is C, Si, Ge, or Sn, by means of Li_3Sb has been previously mentioned in Section VIII-2-A-(4). Similar compounds, $(Et_3E)_3Sb$, where E is Si, Ge, or Sn, have been prepared by the reaction of triethylstibine with hydrides of the group IV elements (335):

$$\mathrm{Et_3Sb + 3Et_3EH \rightarrow (Et_3E)_3Sb + 3EtH}$$

The compounds $(Pr_3Si)_3Sb$ (335) and $(Ph_3Si)_3Sb$ (336) can also be obtained by this method:

$$\mathrm{Et_3Sb + 3Pr_3SiH \rightarrow (Pr_3Si)_3Sb + 3EtH}$$

$$\mathrm{Et_3Sb + 3Ph_3SiH \rightarrow (Ph_3Si)_3Sb + 3EtH}$$

The Et_3Si groups in $(Et_3Si)_3Sb$ can be readily replaced by Ph_3Si or Et_3Ge groups, and the Et_3Ge groups in $(Et_3Ge)_3Sb$ can be replaced by Et_3Sn groups (335):

$$\mathrm{3Ph_3SiH + (Et_3Si)_3Sb \rightarrow 3Et_3SiH + (Ph_3Si)_3Sb}$$

$$\mathrm{3Et_3GeH + (Et_3Si)_3Sb \rightarrow 3Et_3SiH + (Et_3Ge_3)_3Sb}$$

$$\mathrm{3Et_3SnH + (Et_3Ge)_3Sb \rightarrow 3Et_3GeH + (Et_3Sn)_3Sb}$$

The compounds discussed in the above paragraph are very reactive and take part in a number of reactions that involve cleavage of the antimony–metal or antimony–metalloid bond. They are particularly sensitive to oxidation. Thus, both $Bu_2SbSiMe_3$ (37) and $(Me_3Si)_3Sb$ (101) are described as liquids which ignite spontaneously upon exposure to the atmosphere.

Tris(triethylsilyl)stibine (335) reacts with molecular oxygen in hexane at 20° to cleave two of the three Sb—Si bonds:

$$(Et_3Si)_3Sb + O_2 \rightarrow (Et_3SiO)_2SbSiEt_3$$

Benzoyl peroxide in benzene reacts exothermally with the silicon, germanium, and tin compounds to liberate metallic antimony:

$$2(Et_3E)_3Sb + 3(PhCOO)_2 \rightarrow 2Sb + 6Et_3EO_2CPh$$

where E is Si, Ge, or Sn. Tris(triphenylsilyl)stibine reacts with benzoyl peroxide in a similar manner (336). Campbell and co-workers (332) have stated that compounds of the type Ph_2SbSnR_3 react readily with oxygen, but the reaction products could not be obtained in a pure state. It was suggested, however, that polymeric species containing Sb—O—Sn bonds are produced. The oxidation products of other compounds containing Sb—Si, Sb—Sn, Sb—Ge, and Sb—Pb bonds have not been investigated. The compounds $(Ph_3Sn)_2SbPh$ and $(Ph_3Sn)_3Sb$ appear to be unusual in that they are stable in the air for several weeks (333). Tris(triphenylsilyl)stibine has also been described as being relatively stable toward air (336).

Organohalogen compounds are also able to cleave antimony–metal and antimony–metalloid bonds. Thus, alkyl halides react with $(Et_3Ge)_3Sb$ or $(Et_3Sn)_3Sb$ at 100–150° to form tertiary stibines (cf. Section VIII-2-A-(4)):

$$(Et_3E)_3Sb + 3RBr \rightarrow R_3Sb + 3Et_3EBr$$

where E is Ge or Sn, and R is benzyl or cyclopentyl. When irradiated at 30°, bromobenzene and $(Et_3Ge)_3Sb$ react to give triphenylstibine (335):

$$(Et_3Ge)_3Sb + 3PhBr \rightarrow Ph_3Sb + 3Et_3GeBr$$

The reaction of equimolar quantities of benzyl bromide and $(Et_3Ge)_3Sb$ at 100° results in the cleavage of only one Sb—Ge bond:

$$(Et_3Ge)_3Sb + PhCH_2Br \rightarrow (Et_3Ge)_2SbCH_2Ph + Et_3GeBr$$

The interaction of methyl iodide and $Ph_2SbSnBu_3$ yields tributyltin iodide, but the other products of the reaction have not been identified (332). 1,2-Dibromoethane reacts with $(Et_3Si)_3Sb$ (335), $(Ph_3Si)_3Sb$ (336), or $(Et_3Ge)_3Sb$ (335) to give ethylene:

$$2(R_3E)_3Sb + 3BrCH_2CH_2Br \rightarrow 2Sb + 6R_3EBr + C_2H_4$$

where E is Si or Ge.

In some cases at least, the compounds discussed in this section are sensitive to moisture (37, 101) and light (101). The thermal stability of the corresponding silicon, germanium, and tin compounds appears to decrease in that order (101, 335). Catalytic quantities of aluminum bromide markedly accelerate the decomposition of $(Et_3Sn)_3Sb$ at 150°; under these conditions the reaction proceeds as follows (335):

$$4(Et_3Sn)_3Sb \rightarrow 9Et_4Sn + 3Sn + 4Sb$$

The compounds containing Sb—Si, Sb—Ge, Sb—Sn, or Sb—Pb bonds have been described as colorless or pale yellow liquids or solids. They are monomeric in aromatic hydrocarbons (101, 333) and in THF (333). The compounds of the $(Me_3E)_3Sb$ series (where E is Si, Ge, or Sn) have dipole moments in benzene solution in the 1.4–1.7 D range; this result suggests that the E_3Sb skeleton is pyramidal (101).

Compounds containing Sb—P and Sb—B bonds have also been described. Coates and Livingstone (337) have found that the reaction of diphenylchlorostibine and dimethylphenylphosphine yields a crystalline addition compound, which they formulated as the phosphonium salt $[Ph_2SbP(Ph)Me_2]Cl$. Burg and Grant (1) have prepared dimethylstibinoborane, Me_2SbBH_2, in 26% yield by the interaction of tetramethyldistibine and diborane at 100°. Dimethylstibinoborane is formed as a by-product in the reaction of dimethylstibine with diborane and also in the reduction of dimethylbromostibine with sodium borohydride (cf. Section VIII-1). It is stable up to 200° and does not undergo exchange reactions with trimethylborane. Heating with trimethylamine results in the formation of Me_3NBH_3. The relatively unreactive character of dimethylstibinoborane and the fact that it is a monomer have been attributed to the existence of fairly strong π-bonding between the antimony and boron atoms. The compound decomposes at about 210° to give a black solid of approximate composition SbB.

REFERENCES

1. A. B. Burg and L. R. Grant, *J. Amer. Chem. Soc.*, **81**, 1 (1959).
2. K. Issleib and B. Hamann, *Z. Anorg. Allg. Chem.*, **332**, 179 (1964).
3. K. Issleib and B. Hamann, *Z. Anorg. Allg. Chem.*, **339**, 289 (1965).
4. K. Issleib, B. Hamann, and L. Schmidt, *Z. Anorg. Allg. Chem.*, **339**, 298 (1965).
5. E. Wiberg and K. Mödritzer, *Z. Naturforsch.*, *B*, **11**, 753 (1956).
6. E. Wiberg and K. Mödritzer, *Z. Naturforsch.*, *B*, **12**, 128 (1957).
7. E. Wiberg and K. Mödritzer, *Z. Naturforsch.*, *B*, **12**, 131 (1957).
8. A. N. Nesmeyanov, A. E. Borisov, and N. V. Novikova, *Izv. Akad. Nauk SSSR, Otd. Khim. Nauk*, 194 (1963).
9. A. N. Nesmeyanov, A. E. Borisov, and N. V. Novikova, *Izv. Akad. Nauk SSSR, Ser. Khim.*, 815 (1967).
10. A. N. Nesmeyanov, A. E. Borisov, and N. V. Novikova, *Dokl. Akad. Nauk SSSR*, **172**, 1329 (1967).
11. A. N. Nesmeyanov, A. E. Borisov, and N. V. Novikova, *Izv. Akad. Nauk SSSR, Ser. Khim.*, 763 (1965).
12. A. E. Borisov, N. V. Novikova, N. A. Chumaevskii, and E. B. Shkirtil, *Dokl. Akad. Nauk SSSR*, **173**, 855 (1967).

13. L. H. Long and J. F. Sackman, *Res. Correspondence*, **8,** S23 (1955).
14. J. Seifter, *J. Amer. Chem. Soc.*, **61,** 530 (1939).
15. H. Hibbert, *Ber.*, **39,** 160 (1906).
16. G. T. Morgan and G. R. Davies, *Proc. Roy. Soc., Ser. A*, **110,** 523 (1926).
17. T. M. Lowry and J. H. Simons, *Ber.*, **63,** 1595 (1930).
18. W. J. C. Dyke, W. C. Davies, and W. J. Jones, *J. Chem. Soc.*, 463 (1930).
19. H. Hartmann and G. Kühl, *Z. Anorg. Allg. Chem.*, **312,** 186 (1961).
20. C. L. Tseng and W. Y. Shih, *J. Chin. Chem. Soc.*, **4,** 183 (1936); through *C.A.*, **31,** 669 (1937).
21. G. J. O'Donnell, *Iowa State Coll. J. Sci.*, **20,** 34 (1945).
22. L. Maier, D. Seyferth, F. G. A. Stone, and E. G. Rochow, *Z. Naturforsch., B*, **12,** 263 (1957).
23. L. Maier, D. Seyferth, F. G. A. Stone, and E. G. Rochow, *J. Amer. Chem. Soc.*, **79,** 5884 (1957).
24. A. N. Nesmeyanov, A. E. Borisov, and N. V. Novikova, *Izv. Akad. Nauk SSSR, Otd. Khim. Nauk*, 952 (1960).
25. D. Seyferth, *J. Amer. Chem. Soc.*, **80,** 1336 (1958).
26. R. N. Sterlin, R. D. Yatsenko, L. N. Pinkina, and I. L. Knunyants, *Izv. Akad. Nauk SSSR, Otd. Khim. Nauk*, 1991 (1960).
27. I. P. Tsukervanik and D. Smirnov, *Zh. Obshch. Khim.*, **7,** 1527 (1937); through *C.A.*, **31,** 8518 (1937).
28. T. V. Talalaeva and K. A. Kocheshkov, *Izv. Aka. Nauk SSSR, Otd. Khim. Nauk*, 290 (1953).
29. H. E. Ramsden (to Metal and Thermit Corp.), Brit. Pat. 823,958 (Nov. 18, 1959); *C.A.*, **54,** 17239 (1960).
30. W. Voskuil and J. F. Arens, *Rec. Trav. Chim. Pays-Bas*, **83,** 1301 (1964).
31. H. Hartmann, H. Niemöller, W. Reiss, and B. Karbstein, *Naturwissenschaften*, **46,** 321 (1959).
32. Z. M. Manulkin, A. N. Tatarenko, and F. Yu. Yusupov, *Dokl. Akad. Nauk SSSR*, **88,** 687 (1953); through *C.A.*, **48,** 2631 (1954).
33. H. E. Ramsden (to Metal and Thermit Corp.), U.S. Pat. 3,010,983 (Nov. 28, 1961); *C.A.*, **56,** 8750 (1962).
34. D. Seyferth (to Metal and Thermit Corp.), U.S. Pat. 2,964,550 (Dec. 13, 1960); *C.A.*, **55,** 6349 (1961).
35. W. Steinkopf, I. Schubart, and J. Roch, *Ber.*, **65,** 409 (1932).
36. M. E. Brinnand, W. J. C Dyke, W. H. Jones, and W. J. Jones, *J. Chem. Soc.*, 1815 (1932).
37. S. Herbstman, *J. Org. Chem.*, **29,** 986 (1964).
38. G. S. Hiers, *Org. Syn.*, Coll. Vol. 1, 2nd. ed., Wiley, New York, 1941, p. 550.
39. V. P. Glushkova, T. V. Talalaeva, Z. P. Razmanova, G. S. Zhdanov, and K. A. Kocheshkov, *Sb. Statei Obshch. Khim.*, **2,** 992 (1953); through *C.A.*, **49,** 6859 (1955).
40. J. I. Harris, S. T. Bowden, and W. J. Jones, *J. Chem. Soc.*, 1568 (1947).
41. M. Fild, O. Glemser, and G. Christoph, *Angew. Chem., Int. Ed. Engl.*, **3,** 801 (1964).
42. E. Krause and G. Renwanz, *Ber.*, **65,** 777 (1932).
43. E. Ramsden (to Metal and Thermit Corp.), Brit. Pat. 824,944 (Dec. 9, 1959); through *C.A.*, **54,** 17238 (1960).
44. A. N. Tatarenko and Z. M. Manulkin, *Zh. Obshch. Khim.*, **38,** 273 (1968).
45. F. Schulze, *Iowa State Coll. J. Sci.*, **8,** 225 (1933).
46. L. A. Woods, *Iowa State Coll. J. Sci.*, **19,** 61 (1944).
47. F. Yu. Yusunov and Z. M. Manulkin, *Dokl. Akad. Nauk SSSR*, **97,** 267 (1954).
48. F. Yu. Yusunov and Z. M. Manulkin, *Zh. Obshch. Khim.*, **31,** 3757 (1961).

49. A. N. Tatarenko, *Dokl. Akad. Nauk Uzb. SSR*, 35 (1955); through *C.A.*, **52**, 20005 (1958).
50. G. T. Morgan and G. R. Davies, *Proc. Roy. Soc., Ser. A*, **127**, 1 (1930).
51. I. G. M. Campbell, *J. Chem. Soc.*, 4 (1947).
52. I. G. M. Campbell, *J. Chem. Soc.*, 3116 (1955).
53. I. G. M. Campbell and A. W. White, *J. Chem. Soc.*, 1184 (1958).
54. T. V. Talalaeva and K. A. Kocheshkov, *Zh. Obshch. Khim.*, **16**, 777 (1946); through *C.A.*, **41**, 1215 (1947).
55. A. N. Nesmeyanov, A. E. Borisov, and N. V. Novikova, *Izv. Akad. Nauk SSSR, Otd. Khim. Nauk*, 1578 (1961).
56. A. N. Nesmeyanov, A. E. Borisov, and N. V. Novikova, *Tetrahedron Lett.*, No. 8, 23 (1960).
57. A. N. Nesmeyanov, A. E. Borisov, and N. V. Novikova, *Izv. Akad. Nauk SSSR, Otd. Khim. Nauk*, 147 (1960).
58. A. N. Nesmeyanov, A. E. Borisov, and N. V. Novikova, *Izv. Akad. Nauk SSSR, Otd. Khim. Nauk*, 612 (1961).
59. L. I. Zakharkin, V. I. Bregadze, and O. Yu. Okhlobystin, *J. Organometal. Chem.*, **4**, 211 (1965).
60. G. Wittig and D. Hellwinkel, *Chem. Ber.*, **97**, 789 (1964).
61. D. M. Heinekey and I. T. Millar, *J. Chem. Soc.*, 3101 (1959).
62. F. C. Leavitt, T. A. Manuel, and F. Johnson, *J. Amer. Chem. Soc.*, **81**, 3163 (1959) .
63. F. C. Leavitt, T. A. Manuel, F. Johnson, L. U. Matternas, and D. S. Lehman, *J. Amer. Chem. Soc.*, **82**, 5099 (1960).
64. E. H. Braye, W. Hübel, and I. Caplier, *J. Amer. Chem. Soc.*, **83**, 4406 (1961).
65. H. Hartmann and G. Kühl, *Angew. Chem.*, **68**, 619 (1956).
66. H. Hartmann, E. Dietz, K. Komorniczyk, and W. Reiss, *Naturwissenschaften*, **48**, 570 (1961).
67. E. O. Fischer and S. Schreiner, *Chem. Ber.*, **93**, 1417 (1960).
68. A. Michaelis and A. Reese, *Justus Liebigs Ann. Chem.*, **233**, 39 (1886).
69. A. Michaelis and U. Genzken, *Justus Liebigs Ann. Chem.*, **242**, 164 (1887).
70. S. P. Olifirenko, *Visn. L'vivs'k. Berzh. Univ., Ser. Khim.*, No. 6, 100 (1963); through *C.A.*, **63**, 8401 (1965).
71. D. E. Worrall, *J. Amer. Chem. Soc.*, **52**, 2046 (1930).
72. D. E. Worrall, *J. Amer. Chem. Soc.*, **62**, 2514 (1940).
73. A. Étienne, *Bull. Soc. Chim. Fr.*, 50 (1947).
74. D. R. Lyon, F. G. Mann, and G. H. Cookson, *J. Chem. Soc.*, 662 (1947).
75. H. Jenkner, *Z. Naturforsch., B*, **12**, 809 (1957).
76. Kali-Chemie Akt.-Ges., Brit. Pat. 820, 146 (Sept. 16, 1959); *C.A.*, **54**, 6550 (1960).
77. L. I. Zakharkin and O. Yu. Okhlobystin, *Dokl. Akad. Nauk SSSR*, **116**, 236 (1957).
78. W. Stamm and A. Breindel, *Angew. Chem., Int. Ed. Engl.*, **3**, 66 (1964).
79. W. Stamm, *Trans. N.Y. Acad. Sci.*, **28**, 396 (1966).
80. A. N. Nesmeyanov and L. G. Makarova, *Zh. Obshch. Khim.*, **7**, 2649 (1937); through *C.A.*, **32**, 2095 (1938).
81. W. J. Considine and J. J. Ventura, *J. Organometal. Chem.*, **3**, 420 (1965).
82. H. Schumann, H. Köpf, and M. Schmidt, *Z. Anorg. Allg. Chem.*, **331**, 200 (1964).
83. G. T. Morgan and V. E. Yarsley, *Proc. Roy. Soc., Ser. A*, **110**, 534 (1926).
84. C. H. Bamford, D. L. Levi, and D. M. Newitt, *J. Chem. Soc.*, 468 (1946).
85. E. Wiberg and K. Mödritzer, *Z. Naturforsch., B*, **11**, 750 (1956).
86. I. G. M. Campbell, *J. Chem. Soc.*, 3109 (1950).
87. Y. Matsumura, M. Shindo, and R. Okawara, *Inorg. Nucl. Chem. Lett.*, **3**, 219 (1967).
88. I. G. M. Campbell, *J. Chem. Soc.*, 4448 (1952).

89. I. G. M. Campbell and D. J. Morrill, *J. Chem. Soc.*, 1662 (1955).
90. E. J. Kupchik and P. J. Calabretta, *Inorg. Chem.*, **4,** 973 (1965).
91. R. R. Sauers, *Chem. Ind.* (London), 717 (1960).
92. A. N. Nesmeyanov and K. A. Kocheshkov, *Izv. Akad. Nauk SSSR, Otd. Khim. Nauk* 416 (1944); through *C.A.*, **39,** 4320 (1945).
93. R. Nakai and Y. Yamakawa, *Bull. Inst. Chem. Res., Kyoto Univ.*, **24,** 80 (1951).
94. F. B. Makin and W. A. Waters, *J. Chem. Soc.*, 843 (1938).
95. L. A. Woods and H. Gilman, *Proc. Iowa Acad. Sci.*, **48,** 251 (1941).
96. W. Hewertson and H. R. Watson, *J. Chem. Soc.*, 1490 (1962).
97. C. Löwig and E. Schweizer, *Justus Liebigs Ann. Chem.*, **75,** 315 (1850).
98. S. Friedländer, *J. Prakt. Chem.*, **70,** 449 (1857).
99. H. Landolt, *J. Prakt. Chem.*, **84,** 328 (1861).
100. Yu. N. Shlyk, G. M. Bogolyubov, and A. A. Petrov, *Zh. Obshch. Khim.*, **38,** 1199 (1968).
101. E. Amberger and R. W. Salazar, G., *J. Organometal. Chem.*, **8,** 111 (1967).
102. Associated Lead Manufacturers Ltd., and F. B. Lewis, Brit. Pat. 854,776 (Nov. 23, 1960); through *C.A.*, **55,** 11362 (1961).
103. N. S. Vyazankin, O. A. Kruglaya, G. A. Razuvaev, and G. S. Semchikova, *Dokl. Akad. Nauk SSSR*, **166,** 99 (1966).
104. A. Ya. Yakubovich, V. A. Ginsburg, and S. P. Makarov, *Dokl. Akad. Nauk SSSR*, **71,** 303 (1950); through *C.A.*, **44,** 8320 (1950).
105. A. Ya. Yakubovich and S. P. Makarov, *Zh. Obshch. Khim.*, **22,** 1528 (1952).
106. J. W. Dale, H. J. Emeléus, R. N. Haszeldine, and J. H. Moss, *J. Chem. Soc.*, 3708 (1957).
107. F. Nerdel, J. Buddrus, and K. Höher, *Chem. Ber.*, **97,** 124 (1964).
108. G. Wittig and K. Clauss, *Justus Liebigs Ann. Chem.*, **577,** 26 (1952).
109. V. L. Foss, E. A. Besolova, and I. F. Lutsenko, *Zh. Obshch. Khim.*, **35,** 759 (1965).
110. A. N. Nesmeyanov and E. G. Ippolitov, *Vestn. Mosk. Univ.*, **10,** No. 10, *Ser. Fiz. Mat. i Estestv. Nauk*, No. **7,** 87 (1955); through *C.A.*, **50,** 9906 (1956).
111. T. V. Talalaeva and K. A. Kocheshkov, *Zh. Obshch. Khim.*, **8,** 1831 (1938); through *C.A.*, **33,** 5819 (1939).
112. F. A. Paneth and H. Loleit, *J. Chem. Soc.*, 366 (1935).
113. H. Raudnitz, *Ber.*, **60,** 743 (1927).
114. R. R. Holmes and E. F. Bertaut, *J. Amer. Chem. Soc.*, **80,** 2983 (1958).
115. G. T. Morgan and V. E. Yarsley, *J. Chem. Soc.*, **127,** 184 (1925).
116. G. O. Doak, G. G. Long, and M. E. Key, *Inorg. Syn.*, **9,** 92 (1967).
117. R. L. McKenney and H. H. Sisler, *Inorg. Chem.*, **6,** 1178 (1967).
118. W. von E. Doering and A. K. Hoffmann, *J. Amer. Chem. Soc.*, **77,** 521 (1955).
119. G. Wittig and K. Torssell, *Acta Chem. Scand.*, **7,** 1293 (1955).
120. W. J. C. Dyke and W. J. Jones, *J. Chem. Soc.*, 1921 (1930).
121. R. N. Haszeldine and B. O. West, *J. Chem. Soc.*, 3631 (1956).
122. G. Wittig and H. Laib, *Justus Liebigs Ann. Chem.*, **580,** 57 (1953).
123. Y. Matsumura and R. Okawara, *Inorg. Nucl. Chem. Lett.*, **4,** 219 (1968).
124. B. J. Pullman and B. O. West, *J. Inorg. Nucl. Chem.*, **19,** 262 (1961).
125. F. Hewitt and A. K. Holliday, *J. Chem. Soc.*, 530 (1953).
126. V. N. Nefedov, I. S. Kirin, V. M. Zaitsev, G. A. Semenov, and B. E. Dzevitskii, *Zh. Obshch. Khim.*, **33,** 2407 (1963).
127. G. O. Doak, G. G. Long, and L. D. Freedman, *J. Organometal. Chem.*, **4,** 82 (1965).
128. C. Löloff, *Ber.*, **30,** 2834 (1897).
129. Z. M. Manulkin and A. N. Tatarenko, *Zh. Obshch. Khim.*, **21,** 93 (1951).
130. A. E. Goddard, *J. Chem. Soc.*, **121,** 36 (1922).

131. A. E. Goddard, *J. Chem. Soc.*, **123,** 1161 (1923).
132. V. I. Lodochnikova, E. M. Panov, and K. A. Kocheshkov, *Zh. Obshch. Khim.*, **34,** 946 (1964).
133. G. A. Razuvaev and M. A. Shubenko, *Dokl. Akad. Nauk SSSR*, **67**, 1049 (1949); through *C.A.*, **44,** 1435 (1950).
134. M. A. Shubenko, *Sb. Statei Obshch. Khim.*, **2,** 1043 (1953); through *C.A.*, **49,** 6856 (1955).
135. A. N. Nesmeyanov and L. G. Makarova, *Uch. Zap. Mosk. Gos. Univ.*, No. 132, *Org. Khim.*, **7,** 109 (1950); through *C.A.*, **49,** 3903 (1955).
136. M. C. Henry and G. Wittig, *J. Amer. Chem. Soc.*, **82,** 563 (1960).
137. G. Wittig and K. Schwarzenbach, *Justus Liebigs Ann. Chem.*, **650,** 1 (1961).
138. L. P. Petrenko, *Zh. Obshch. Khim.*, **24,** 520 (1954).
139. L. P. Petrenko, *Tr. Voronezhsk. Gos. Univ.*, **57,** 145 (1959); through *C.A.*, **55,** 6425 (1961).
140. G. Wittig and D. Hellwinkel, *Angew. Chem., Int. Ed. Engl.*, **1,** 53 (1962).
141. P. May, *J. Chem. Soc.*, **97,** 1956 (1910).
142. M. Becke-Goehring and H. Thielemann, *Z. Anorg. Allg. Chem.*, **308,** 33 (1961).
143. A. B. Bruker and N. M. Nikiforova, *Zh. Obshch. Khim.*, **18,** 1133 (1948); through *C.A.*, **43,** 1737 (1949).
144. O. A. Reutov, O. A. Ptitsyna, T. P. Karpov, and T. A. Smolina, *Nauch. Dokl. Vysshei Shkoly, Khim. Khim. Tekhnol.*, No. **1,** 115 (1958); through *C.A.*, **52,** 17912 (1958).
145. J. Seifter, *J. Pharmacol. Exp. Ther.*, **66,** 366 (1939).
146. C. H. Bamford and D. M. Newitt, *J. Chem. Soc.*, 695 (1946).
147. W. Ipatiew and G. Rasuwajew, *Ber.*, **63,** 1110 (1930).
148. C. J. Thompson, H. J. Coleman, R. L. Hopkins, and H. T. Rall, *Anal. Chem.*, **37,** 1042 (1965).
149. P. B. Ayscough and H. J. Emeléus, *J. Chem. Soc.*, 3381 (1954).
150. H. Schmidt, *Justus Liebigs Ann. Chem.*, **429,** 123 (1922).
151. A. E. Goddard and V. E. Yarsley, *J. Chem. Soc.*, 719 (1928).
152. R. N. Sterlin, S. S. Dubov, W. K. Li, L. P. Vakhomchik, and I. L. Knunyants, *Zh. Vses. Khim. Obshchest.*, **6,** 110 (1961); through *C.A.*, **55,** 15336 (1961).
153. R. N. Haszeldine and B. O. West, *J. Chem. Soc.*, 3880 (1957).
154. D. Wittenberg and H. Gilman, *J. Org. Chem.*, **23,** 1063 (1958).
155. A. D. Britt and E. T. Kaiser, *J. Phys. Chem.*, **69,** 2775 (1965).
156. A. D. Britt and E. T. Kaiser, *J. Org. Chem.*, **31,** 112 (1966).
157. M. S. Malinovsky and S. P. Olifirenko, *Zh. Obshch. Khim.*, **25,** 122 (1955).
158. M. S. Malinovsky and S. P. Olifirenko, *Zh. Obshch. Khim.*, **26,** 1402 (1956).
159. A. E. Borisov, *Izv. Akad. Nauk SSSR, Otd. Khim. Nauk*, 402 (1951); through *C.A.* **46,** 2995 (1951).
160. O. Schmitz-DuMont and B. Ross, *Z. Naturforsch., B*, **20,** 72 (1965).
161. J. Chatt, *J. Chem. Soc.*, 652 (1951).
162. J. Chatt and R. G. Wilkins, *J. Chem. Soc.*, 4300 (1952).
163. K. A. Jensen, *Z. Anorg. Allg. Chem.*, **229,** 225 (1936).
164. J. Chatt and R. G. Wilkins, *J. Chem. Soc.*, 2532 (1951).
165. J. Chatt and R. G. Wilkins, *J. Chem. Soc.*, 525 (1956).
166. J. Chatt and L. M. Venanzi, *J. Chem. Soc.*, 3858 (1955).
167. J. Chatt, L. A. Duncanson, and L. M. Venanzi, *J. Chem. Soc.*, 4461 (1955).
168. J. Chatt, L. A. Duncanson, and L. M. Venanzi, *J. Chem. Soc.*, 2712 (1956).
169. J. Chatt and L. M. Venanzi, *J. Chem. Soc.*, 2351 (1957).
170. D. Benlian and M. Bigorgne, *Bull. Soc. Chim. Fr.*, 1583 (1963).

171. G. E. Coates, *J. Chem. Soc.*, 2003 (1951).
172. M. Angoletta, *Gazz. Chim. Ital*,. **89,** 2359 (1959).
173. R. P. M. Werner, *Z. Naturforsch., B*, **16,** 477 (1961).
174. A. S. Kasenally, J. Lewis, A. R. Manning, J. R. Miller, R. S. Nyholm, and M. H. B. Stiddard, *J. Chem. Soc.*, 3407 (1965).
175. W. Hieber and H. Heusinger, *J. Inorg. Nucl. Chem.*, **4,** 178 (1957).
176. M. A. Bennett and G. Wilkinson, *J. Chem. Soc.*, 1418 (1961).
177. C. N. Matthews, T. A. Magee, and J. H. Wotiz, *J. Amer. Chem. Soc.*, **81,** 2273 (1959).
178. T. A. Manuel and F. G. A. Stone, *J. Amer. Chem. Soc.*, **82,** 366 (1960).
179. W. Reppe and W. J. Schweckendiek, *Justus Liebigs Ann. Chem.*, **560,** 104 (1948).
180. W. Hieber and W. Freyer, *Chem. Ber.*, **93,** 462 (1960).
181. W. Hieber and W. Schropp, Jr., *Z. Naturforsch., B*, **14,** 460 (1959).
182. E. W. Abel, M. A. Bennett, and G. Wilkinson, *J. Chem. Soc.*, 2323 (1959).
183. L. Malatesta and A. Aràneo, *J. Chem. Soc.*, 3803 (1957).
184. L. Vaska, *J. Amer. Chem. Soc.*, **83,** 756 (1961).
185. L. Vallarino, *J. Inorg. Nucl. Chem.*, **8,** 288 (1958).
186. A. Davison, M. L. H. Green, and G. Wilkinson, *J. Chem. Soc.*, 3172 (1961).
187. T. A. Magee, C. N. Matthews, T. S. Wang, and J. H. Wotiz, *J. Amer. Chem. Soc.*, **83,** 3200 (1961).
188. R. A. Shaw, B. C. Smith, and C. P. Thakur, *Justus Liebigs Ann. Chem.*, **713,** 30 (1968).
189. R. E. Weston, Jr., *J. Amer. Chem. Soc.*, **76,** 2645 (1954).
190. K. Mislow, A. Zimmerman, and J. T. Melillo, *J. Amer. Chem. Soc.*, **85,** 594 (1963).
191. I. G. M. Campbell and R. C. Poller, *Chem. Ind.* (London), 1126 (1953).
192. K. Bowden and E. A. Braude, *J. Chem. Soc.*, 1068 (1952).
193. H. H. Jaffé, *J. Chem. Phys.*, **22,** 1430 (1954).
194. C. N. R. Rao, J. Ramachandran, M. S. C. Iah, S. Somasekhara, and T. V. Rajakumar, *Nature*, **183,** 1475 (1959).
195. C. N. R. Rao, J. Ramachandran, and A. Balasubramian, *Can. J. Chem.*, **39,** 171 (1961).
196. O. V. Kolninov and Z. V. Zvonkova, *Zh. Fiz. Khim.*, **36,** 2228 (1962).
197. H. Zorn, H. Schindlbauer, and D. Hammer, *Monatsh. Chem.*, **98,** 731 (1967).
198. E. J. Rosenbaum and T. A. Ashford, *J. Chem. Phys.*, **7,** 554 (1939).
199. E. J. Rosenbaum, D. J. Rubin, and C. R. Sandberg, *J. Chem. Phys.*, **8,** 366 (1940).
200. R. K. Sheline, *J. Chem. Phys.*, **18,** 602 (1950).
201. H. Siebert, *Z. Anorg. Allg. Chem.*, **273,** 161 (1953).
202. A. N. Nesmeyanov, M. E. Batuev, and A. E. Borisov, *Izv. Akad. Nauk SSSR, Otd. Khim. Nauk* 567 (1949); through *C.A.*, **44,** 2374 (1950).
203. F. A. Miller and D. H. Lemmon, *Spectochim. Acta, A*, **23,** 1099 (1967).
204. F. Oswald, *Z. Anal. Chem.*, **197,** 309 (1963).
205. A. E. Borisov, N. V. Novikova, N. A. Chumaevskii, and E. B. Shkirtil, *Ukr. Fiz. Zh.*, **13,** 75 (1968); through *C.A.*, **69,** 1359 (1968).
206. K. A. Jensen and P. H. Nielsen, *Acta Chem. Scand.*, **17,** 1875 (1963).
207. B. J. Pullman and B. O. West, *Aust. J. Chem.*, **17,** 30 (1964).
208. A. E. Borisov, N. V. Novikova, and N. A. Chumaevskii, *Dokl. Akad. Nauk SSSR* **136,** 129 (1961).
209. G. Bouquet, A. Loutellier, and M. Bigorgne, *J. Mol. Structure*, **1,** 211 (1967–68).
210. R. N. Sterlin and S. S. Dubov, *Zh. Vses. Khim. Obshchest.*, **7,** No. 1, 117 (1962); through *C.A.*, **57,** 294 (1962).
211. M. Margoshes and V. A. Fassel, *Spectrochim. Acta*, **7,** 14 (1955).
212. R. D. Kross and V. A. Fassell, *J. Amer. Chem. Soc.*, **77,** 5858 (1955).
213. L. A. Harrah, M. T. Ryan, and C. Tamborski, *Spectrochim. Acta*, **18,** 21 (1962).

214. K. M. Mackay, D. B. Sowerby, and W. C. Young, *Spectrochim. Acta, A*, **24,** 611 (1968).
215. A. L. Allred and A. L. Hensley, Jr., *J. Inorg. Nucl. Chem.*, **17,** 43 (1961).
216. C. R. McCoy and A. L. Allred, *J. Inorg. Nucl. Chem.*, **25,** 1219 (1963).
217. R. G. Kostyanovskii, I. I. Chervin, V. V. Yakshin, and A. U. Stepanyants, *Izv. Akad. Nauk SSSR, Ser. Khim.*, 1629 (1967).
218. A. G. Massey, E. W. Randall, and D. Shaw, *Spectrochim. Acta*, **21,** 263 (1965).
219. R. J. Chuck, A. G. Massey, E. W. Randall, and D. Shaw, in B. Pesce, ed., *Nuclear Magnetic Resonance in Chemistry*, Academic Press, New York, 1965, p. 189.
220. W. Brügel, T. Ankel, and F. Krückeberg, *Z. Electrochem.*, **64,** 1121 (1960).
221. R. G. Barnes and P. J. Bray, *J. Chem. Phys.*, **23,** 1177 (1955).
222. S. S. Dubov, B. I. Tetel'baum, and R. N. Sterlin, *Zh. Vses. Khim. Obshchest.*, **7,** 691 (1962); through *C.A.*, **58,** 8538 (1963).
223. H. P. Fritz and K. E. Schwarzhans, *J. Organometal. Chem.*, **5,** 103 (1966).
224. S. E. Gukasyan and V. S. Shpinel, *Phys. Status Solidi*, **29,** 49 (1968).
225. H. J. M. Bowen, *Trans. Faraday Soc.*, **50,** 463 (1954).
226. J. Wetzel, *Z. Kristallogr.*, **104,** 305 (1942).
227. V. I. Iveronova and F. M. Roitburd, *Zh. Fiz. Khim.*, **26,** 810 (1952); through *C.A.*, **46,** 10767 (1952).
228. L. S. Birks and J. M. Siomkajlo, *Spectrochim. Acta*, **18,** 363 (1962).
229. D. Mootz, P. Holst, L. Berg, and K. Drews, *Z. Kristallogr.*, **117,** 233 (1962).
230. L. H. Long and J. F. Sackman, *Trans. Faraday Soc.*, **51,** 1062 (1955).
231. H. A. Skinner, "The Strengths of Metal-to-Carbon Bonds," in F. G. A. Stone and R. West, eds., *Advances in Organometallic Chemistry*, Vol. 2, Academic Press, New York, 1964, p. 98.
232. S. J. W. Price and A. F. Trotman-Dickenson, *Trans. Faraday Soc.*, **54,** 1630 (1958).
233. W. F. Lautsch, *Chem. Tech.* (Berlin), 10, 419 (1958); through *C.A.*, **53,** 43 (1959).
234. W. F. Lautsch, P. Erzberger, and A. Tröber, *Wiss. Z. Tech. Hochsch. Chem. Leuna-Merseburg*, **1,** 31 (1958–59); through *C.A.*, **54,** 13845 (1960).
235. K. H. Birr, *Z. Anorg. Allg. Chem.*, **306,** 21 (1960).
236. K. H. Birr, *Z. Anorg. Allg. Chem.*, **311,** 92 (1961).
237. R. E. Winters and R. W. Kiser, *J. Organometal. Chem.*, **10,** 7 (1967).
238. F. I. Vilesov and V. M. Zaitsev, *Dokl. Akad. Nauk SSSR*, **154,** 886 (1964).
239. E. J. Rosenbaum and C. R. Sandberg, *J. Amer. Chem. Soc.*, **62,** 1622 (1940).
240. M. V. Forward, S. T. Bowden, and W. J. Jones, *J. Chem. Soc.*, S121 (1949).
241. E. Bergmann and W. Schütz, *Z. Phys. Chem.* (Leipzig), *B*, **19,** 401 (1932).
242. C. P. Smyth, *J. Org. Chem.*, **6,** 421 (1941).
243. F. Hein, A. Schleede, and H. Kallmeyer, *Z. Anorg. Allg. Chem.*, **311,** 260 (1961).
244. M. J. Aroney, R. J. W. LeFèvre, and J. D. Saxby, *J. Chem. Soc.*, 1739 (1963).
245. P. Pascal, *C. R. Acad. Sci., Paris*, **218,** 57 (1944).
246. N. K. Parab and D. M. Desai, *Curr. Sci.*, **26,** 389 (1957); through *C.A.*, **52,** 16856 (1958).
247. N. K. Parab and D. M. Desai, *J. Indian Chem. Soc.*, **35,** 569 (1958).
248. E. O. Fischer, G. Joos, and W. Meer, *Z. Naturforsch., B*, **13,** 456 (1958).
249. P. Bothorel, *Ann. Chim.* (Paris), **4,** 669 (1959).
250. P. Bothorel, *C. R. Acad. Sci., Paris*, **251,** 1628 (1960).
251. A. E. Goddard, J. N. Ashley, and R. B. Evans, *J. Chem. Soc.*, **121,** 978 (1922).
252. F. Kh. Solomakhina, *Tr. Tashkentsk. Farm. Inst.*, **1,** 321 (1957); through *C.A.*, **55,** 15389 (1961).
253. M. S. Kharasch, E. V. Jensen, and S. Weinhouse, *J. Org. Chem.*, **4,** 429 (1949).
254. L. Maier, *Tetrahedron Lett.*, No. 6, 1 (1959).

255. R. Müller and C. Dathe, *Chem. Ber.*, **99**, 1609 (1966).
256. R. Müller, *Organometal. Chem. Rev.*, **1**, 359 (1966).
257. R. Müller and C. Dathe, East Ger. Pat. 49,607 (Aug. 20, 1966); through *C.A.*, **66**, 2760 (1967).
258. R. Müller and C. Dathe (to Institut für Silikon- und Fluorkarbon-Chemie), Ger. Pat. 1,249,865 (Sept. 14, 1967); through *C.A.*, **68**, 4829 (1968).
259. E. A. Besolova, V. L. Foss, and I. F. Lutsenko, *Zh. Obshch. Khim.*, **38**, 267 (1968).
260. F. F. Blicke, U. O. Oakdale, and F. D. Smith, *J. Amer. Chem. Soc.*, **53**, 1025 (1931).
261. G. O. Doak, G. G. Long, L. D. Freedman, and M. O'Brien, unpublished results.
262. M and T Chemicals Inc., Neth. Appl. 6,505,216 (Oct. 25, 1965); through *C.A.*, **64**, 9766 (1966).
263. H. I. Weingarten and W. A. White (to Monsanto Co.), U.S. Pat. 3,366,655 (Jan. 30, 1968); through *C.A.*, **68**, 9268 (1968).
264. E. A. Besolova, V. L. Foss, and I. F. Lutsenko, *Zh. Obshch. Khim.*, **38**, 1574 (1968).
265. P. G. Sergeev and A. B. Bruker, *Zh. Obshch. Khim.*, **27**, 2220 (1957).
266. P. Pfeiffer and P. Schmidt, *J. Prakt. Chem.*, **152**, 27 (1939).
267. F. F. Blicke and U. O. Oakdale, *J. Amer. Chem. Soc.*, **55**, 1198 (1933).
268. D. B. Meyers and J. W. Jones, *J. Amer. Pharm. Ass.*, **39**, 401 (1950).
269. G. O. Doak and H. H. Jaffé, *J. Amer. Chem. Soc.*, **72**, 3025 (1950).
270. M. S. Kharasch (to Eli Lilly and Co.), U.S. Pat. 1,684,920 (Sept. 18, 1928); *C.A.*, **22**, 4538 (1928).
271. T. C. Sun and J. Y. Chi, *Yao Hsueh Hsueh Pao*, **7**, 266 (1959); through *C.A.*, **54**, 10915 (1960).
272. G. T. Morgan and G. R. Davies, *Proc. Roy. Soc., Ser. A*, **143**, 38 (1933).
273. H. Schmidt, *Justus Liebigs Ann. Chem.*, **421**, 174 (1920).
274. P. Pfeiffer and K. Schneider, *Ber.*, **68**, 50 (1935).
275. A. N. Nesmeyanov, O. A. Reutov, O. A. Ptitsyna, and P. A. Tsurkan, *Izv. Akad. Nauk SSSR, Otd. Khim. Nauk*, 1435 (1958).
276. O. A. Reutov, O. A. Ptitsyna, and G. Ertel, *Chem. Tech.* (Berlin), **10**, 201 (1958).
277. A. N. Nesmeyanov, O. A. Reutov, T. P. Tolstaya, O. A. Ptitsyna, L. S. Isaeva, M. F. Turchinskii, and G. P. Bochkareva, *Dokl. Akad. Nauk SSSR*, **125**, 1265 (1959).
278. O. A. Reutov and O. A. Ptitsyna, *Dokl. Akad. Nauk SSSR*, **79**, 819 (1951); through *C.A.*, **46**, 6093 (1952).
279. L. Maier, E. G. Rochow, and W. C. Fernelius, *J. Inorg. Nucl. Chem.*, **116**, 213 (1960).
280. C. J. Wilkins, *J. Chem. Soc.*, 2726 (1951).
281. A. B. Bruker, *Zh. Obshch. Khim.*, **6**, 1823 (1936); through *C.A.*, **31**, 4291 (1937).
282. A. B. Bruker and E. S. Makhlis, *Zh. Obshch. Khim.*, **7**, 1880 (1937); through *C.A.*, **32**, 72 (1938).
283. A. B. Bruker, *Zh. Obshch. Khim.*, **18**, 1297 (1948); through *C.A.*, **43**, 4647 (1949).
284. H. Schmidt and F. Hoffmann, *Ber.*, **59**, 555 (1926).
285. O. A. Reutov, O. A. Ptitsyna, A. N. Lovstova, and V. F. Petrova, *Zh. Obshch. Khim.*, **29**, 3888 (1959).
286. K. A. Kocheshkov and A. N. Nesmeyanov, *Zh. Obshch. Khim.*, **4**, 1102 (1934); through *C.A.*, **29**, 3993 (1935).
287. A. N. Nesmejanow and K. A. Kozeschkow, *Ber.*, **67**, 317 (1934).
288. M. S. Malinovsky and S. P. Olifirenko, *Zh. Obshch. Khim.*, **26**, 118 (1956).
289. O. J. Scherer, J. F. Schmidt, and M. Schmidt, *Z. Naturforsch., B*, **19**, 447 (1964).
290. F. F. Blicke and U. O. Oakdale, *J. Amer. Chem. Soc.*, **54**, 2993 (1932).
291. M and T Chemicals Inc., Neth. Appl. 6,505,218 (Oct. 25, 1965); through *C.A.*, **64**, 9766 (1966).
292. E. J. Kupchik and C. T. Theisen, *J. Organometal. Chem.*, **11**, 627 (1968).

293. W. T. Reichle, *J. Organometal. Chem.*, **13,** 529 (1968).
294. M and T Chemicals Inc., Neth. Appl. 6,505,219 (Oct. 25, 1965); through *C.A.*, **64,** 9766 (1966).
295. N. L. Remes and J. J. Ventura (to M and T Chemicals Inc.), Brit. Pat. 1,079,658 (Aug. 16, 1967); *C.A.*, **68,** 4829 (1968).
296. D. C. Evans (to M & T Chemicals Inc.), Brit. Pat. 1,079,659 (Aug. 16, 1967); through *C.A.*, **68,** 4829 (1968).
297. J. R. Leebrick and N. L. Remes (to M & T Chemicals Inc.), U.S. Pat. 3,367,954 (Feb. 6, 1968); *C.A.*, **68,** 10179 (1968).
298. H. Schmidbaur and M. Schmidt, *Angew. Chem.*, **73,** 655 (1961).
299. H. Schmidbaur, H. S. Arnold, and E. Beinhofer, *Chem. Ber.*, **97,** 449 (1964).
300. M. K. Saikini, *Uch. Zap. Kazansk. Gos. Univ.*, **116,** No. 2, 129 (1956); through *C.A.* **51,** 7191 (1957).
301. R. E. Dessy, T. Chivers, and W. Kitching, *J. Amer. Chem. Soc.*, **88,** 467 (1966).
302. Y. L. Chi and S. Y. Moh, *K'e Hsueh Tung Pao*, No. 2, 50 (1957); through *C.A.*, **53,** 18896 (1959).
303. A. N. Nesmeyanov, N. K. Gipp, L. G. Makarova, and K. K. Mozgova, *Izv. Akad. Nauk SSSR, Otd. Khim. Nauk*, 298 (1953).
304. H. H. Jaffé and G. O. Doak, *J. Amer. Chem. Soc.*, **71,** 602 (1949).
305. O. A. Reutov and O. A. Ptitsyna, *Izv. Akad. Nauk SSSR, Otd. Khim. Nauk*, 93 (1952); *C.A.*, **47,** 1631 (1953).
306. S. M. Mistry and P. C. Guha, *J. Indian Inst. Sci., A*, **15,** 25 (1932); *C.A.*, **26,** 4589 (1932).
307. H. H. Jaffé and G. O. Doak, *J. Amer. Chem. Soc.*, **72,** 3027 (1950).
308. K. Hiratsuka, *Nippon Kagaku Zasshi*, **58,** 1163 (1937); *C.A.*, **33,** 158 (1939).
309. M. M. Koton and F. S. Florinskii, *Dokl. Akad. Nauk SSSR*, **137,** 1368 (1961).
310. M. M. Koton and T. M. Kiseleva, *Izv. Akad. Nauk SSSR, Otd. Khim. Nauk*, 1783 (1961).
311. M and T Chemicals Inc., Neth. Appl. 6,505,215 (Oct. 25, 1965); through *C.A.*, **64,** 9766 (1966).
312. M and T Chemicals Inc., Neth. Appl. 6,505,217 (Oct. 25, 1965); through *C.A.*, **64,** 9767 (1966).
313. J. Hasenbäumer, *Ber.*, **31,** 2910 (1898).
314. A. Michaelis and A. Günther, *Ber.*, **44,** 2316 (1911).
315. E. A. H. Friedheim, U.S. Pat. 2,430,461 (Nov. 11, 1947); through *C.A.*, **42,** 1973 (1948).
316. E. A. H. Friedheim, Brit. Pat. 655,435 (July 18, 1951); through *C.A.*, **47,** 144 (1953).
317. E. A. H. Friedheim, U.S. Pat. 2,659,723 (Nov. 17, 1953); *C.A.*, **49,** 1815 (1955).
318. E. A. H. Friedheim, U.S. Pat. 2,880,222 (March 31, 1959); through *C.A.*, **53,** 16158 (1959).
319. M and T Chemicals Inc., Neth. Appl. 6,505,233 (Oct. 25, 1965); through *C.A.*, **64,** 14220 (1966).
320. D. C. Evans (to M & T Chemicals Inc.), Brit. Pat. 1,106,035 (March 13, 1968); *C.A.*, **68,** 10179 (1968).
321. E. J. Kupchik and E. F. McInerney, *J. Organometal. Chem.*, **11,** 291 (1968).
322. Chem. Fabrik von Heyden, Ger. Pat. 268,451 (Dec. 16, 1913); through *Chem. Zentr.*, **85,** I, 309 (1914).
323. Farbwerke vorm. Meister Lucius & Brüning, Ger. Pat. 270,255 (Feb. 13, 1914); through *Chem. Zentr.*, **85,** I, 829 (1914).
324. Chem. Fabrik von Heyden, Ger. Pat. 396,697 (June 10, 1924); through *Chem. Zentr.*, **95,** II, 760 (1924).

325. P. Ehrlich and P. Karrer, *Ber.*, **46**, 3564 (1913).
326. W. Riddell and S. Basterfield, *Trans. Roy. Soc. Can.*, **23**, Sect. 3, 45 (1929).
327. H. Lecoq, *Bull. Soc. Chim. Belges*, **42**, 199 (1933).
328. H. Lecoq, *J. Pharm. Belg.*, **19**, 173 (1937).
329. F. Klages and W. Rapp, *Chem. Ber.*, **88**, 384 (1955).
330. W. Kuchen and H. Ecke, *Z. Anorg. Allg. Chem.*, **321**, 138 (1963).
331. F. A. Paneth, *Trans. Faraday Soc.*, **30**, 179 (1934).
332. I. G. M. Campbell, G. W. A. Fowles, and L. A. Nixon, *J. Chem. Soc.*, 3026 (1964).
333. H. Schumann, T. Östermann, and M. Schmidt, *J. Organometal. Chem.*, **8**, 105 (1967).
334. H. Schumann and M. Schmidt, *Inorg. Nucl. Chem. Lett.*, **1**, 1 (1965).
335. N. S. Vyazankin, G. A. Razuvaev, O. A. Kruglaya, and G. S. Semchikova, *J. Organometal. Chem.*, **6**, 474 (1966).
336. N. S. Vyazankin, G. S. Kalinina, O. A. Kruglaya, and G. A. Razuvaev, *Zh. Obsnch. Khim.*, **38**, 205 (1968).
337. G. E. Coates and J. G. Livingstone, *Chem. Ind.* (London), 1366 (1958).

IX Organobismuth compounds

1. TRIALKYL- AND TRIARYLBISMUTH COMPOUNDS

A. Preparation

The study of organobismuth compounds was begun in 1850 by Löwig and Schweizer (1) who prepared triethylbismuth by the reaction between ethyl iodide and an alloy of bismuth and potassium. Similar methods of synthesizing triethylbismuth were also used by Breed (2, 3) and by Dünhaupt (4). It was later shown (5–8) that triarylbismuth compounds can be satisfactorily obtained by the interaction of aryl halides and bismuth–sodium alloys.

Another early method of obtaining tertiary bismuth compounds involved the reaction of organic mercurials with bismuth metal or with bismuth trihalides. Thus, Frankland and Duppa (9) prepared triethylbismuth by heating a mixture of diethylmercury and powdered bismuth at 120–140°. Similarly, Hilpert and Grüttner (10) obtained a 41% yield of triphenylbismuth by heating diphenylmercury and bismuth metal in a stream of hydrogen at 250° for ten minutes. Challenger and Allpress (11) found that triphenylbismuth can be prepared in almost quantitative yield by the interaction of diphenylmercury and bismuth tribromide in ethereal solution. The high toxicity and cost of organic mercurials have discouraged the use of these compounds in synthetic work.

Organozinc compounds were also used at one time for the synthesis of trialkylbismuth compounds (12, 13). This method was never widely employed because of the difficulties involved in the preparation and handling of the spontaneously inflammable alkylzinc compounds. It became completely obsolete in 1904 when Pfeiffer and Pietsch (14) showed that the more easily handled Grignard reagents are equally effective for the preparation of tertiary bismuth compounds. A considerable number of trialkyl- and triarylbismuth compounds have been prepared from the corresponding Grignard reagents.

Although the Grignard reaction is at present the best method for establishing the carbon–bismuth bond, several other good methods are also available. Excellent yields of trialkylbismuth compounds have been obtained by the reaction between aluminum alkyls and bismuth trichloride (15–17). It has also been found that lithium tetraalkylaluminates (18) can be substituted for the aluminum alkyls and that bismuth trioxide (19) can be used in place of the bismuth trihalide. In the latter case, however, only low yields of trialkylbismuth compounds were obtained. The use of organolithium compounds for the preparation of tertiary bismuth compounds has not been thoroughly investigated, although the following compounds have been synthesized by the reaction of the appropriate lithium reagent with bismuth trichloride or tribromide: triphenylbismuth (20), tris(*p*-dimethylaminophenyl)bismuth (21), triisopropenylbismuth (22), and the *cis* and *trans* isomers of tripropenylbismuth (22). Tricyclopentadienylbismuth (23, 24) and tris(methylcyclopentadienyl)bismuth (24) have been prepared in goods yields by the interaction of the corresponding organosodium compounds with bismuth trichloride. An attempt to prepare tri-*p*-biphenylylbismuth by the Wurtz-Fittig type reaction (employing sodium) was, however, unsuccessful (25). Recently, Müller and Dathe (26–29) have shown that alkyl- and aryltrifluorosilanes can be converted in aqueous solution to a variety of organometallic compounds. For example, triphenylbismuth is formed in 50% yield by the addition of phenyltrifluorosilane and ammonium fluoride to a solution of bismuth hydroxide in hydrofluoric acid.

Trialkylbismuth compounds can also be synthesized by the electrochemical reaction of organometallic reagents at a sacrificial anode of bismuth. Thus, triethylbismuth has been obtained in high yields by the electrolysis of sodium alkoxytriethylaluminates (30), sodium tetraethylborate (31), sodium tetraethylaluminate (32), or potassium tetraethylaluminate (32). Electrolysis at a bismuth anode has also been used to convert silver cyanide in pyridine solution to bismuth(III) cyanide (33). The electrochemical synthesis of organometallic compounds has been reviewed by Marlett (34).

Gilman and co-workers (35, 36) were the first to describe the conversion of diazonium salts to triarylbismuth compounds. Their technique consisted of treating a suspension of an aromatic diazonium chlorobismuthate (ArN_2BiCl_4) with copper powder and then adding aqueous ammonia or hydrazine to the reaction mixture. They showed that the copper decomposes the diazonium salt to arylbismuth halides of the types $ArBiCl_2$ and Ar_2BiCl; these halides, in turn, are converted by the ammonia or hydrazine to the triarylbismuth compound. The yields reported by Gilman and co-workers (35, 36) were rather low, ranging from a "trace" of tri-α-naphthylbismuth to a 22% yield of triphenylbismuth. Attempts to prepare tris(*p*-sulfophenyl)-bismuth by their procedure were unsuccessful. Later workers (37, 38) obtained somewhat better results by treatment of the chlorobismuthate with metallic bismuth:

$$3ArN_2BiCl_4 + 2Bi \rightarrow Ar_3Bi + 4BiCl_3 + 3N_2$$

For example, tri-α-naphthylbismuth was prepared in 15% yield by this method, and triphenylbismuth was obtained in 50% yield. The diazonium method was further improved by the use of tetrafluoroborates (39) in place of the chlorobismuthates. Yields of triarylbismuth compounds as high as 69% were reported. Reutov (40) has described the preparation of triarylbismuth compounds by the reaction between potassium arylazoformates (ArN_2CO_2K), bismuth trichloride, and bismuth metal; but this procedure appears to have no advantages over the employment of diazonium salts. Triphenylbismuth has also been obtained by the treatment of a hydrochloric acid solution of bismuth trichloride and phenylhydrazine with cupric and ferric chlorides (41). This reaction probably involves the intermediate formation of a benzenediazonium salt.

The reaction of diaryliodonium salts with bismuth trichloride and powdered bismuth is another method of preparing triarylbismuth compounds (42, 43):

$$3Ar_2ICl + 2Bi \xrightarrow{BiCl_3} Ar_3Bi + BiCl_3 + 3ArI$$

Experiments with unsymmetrical iodonium salts indicate that the bismuth atom is preferentially arylated by the more electron-attracting group. For example, (*p*-chlorophenyl)phenyliodonium chloride gives 13% tris(*p*-chlorophenyl)bismuth and no triphenylbismuth. Reutov and co-workers believe that these facts provide evidence for heterolytic cleavage of iodonium salts. It has also been shown (44) that diphenylchloronium and diphenylbromonium salts can be used to arylate bismuth and give good yields of triphenylbismuth.

There have been several reports of the preparation of trialkylbismuth compounds by the action of free radicals on bismuth mirrors. Paneth and his co-workers (45–47) were the first to investigate this reaction. They found that methyl and ethyl radicals readily remove cold bismuth mirrors and yield the corresponding trialkylbismuth compounds. With heated mirrors, there was evidence for the possible formation of tetramethyldibismuth and tetraethyldibismuth. More recently, Bell, Pullman, and West (48) have reported that trifluoromethyl radicals (generated by the pyrolysis of hexafluoroacetone) react with a heated bismuth mirror to form a colorless, rather involatile liquid, which was tentatively identified as tris(trifluoromethyl)bismuth. The quantity of liquid obtained was too small (around five mg) to permit full chemical analysis, but its infrared spectrum suggested the presence of trifluoromethyl groups. Furthermore, alkaline hydrolysis of the liquid yielded fluoroform, while pyrolysis gave metallic bismuth.

Unsymmetrical tertiary bismuth compounds can be obtained by the reaction of Grignard or lithium reagents with alkyl- or arylbismuth halides of the type $RBiX_2$ or R_2BiX. The method was first used by Challenger (49) to prepare diphenyl-α-naphthylbismuth:

$$(C_6H_5)_2BiBr + \alpha\text{-}C_{10}H_7MgBr \rightarrow \alpha\text{-}C_{10}H_7Bi(C_6H_5)_2 + MgBr_2$$

Similarly, the Grignard reaction has been employed by Gilman and Yablunky (21) for the preparation of a considerable number of other unsymmetrical triarylbismuth compounds. The aliphatic compound diethylamylbismuth has been synthesized by the interaction of amylbismuth dichloride and ethylmagnesium bromide (50). In a similar way, 1,4-phenylenebis(diphenylbismuth) has been prepared by the reaction of two moles of diphenylbismuth chloride with one mole of 1,4-phenylenedilithium (51). Compounds containing the bismuth atom in a ring system have also been obtained via Grignard or lithium reagents. Thus, ethylcyclopentamethylenebismuth (52) was prepared from ethylbismuth dibromide and the di-Grignard reagent obtained from 1,5-dibromopentane, while 5-phenyl-5*H*-dibenzobismole (53) has been synthesized by the reaction between 2,2′-biphenylylenedilithium and phenylbismuth diiodide. A polymeric tertiary bismuth compound (54) has been obtained by the reaction of poly(lithiostyrene) with diphenylbismuth chloride.

Gilman and Yablunky (55) have described another method for converting diarylbismuth halides to unsymmetrical compounds. They found that these halides react in liquid ammonia with lithium, sodium, potassium, calcium, and barium to form deeply colored, unstable, and highly reactive compounds of the type Ar_2BiM or $(Ar_2Bi)_2M$. Treatment of these compounds with aryl halides leads to the formation of fair yields of unsymmetrical triarylbismuth compounds.

In spite of numerous attempts (49, 56, 57), no compound has yet been prepared in which a simple alkyl group and an aryl group are bonded to the same bismuth atom. Hartmann and co-workers (58), however, have prepared several mixed alkynyl-aryl compounds of the types $Ar_2BiC{\equiv}CH$, $Ar_2BiC{\equiv}CC_6H_5$, and $Ar_2BiC{\equiv}CBiAr_2$. The following reactions were employed:

$$Ar_2BiCl + NaC{\equiv}CH \rightarrow Ar_2BiC{\equiv}CH + NaCl$$

$$Ar_2BiCl + NaC{\equiv}CC_6H_5 \rightarrow Ar_2BiC{\equiv}CC_6H_5 + NaCl$$

$$Ar_2BiCl + 2NaC{\equiv}CH \rightarrow Ar_2BiC{\equiv}CBiAr_2 + 2NaCl + C_2H_2$$

In a number of cases, acetylene Grignard reagents could be used instead of the sodium acetylides. Liquid ammonia or tetrahydrofuran were used as solvents in the sodium acetylide reactions; only the latter solvent was used with the Grignard reagents.

Unsymmetrical tertiary bismuth compounds containing alkyl and perfluoroalkyl groups have been prepared by the reaction of perfluoroalkyl iodides with trialkylbismuth compounds at 100° (59, 60). For example, trimethylbismuth reacts completely with a moderate excess of trifluoroiodomethane to give dimethyl(trifluoromethyl)bismuth (82% conversion) and methylbis(trifluoromethyl)bismuth (18% conversion). No evidence for the formation of tris(trifluoromethyl)bismuth could be obtained. Tris(trifluoromethyl)phosphine can be used instead of the trifluoroiodomethane:

$$(CF_3)_3P + (CH_3)_3Bi \rightarrow (CH_3)_2BiCF_3 + CH_3P(CF_3)_2$$

$$(CF_3)_3P + (CH_3)_2BiCF_3 \rightarrow CH_3Bi(CF_3)_2 + CH_3P(CF_3)_2$$

Despite the relative ease with which it is possible to replace one or two alkyl groups by perfluoroalkyl groups, conversion of a trialkylbismuth compound to a tris(perfluoroalkyl)bismuth compound has not yet been accomplished.

In addition to the reactions discussed above, trialkyl- and triarylbismuth compounds have been synthesized by a number of methods that are probably now of only historical interest. These methods are included in a review by Gilman and Yale (61).

B. Reactions

(*1*) *Trialkylbismuth compounds*

With the exception of a few Lewis acid–base reactions, which are discussed below, there are virtually no reactions of trialkylbismuth compounds which do not involve cleavage of the carbon–bismuth bond. Most of these compounds are spontaneously inflammable in air; except for trimethylbismuth, they cannot be distilled at ordinary pressures without decomposition (60, 61). The controlled oxidation of triethylbismuth at −50 to −60° yields a white solid that has been identified as diethylbismuth ethoxide, $(C_2H_5)_2BiOC_2H_5$ (62). This compound is relatively stable to further oxidation at low temperatures. Between 5 and 25° the products of oxidation of triethylbismuth include diethyl peroxide, ethyl ether, ethyl alcohol, ethylene, and a solid whose composition approximates $C_2H_5BiO_2$. When tribenzylbismuth is exposed to the air, clouds of smoke are formed immediately (63). It has been suggested (64) that the oxidation of this compound yields benzaldehyde and inorganic bismuth. Treatment of tribenzylbismuth with an acetone solution of potassium permanganate gives an immediate precipitate of manganese dioxide (63).

Even at 0° trialkylbismuth compounds are cleaved by chlorine or bromine in accordance with the following equation (12, 13):

$$R_3Bi + X_2 \rightarrow R_2BiX + RX$$

It has been reported (22), however, that the *cis* and *trans* isomers of triisopropenylbismuth are converted on treatment with bromine at $-55°$ to the corresponding triisopropenylbismuth dibromides. The reaction of mixed alkylperfluoroalkylbismuth compounds with bromine or iodine at $-26°$ leads to cleavage of C—Bi bonds and the formation of mixtures of alkyl and perfluoroalkyl halides (59, 60). No evidence of preferential cleavage of either alkylbismuth or perfluoroalkylbismuth bonds has been obtained. Tribenzylbismuth reacts with a slight excess of iodine in benzene to yield bismuth triiodide (63).

In general, trialkylbismuth compounds are not affected by water or aqueous bases. Compounds containing perfluoroalkyl groups, however, are rapidly hydrolyzed by 5*N* sodium hydroxide (59, 60). For example, dimethyl(trifluoromethyl)bismuth reacts with base to give a quantitative yield of fluoroform. Trialkylbismuth compounds are hydrolyzed by both inorganic and organic acids. Hydrogen sulfide, mercaptans, and selenols (65, 66) appear to be particularly effective in cleaving the carbon–bismuth bond.

Some metals and inorganic salts can also cause rupture of the carbon–bismuth bond. Thus, dialkylbismuth halides can be prepared by the reaction between trialkylbismuth compounds and bismuth trichloride or tribromide (12, 13). Silver nitrate and mercuric chloride readily cleave trialkylbismuth compounds and are in turn reduced to the free metals (61). Treatment of tribenzylbismuth with sodium yields free bismuth (63). The reaction of triisopropenylbismuth with thallium(III) chloride in ether at $-40°$ has been found (22) to give a 98% yield of diisopropenylthallium chloride. The electrolysis of triethylbismuth at a lead anode has been used to prepare tetraethyllead (67).

When mixtures of triethylbismuth and triethylgermane are heated under nitrogen at 130–150°, compounds containing Ge—Bi bonds are formed (68). These reactions can be represented by the following general equation (where $n = 1, 2,$ or 3):

$$n(C_2H_5)_3GeH + (C_2H_5)_3Bi \rightarrow nC_2H_6 + [(C_2H_5)_3Ge]_nBi(C_2H_5)_{3-n}$$

All three compounds are easily oxidized, but are thermally stable under nitrogen up to 200°. At higher temperatures they decompose to metallic bismuth and hexaethyldigermane. The reaction of triethylbismuth with triethylsilane or triethyltin hydride has also been investigated (69, 70). In these cases only compounds of the type $[(C_2H_5)_3M]_3Bi$ (where M = Si or Sn) were isolated.

It has been generally found (12, 13, 59, 60, 71–73) that trialkylbismuth compounds have virtually no basic or donor properties. Thus, the reaction of trimethylbismuth with methyl iodide does not yield a bismuthonium salt even at elevated temperatures. Instead, the following reaction is observed (12, 13):

$$(CH_3)_3Bi + 2CH_3I \rightarrow CH_3BiI_2 + 2C_2H_6$$

Perfluoroalkyl iodides undergo an exchange reaction with trialkylbismuth compounds at 100° to form dialkyl(perfluoroalkyl)bismuth and alkylbis-(perfluoroalkyl)bismuth (59, 60). The latter two classes of compounds have been shown (60) to be devoid of donor properties; thus, they do not form adducts with the strong Lewis acid, boron trifluoride. In contrast, Benlian and Bigorgne (74) in 1963 reported that triethylbismuth reacts with nickel tetracarbonyl and molybdenum hexacarbonyl to form a nickel complex, $Ni(CO)_3(BiEt_3)$, and two molybdenum complexes, $Mo(CO)_5(BiEt_3)$ and *cis*-$Mo(CO)_4(BiEt_3)_2$. The authors noted that these compounds were difficult to prepare and that triethylbismuth is a much weaker ligand than the corresponding compounds of antimony, arsenic, and phosphorus.

It has been recently found (60) that dimethyl(trifluoromethyl)bismuth and methylbis(trifluoromethyl)bismuth have acceptor properties and form adducts at low temperatures with dimethylamine. Trimethylbismuth and the amine do not react under these conditions. The ability of the trifluoromethylbismuth derivatives to react as Lewis acids is not altogether unexpected, since tris(trifluoromethyl)stibine forms a relatively stable adduct with dimethylamine. It has also been reported (75) that triphenyl derivatives of phosphorus, arsenic, and antimony (but not bismuth) form solid 1:1 complexes with the π-donor, hexamethylbenzene.

The compound $(C_2H_5)_3Bi \cdot Tl(C_6H_5)_3$ has been mentioned in the patent literature (76), but no information about the preparation or structure of this material is available. Thermal decomposition of the substance at 395° is said to yield a bismuth–thallium alloy.

(2) *Triarylbismuth compounds*

Triarylbismuth compounds are relatively stable. They are usually unaffected by oxygen or water at ordinary temperatures and are generally much less reactive than the alkyl compounds. Unlike the latter substances the triarylbismuth compounds can be readily converted to a variety of derivatives of the type Ar_3BiX_2 (X is an electronegative group). These reactions are discussed in Section IX-3-A. It has been stated that tri-α-thienylbismuth (77, 78) slowly undergoes oxidation to bismuth oxide and di-α-thienyl in the presence of air, but the stoichiometry of this reaction has not been established. Triphenylbismuth can be readily distilled (bp 242° at 14 mm) without decomposition and has been obtained so pure that it has been used in measurements of the atomic weight of bismuth (79–81).

The carbon–bismuth bond in triarylbismuth compounds is able to survive the conditions of free-radical initiated polymerization. Thus, diphenyl(*p*-vinylphenyl)bismuth can be polymerized in toluene at 65° in the presence of 2,2′-azobisisobutyronitrile to a white powder of composition $(C_{20}H_{17}Bi)_n$ (54). Under similar reaction conditions, diphenyl(*p*-vinylphenyl)bismuth and styrene copolymerize to a material containing carbon–bismuth bonds. The polymerization of tris(*p*-vinylphenyl)bismuth has been carried out by heating at 100° for six hours without the addition of any initiator (82). The monomer has also been copolymerized with styrene or methyl methacrylate both in the presence of 0.02% di-*tert*-butyl peroxide and in the absence of any initiator.

Triarylbismuth compounds are usually cleaved quantitatively by strong mineral acids to inorganic bismuth salts (5, 49):

$$Ar_3Bi + HY \rightarrow 3ArH + BiY_3$$

When weak acids (83) or halogens (53, 56) are used, it is often possible to obtain intermediate products of the type Ar_2BiY or $ArBiY_2$. The reaction of triarylbismuth compounds with thiols or carboxylic acids generally results in the cleavage of two aryl groups (84). Tri-α-naphthylbismuth, however, is unaffected by thiophenol even in boiling xylene; in contrast, the phenyl groups of diphenyl-α-naphthylbismuth are readily cleaved by the thiol. No reason for the anomalous behavior of tri-α-naphthylbismuth has been suggested. When triphenylbismuth is allowed to react with thiosalicylic acid, the following cyclic derivative is obtained (84):

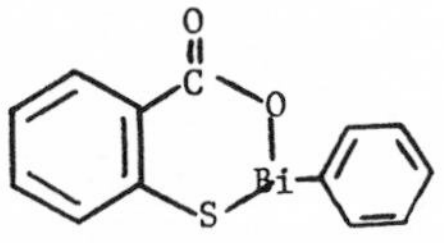

Heating triarylbismuth compounds with phenols leads to essentially complete cleavage of the carbon–bismuth bonds (85).

A wide variety of inorganic substances are able to cleave the carbon–bismuth bonds of triarylbismuth compounds. Halogen halides and cyanogen halides give compounds of the type Ar_2BiX in which X is the more electronegative component of the halide, e.g. (11, 56, 86):

$$Ph_3Bi + ICl \rightarrow Ph_2BiCl + PhI$$

$$Ph_3Bi + BrCN \rightarrow Ph_2BiCN + PhBr$$

Reaction of triarylbismuth compounds with thiocyanogen yields diarylbismuth thiocyanates and aryl thiocyanates (87). Some metallic and metalloid halides are also arylated by triarylbismuth compounds (20, 64, 88):

$$Ar_3Bi + 2BiCl_3 \rightarrow 3ArBiCl_2$$

$$Ar_3Bi + BiCl_3 \rightarrow Ar_2BiCl + ArBiCl_2$$

$$2Ar_3Bi + BiCl_3 \rightarrow 3Ar_2BiCl$$

$$Ar_3Bi + AsCl_3 \rightarrow Ar_2BiCl + ArAsCl_2$$

$$Ar_3Bi + PCl_3 \rightarrow Ar_2BiCl + ArPCl_2$$

$$Ar_3Bi + HgCl_2 \rightarrow Ar_2BiCl + ArHgCl$$

$$2Ar_3Bi + TlCl_3 \rightarrow 2Ar_2BiCl + Ar_2TlCl$$

The reaction of triphenylbismuth with antimony trichloride yields diphenylbismuth chloride and triphenylantimony dichloride; i.e., the antimony trichloride is converted to a pentavalent antimony compound (64). The stoichiometry and mechanism of this reaction have not been explained. It has been reported (88), however, that when a chloroform solution of triphenylbismuth and antimony trichloride is heated in a carbon dioxide atmosphere, the products include diphenylchlorostibine, bismuth trichloride, and traces of benzene. The reaction of triphenylbismuth with tin tetrachloride gives a complex mixture from which diphenyltin dichloride can be isolated (88, 89). Some inorganic halides, e.g., aluminum chloride (90), ferric chloride (90), silicon tetrachloride (91), titanium tetrachloride (91), cupric chloride (64, 88), and zinc chloride (88), react with triphenylbismuth in ether or chloroform to form diphenylbismuth chloride or bismuth trichloride but are apparently not phenylated in the course of the reaction; in these cases the cleaved phenyl groups are converted to benzene or biphenyl (88, 90, 91). It has been stated (91, 92) that the interaction of triphenylbismuth and silver nitrate in chloroform gives $(C_6H_5Ag)_2 \cdot AgNO_3$ and bismuth trichloride, but little information about this reaction is available.

Koton (93) has studied the cleavage of organometallic compounds by trimethylamine hydrochloride. He found that when an alcoholic solution of this hydrochloride and triphenylbismuth is heated to 130° in a sealed tube, a quantitative yield of benzene and bismuth trichloride is obtained. Under the same conditions, tri-α-naphthylbismuth gives di-α-naphthylbismuth chloride as well as naphthalene and bismuth trichloride.

Unlike more reactive organometallic compounds, organobismuth compounds do not add to aldehydes or ketones (61). Triarylbismuth compounds, however, can react with acyl halides to give ketones (64). Thus, triphenylbismuth reacts with acetyl or benzoyl chloride to give a low yield of acetophenone or benzophenone:

$$(C_6H_5)_3Bi + RCOCl \rightarrow RCOC_6H_5 + (C_6H_5)_2BiCl$$

Similarly, α-naphthyl phenyl ketone has been isolated in low yield from the reaction between tri-α-naphthylbismuth and benzoyl chloride in the presence of either aluminum or ferric chloride (94).

Schmitz-DuMont and Ross (95) have investigated the ammonolysis of a large number of organometallic compounds. On treatment of the triphenyl derivatives of arsenic, antimony, and bismuth with a solution of potassium amide in liquid ammonia, it was found that all the phenyl groups were removed as benzene to yield $KAs(NH_2)_2$, $KSb(NH_2)_2$, and BiN, respectively. The bismuth nitride, a black, highly explosive substance, is formed from a brown, short-lived intermediate, which may be $Bi(NH_2)_3$. The following reaction sequence was suggested:

$$(C_6H_5)_3Bi + 3NH_3 \rightarrow Bi(NH_2)_3 + 3C_6H_6$$

$$Bi(NH_2)_3 \rightarrow BiN + 2NH_3$$

The authors (95) concluded that the ease of ammonolysis increases as the heteroatom becomes more metallic. Thus, triphenylbismuth is cleaved at $-60°$ by as little as 0.2 mole of potassium amide per mole of bismuth compound, while triphenylstibine, -arsine, and -phosphine require increasingly more vigorous conditions.

Gilman and Yablunky (55) have noted that the C—Bi bond in triphenylbismuth is cleaved by sodium in liquid ammonia, but the course of this reaction has not been elucidated.

In accordance with one of the general principles of organometallic chemistry (96), the bismuth in triarylbismuth compounds can sometimes be displaced by other metals, especially by more electropositive ones. For example, when a solution of triphenylbismuth is treated with a large excess of lithium, a 22% yield of phenyllithium is obtained (97). The reaction of triphenylbismuth with antimony at 300° gives about an 85% yield of triphenylstibine (98). Mercury reacts with triarylbismuth compounds to a limited extent. Thus, when 0.1 mole of triphenylbismuth and 150 g of mercury were heated with boiling chloroform, only 0.5 g of bismuth and 1.5 g of diphenylmercury could be isolated (88). Gilman and Yale (61) refer to unpublished work which indicates that mercury does not react with either triphenylbismuth or tri-*p*-tolylbismuth in refluxing benzene. Hilpert and Grüttner (10) have attempted to determine the equilibrium position of the following reaction at 250°:

$$2(C_6H_5)_3Bi + 3Hg \rightleftharpoons 3(C_6H_5)_2Hg + 2Bi$$

They found that the reaction of triphenylbismuth and mercury at this temperature gives a 24% yield of diphenylmercury; under the same conditions diphenylmercury and bismuth yield 41% triphenylbismuth. In both reactions a 2% yield of biphenyl was obtained.

Gilman and co-workers (99, 100) have observed that the treatment of triarylbismuth compounds with *n*-butyllithium or *n*-butylsodium results in the following type of metal–metal exchange:

$$Ar_3Bi + 3n\text{-}C_4H_9Li \rightarrow 3ArLi + (n\text{-}C_4H_9)_3Bi$$

This reaction is usually aided by electron-attracting substituents on the aryl groups and retarded by electron-repelling substituents. Steric effects are also important. Thus, tri-*o*-tolylbismuth, tris(*o*-chlorophenyl)bismuth, and trimesitylbismuth are not cleaved at all by *n*-butyllithium.

More recently, Wittig and Maercker (101) have reported that triphenylbismuth reacts slowly with *p*-tolyllithium in ether to yield phenyllithium. An "ate-complex" was suggested as an intermediate:

$$(C_6H_5)_3Bi + p\text{-}CH_3C_6H_4Li \rightleftharpoons$$
$$[p\text{-}CH_3C_6H_4Bi(C_6H_5)_3]Li \rightleftharpoons p\text{-}CH_3C_6H_4Bi(C_6H_5)_2 + C_6H_5Li$$

The triphenyl derivatives of phosphorus, arsenic, and antimony were also found to undergo this reaction; triphenylamine, however, was not cleaved by *p*-tolyllithium.

The reaction of triphenylbismuth with hydrogen has been studied under various conditions. When the bismuth compound is dissolved in xylene and heated at 250° with hydrogen at 60 atmospheres, the following reaction requires about 24 hours to go to completion (102):

$$2(C_6H_5)_3Bi + 3H_2 \rightarrow 6C_6H_6 + 2Bi$$

No reduction is observed at room temperature and four atmospheres of hydrogen, even in the presence of a rhodium-on-alumina catalyst that is usually effective for the hydrogenation of aromatic compounds; this result has been attributed to a poisoning of the catalyst by the trivalent bismuth compound (103). Jackson and Sasse (104) have shown that triphenylbismuth inhibits the hydrogenation of crotonic acid with W-7 Raney nickel; nevertheless, when this catalyst is refluxed for two hours with triphenylbismuth in methanol, an 85% yield of benzene is obtained. More recently, Thompson and co-workers (105) have reported that the vapor-phase catalytic hydrogenation of triphenylbismuth (as well as triphenylarsine and triphenylstibine) yields cyclohexane.

There have been several studies (106, 107) of the radiation-induced decomposition of triarylbismuth compounds. When carried out in aromatic solvents, the photolysis of triphenylbismuth by ultraviolet light proceeds by a homolytic reaction in which metallic bismuth and phenyl radicals are formed. About half of the available phenyl radicals add to the solvent molecules to give substituted phenylcyclohexadienyl radicals; these are then oxidized to biaryls by the remainder of the phenyl radicals, which are in turn reduced to benzene. The three phenyl radicals are probably liberated from the triphenylbismuth in a stepwise manner; i.e., both $(C_6H_5)_2Bi\cdot$ and the diradical $C_6H_5\dot{B}i\cdot$ may be present as intermediates during the photolysis.

However, neither the dimer $(C_6H_5)_2BiBi(C_6H_5)_2$ nor polymers of the type $(C_6H_5Bi)_n$ have been identified as decomposition products, although a non-volatile bismuth-containing residue is formed. The decomposition of triphenylbismuth in benzene or chloroform can also be induced by γ-rays from a ^{60}Co source (108, 109). In benzene the radiolysis products include biphenyl, phenylcyclohexadienes, hydrogen, metallic bismuth, and an unidentified solid containing about 60% bismuth. In chloroform, γ-radiation results in the formation of triphenylbismuth dichloride, probably via radiolysis of the solvent. Neutron irradiation of triphenylbismuth has been used as a method for preparing enriched ^{210}Bi (110). Triphenylbismuth prepared from an alpha-emitting ^{210}Bi isomer has been used as a source of carrier-free ^{206}Tl (111).

It has been reported (112–115) that organic compounds of ^{210}Po are formed as the result of the β-decay of ^{210}Bi incorporated in triarylbismuth compounds. The ^{210}Po compounds isolated by means of paper or thin-layer chromatography were tentatively identified as being of the types Ar_2Po, Ar_2PoCl_2, and Ar_3PoCl.

Like their aliphatic analogues, the triarylbismuth compounds appear to be almost devoid of nucleophilic or donor character. Thus, numerous attempts at the direct quaternization of triarylbismuth compounds have failed (64, 116–118). For example, Challenger and Ridgway (64) found that when triphenylbismuth and benzyl chloride are dissolved in ether and allowed to sit at room temperature, there is virtually no reaction even after eight weeks. In the absence of a solvent, triphenylbismuth and benzyl chloride react vigorously to form hydrogen chloride, diphenylmethane, and inorganic bismuth. Chatt and Mann (117) have reported that they were unable to obtain tetraphenylbismuthonium bromide by the reaction between triphenylbismuth and bromobenzene in the presence of aluminum chloride, even though this type of reaction gives satisfactory yields of phosphonium, arsonium, and stibonium salts. In sharp contrast to the many failures to quaternize triarylbismuth compounds, it has been reported (119) that hexachloro-*s*-trithiane 1,1,3,3,5,5-hexaoxide reacts with an ethereal solution of triphenylbismuth to form a di-quaternary salt:

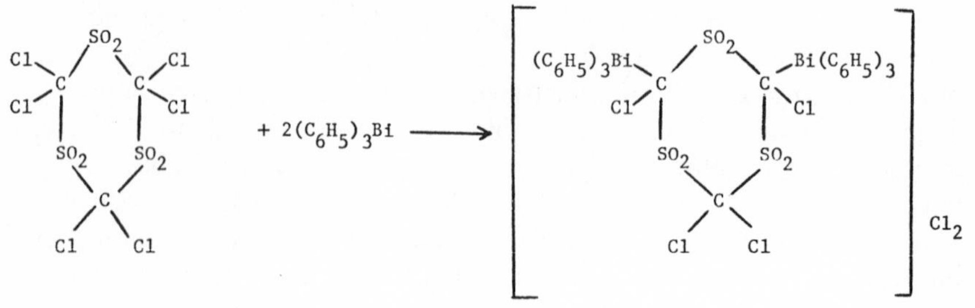

$$\text{(cyclo-}[C(Cl)_2SO_2]_3) + 2(C_6H_5)_3Bi \longrightarrow \left[\text{cyclo-}\{(C_6H_5)_3Bi(Cl)C\text{-}SO_2\text{-}C(Cl)Bi(C_6H_5)_3\text{-}SO_2\text{-}C(Cl)_2\text{-}SO_2\}\right]Cl_2$$

The preparation of bismuthonium compounds by indirect methods is discussed in Section IX-4.

The reported coordination chemistry of triarylbismuth compounds is rather meager. For example, attempts to form double salts of mercuric chloride and triarylbismuth compounds have in general been unsuccessful (7, 64); Stilp (120), however, reported in his inaugural dissertation that such double salts can be prepared and are, in fact, stable enough to be recrystallized from acetic acid without decomposition. In recent years there have been several reports that triarylbismuth compounds can form complexes with Lewis acids. Becke-Goehring and Thielemann (121) have found that triphenylbismuth reacts readily with sulfur trioxide to yield a mixture consisting largely of an inorganic bismuth sulfate as well as a small amount of a substance formulated as $(C_6H_5)_3Bi{\cdot}SO_3$. This material is reasonably stable; it melts with decomposition at 236° and can be precipitated from alcohol with petroleum ether. Hydrolysis of the adduct with hot water yields triphenylbismuth and sulfuric acid. Nuttall and co-workers (122) have described the preparation of an adduct, $(C_6H_5)_3Bi{\cdot}AgClO_4$, by mixing ethereal solutions of triphenylbismuth and silver perchlorate. The infrared spectrum of the solid adduct indicates that the perchlorate ion is free and uncoordinated. The authors (122) point out that the silver may be coordinated to the triphenylbismuth both through the bismuth atom and the π-electron systems of the aromatic rings. Benlian and Bigorgne (74) have reported that triphenylbismuth reacts with molybdenum hexacarbonyl to form $Mo(CO)_5(Ph_3Bi)$. Except for the infrared spectrum of this substance, no chemical or physical properties are described. A sluggish reaction between triphenylbismuth and iron pentacarbonyl has also been observed (123). The pyrophoric reaction product has been formulated as $(Ph_3Bi)_2Fe(CO)_4$. Attempts to prepare triphenylbismuth-substituted nickel (124) and rhodium (125) carbonyls have not been successful. The formation of an orange and a red coordination complex of niobium pentachloride with triphenylbismuth has been recently described (126). The orange complex, $(C_6H_5)_3Bi{\cdot}NbCl_5$, precipitates when niobium pentachloride and triphenylbismuth are mixed in carbon tetrachloride, *n*-hexane, or cyclohexane. The red complex, $(C_6H_5)_3Bi{\cdot}2NbCl_5$, is formed in benzene and can be isolated by evaporating the solution to dryness. Similar complexes of niobium pentachloride with the triphenyl derivatives of the other group V representative elements can also be obtained. No information about the structure of any of these coordination compounds has been published.

A substance with the formula $[(C_6H_5)_3BiMn(CO)_4]_2$ is mentioned in the patent literature (127), but no description of its preparation or properties is given.

The conversion of triphenylbismuth to an ylid has been recently reported (128). When a mixture of diazotetraphenylcyclopentadiene and triphenylbismuth is heated to 140°, the tetraphenylcyclopentadiene moiety (presumably generated as a carbene) reacts with the lone pair of electrons on the bismuth atom and forms a cyclopentadienylide:

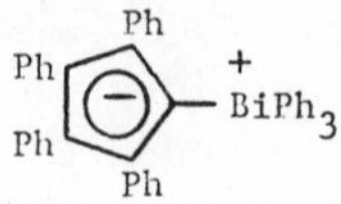

Ph Ph Ph Ph $\overset{+}{\text{BiPh}_3}$

A similar process has been used to prepare cyclopentadienylide derivatives of sulfur, nitrogen, phosphorus, arsenic, and antimony. The bismuthonium ylid (like the corresponding pyridinium ylid) is deep blue, while the phosphonium, arsonium, and stibonium ylids are yellow. In benzene solution the bismuth compound exhibits ultraviolet absorption maxima at 280, 345, and 596 mμ; in methanol the compound is red-purple and has maxima at 240, 335, and 528 mμ. These maxima are almost the same as those shown by pyridinium tetraphenylcyclopentadienylide in the same solvents. Lloyd and Singer suggest that the vacant 6*d*-orbitals of bismuth, unlike the 4*d*- or 5*d*-orbitals of arsenic or antimony, cannot effectively overlap the 2*p*-orbitals of the anionic ring (128). The bismuthonium ylid appears to be a very unstable compound, and an analytically pure sample could not be prepared by recrystallization or by chromatographic methods. Similarly, attempts to convert the ylid to a picrate or perchlorate resulted in decomposition. The compound also decomposed rapidly in methanolic sodium hydroxide and yielded tetraphenylcyclopentadiene.

The reaction of triphenylbismuth with chloramine-T to yield a bismuthonium-N-tosylimine is discussed (Section IX-5) in connection with the preparation of pentaphenylbismuth. Attempts to form a tosylimine derivative of the heterocyclic compound 5-phenyl-5*H*-dibenzobismole were, however, unsuccessful (53).

C. Physical Properties

The trialkylbismuth compounds are colorless or pale yellow liquids; and, except for the nitro derivatives, triarylbismuth compounds are colorless solids. Tribenzylbismuth has been described (63) as a solid having a faint tinge of green, while tris(*o*-chlorobenzyl)bismuth and tris(*o*-bromobenzyl)-bismuth are lemon-yellow and orange-yellow solids, respectively. Tricyclopentadienylbismuth and tris(methylcyclopentadienyl)bismuth both exist in two solid modifications, one orange and one black (24).

In contrast to the ultraviolet absorption spectra of triphenylphosphine, -arsine, and -stibine, the spectrum of triphenylbismuth does not show a distinct maximum near 250 mμ (129–131). There are, however, inflection points at 248 mμ ($\varepsilon = 1.29 \times 10^4$) and 280 m$\mu$ ($\varepsilon = 4.26 \times 10^3$). Jaffé (129) has suggested that the intense absorption in the 250 mμ region arises from transitions of electrons that are delocalized over the entire molecule; i.e., the three phenyl groups attached to a trivalent phosphorus, arsenic, antimony, or bismuth atom are conjugated with one another. The shoulder at 280 mμ probably represents an $n \rightarrow \pi^*$ transition of one of the lone-pair electrons on the bismuth atom. Like the ultraviolet absorption spectrum of triphenylbismuth, the spectrum of 1,4-phenylenebis(diphenylbismuth) has no distinct maxima but two shoulders, with inflection points at 248 mμ (log $\varepsilon = 4.60$) and 279 mμ (log $\varepsilon = 4.32$) (51).

The facile photochemical cleavage of the carbon–bismuth bond has made it difficult to obtain Raman spectra of organobismuth compounds; and, accordingly, only a few such spectra have been reported. Pai (132) determined the Raman spectrum of trimethylbismuth and found that the low-frequency region consisted of only two very strong, broad lines at 460 and 171 cm^{-1}. His analysis of the spectrum indicates that the compound possesses a pyramidal structure with the bismuth atom at the apex and the three methyl groups forming a triangular base. A number of investigators (133–135) have used Pai's data for calculations of the force constants of the Bi—C bond. Relatively recent values obtained by Siebert (135) are 1.82×10^5 dynes/cm for the stretching force constant and 0.120×10^5 dynes/cm for the bending force constant.

Recently, a Raman technique has been developed which involves the irradiation of unstable compounds with intense 4358 A mercury light at temperatures low enough to inhibit photochemical decomposition (136). This technique has been used to obtain a weak Raman spectrum of triethylbismuth at $-120°$. A very strong, barely resolved triplet at about 440, 450, and 460 cm^{-1} was assigned to Bi—C stretching; the rest of the spectrum is almost identical with that of tetraethyllead obtained under the same conditions.

The infrared spectra of trimethylbismuth and a number of dialkyl(perfluoroalkyl)- and alkylbis(perfluoroalkyl)bismuth compounds have been determined in the 700–4000 cm^{-1} region (60). Almost all the bands observed were assigned to various C—H and C—F vibrations. Weak bands in the 1253–1262 and 907–915 cm^{-1} regions were tentatively assigned to first overtones of the asymmetric and symmetric Bi—C stretch, respectively. The infrared spectrum of triphenylbismuth has been reported by a number of investigators (131, 137–141). A weak peak at 435 cm^{-1} has been tentatively assigned to the asymmetric Bi—C vibration; three sharp peaks at

216, 224, and 236 cm^{-1} are probably associated with phenyl-Bi vibrations (140). The infrared spectrum of tricyclopentadienylbismuth indicates that this molecule is pyramidal with the bismuth atom bound by essentially ionic bonds to all three rings (142). The infrared spectra of 1,4-phenylenebis-(diphenylbismuth) (51), tris(perfluorovinyl)bismuth (143), and the *cis* and *trans* isomers of tripropenylbismuth (22) have also been reported.

Allred and co-workers (144, 145) have studied the proton NMR spectra of trimethylbismuth and a large number of related compounds and concluded that the proton chemical shift is related to the electronegativity of the central atom.

Triphenylbismuth has a zero dipole moment (146–148) and, accordingly, has been assumed to be planar. An X-ray diffraction study by Wetzel (149) in 1942 appeared to agree with a planar distribution of the Bi—C bonds, and this conclusion has been widely quoted. Recent work by Hawley, Ferguson, and Harris (150), however, has shown that Wetzel's data are almost certainly in error, and hence the geometry of triphenylbismuth must be regarded as an open question. Tricyclopentadienylbismuth has a dipole moment of 1.17 D (24); this result is consistent with the pyramidal structure deduced from the infrared spectrum of the compound. Very recently, the dipole moment of 1,4-phenylenebis(diphenylbismuth) has been found (51) to be zero. This result suggests, but does not prove, that there is a planar arrangement of the carbon atoms bonded to each bismuth atom.

In sharp contrast to the low dipole moments of the triarylbismuth compounds, the dipole moment of tris(perfluorovinyl)bismuth has been found (151) to be 3.99 D. Furthermore, the dipole moment of this bismuth compound is appreciably larger than the moments of the tris(perfluorovinyl) derivatives of the other group V elements; the order of dipole moments in this series is Bi $>$ Sb $>$ As $>$ P. In the absence of structural information about these compounds, it is difficult to explain the observed trend.

Only a few thermochemical measurements have been made on organobismuth compounds. The heat of combustion of trimethylbismuth was determined by Long and Sackman (152) with a static bomb calorimeter. After corrections were made to allow for the formation of some free bismuth and carbon in the combustion, the heat of formation (ΔH_f°) of liquid trimethylbismuth at 25° was found to be 37.5 kcal/mole; i.e., the reaction between the elements to form trimethylbismuth is definitely endothermic. Since the heat of vaporization of trimethylbismuth is 8.3 kcal/mole at 25°, the heat of formation of the gaseous species is 45.8 kcal/mole. Subtracting this quantity from the sum of the heat of atomization of bismuth plus the heat of formation of three moles of methyl radicals yields 101.2 kcal/mole as the energy required to remove all the methyl radicals from one mole of gaseous trimethylbismuth. The mean dissociation energy of the Bi—C bond in trimethylbismuth is, therefore, 33.7 kcal/mole (153). Kinetic studies (154) on the thermal decomposition of trimethylbismuth indicate that the dissociation energy of the *first* Bi—C bond in this compound is 44.0 kcal/mole. It is generally found that the energy required to cleave the first metal–carbon bond in an organometallic compound (in which the valence of the metal is 2, 3, or 4) is significantly larger than the mean dissociation energy.

The heat of combustion of triethylbismuth in a Berthelot bomb calorimeter has been found by Lautsch and co-workers (155, 156) to be −1185.5 kcal/mole. This value leads to a C—Bi bond energy of 23.8 kcal/mole.

Using a static bomb calorimeter, Birr (157) has measured the heat of combustion of triphenylbismuth and has used this result to obtain 46.8 kcal/mole as the mean dissociation energy of the Bi—C bond in triphenylbismuth. There are errors in Birr's calculations, however, and Skinner (158) has recalculated the value of the dissociation energy to be 42.2 kcal/mole. The fact that the phenyl compound has a higher dissociation energy than the methyl compound has been attributed (157) to π-bonding between the bismuth atom and the benzene rings.

Long and Sackman (159) have measured the vapor pressure, melting point, and liquid density of a carefully purified sample of trimethylbismuth. The vapor pressure results were obtained over the range −25 to +15° and lead to an extrapolated normal boiling point of 109.3° and a Trouton's constant of 21.7. These data are in good agreement with some earlier work of Bamford and co-workers (160), who reported an extrapolated normal boiling point of 107.1°. The extrapolated boiling points, in turn, do not differ appreciably from several directly determined values. Long and Sackman (159) reported the melting point as −107.7°; this value is about 22° lower than the mp of −85.8° reported by Bamford and co-workers (160). No explanation for this discrepancy has been given. It is interesting, however, that Long and Sackman (159) have noted that their melting point value for trimethylstibine is about 25° lower than the corresponding value reported by Bamford and co-workers (160).

More recently, Amberger (161) has reported vapor pressure measurements on trimethylbismuth over the temperature range −58 to +107°. The data were found to fit the equation,

$$\log p = -A/T - B \log T + C$$

where p is in torr, $A = 2225.7$, $B = 2.749$, and $C = 15.8011$. Amberger's results lead to an extrapolated normal boiling point of 108.8° and a molar enthalpy of vaporization of 8.3768 kcal/mole at 107°.

The physical properties of the trivinyl derivatives of phosphorus, arsenic, antimony, and bismuth have been studied by Maier and co-workers (72, 162). Trivinylbismuth was found to have a freezing point of −124.5°, an extrapolated normal boiling point of 158.1°, and a Trouton's constant of 26.5;

the vapor pressure data fit the equation, $\log p = B - A/T$, where $A = 2499$ and $B = 8.674$.

Smith and Andrews (163) have measured the heat capacities from 100 to 320°K of triphenylbismuth and a number of other compounds containing phenyl–metal bonds. It was concluded that there is a weakening of the binding force between the phenyl group and the metal atom as the mass of the latter increases.

By means of a photoionization method, Vilesov and Zaitsev (164) have found that the first ionization potential of triphenylbismuth is 7.3 eV. This value is not significantly different from the ionization potentials of the corresponding compounds of phosphorus, arsenic, and antimony; it is, however, appreciably larger than the ionization potential of triphenylamine. It was concluded that these potentials are associated with the removal of one of the electrons of the central atom's unshared pair. The mass spectrum of triphenylbismuth is discussed in Section IV-6-C.

Forward, Bowden, and Jones (165) measured the density and surface tension of liquid triphenylbismuth and calculated the parachor of this compound to be 650.4. Since this value was found to be essentially independent of temperature, it was concluded that liquid triphenylbismuth is normal, i.e., unassociated. If Sugden's value of 190.0 for the phenyl group is used, the atomic parachor of bismuth comes out to be 80.4. This is in reasonable agreement with other values found in the literature (166). Forward and coworkers (165) also studied the phase equilibria between triphenylbismuth and each of the following compounds: triphenylmethane, triphenylamine, triphenylstibine, and tetraphenyltin.

Cryoscopic measurements in benzene have yielded normal molecular weights for triphenylbismuth (167, 168), tri-*p*-tolylbismuth (167), tri-α-naphthylbismuth (167), tris(*o*-chlorobenzyl)bismuth (63), tricyclopentadienylbismuth (24), and a number of mixed alkynyl-aryl compounds of the types $Ar_2BiC{\equiv}CH$, $Ar_2BiC{\equiv}C{-}C_6H_5$, and $Ar_2BiC{\equiv}CBiAr_2$ (58).

Parab and Desai (169, 170) have determined the diamagnetic susceptibility of several triarylbismuth compounds, and calculated the average susceptibility of the trivalent bismuth atom to be -36.98×10^{-6} cgs unit. A comparison of this result with those obtained with organoantimony compounds led to the conclusion that the Bi—C bond is more ionic than the Sb—C bond.

The nuclear quadrupole spectrum of ^{209}Bi in triphenylbismuth has been reported (171). As expected for a nucleus with a spin value of $\frac{9}{2}$, four lines were found in the spectrum. The observed frequencies were in excellent agreement with those calculated from theory.

The existence of a correlation between the molar volumes and molar polarizabilities of the triethyl derivatives of nitrogen, phosphorus, arsenic, antimony, and bismuth is mentioned in Section IV-6-C. Several methods that have been used for investigating the conformations of the triphenyl derivatives of the group V elements are discussed in Section VIII-2-C.

2. COMPOUNDS OF THE TYPE R_2BiX AND $RBiX_2$

A. Preparation

Trivalent organobismuth mono- and dihalides are known in both the alkyl and aryl series. The chlorides and bromides are readily prepared by the reaction of a trialkyl- or triarylbismuth compound with bismuth chloride or bismuth bromide, respectively (11–13, 20, 21, 56, 58). The monohalides are formed in good yields when one mole of bismuth chloride or bromide in dry ether is added to two moles of an ethereal solution of the tertiary bismuth compound:

$$2R_3Bi + BiX_3 \rightarrow 3R_2BiX$$

If the molar ratio of inorganic bismuth halide and organobismuth compound is reversed, dihalides are obtained:

$$R_3Bi + 2BiX_3 \rightarrow 3RBiX_2$$

Compounds prepared by these reactions include methylbismuth dichloride (12) and dibromide (12), ethylbismuth dichloride (12) and dibromide (12), isobutylbismuth dibromide (13), isoamylbismuth dibromide (13), phenylbismuth dibromide (11), diphenylbismuth chloride (20, 21, 56), *o*-tolylbismuth dibromide (21), α-naphthylbismuth dibromide (11, 56), di-*p*-tolylbismuth chloride (11), di-α-naphthylbismuth chloride (11), bis(*p*-methoxyphenyl)bismuth chloride (58), and bis(*p*-dimethylaminophenyl)bismuth chloride (58). The iodides can be prepared by the reaction of the corresponding chlorides or bromides with sodium or potassium iodide (4, 21, 172). Alkylbismuth diiodides can also be obtained by heating a trialkylbismuth compound with an alkyl iodide (12, 13). Another method for preparing the mono- and dihalides involves the reaction of bismuth chloride or bromide with Grignard reagents. For example, methylbismuth dibromide (173) and ethylbismuth dibromide (52) have been obtained in low yields by the addition of the corresponding Grignard reagent to an ether solution of bismuth bromide; and the patent literature (174) mentions the formation of bis(vinylphenyl)bismuth halides by the reaction of two moles of Grignard reagent with one mole of an inorganic bismuth compound. There is also a report (175) of the preparation of phenylbismuth dichloride by the action of phenylmagnesium bromide on an excess of bismuth chloride. The reported melting point (74°) of the dihalide is, however, much lower than the melting points of other arylbismuth dichlorides and is, in fact, rather close to the melting point of triphenylbismuth (77.6°). Accordingly, the preparation of phenylbismuth dichloride by means of the Grignard reaction must be regarded as uncertain.

Organometallic compounds other than Grignard reagents have also been used for the preparation of trivalent organobismuth halides. Thus, methylbismuth dibromide has been prepared by the reaction of dimethylzinc with bismuth bromide (12); tetraphenyllead has been found (176) to react with bismuth bromide to yield diphenylbismuth bromide and diphenyllead dibromide. Diethylbismuth chloride has probably been formed by the interaction of tetraethyllead and bismuth chloride, but the spontaneously inflammable product was not analyzed (177).

The cleavage of trialkyl- and triarylbismuth compounds by a variety of inorganic and organic reagents often leads to the formation of trivalent organobismuth halides. Reactions of this type are reviewed in Section IX-1-B. The decomposition of triarylbismuth dihalides to diarylbismuth halides and arylbismuth dihalides is discussed in Section IX-3-B. The formation of trivalent mono- and dihalides as intermediates in the preparation of triarylbismuth compounds from diazonium salts is mentioned in Section IX-1-A. The decomposition of tetraphenylbismuthonium tribromide at room temperature has been shown to yield diphenylbismuth bromide (cf. Section IX-4).

It has been found possible to prepare arylbismuth dihalides by the reaction of a diarylbismuth halide with a halogen or halogen halide. Thus, phenylbismuth dibromide has been obtained by the interaction of either diphenylbismuth bromide (49) or diphenylbismuth iodide (178) with bromine. Treatment of diphenylbismuth bromide with iodine monochloride yields phenylbismuth chlorobromide, $C_6H_5BiClBr$ (56).

Methylbismuth dibromide has been prepared in low yield by the copper-catalyzed reaction of bismuth metal and methyl bromide at 250° (179). It has been suggested (60) that trifluoromethylbismuth diodide and bis(trifluoromethyl)bismuth iodide are formed in trace amounts by heating trifluoroiodomethane with bismuth metal at 245° for 65 hours.

Yakubovich and co-workers (180, 181) have prepared a variety of organometallic compounds by the reaction of diazoalkanes with metal halides. When diazomethane was allowed to react with bismuth trichloride in ether or benzene, a substance was isolated that appeared to be chloromethylbismuth oxide. Even though an effort had been made to exclude moisture from the reaction vessel, it was suggested that the oxide resulted from traces of water inadvertently introduced:

$$BiCl_3 + CH_2N_2 \longrightarrow ClCH_2BiCl_2 \xrightarrow{H_2O} ClCH_2BiO$$

On vacuum evaporation of a hydrochloric acid solution of the oxide, an impure sample of chloromethylbismuth dichloride was obtained. Attempts to purify this material by recrystallization were unsuccessful.

A few compounds of the type Ar_2BiX have been prepared where X is azide, cyanide, thiocyanate, or selenocyanate. The formation of these substances is discussed in connection with the reactions of either triarylbismuth compounds (Section IX-1-B-(2)) or compounds of the types Ar_3BiX_2 and $Ar_3Bi(OH)X$ (Section IX-3-B). Diphenylbismuth cyanide and diphenybismuth thiocyanate have also been prepared by metathetical reactions between diphenylbismuth bromide and potassium cyanide and lead thiocyanate, respectively (87). Only one trivalent organobismuth nitrate has been characterized. This is ethylbismuth dinitrate, which was prepared by the reaction of alcoholic silver nitrate with ethylbismuth diiodide (182).

B. Reactions

The trivalent organobismuth halides are, in general, reactive and unstable solids. They are usually decomposed by water or alcohol, although several *ortho*-substituted aryl compounds have been obtained which appear to be unaffected by moisture and which can be recrystallized from alcohol (36). Alkyl compounds of the type R_2BiX are particularly sensitive substances; they are spontaneously inflammable in air and may decompose on standing even when precautions are taken to exclude water and oxygen. Diarylbismuth compounds should be handled with caution, since some of them are known to be powerful sternutators (183). Thus, diphenylbismuth chloride, bromide, and cyanide are probably more powerful than the chemical warfare agent, diphenylchloroarsine. Diphenylbismuth propionate, bromoacetate, and iodoacetate are barely tolerable at a concentration of 1 part in 2.5×10^7 parts of air, and produce unpleasant after-effects.

The trivalent arylbismuth iodides appear to be much less reactive than the corresponding bromides or chlorides. Thus, the preparation of diphenylbismuth chloride and bromide must be carried out under strictly anhydrous conditions, while one method for crystallizing diphenylbismuth iodide involves pouring an alcoholic solution of the compound into boiling water (172). Treatment of the iodide with liquid ammonia for three hours yields 21% triphenylbismuth, but 65% of the iodide can be recovered unchanged (55). In contrast, a similar experiment with diphenylbismuth bromide results in considerable decomposition to inorganic bismuth; no triphenylbismuth is obtained, and only 38.2% of the starting material can be recovered. Also, the iodide is reduced only slowly with hydrazine hydrate to yield 61.2% of triphenylbismuth; diphenylbismuth chloride is converted under the same conditions almost instantaneously to a 93.5% yield of the tertiary compound.

There are several reports of the conversion of the mono- and dihalides to oxides or hydroxides. Marquardt (12) found that the latter types of compounds could be obtained from the zinc bromide double salts of methylbismuth dibromide and dimethylbismuth bromide. Thus, when an alcoholic solution of the methylbismuth dibromide double salt is treated with ammonia, an amorphous solid, CH_3BiO, is formed. This oxide appears to be amphoteric; i.e., it is insoluble in water or aqueous ammonia but soluble in both dilute sodium hydroxide and dilute nitric acid. It is very easily oxidized by air. Dimethylbismuth hydroxide, $(CH_3)_2BiOH$, has been prepared by the hydrolysis of the double salt of dimethylbismuth bromide and zinc bromide. The hydroxide is a spontaneously inflammable solid. On treatment with hydrochloric acid it yields methane; with methyl iodide it forms methylbismuth diiodide. The action of ammonia on an alcoholic solution of diphenylbismuth bromide appears to give an unstable diphenylbismuth hydroxide, which rapidly decomposes to give triphenylbismuth and bismuth hydroxide (6). Lecoq (184) has reported that phenylbismuth dichloride (Section IX-2-A) reacts instantly with water to give a white, granular precipitate, C_6H_5BiO, which is soluble in alcohol and alkalies and which gives a normal molecular weight. The possible conversion of chloromethylbismuth dichloride to the corresponding oxide has been previously discussed (cf. Section IX-2-A).

Diethylbismuth ethoxide has been prepared by the reaction of diethylbismuth bromide with sodium ethoxide in absolute alcohol; the product was purified by sublimation *in vacuo* (62). A trivalent organobismuth mercaptide, $(C_6H_5)_2BiSC_6H_5$, has also been reported (84). This was obtained by adding diphenylbismuth chloride to an ethereal solution of thiophenol and recrystallizing the product from benzene.

Kupchik and Theisen (185) have recently reported that phenylbismuth bis(dialkyldithiocarbamates) can be prepared by allowing a stoichiometric quantity of phenylbismuth dibromide to react with a sodium dialkyldithiocarbamate in chloroform at 25°. When diphenylbismuth chloride was used, however, the expected diphenylbismuth dialkyldithiocarbamates could not be isolated; instead, phenylbismuth bis(dialkyldithiocarbamates) were again obtained. This result illustrates the ease with which the C—Bi bond is sometimes cleaved. The molecular weights of the bis(dialkyldithiocarbamates) in benzene, chlorobenzene, or camphor were found to be somewhat higher than calculated, but the authors (185) did not suggest that the compounds were associated. The infrared spectra contain a strong band near 1500 cm^{-1} that was attributed to the stretching frequency of the partial C=N bond of the following canonical form:

$$C_6H_5Bi(S\underset{\substack{|\\ S^-}}{C}{=}\overset{+}{N}R_2)_2$$

The ultraviolet spectra show intense absorption near 260 mμ (ε, 6×10^4). This band is apparently not associated with the phenyl group but with the dialkyldithiocarbamate residue. Thus, bismuth tris(diethyldithiocarbamate) absorbs at 260 mμ (ε, 7.06×10^4).

Diphenylbismuth halides in liquid ammonia react with two equivalents of lithium, sodium, potassium, calcium, or barium to form deep red, highly reactive compounds of the type $(C_6H_5)_2BiM$ or $(C_6H_5)_2BiM_{1/2}$ (55). When only one equivalent of metal is used, a deep-green solution results, which changes to red when more metal is added. It has been suggested (55) that the green color is due either to the diphenylbismuth radical or to tetraphenyldibismuth, $(C_6H_5)_2BiBi(C_6H_5)_2$. The electrolytic reduction of diphenylbismuth chloride in anhydrous glyme (1, 2-dimethoxyethane) has also been investigated (186). When the electrolysis is conducted at -1.2 volts, one electron is involved and a pale yellow solution is obtained. This result has been attributed (186) to the formation of tetraphenyldibismuth by the rapid coupling of diphenylbismuth radicals. Further reduction of the solution involves two electrons per mole of tetraphenyldibismuth and gives rise to a green solution. This solution presumably contains the diphenylbismuth anion, which is unstable and deposits bismuth metal.

It is of interest that all attempts to isolate tetraphenyldibismuth have failed, although the corresponding arsenic and antimony compounds are well known. Thus, Blicke and his co-workers (172) were unsuccessful in obtaining the dibismuth compound by the reaction of diphenylbismuth iodide with silver, mercury, zinc, copper bronze, or sodium hypophosphite. In the case of the first two metals, black precipitates were formed. Sodium hypophosphite at room temperature did not seem to react with an alcoholic solution of the bismuth iodide, while a black product was produced on warming. None of these black materials was characterized. The apparent instability of the dibismuth compounds may be associated with the fact that the Bi—Bi bond strength is only 25 kcal/mole while the As—As and Sb—Sb bond strengths are 32.1 kcal/mole and 30.2 kcal/mole, respectively (187). Dessy and co-workers (188) have reported that reductive cleavage of the metal–metal bond requires lower cathodic potentials as one goes from arsenic to antimony to bismuth.

The polarographic reduction of diphenylbismuth iodide in aqueous alcohol produces a single wave with the diffusion current directly proportional to the concentration of the bismuth compound (189). The observed half-wave potential is a function of the pH; when the solution is $1N$ in hydrochloric acid, the potential at 25° is -0.166 volt (relative to the saturated calomel electrode) (190).

The reaction of phenylbismuth dibromide or diphenylbismuth chloride in ether with lithium borohydride or lithium aluminum hydride at low temperatures has been found (191) to yield black, ether-insoluble, polymeric substances the composition of which corresponds to $(C_6H_5Bi)_n$; biphenyl and triphenylbismuth were also formed in the reaction between the diphenylbismuth halide and either reducing agent. The black material is easily oxidized. It reacts with air or oxygen to yield a white solid formulated as $(C_6H_5BiO)_n$; this substance is presumably not the same as the monomeric oxide described by Lecoq (cf. Section IX-2-A). The black material reacts with bromine vapor to give a quantitative yield of yellow, crystalline phenylbismuth dibromide (identified by its bismuth analysis and melting point).

The reduction of methylbismuth dichloride or dimethylbismuth chloride with lithium aluminum hydride at low temperatures gives methylbismuthine, CH_3BiH_2, and dimethylbismuthine, $(CH_3)_2BiH$, respectively (161). Both hydrides are unstable at room temperature and disproportionate to bismuthine, BiH_3, and trimethylbismuth.

Nesmeyanov and co-workers (192) have prepared several compounds containing rhenium-metal bonds by the reaction of sodium rhenium pentacarbonyl with organometallic halides. Included in their study was the following reaction:

$$(C_6H_5)_2BiCl + NaRe(CO)_5 \rightarrow (C_6H_5)_2BiRe(CO)_5$$

The yield of diphenylbismuth rhenium pentacarbonyl thus obtained was 66%. The compound proved stable enough to be sublimed (40° at 0.01 mm), but it decomposed when dissolved in polar organic solvents.

A large number of compounds containing cobalt–metal bonds have been prepared by the treatment of organometallic halides with bis(dimethylglyoximato)cobalt compounds of the type $HCo(D_2H_2)B$, where B is a base and D is the dianion of dimethylglyoxime (193). Among these compounds is a pyridine complex of $(C_6H_5)_2Bi(CoD_2H_2)$. This material is stable toward oxygen, soluble in several common organic solvents, and resistant to alkali. The bismuth–cobalt bond can be cleaved by oxidation with bromine or reduction with sodium borohydride; the compound is also decomposed by strong acids. It has been stated (193) that the thermal stability of the compounds containing elements of the fifth group increases with increasing atomic weight of the element, but no experimental evidence has been presented.

Makarova (194) has investigated the reaction of mercuric chloride with diphenylbismuth chloride. When an aqueous alcoholic solution of the two compounds is boiled for about 30 minutes, an almost quantitative yield of phenylmercuric chloride is obtained:

$$(C_6H_5)_2BiCl + HgCl_2 + H_2O \rightarrow C_6H_5HgCl + BiOCl + C_6H_6 + HCl$$

In alkaline medium the reaction yields some triphenylbismuth and a little phenylmercuric chloride. In no case was any diphenylmercury detected as a reaction product.

Okawara and co-workers (195) have described the preparation of pyridine adducts of phenylbismuth dihalides by means of two different methods. One method simply consists of the addition of pyridine to the dihalide:

$$C_6H_5BiX_2 + 2C_6H_5N \rightarrow C_6H_5BiX_2{\cdot}2C_6H_5N$$

The other method involves the cleavage of a phenyl group from a diphenylbismuth halide:

$$2(C_6H_5)_2BiX + 2C_6H_5N \rightarrow C_6H_5BiX_2{\cdot}2C_6H_5N + (C_6H_5)_3Bi$$

Roper and Wilkins (196) have investigated the conductivity of a considerable number of 2,2′-dipyridyl adducts of group V halides. Included in this study is a 1:1 adduct of 2,2′-dipyridyl and phenylbismuth dibromide. This adduct, which was prepared by mixing benzene solutions of the ligand and the dibromide, was found to be a relatively weak electrolyte in nitrobenzene. The authors (196) also characterized a dipyridinium salt of the complex anion $C_6H_5BiBr_3^-$. Attempts to obtain a reaction between 2,2′-dipyridyl and diphenylbismuth chloride or iodide were unsuccessful.

Silicon-containing amino derivatives of bismuth have been recently prepared via trivalent organobismuth halides. These reactions are discussed in Section III-5.

The conversion of trivalent organobismuth halides to unsymmetrical tertiary compounds is described in Section IX-1-A.

3. COMPOUNDS OF THE TYPE R_3BiX_2 AND $R_3Bi(OH)X$

A. Preparation

Triarylbismuth dihalides are among the most readily available and best known organic compounds of bismuth. The bromides and chlorides are usually prepared by the addition of the stoichiometric quantity of bromine or chlorine to an ice-cold solution of the triarylbismuth in chloroform (49), carbon tetrachloride (25), or petroleum ether (20). The yields are often greater than 90%. Sulfuryl chloride (11, 197), sulfur monochloride (197), thionyl chloride (197), and iodine trichloride (91, 167) also yield dichlorides upon reaction with the trivalent bismuth compounds, but some cleavage of the carbon–bismuth bond occurs simultaneously. Triphenylbismuth difluoride has been prepared by the interaction of the dichloride and potassium fluoride (87). No triarylbismuth diiodide has ever been isolated. Triphenylbismuth diiodide is probably formed at $-78°$ by the reaction of iodine with triphenylbismuth in ether solution (56, 86); when the mixture is warmed, however, the diiodide decomposes to give diphenylbismuth iodide and iodobenzene. Feigl and co-workers (198, 199) have shown that triphenylbismuth reacts with iodine vapor to form a brown substance and that this reaction can be made the basis of a sensitive spot test for triphenylbismuth. The brown substance, they suggested, is a polyiodide of the type $(C_6H_5)_3$-$BiI_2 \cdot xI_2$. No evidence in support of this suggestion, however, has yet been presented. The results of Beveridge, Harris, and Inglis (200) indicate that the reaction of triphenylbismuth with excess iodine in acetonitrile does not produce a polyiodide (cf. Section IX-3-C).

The dichlorides and dibromides react with the appropriate silver salts to give a variety of triaryl compounds of the type Ar_3BiX_2, where X is an inorganic or organic anion. In this way, dinitrates (6, 25, 57), dicyanates (201), diazides (92), dihydroxides (92), and dibenzoates (55) have been prepared. Several triarylbismuth diacetates have been prepared by the reaction between a dihalide and lead(II) acetate (57). Sodium salts of organic acids can be used in place of silver or lead salts (202). Triphenylbismuth carbonate has been obtained by treating a suspension of the dibromide in alcohol with an excess of solid sodium hydroxide and then passing carbon dioxide into the reaction mixture (6). The sulfate has been prepared by dissolving the dichloride in cold, concentrated sulfuric acid, drawing air through the solution to remove hydrogen chloride, and then pouring the mixture on ice (167, 197). Attempts to prepare triphenylbismuth diperchlorate by the reaction between the dichloride (or dibromide) and two equivalents of silver perchlorate have resulted in the formation of tetraphenylbismuthonium perchlorate (203). The mechanism of this rearrangement has not been elucidated.

The dihydroxides and carbonates prepared by the methods described above can be readily converted into other compounds of the type Ar_3BiY_2 by treatment with the appropriate organic or inorganic acid. Oxidation methods are often useful for preparing diacetates, dibenzoates, and dinitrates. Thus, triphenylbismuth diacetate has been prepared by the reaction of triphenylbismuth with lead tetraacetate (20, 53); triphenylbismuth dibenzoate, by the reaction of triphenylbismuth with benzoyl peroxide (201); and triphenylbismuth dinitrate, by the reaction of triphenylbismuth with benzoyl nitrate (86). Attempts to prepare triphenylbismuth oxide or dihydroxide by the oxidation of triphenylbismuth have failed. Thus, when triphenylbismuth was treated with hydrogen peroxide, an exothermic reaction occurred and an unidentified, infusible solid was obtained (204). Similarly, the reaction of 2,4,4-trimethyl-1-pyrroline-1-oxide with triphenylbismuth resulted in extensive decomposition of the bismuth compound and the formation of a considerable amount of benzene (205). In contrast, triphenylphosphine and triphenylarsine were converted to oxides by the pyrroline-1-oxide; triphenylstibine, like triphenylbismuth, was not effective as an oxygen acceptor in this reaction.

Challenger and co-workers have reported the preparation of compounds of the type $Ar_3Bi(OH)X$, where X is chloride (92, 167), bromide (167), or cyanide (92). The hydroxychlorides and hydroxybromides were obtained by several methods: by passing moist ammonia gas into a chloroform solution of the triarylbismuth dihalide; by warming sodium with a solution of the dihalide in moist ether; by the action of aqueous ammonia on the dihalide; and by dissolving a triarylbismuth sulfate in ammonia and treating the solution with hydrochloric acid. The hydroxycyanides were prepared by treatment of the halide with aqueous potassium cyanide and recrystallization of the product from a mixture of chloroform and petroleum ether. Solomakhina (91) has isolated triphenylbismuth hydroxychloride as a by-product from the reaction of triphenylbismuth with iodine trichloride.

B. Reactions

Except for the metathetical reactions discussed above, compounds of the R_3BiX_2 type undergo few reactions that do not involve cleavage of the carbon–bismuth bond. Pentaphenylbismuth was first prepared by the reaction of triphenylbismuth dichloride with phenyllithium at $-75°$ (206). The nuclear nitration of triarylbismuth dinitrates has been carried out successfully at relatively low temperatures. Thus, triphenylbismuth dinitrate on treatment with fuming nitric acid at 0° yields a mixture of tris(nitrophenyl)bismuth dinitrates (207). The orientation of the nitro groups is approximately 86% *meta*, 12% *ortho*, and 2% *para*. In other work (86) dinitro, tetranitro, and hexanitro derivatives of triphenylbismuth dinitrate have been reported; no information concerning the structure of these compounds has been given. Trinitro derivatives of tri-*o*-tolyl- (57), tri-*p*-tolyl- (57, 208), and tris(*p*-carbomethoxyphenyl)bismuth dinitrates (57) have also been prepared by nitration.

Supniewski (209) has reported the nuclear sulfonation of tri-*o*-tolylbismuth dichloride. He isolated a mono- and a disulfo derivative of tri-*o*-tolylbismuth sulfate. Adams and Supniewski (57) obtained a trisulfo derivative which was converted to tris(hydroxy-2-tolyl)bismuth hydroxychloride. The methyl groups of tri-*o*-tolyl- and tri-*p*-tolylbismuth dichlorides can be oxidized to carboxy groups by means of potassium permanganate, lead tetraacetate, or chromic acid.

Attempts to cleave the ether linkage in tris(*p*-methoxyphenyl)bismuth dibromide with either hydriodic acid or aluminum chloride have resulted either in extensive decomposition or recovery of the unchanged starting material (57). The bromination of tri-*o*-tolyl- or tri-*p*-tolylbismuth dibromide in the presence of ultraviolet light caused cleavage of the carbon–bismuth bond and the formation of bismuth bromide and *o*-bromo- or *p*-bromobenzyl bromide (57). Treatment of triphenylbismuth hydroxychloride with hydrobromic acid did not give the expected triphenylbismuth chlorobromide (178). Instead, equimolar quantities of triphenylbismuth dichloride and dibromide were isolated.

The reaction between triarylbismuth dihalides and Grignard reagents was first investigated by Challenger and co-workers (49, 167) in attempts to prepare tetraarylbismuthonium halides. The desired products, however, were never isolated. In the case of the reaction between triphenylbismuth dibromide and phenylmagnesium bromide, the only compounds obtained were triphenylbismuth, diphenylbismuth bromide, phenylbismuth dibromide, and bromobenzene. It was noted by Challenger (49) that an intense but transitory purple color developed during the course of these Grignard reactions. Gilman and Yablunky (210) showed that this purple color could be made the basis of a sensitive test for aryl Grignard reagents and other active arylmetallics such as aryllithiums and arylpotassiums. The purple color can probably be attributed to the formation of an unstable, purple pentaarylbismuth compound (cf. Section IX-5). No color is given by any alkylmetallic compound or by relatively unreactive arylmetallic compounds, e.g., tetraphenyllead. Steric factors either in the bismuth compound or the reactive organometallic reagent decrease the intensity of the purple color or may even prevent its formation. For example, tri-*o*-tolylbismuth dichloride gives a weak purple color with phenylmagnesium bromide or with *p*-tolylmagnesium bromide and a negative test with α-naphthylmagnesium bromide. Trimesitylbismuth dichloride and tri-α-naphthylbismuth dibromide give negative tests with Grignard reagents. Trivalent organobismuth halides (e.g., diarylbismuth halides and arylbismuth dihalides) also fail to give the color test.

The reduction of triarylbismuth dihalides to the corresponding triarylbismuth compounds has been carefully studied (57, 211). Hydrazine hydrate appears to be the reagent of choice for this reaction (211). Another good procedure involves solution of the dihalide in acetone and treatment of the cooled mixture with concentrated, aqueous sodium hydrosulfite (57). This method has been used to prepare tris(nitrophenyl)bismuth and tris(nitro-4-tolyl)bismuth. It has also proved useful in obtaining very pure tri-*m*-tolylbismuth. When this compound is prepared by the Grignard reaction, it is contaminated with traces of toluene and *m*-bromotoluene which are difficult to remove. The crude compound is converted in the usual manner into tri-*m*-tolylbismuth dichloride, which is purified by recrystallization from acetone and is then reduced with sodium hydrosulfite to tri-*m*-tolylbismuth. Attempts to prepare the pentavalent hydride $(C_6H_5)_3BiH_2$ by the reduction of triphenylbismuth dichloride with either lithium borohydride or lithium aluminum hydride have been unsuccessful (212). The only organic product isolated was triphenylbismuth. It has recently been found that the reaction of triphenylbismuth dichloride and a sodium dialkyldithiocarbamate gives triphenylbismuth and a high yield of a tetraalkylthiuram disulfide (185). The electrolytic reduction of triphenylbismuth dibromide has been found to be a one-step, two-electron process which involves loss of bromine as bromide ion and formation of triphenylbismuth (186):

$$Ph_3BiBr_2 + 2e^- \rightarrow Ph_3Bi + 2Br^-$$

Nefedov and co-workers (213, 214) have studied the rate of isotope exchange between triphenylbismuth (containing radioactive bismuth) and triphenylbismuth dichloride in alcoholic solution. They found that the exchange is first order with respect to each component and that the activation energy is 15.9 kcal/mole. Since the analogous antimony reaction has a similar rate, it was concluded (214) that the bond energy of 5*s* electrons in antimony is close to that of 6*s* electrons in bismuth.

The carbon–bismuth bond in the triarylbismuth dihalides is easily cleaved. The stability of the dihalides is in the order: difluorides > dichlorides > dibromides > diiodides (87). Two types of decomposition appear to be possible (87):

$$(1) \quad Ar_3BiX_2 \rightarrow Ar_2BiX + ArX$$

$$(2) \quad 2Ar_3BiX_2 \rightarrow Ar_3Bi + ArBiX_2 + 2ArX$$

At higher temperatures the triarylbismuth formed in the second type of reaction may decompose in the following way (61, 87):

$$2Ar_3Bi \rightarrow 3Ar\text{—}Ar + 2Bi$$

As previously mentioned, triarylbismuth diiodides are not stable at room temperature but decompose spontaneously to diarylbismuth iodides and aryl iodides. Triarylbismuth dibromides are more stable but apparently undergo a similar type of decomposition on heating. Thus, triphenylbismuth dibromide on boiling with benzene gives a good yield of diphenylbismuth bromide (49). If the pure dibromide is heated for a few minutes at 100°, it decomposes to a yellow, viscous material which apparently contains bromobenzene (49). Similarly, tri-α-naphthylbismuth dibromide on being heated at 100° for one-half hour yields α-bromonaphthalene, tri-α-naphthylbismuth, α-naphthylbismuth dibromide, and inorganic bismuth which was isolated as the oxybromide (11). Triphenylbismuth dichloride decomposes only slowly in boiling benzene (56, 87); the pure dichloride melts at 141° and decomposes at 150° (87). When the dichloride is refluxed with an aqueous alcoholic solution of mercuric chloride for 25–35 minutes, a nearly quantitative yield of phenylmercuric chloride is obtained (194). The reaction can apparently be represented by the following equation:

$$(C_6H_5)_3BiCl_2 + HgCl_2 + H_2O \rightarrow C_6H_5HgCl + BiOCl + 2C_6H_6 + Cl_2$$

In alkaline medium the reaction of the dichloride with mercuric oxide gives triphenylbismuth. In no case was the formation of diphenylmercury observed. By contrast, tris(*p*-bromophenyl)antimony dichloride fails to react with mercuric chloride in neutral, alcoholic solution, while in alkaline medium it gives a 75% yield of bis(*p*-bromophenyl)mercury. Triphenylbismuth difluoride is stable up to 200° (87). At 250–260° it decomposes rapidly to form fluorobenzene, triphenylbismuth, biphenyl, and, presumably, inorganic bismuth.

Like the dihalides other compounds of the type Ar_3BiX_2 are also thermally unstable. Triphenylbismuth dihydroxide decomposes violently at 100–120° and deflagrates when heated in a flame (92). In solvents such as water, acetone, ether, petroleum ether, chloroform, and alcohols the dihydroxide decomposes even in the cold to yield bismuth hydroxide and triphenylbismuth. With ethyl, *n*-propyl, or isopropyl alcohols, oxidation apparently takes place, and the corresponding carbonyl compound as well as benzene can be detected in each case (92). Triphenylbismuth diazide decomposes at 100° to yield phenyl azide, triphenylbismuth, and diphenylbismuth azide (92); the latter compound is also obtained from hot solutions of the hydroxyazide in organic solvents. Treatment of the diazide with hydrochloric acid gives hydrazoic acid, benzene, and bismuth chloride. The carbon–bismuth bond in triphenylbismuth hydroxycyanide is also easily cleaved. In hot alcohol the hydroxycyanide gives triphenylbismuth, diphenylbismuth cyanide, benzonitrile, and inorganic matter; in boiling water the compound also loses benzonitrile readily (92).

Attempts to prepare triarylbismuth dithiocyanates or diselenocyanates by metathetical reactions of triarylbismuth dihalides have yielded diarylbismuth thiocyanates (87) or selenocyanates (215), respectively. From these results Challenger and co-workers (87, 215) have concluded that the dithiocyanates and diselenocyanates must be extremely unstable compounds. Triarylbismuth dicyanides have also never been isolated. When triphenylbismuth dichloride was heated with an alcoholic solution of potassium cyanide, diphenylbismuth cyanide was obtained (87). When aqueous solutions of potassium cyanide were allowed to react with either triphenylbismuth dichloride or dihydroxide, triphenylbismuth hydroxycyanide was formed (92). Heating a dry mixture of the dichloride with two molecular proportions of potassium cyanide produced benzonitrile at temperatures above 130°.

Radioactive ^{210}Bi has been obtained by irradiating triphenylbismuth dichloride with slow neutrons, dissolving the irradiated product in ether, and extracting with dilute hydrochloric acid containing bismuth chloride as a carrier (216). The yield of radioactive bismuth was 80%.

It has been reported (112–115) that organic compounds of ^{210}Po are formed as the result of the β-decay of ^{210}Bi incorporated in triarylbismuth dihalides (prepared by utilizing the β-decay of ^{210}Pb in tetraaryllead compounds). The ^{210}Po compounds isolated by means of paper or thin-layer chromatography were tentatively identified as diarylpolonium dichlorides, triarylpolonium chlorides, and, in one case, a diarylpolonium, $(p\text{-}CH_3C_6H_4)_2Po$.

Triphenylbismuth dichloride prepared from an alpha-emitting ^{210}Bi isomer has been used as a source of carrier-free ^{206}Tl (111).

C. Physical properties

In general, the pentavalent bismuth compounds discussed in this section are relatively stable crystalline solids which are usually soluble in chloroform, dioxane, benzene, and acetone, and insoluble in water, alcohol, and ether.

The crystal structure of triphenylbismuth dichloride has been studied by means of X-ray diffraction by a number of investigators (150, 217, 218). The results obtained by Hawley, Ferguson, and Harris (150) clearly show the trigonal bipyramidal arrangement of the molecule. The bismuth atom and the three carbon atoms to which it is bonded are coplanar, and the Bi—Cl bonds are normal to this plane. The inclinations of the benzene rings to this plane are 31°, 70°, and 79°. The mean Bi—Cl distance is 2.605 A, which is significantly longer than the Bi—Cl bond in bismuth trichloride but comparable with the Bi—Cl distances in the $BiCl_5^{-2}$ ion. The mean Bi—C distance is 2.24 A. The unit cell contains eight molecules of the dichloride.

Beveridge, Harris, and Inglis (200) have made a thorough study of the electrolytic conductance in acetonitrile of compounds of the type Ph_3MX_2 where X is a halogen and M is P, As, Sb, or Bi. The molar conductance values for triphenylbismuth dichloride and dibromide were found to be very low ($\Lambda_m < 1.0$) and to vary little with concentration. When bromine was added to a solution of triphenylbismuth in acetonitrile, there was little change in the conductance (which was about zero) until the molar ratio of bromine to triphenylbismuth was slightly greater than 1:1. At this point the solution acquired a yellow color, and there began a regular but very slight rise in conductance up to and beyond the 2:1 ratio without break. These results were attributed to the conductance of bromine in acetonitrile; the experiment provided no evidence for the formation of triphenylbismuth tetrabromide.

When iodine was added at room temperature to a solution of triphenylbismuth in acetonitrile, the conductance rose hardly at all above that of the initial solution. From the beginning of the titration, the solution acquired a yellow color which deepened progressively; when the molar ratio of iodine to triphenylbismuth reached 3:1, black bismuth triiodide precipitated. The overall equation is probably as follows:

$$Ph_3Bi + 3I_2 \rightarrow BiI_3 + 3PhI$$

A similar reaction in ether has been shown (86) to go through the intermediate formation of diphenylbismuth iodide and phenylbismuth diiodide.

When the titration with iodine was repeated at −35°, the system behaved differently. The solution remained colorless until the molar ratio of iodine to triphenylbismuth was 1:1, and the conductometric titration pattern was similar to that described above for triphenylbismuth plus bromine. It was concluded that triphenylbismuth diiodide is formed in acetonitrile at −35° and possesses a very low molar conductance. The results also indicate that the iodide does not react with further quantities of iodine to produce the tetraiodide.

The reaction of triphenylbismuth with iodine bromide in acetonitrile has also been investigated conductometrically. The solution of triphenylbismuth showed virtually no change in conductance until the 2:1 (IBr to Ph_3Bi) ratio was reached at which point the conductance began to rise slowly. The color of the solution was pale yellow from the beginning of the titration, and became markedly darker after the 2:1 ratio was reached. When the molar ratio of iodine bromide to triphenylbismuth was 1:1, diphenylbismuth bromide could be isolated from the solution; when the ratio was 2:1, phenylbismuth dibromide could be isolated. It was concluded that the reaction between triphenylbismuth and iodine bromide occurs in two stages:

$$(1) \quad Ph_3Bi + IBr \rightarrow Ph_2BiBr + PhI$$

$$(2) \quad Ph_2BiBr + IBr \rightarrow PhBiBr_2 + PhI$$

The increase in conductance after the 2:1 ratio was attributed to the conductance of excess iodine bromide in acetonitrile.

Jensen (219) has found that the molar polarization of triphenylbismuth dichloride in benzene is independent of temperature; accordingly, the compound has no dipole moment. A previous report (220) that the dichloride has a dipole moment of 1.17 D is apparently in error caused by the extraordinarily large (about 30 cm^3 per mole) atomic polarization of the compound. Jensen's result (219) is consistent with the assumption that the geometry of the dichloride in benzene solution is similar to that found in the solid state.

The literature contains relatively little information concerning the spectral properties of pentavalent bismuth compounds. The infrared spectra of triphenylbismuth dichloride and dibromide have been reported (131, 140). The spectrum of the latter compound contains strong, sharp bands at 219 cm^{-1} and 242 cm^{-1} which were assigned to Bi—Br vibrations. The ^{19}F NMR spectrum of triphenylbismuth difluoride shows that the fluorines are spectroscopically equivalent (221). The spectrum was found to be essentially temperature independent and relatively insensitive to solvent and concentration. The ^{19}F chemical shift is 81 ppm upfield with respect to trifluoroacetic acid used as an external reference.

Two groups of investigators (167, 168) have shown by means of cryoscopic measurements that triphenylbismuth dichloride is monomolecular in benzene. Triphenylbismuth difluoride is also monomeric in this solvent (221).

The diamagnetic susceptibilities of several triarylbismuth dihalides have been determined by Parab and Desai (169, 170). The average molar susceptibility of the pentavalent bismuth atom was found to be -30.48×10^{-6} cgs unit. Comparison of this value with the corresponding value (-15.50×10^{-6} cgs unit) for the pentavalent antimony atom led the authors to conclude that the Bi—C bond is more ionic than the Sb—C bond.

The optic axial angles of triphenylbismuth dichloride have been measured as a function of both wavelength and temperature by Greenwood (222) and by Bryant (223). The former author (222) reported that the crystalline dichloride can rotate the plane of polarization. Bryant (223), however, was unable to confirm this observation.

4. BISMUTHONIUM COMPOUNDS

In spite of numerous attempts by earlier investigators (11–13, 49, 64, 116, 117, 167, 224), it was not until 1952 that the first bismuthonium compounds were prepared by Wittig and Clauss (206). These workers found that treatment of pentaphenylbismuth in ether at $-70°$ with one mole of bromine in carbon tetrachloride gave the highly unstable, colorless, crystalline tetraphenylbismuthonium bromide:

$$(C_6H_5)_5Bi + Br_2 \rightarrow (C_6H_5)_4BiBr + C_6H_5Br$$

With two moles of bromine, the reaction yielded tetraphenylbismuthonium tribromide as an orange-colored, crystalline powder, which was quickly isolated and gave a good analysis for bismuth.

Tetraphenylbismuthonium chloride was prepared by treatment of pentaphenylbismuth at $-70°$ with the stoichiometric quantity of dry hydrogen chloride in ether:

$$(C_6H_5)_5Bi + HCl \rightarrow (C_6H_5)_4BiCl + C_6H_6$$

The chloride was colorless and gave good analytical results when it was filtered off at low temperatures, washed with cold ether, and dried *in vacuo* at −30°. Although the solid chloride decomposed in a few minutes at room temperature, it was reasonably stable in water for several days at 20°. Treatment of an aqueous solution with sodium nitrate, sodium perchlorate, or sodium tetraphenylborate yielded the relatively stable tetraphenylbismuthonium nitrate, perchlorate, and tetraphenylborate. The latter compound was also prepared by the reaction of pentaphenylbismuth and triphenylboron:

$$(C_6H_5)_5Bi + (C_6H_5)_3B \rightarrow [(C_6H_5)_4Bi][(C_6H_5)_4B]$$

The nitrate is moderately soluble in water and, because of its stability, may be a convenient source of an aqueous solution of the tetraphenylbismuthonium cation. The sulfate appears to be very soluble in water and has not been isolated in the solid state.

Tetraphenylbismuthonium chloride could not be converted by metathesis into the corresponding cyanide or nitrite. Treatment of an aqueous solution of the chloride with potassium cyanide gave benzonitrile and triphenylbismuth; similarly, the chloride plus sodium nitrite yielded nitrobenzene and triphenylbismuth.

It has been recently found (203) that the reaction between triphenylbismuth dichloride (or dibromide) and two equivalents of silver perchlorate results in the formation of tetraphenylbismuthonium perchlorate. This reaction is remarkable since metathesis between triarylbismuth dihalides and other silver salts has been used many times previously without any rearrangement being noted. It is also of interest since it offers a comparatively simple route to the preparation of the tetraphenylbismuthonium ion. This cation, by analogy with the corresponding arsonium and stibonium cations, may prove to be valuable as a precipitant in analytical chemistry. When tetraphenylbismuthonium perchlorate is dissolved in absolute alcohol and warmed with an equimolar amount of sodium tetraphenylborate, a voluminous white precipitate of tetraphenylbismuthonium tetraphenylborate is obtained.

The preparation of a di-quaternary salt by the direct quaternization of triphenylbismuth by hexachloro-*s*-trithiane 1,1,3,3,5,5-hexaoxide has been previously mentioned (Section IX-1-B-(2)). This salt is described (119) as a colorless, amorphous substance, mp 110–112°. No other information about its chemical and physical properties is available.

Dötzer (225) has recently reported the synthesis of a tetraalkylbismuthonium salt via a complex of triethylbismuth and triethylaluminum:

$$(C_2H_5)_3Bi{:}Al(C_2H_5)_3 + CH_3Cl \rightarrow [(CH_3)(C_2H_5)_3Bi][(C_2H_5)_3AlCl]$$

When the bismuthonium salt thus prepared was treated with a toluene suspension of sodium tetraphenylborate, a 48% yield of methyltriethylbismuthonium tetraphenylborate was obtained. The latter substance was colorless and could be recrystallized from aqueous acetone. The use of aluminum alkyls and other Lewis acids for the preparation of onium salts has also been claimed in the patent literature (226), but no examples involving bismuth were given.

Except for the metathetical reactions discussed above, little is known about the chemical properties of bismuthonium compounds. It has been noted (206) that these compounds are less stable than the corresponding phosphorus, arsenic, and antimony analogs. Thus, tetraphenylbismuthonium chloride and bromide in the solid state decompose rapidly at room temperature as follows:

$$(C_6H_5)_4BiX \rightarrow (C_6H_5)_3Bi + C_6H_5X$$

The cyanide and nitrite have not been isolated but apparently undergo a similar type of decomposition. The tribromide is stable at $-30°$ for several hours. When warmed to $+5°$ in the course of fifteen minutes, it gradually decomposes into triphenylbismuth dibromide and bromobenzene:

$$(C_6H_5)_4BiBr_3 \rightarrow (C_6H_5)_3BiBr_2 + C_6H_5Br$$

When quickly warmed to room temperature, the tribromide undergoes an exothermic reaction giving diphenylbismuth bromide:

$$(C_6H_5)_4BiBr_3 \rightarrow (C_6H_5)_2BiBr + 2C_6H_5Br$$

The perchlorate, as noted above, is relatively stable; it explodes, however, if heated strongly.

The reaction of tetraphenylbismuthonium chloride with phenyllithium at $-70°$ in ether gives a 67% yield of pentaphenylbismuth and a 23% yield of triphenylbismuth (206).

The infrared spectrum of tetraphenylbismuthonium perchlorate has been determined in the region between 4000 and 250 cm^{-1} (203). The observed bands may be divided into three groups: (a) fifteen bands which can be assigned to the phenyl groups; (b) five bands assigned to the perchlorate ion; and (c) three unidentified bands.

5. PENTAARYLBISMUTH COMPOUNDS

After Wittig and Clauss (206) found that the thermal stability of pentaphenylphosphorus, -arsenic, and -antimony increases in this order, it was natural for them to attempt the preparation of the corresponding bismuth compound. Since no bismuthonium compound had been previously described, they were unable to use the general method employed for the other pentaaryl compounds:

$$(C_6H_5)_4MX + C_6H_5Li \rightarrow (C_6H_5)_5M + LiX$$

Pentaphenylantimony, however, had also been made by the reaction of phenyllithium with triphenylantimony dihalides, and Wittig and Clauss (206) found that this method could be extended to the bismuth analog:

$$(C_6H_5)_3BiCl_2 + 2C_6H_5Li \rightarrow (C_6H_5)_5Bi + 2LiCl$$

At room temperature, only decomposition products such as triphenylbismuth could be isolated. At $-75°$, however, a voluminous, yellow precipitate was obtained, which changed to a violet powder upon being slowly warmed to room temperature and treated with a slight excess of phenyllithium. When this powder was dissolved in acetone and then precipitated with water, violet needles were obtained which gave good analytical figures for pentaphenylbismuth. The yield was 81%; a small amount of triphenylbismuth was also obtained. Hellwinkel and Kilthau (227) have recently reported that recrystallization of pentaphenylbismuth from aqueous tetrahydrofuran (under nitrogen) yields a very pure product as deep-purple crystals, mp 90–100°; the observed melting point depended on the rate of heating.

In 1964 Wittig and Hellwinkel (53) described a new method for the preparation of pentaphenylbismuth by the reaction of phenyllithium with the N-triphenylbismuth derivative of *p*-toluenesulfonamide:

$$2C_6H_5Li + p\text{-}CH_3C_6H_4SO_2N{=}Bi(C_6H_5)_3 \rightarrow (C_6H_5)_5Bi + p\text{-}CH_3C_6H_4SO_2NLi_2$$

Triphenylbismuth is also formed in this reaction. The yield of pentaphenylbismuth is appreciably lower than in the original method of Wittig and Clauss (206). The required N-triphenylbismuth derivative is made by the reaction of triphenylbismuth with chloramine-T in acetonitrile:

$$(C_6H_5)_3Bi + p\text{-}CH_3C_6H_4SO_2NClNa \rightarrow p\text{-}CH_3C_6H_4SO_2N{=}Bi(C_6H_5)_3 + NaCl$$

The acetonitrile mother liquor on standing in the air yielded a deposit which could be converted by treatment with acetic acid into triphenylbismuth diacetate. The deposit was not analyzed or otherwise characterized. It is possible that it may have been triphenylbismuth dihydroxide.

As previously noted, pentaphenylbismuth may be prepared by reaction of tetraphenylbismuthonium chloride with phenyllithium. This reaction has at present no synthetic utility since the bismuthonium chloride has been prepared only by the treatment of pentaphenylbismuth with hydrogen chloride.

Pentaphenylbismuth is significantly less stable than its phosphorus, arsenic, and antimony analogs (206). When heated between 100 and 105° in a nitrogen atmosphere, the bismuth compound decomposes exothermally with the evolution of a gas. The products of the decomposition include benzene, biphenyl, and triphenylbismuth. At room temperature pentaphenylbismuth is stable for several days under nitrogen. In air the compound decomposes in a short time to a yellow-brown material. Attempts to convert pentaphenylbismuth into a hexaphenylbismuthate complex have been unsuccessful (227). The reactions of pentaphenylbismuth with chlorine, bromine, hydrogen chloride, and triphenylboron yield bismuthonium compounds and have been previously discussed (Section IX-4).

Nefedov and co-workers (228) have investigated the chemical changes accompanying the β-decay of ^{210}Bi in solid pentaphenylbismuth. The tagged bismuth compound was obtained by the arylation of triphenylbismuth dichloride which had, in turn, been synthesized from neutron-irradiated bismuth. After the radioactive pentaphenylbismuth was allowed to stand for a time equal to five half-lives of ^{210}Bi, it was possible to detect in the crystals various phenylpolonium compounds.

REFERENCES

1. C. Löwig and E. Schweizer, *Justus Liebigs Ann. Chem.*, **75**, 315 (1850).
2. Breed, *Justus Liebigs Ann. Chem.*, **82**, 106 (1852).
3. Breed, *J. Prakt. Chem.*, **56**, 341 (1852).
4. F. Dünhaupt, *J. Prakt. Chem.*, **61**, 399 (1854).
5. A. Michaelis and A. Polis, *Ber.*, **20**, 54 (1887).
6. A. Michaelis and A. Marquardt, *Justus Liebigs Ann. Chem.*, **251**, 323 (1889).
7. A. Gillmeister, *Ber.*, **30**, 2843 (1897).
8. St. Rozenblumówna and St. Weil, *Bull. Trav. Inst. Pharm. État*, No. 1, 3 (1927); through *C.A.*, **21**, 1449 (1927).
9. E. Frankland and B. F. Duppa, *J. Chem. Soc.*, **17**, 29 (1864).
10. S. Hilpert and G. Grüttner, *Ber.*, **46**, 1675 (1913).
11. F. Challenger and C. F. Allpress, *J. Chem. Soc.*, **119**, 913 (1921).
12. A. Marquardt, *Ber.*, **20**, 1516 (1887).
13. A. Marquardt, *Ber.*, **21**, 2035 (1888).
14. P. Pfeiffer and H. Pietsch, *Ber.*, **37**, 4620 (1904).
15. L. I. Zakharkin and O. Yu. Okhlobystin, *Izv. Akad. Nauk SSSR, Otd. Khim. Nauk*, 1942 (1959).
16. Kali-Chemie Aktiengesellschaft, Brit. Pat. 820,146 (Sept. 16, 1959); through *C.A.*, **54**, 6550 (1960).
17. Farbwerke Hoechst Aktiengesellschaft, Brit. Pat. 839,370 (June 29, 1960); through *C.A.*, **55**, 3435 (1961).
18. R. S. Dickson and B. O. West, *Aust. J. Chem.*, **15**, 710 (1962).
19. W. Stamm and A. Breindel, *Angew. Chem.*, **76**, 99 (1964).
20. L. A. Zhitkova, N. I. Sheverdina, and K. A. Kocheshkov, *Zh. Obshch. Khim.*, **8**, 1839 (1938); *C.A.*, **33**, 5819 (1939).
21. H. Gilman and H. L. Yablunky, *J. Amer. Chem. Soc.*, **63**, 207 (1941).
22. A. E. Borisov, M. A. Osipova, and A. N. Nesmeyanov, *Izv. Akad. Nauk SSSR, Ser. Khim.*, 1507 (1963).
23. E. O. Fischer and S. Schreiner, *Angew. Chem.*, **69**, 205 (1957).
24. E. O. Fischer and S. Schreiner, *Chem. Ber.*, **93**, 1417 (1960).

25. D. E. Worrall, *J. Amer. Chem. Soc.*, **58,** 1820 (1936).
26. R. Müller and C. Dathe, *Chem. Ber.*, **99,** 1609 (1966).
27. R. Müller and C. Dathe, East Ger. Pat. 49,607 (Aug. 20, 1966); through *C.A.*, **66,** 2760 (1967).
28. R. Müller, *Organometal. Chem. Rev.*, **1,** 359 (1966).
29. R. Müller and C. Dathe (to Institut für Silikon- und Fluorkarbon-Chemie), Ger. Pat. 1,249,865 (Sept. 14, 1967); through *C.A.*, **68,** 4829 (1968).
30. K. Ziegler and H. Lehmkuhl (to K. Ziegler), Ger. Pat. 1,127,900 (Apr. 19, 1962); *C.A.*, **57,** 11235 (1962).
31. K. Ziegler and O. W. Steudel, *Justus Liebigs Ann. Chem.*, **652,** 1 (1962).
32. K. Ziegler and H. Lehmkuhl (to K. Ziegler), Ger. Pat. 1,161, 562 (Jan. 23,1964); through *C.A.*, **60,** 11623 (1964).
33. H. Schmidt and H. Meinert, *Z. Anorg. Allg. Chem.*, **295,** 173 (1958).
34. E. M. Marlett, *Ann. N.Y. Acad. Sci.*, **125,** Art. 1, 12 (1965).
35. H. Gilman and A. C. Svigoon, *J. Amer. Chem. Soc.*, **61,** 3586 (1939).
36. H. Gilman and H. L. Yablunky, *J. Amer. Chem. Soc.*, **63,** 949 (1941).
37. T. K. Kozminskaya, M. M. Nad, and K. A. Kocheshkov, *Zh. Obshch. Khim.*, **16,** 891 (1946); *C.A.*, **41,** 2014 (1947).
38. M. M. Nad, T. K. Kozminskaya, and K. A. Kocheshkov, *Zh. Obshch. Khim.*, **16,** 897 (1946); *C.A.*, **41,** 2014 (1947).
39. A. N. Nesmeyanov, T. P. Tolstaya, and L. S. Isaeva, *Dokl. Akad. Nauk SSSR*, **122,** 614 (1958).
40. O. A. Reutov, *Vestn. Mosk. Univ.*, **8,** No. 3, *Ser. Fiz. Mat. i Estestv. Nauk*, No. **2,** 119 (1953); *C.A.*, **49,** 3867 (1955).
41. A. B. Bruker and K. M. Malkov, *Dokl. Akad. Nauk SSSR*, **128,** 948 (1959).
42. O. A. Reutov, O. A. Ptitsyna, and N. B. Styazhkina, *Dokl. Akad. Nauk SSSR*, **122,** 1032 (1958).
43. O. A. Ptitsyna, O. A. Reutov, and Yu. S. Ovodov, *Izv. Akad. Nauk SSSR, Otd. Khim. Nauk*, 638 (1962).
44. A. N. Nesmeyanov, O. A. Reutov, T. P. Tolstaya, O. A. Ptitsyna, L. S. Isaeva, M. F. Turchinskii, and G. P. Bochkareva, *Dokl. Akad. Nauk SSSR*, **125,** 1265 (1959).
45. F. Paneth and W. Hofeditz, *Ber.*, **62,** 1335 (1929).
46. F. Paneth, *Trans. Faraday Soc.*, **30,** 179 (1934).
47. F. A. Paneth and H. Loleit, *J. Chem. Soc.*, 366 (1935).
48. T. N. Bell, B. J. Pullman, and B. O. West, *Aust. J. Chem.*, **16,** 722 (1963).
49. F. Challenger, *J. Chem. Soc.*, **105,** 2210 (1914).
50. I. Norvick, *Nature*, **135,** 1038 (1935).
51. H. Zorn, H. Schindlbauer, and D. Hammer, *Monatsh. Chem.*, **98,** 731 (1967).
52. G. Grüttner and M. Wiernik, *Ber.*, **48,** 1473 (1915).
53. G. Wittig and D. Hellwinkel, *Chem. Ber.*, **97,** 789 (1964).
54. D. Braun, H. Daimon, and G. Becker, *Makromol. Chem.*, **62,** 183 (1963).
55. H. Gilman and H. L. Yablunky, *J. Amer. Chem. Soc.*, **63,** 212 (1941).
56. F. Challenger and C. F. Allpress, *Proc. Chem. Soc.*, **30,** 292 (1914).
57. J. V. Supniewski and R. Adams, *J. Amer. Chem. Soc.*, **48,** 507 (1926).
58. H. Hartmann, G. Habenicht, and W. Reiss, *Z. Anorg. Allg. Chem.*, **317,** 54 (1962).
59. T. N. Bell, B. J. Pullman, and B. O. West, *Proc. Chem. Soc.*, 224 (1962).
60. T. N. Bell, B. J. Pullman, and B. O. West, *Aust. J. Chem.*, **16,** 636 (1963).
61. H. Gilman and H. L. Yale, *Chem. Rev.*, **30,** 281 (1942).
62. G. Calingaert, H. Soroos, and V. Hnizda, *J. Amer. Chem. Soc.*, **64,** 392 (1942).
63. G. Bähr and G. Zoche, *Chem. Ber.*, **90,** 1176 (1957).
64. F. Challenger and L. R. Ridgway, *J. Chem. Soc.*, **121,** 104 (1922).

65. H. Gilman and J. F. Nelson, *J. Amer. Chem. Soc.*, **59,** 935 (1937).
66. J. F. Nelson, *Iowa State Coll. J. Sci.*, **12,** 145 (1937).
67. A. P. Giraitis (to Ethyl Corp.), Ger. Pat. 1.046,617 (Dec. 18, 1958); through *C.A.*, **55,** 383 (1961).
68. O. A. Kruglaya, N. S. Vyazankin, and G. A. Razuvaev, *Zh. Obshch. Khim.*, **35,** 394 (1965).
69. N. S. Vyazankin, O. A. Kruglaya, G. A. Razuvaev, and G. S. Semchikova, *Dokl. Akad. Nauk SSSR*, **166,** 99 (1966).
70. N. S. Vyazankin, G. A. Razuvaev, O. A. Kruglaya, and G. S. Semchikova, *J. Organometal. Chem.*, **6,** 474 (1966).
71. G. E. Coates, *J. Chem. Soc.*, 2003 (1951).
72. L. Maier, D. Seyferth, F. G. A. Stone, and E. G. Rochow, *J. Amer. Chem. Soc.*, **79,** 5884 (1957).
73. D. Seyferth, *J. Amer. Chem. Soc.*, **80,** 1336 (1958).
74. D. Benlian and M. Bigorgne, *Bull. Soc. Chim. Fr.*, 1583 (1963).
75. R. A. Shaw, B. C. Smith, and C. P. Thakur, *Chem. Commun.*, 228 (1966).
76. T. P. Whaley and V. Norman (to Ethyl Corp.), U.S. Pat. 3,071,493 (Jan. 1, 1963); *C.A.*, **58,** 4235 (1963).
77. E. Krause and G. Renwanz, *Ber.*, **62,** 1710 (1929).
78. E. Krause and G. Renwanz, *Ber.*, **65,** 777 (1932).
79. A. Classen and O. Ney, *Ber.*, **53,** 2267 (1920).
80. A. Classen and O. Ney, *Z. Anorg. Allg. Chem.*, **115,** 253 (1921).
81. A. Classen and G. Strauch, *Z. Anorg. Allg. Chem.*, **141,** 82 (1924).
82. M. M. Koton and F. S. Florinskii, *Dokl. Akad. Nauk. SSSR*, **169,** 598 (1966).
83. F. Challenger, A. L. Smith, and F. J. Paton, *J. Chem. Soc.*, **123,** 1046 (1923).
84. H. Gilman and H. L. Yale, *J. Amer. Chem. Soc.*, **73,** 2880 (1951).
85. M. M. Koton, *Zh. Obshch. Khim.*, **17,** 1307 (1947); *C.A.*, **42,** 1903 (1948).
86. J. F. Wilkinson and F. Challenger, *J. Chem. Soc.*, **125,** 854 (1924).
87. F. Challenger and J. F. Wilkinson, *J. Chem. Soc.*, **121,** 91 (1922).
88. F. Kh. Solomakhina, *Tr. Tashkentsk. Farm. Inst.*, **1,** 321 (1957); through *C.A.*, **55,** 15389 (1961).
89. F. Challenger and F. Pritchard, *J. Chem. Soc.*, **125,** 864 (1924).
90. Z. M. Manulkin and A. N. Tatarenko, *Zh. Obshch. Khim.*, **21,** 93 (1951); through *C.A.*, **45,** 7038 (1951).
91. F. Kh. Solomakhina, *Tr. Tashkentsk. Farm. Inst.*, **2,** 317 (1960); through *C.A.*, **57,** 11230 (1962).
92. F. Challenger and O. V. Richards, *J. Chem. Soc.*, 405 (1934).
93. M. M. Koton, *Zh. Obshch. Khim.*, **18,** 936 (1948); through *C.A.*, **43,** 559 (1949).
94. F. Kh. Solomakhina and Z. M. Manulkin, *Tr. Tashkentsk. Farm. Inst.*, **3,** 390 (1962); through *C.A.*, **61,** 3143 (1964).
95. O. Schmitz-DuMont and B. Ross, *Z. Anorg. Allg. Chem.*, **349,** 328 (1967).
96. E. G. Rochow, D. T. Hurd, and R. N. Lewis, *The Chemistry of Organometallic Compounds*, Wiley, New York, 1957, p. 48.
97. T. V. Talalaeva and K. A. Kocheshkov, *Zh. Obshch. Khim.*, **8,** 1831 (1938); through *C.A.*, **33,** 5819 (1939).
98. W. J. Considine and J. J. Ventura, *J. Organometal. Chem.*, **3,** 420 (1965).
99. H. Gilman, H. L. Yablunky, and A. C. Svigoon, *J. Amer. Chem. Soc.*, **61,** 1170 (1939).
100. H. Gilman and H. L. Yale, *J. Amer. Chem. Soc.*, **72,** 8 (1950).
101. G. Wittig and A. Maercker, *J. Organometal. Chem.*, **8,** 491 (1967).
102. W. Ipatiew and G. Rasuwajew, *Ber.*, **63,** 1110 (1930).
103. L. D. Freedman, G. O. Doak, and E. L. Petit, *J. Amer. Chem. Soc.*, **77,** 4262 (1955).

104. G. D. F. Jackson and W. H. F. Sasse, *J. Chem. Soc.*, 3746 (1962).
105. C. J. Thompson, H. J. Coleman, R. L. Hopkins, and H. T. Rall, *Anal. Chem.*, **37,** 1042 (1965).
106. D. H. Hey, D. A. Shingleton, and G. H. Williams, *J. Chem. Soc.*, 5612 (1963).
107. G. A. Razuvaev, G. G. Petukhov, V. A. Titov, and O. N. Druzhkov, *Zh. Obshch. Khim.*, **35,** 481 (1965).
108. C. Heitz and J. P. Adloff, *J. Organometal. Chem.*, **2,** 59 (1964).
109. D. B. Peterson, T. Arakawa, D. A. G. Walmsley, and M. Burton, *J. Phys. Chem.*, **69,** 2880 (1965).
110. D. S. Popplewell, *J. Inorg. Nucl. Chem.*, **25,** 318 (1963).
111. V. D. Nefedov and O. V. Larionov, *Radiokhimiya*, **3,** 639 (1961); *C.A.*, **56,** 4341 (1962).
112. A. N. Murin, V. D. Nefedov, V. M. Zaitsev, and S. A. Grachev, *Dokl. Akad. Nauk SSSR*, **133,** 123 (1960).
113. V. D. Nefedov, M. A. Toropova, S. A. Grachev, and Z. A. Grant, *Zh. Obshch. Khim.*, **33,** 15 (1963).
114. V. D. Nefedov, S. A. Grachev, and S. Gluvka, *Zh. Obshch. Khim.*, **33,** 333 (1963).
115. V. D. Nefedov, M. Vobecky, and I. Borak, *Radiokhimiya*, **7,** 628 (1965); through *C.A.*, **64,** 17630 (1966).
116. W. C. Davies and W. P. G. Lewis, *J. Chem. Soc.*, 1599 (1934).
117. J. Chatt and F. G. Mann, *J. Chem. Soc.*, 1192 (1940).
118. M. C. Henry and G. Wittig, *J. Amer. Chem. Soc.*, **82,** 563 (1960).
119. Z. El-Hewehi and D. Hempel, *J. Prakt. Chem.*, **22,** 1 (1963).
120. K. Stilp, Inaugural Dissertation, University of Rostock, 1910; quoted by H. Gilman and H. L. Yale, ref. 61.
121. M. Becke-Goehring and H. Thielemann, *Z. Anorg. Allg. Chem.*, **308,** 33 (1961).
122. R. H. Nuttall, E. R. Roberts, and D. W. A. Sharp, *J. Chem. Soc.*, 2854 (1962).
123. F. Hein and H. Pobloth, *Z. Anorg. Allg. Chem.*, **248,** 84 (1941).
124. W. Reppe and W. J. Schweckendiek, *Justus Liebigs Ann. Chem.*, **560,** 104 (1948).
125. L. Vallarino, *J. Inorg. Nucl. Chem.*, **8,** 290 (1958).
126. J. Desnoyers and R. Rivest, *Can. J. Chem.*, **43,** 1879 (1965).
127. T. H. Coffield (to Ethyl Corp.), U.S. Pat. 3,100,214 (Aug. 6, 1963).
128. D. Lloyd and M. I. C. Singer, *Chem. Commun.*, 1042 (1967).
129. H. H. Jaffé, *J. Chem. Phys.*, **22,** 1430 (1954).
130. C. N. R. Rao, J. Ramachandran, M. S. C. Iah, S. Somasekhara, and T. V. Rajakumar, *Nature*, **183,** 1475 (1959).
131. C. N. R. Rao, J. Ramachandran, and A. Balasubramanian, *Can. J. Chem.*, **39,** 171 (1961).
132. N. G. Pai, *Proc. Roy. Soc., Ser. A*, **149,** 29 (1935).
133. E. J. Rosenbaum, D. J. Rubin, and C. R. Sandberg, *J. Chem. Phys.*, **8,** 366 (1940).
134. R. K. Sheline, *J. Chem. Phys.*, **18,** 602 (1950).
135. H. Siebert, *Z. Anorg. Allg. Chem.*, **273,** 161 (1953).
136. J. A. Jackson and J. R. Nielsen, *J. Mol. Spectrosc.*, **14,** 320 (1964).
137. R. D. Kross and V. A. Fassel, *J. Amer. Chem. Soc.*, **77,** 5858 (1955).
138. M. Margoshes and V. A. Fassel, *Spectrochim. Acta*, **7,** 14 (1955).
139. L. A. Harrah, M. T. Ryan, and C. Tamborski, *Spectrochim. Acta*, **18,** 21 (1962).
140. K. A. Jensen and P. H. Nielsen, *Acta Chem. Scand.*, **17,** 1875 (1963).
141. D. H. Brown, A. Mohammed, and D. W. A. Sharp, *Spectrochim. Acta*, **21,** 659 (1965).
142. H. P. Fritz, *Chem. Ber.*, **92,** 780 (1959).
143. R. N. Sterlin and S. S. Dubov, *Zh. Vses. Khim. Obshchest.*, **7,** No. 1, 117 (1962); throughout *C.A.*, **57,** 294 (1962).

144. A. L. Allred and A. L. Hensley, Jr., *J. Inorg. Nucl. Chem.*, **17**, 43 (1961).
145. C. R. McCoy and A. L. Allred, *J. Inorg. Nucl. Chem.*, **25**, 1219 (1963).
146. E. Bergmann and W. Schütz, *Z. Phys. Chem.* (Leipzig), *B*, **19**, 401 (1932).
147. C. P. Smyth, *J. Org. Chem.*, **6**, 421 (1941).
148. M. J. Aroney, R. J. W. LeFèvre, and J. D. Saxby, *J. Chem. Soc.*, 1739 (1963).
149. J. Wetzel, *Z. Kristallogr.*, **104**, 305 (1942).
150. D. M. Hawley, G. Ferguson, and G. S. Harris, *Chem. Commun.*, 111 (1966).
151. R. N. Sterlin, S. S. Dubov, W. K. Li, L. P. Vakhomchik, and I. L. Knunyants, *Zh. Vses. Khim. Obshchest.*, **6**, No. 1, 110 (1961); through *C.A*, **55**, 15336 (1961).
152. L. H. Long and J. F. Sackman, *Trans. Faraday Soc.*, **50**, 1177 (1954).
153. L. H. Long, *Pure Appl. Chem.*, **2**, 61 (1961).
154. S. J. W. Price and A. F. Trotman-Dickenson, *Trans. Faraday Soc.*, **54**, 1630 (1958).
155. W. F. Lautsch, *Chem. Tech.* (Berlin), 10, 419 (1958); through *C.A.*, **53**, 43 (1959).
156. W. F. Lautsch, P. Erzberger, and A. Tröber, *Wiss. Z. Tech. Hochsch. Chem. Leuna-Merseburg*, **1**, 31 (1958–59); through *C.A.*, **54**, 13845 (1960).
157. K. H. Birr, *Z. Anorg. Allg. Chem.*, **311**, 92 (1961).
158. H. A. Skinner, "The Strengths of Metal-to-Carbon Bonds," in F. G. A. Stone and R. West, eds., *Advances in Organometallic Chemistry*, Vol. 2, Academic Press, New York, 1964, p. 98.
159. L. H. Long and J. F. Sackman, *Res. Correspondence*, Suppl. to *Research* (London), **8**, S23 (1955).
160. C. H. Bamford, D. L. Levi, and D. M. Newitt, *J. Chem. Soc.*, 468 (1946).
161. E. Amberger, *Chem. Ber.*, **94**, 1447 (1961).
162. L. Maier, D. Seyferth, F. G. A. Stone, and E. G. Rochow, *Z. Naturforsch.*, *B*, **12**, 263 (1957).
163. R. H. Smith and D. H. Andrews, *J. Amer. Chem. Soc.*, **53**, 3661 (1931).
164. F. I. Vilesov and V. M. Zaitsev, *Dokl. Akad. Nauk SSSR*, **154**, 886 (1964).
165. M. V. Forward, S. T. Bowden, and W. J. Jones, *J. Chem. Soc.*, S121 (1949).
166. S. Sugden, *The Parachor and Valency*, Routledge, London, 1930, p. 181.
167. F. Challenger and A. E. Goddard, *J. Chem. Soc.*, **117**, 762 (1920).
168. W. J. Lile and R. J. Menzies, *J. Chem. Soc.*, 617 (1950).
169. N. K. Parab and D. M. Desai, *J. Indian Chem. Soc.*, **35**, 573 (1958).
170. N. K. Parab and D. M. Desai, *Sci. Cult.* (Calcutta), **23**, 430 (1958); *C.A.*, **52**, 15988 (1958).
171. H. G. Robinson, H. G. Dehmelt, and W. Gordy, *Phys. Rev.*, **89**, 1305 (1953).
172. F. F. Blicke, U. O. Oakdale, and F. D. Smith, *J. Amer. Chem. Soc.*, **53**, 1025 (1931).
173. Volmar and Chardeyron, *Bull. Soc. Chim. Fr.*, **31**, 545 (1922).
174. H. E. Ramsden (to M & T Chemicals Inc.), U.S. Pat. 3,109,851 (Nov. 5, 1963); *C.A.*, **60**, 3015 (1964).
175. H. Lecoq, *J. Pharm. Belg.*, **19**, 155 (1937).
176. A. E. Goddard, J. N. Ashley, and R. B. Evans, *J. Chem. Soc.*, **121**, 978 (1922).
177. H. Gilman and L. D. Apperson, *J. Org. Chem.*, **4**, 162 (1939).
178. F. Challenger and C. F. Allpress, *J. Chem. Soc.*, **107**, 16 (1915).
179. L. Maier, E. G. Rochow, and W. C. Fernelius, *J. Inorg. Nucl. Chem.*, **16**, 213 (1961).
180. A. Ya. Yakubovich, V. A. Ginsburg, and S. P. Makarov, *Dokl. Akad. Nauk SSSR*, **71**, 303 (1950); through *C.A.*, **44**, 8320 (1950).
181. A. Ya. Yakubovich and S. P. Makarov, *Zh. Obshch. Khim.*, **22**, 1528 (1952).
182. F. Dünhaupt, *Justus Liebigs Ann. Chem.*, **92**, 371 (1854).
183. H. McCombie and B. C. Saunders, *Nature*, **159**, 491 (1947).
184. H. Lecoq, *J. Pharm. Belg.*, **19**, 173 (1937).
185. E. J. Kupchik and C. T. Theisen, *J. Organometal. Chem.*, **11**, 627 (1968).

186. R. E. Dessy, T. Chivers, and W. Kitching, *J. Amer. Chem. Soc.*, **88**, 467 (1966).
187. L. Pauling, *The Nature of the Chemical Bond*, 3rd ed., Cornell University Press, Ithaca, 1960, p. 85.
188. R. E. Dessy, P. M. Weissman, and R. L. Pohl, *J. Amer. Chem. Soc.*, **88**, 5117 (1966).
189. V. F. Toropova and M. K. Saikina, *Sb. Statei Obshch. Khim.*, **1**, 210 (1953); through *C.A.*, **48**, 12579 (1954).
190. M. K. Saikina, *Uch. Zap., Kazansk. Gos. Univ.*, **116**, No. 2, 129 (1956); through *C.A.*, **51**, 7191 (1957).
191. E. Wiberg and K. Mödritzer, *Z. Naturforsch., B*, **12**, 132 (1957).
192. A. N. Nesmeyanov, K. N. Anisimov, N. E. Kolobova, and V. N. Khandozhko, *Dokl. Akad. Nauk SSSR*, **156**, 383 (1964).
193. G. N. Schrauzer and G. Kratel, *Angew. Chem.*, **77**, 130 (1965).
194. L. G. Makarova, *Zh. Obshch. Khim.*, **7**, 143 (1937); through *C.A.*, **31**, 4290 (1937).
195. R. Okawara, K. Yasuda, and M. Inoue, *Bull. Chem. Soc. Jap.*, **39**, 1823 (1966).
196. W. R. Roper and C. J. Wilkins, *Inorg. Chem.*, **3**, 500 (1964).
197. F. Challenger, *J. Chem. Soc.*, **109**, 250 (1916).
198. F. Feigl, D. Goldstein, and D. Haguenauer-Castro, *Z. Anal. Chem.*, **178**, 419 (1961).
199. F. Feigl and D. Goldstein, *Mikrochim. Acta*, 1 (1966).
200. A. D. Beveridge, G. S. Harris, and F. Inglis, *J. Chem. Soc., A*, 520 (1966).
201. F. Challenger and V. K. Wilson, *J. Chem. Soc.*, 209 (1927).
202. H. Gilman and H. L. Yale, *J. Amer. Chem. Soc.*, **73**, 4470 (1951).
203. G. O. Doak, G. G. Long, S. K. Kakar, and L. D. Freedman, *J. Amer. Chem. Soc.*, **88**, 2342 (1966).
204. J. J. Monagle, *J. Org. Chem.*, **27**, 3851 (1962).
205. F. Agolini and R. Bonnett, *Can. J. Chem.*, **40**, 181 (1962).
206. G. Wittig and K. Clauss, *Justus Liebigs Ann. Chem.*, **578**, 136 (1952).
207. F. Challenger and E. Rothstein, *J. Chem. Soc.*, 1258 (1934).
208. J. Supniewski, *Rocz. Chem.*, **5**, 298 (1925); *C.A.*, **20**, 1984 (1926).
209. J. Supniewski, *Rocz. Chem.*, **6**, 97 (1926); *C.A.*, **21**, 2466 (1927).
210. H. Gilman and H. L. Yablunky, *J. Amer. Chem. Soc.*, **63**, 839 (1941).
211. H. Gilman and H. L. Yablunky, *J. Amer. Chem. Soc.*, **62**, 665 (1940).
212. E. Wiberg and K. Mödritzer, *Z. Naturforsch., B*, **11**, 755 (1956).
213. V. D. Nefedov and V. I. Andreev, *Zh. Fiz. Khim.*, **31**, 563 (1957).
214. V. D. Nefedov, E. N. Sinotova, and V. D. Trenin, *Radiokhimiya*, **2**, 739 (1960); *C.A.*, **55**, 15169 (1961).
215. F. Challenger, A. T. Peters, and J. Halévy, *J. Chem. Soc.*, 1648 (1926).
216. A. N. Murin and V. D Nefedov, *Primenenie Mechenykh At. v Anal. Khim., Akad. Nauk SSSR, Inst. Geokhim. i Anal. Khim.*, 75 (1955); through *C.A.*, **50**, 3915 (1956),
217. G. Greenwood, *Amer. Mineral.*, **16**, 473 (1931).
218. E. V. Stroganov, *Vestn. Leningrad. Univ.*, **14**, No. 4, *Ser. Fiz. Khim.*, No. 1, 103 (1959); through *C.A.*, **53**, 14631 (1959).
219. K. A. Jensen, *Z. Anorg. Allg. Chem.*, **250**, 257 (1943).
220. P. F. Oesper and C. P. Smyth, *J. Amer. Chem. Soc.*, **64**, 173 (1942).
221. E. L. Muetterties, W. Mahler, K. J. Packer, and R. Schmutzler, *Inorg. Chem.*, **3**, 1298 (1964).
222. G. Greenwood, *Mineral. Mag.*, **20**, 123 (1923).
223. W. M. D. Bryant, *Amer. Mineral.*, **20**, 281 (1935).
224. L. G. Makarova and A. N. Nesmeyanov, *Izv. Akad. Nauk SSSR, Otd. Khim. Nauk*, 617 (1945).
225. R. Dötzer, Abstracts of papers of the 3rd International Symposium on Organometallic Chemistry, August 28–September 1, 1967, p. 196.

226. R. Dötzer (to Siemens-Schuckertwerke Akt.-Ges.), Ger. Pat. 1,200,817 (Sept. 16, 1965); *C.A.*, **63**, 15896 (1965).
227. D. Hellwinkel and G. Kilthau, *Justus Liebigs Ann. Chem.*, **705**, 66 (1967).
228. V. D. Nefedov, V. E. Zhuravlev, M. A. Toropova, and A. V. Levchenko, *Radiokhimiya*, **6**, 632 (1964); *C.A.*, **62**, 3600 (1965).

Author Index

Subject Index